T0358349

The Mathematics of Large-Scale Atmosphere and Ocean

The Mathematics of Large-Scale Atmosphere and Ocean

Mike Cullen

Met Office, Exeter, UK

NEW JERSEY • LONDON • SINGAPORE • BEIJING • SHANGHAI • HONG KONG • TAIPEI • CHENNAI • TOKYO

Published by

World Scientific Publishing Co. Pte. Ltd.
5 Toh Tuck Link, Singapore 596224
USA office: 27 Warren Street, Suite 401-402, Hackensack, NJ 07601
UK office: 57 Shelton Street, Covent Garden, London WC2H 9HE

British Library Cataloguing-in-Publication Data
A catalogue record for this book is available from the British Library.

THE MATHEMATICS OF LARGE-SCALE ATMOSPHERE AND OCEAN

ISBN 978-981-124-014-0 (hardcover)
ISBN 978-981-124-015-7 (ebook for institutions)
ISBN 978-981-124-016-4 (ebook for individuals)

For any available supplementary material, please visit
https://www.worldscientific.com/worldscibooks/10.1142/12367#t=suppl

Typeset by Stallion Press
Email: enquiries@stallionpress.com

Printed in Singapore

Dedication

I would like to thank the many people who have inspired this work and have helped to develop it further, as expressed in this second edition. First James Glimm and Alexandre Chorin, who showed me that nonlinear partial differential equations could be solved, and Brian Hoskins for his inspirational work on developing semi-geostrophic theory. Then those who collaborated with me on developing this theory over many years. In particular Jim Purser, Glenn Shutts, Simon Chynoweth, Martin Holt, John Norbury, Mike Sewell, Mark Mawson, Rob Douglas, Ian Roulstone, John Thuburn, Andrew Wheadon, Bob Beare, Colin Cotter, Abeed Visram, Hiroe Yamasaki, John Methven, Laura Mansfield, Michael Haigh and Matt Turner. As one who sits between physics and mathematics, I would then like to thank the mathematicians who proved the results: Yann Brenier, Jean-David Benamou, Gregoire Loeper, Wilfrid Gangbo, Robert McCann, Hamed Maroofi, Giovanni Pisante, Marc Sedjro, Luigi Ambrosio, Maria Colombo, Guido de Philippis, Alessio Figalli, Josiane Faria, Milton Lopes Filho, Helena Nussenveig Lopes, Misha Feldman, Jingrui Cheng, Adrian Tudorascu, Bin Cheng, Tom O'Neill, Beatrice Pelloni, David Gilbert, Charlie Egan and Mark Wilkinson. I would also like to thank Misha Feldman, Bob Beare and Marc Sedjro for reviewing sections of the text.

Preface

Accurate extended-range weather predictions are now routinely available for up to a week ahead, and represent a major achievement of the meteorological community over the last 70 years. The reason that this is possible is the highly predictable nature of large-scale atmospheric circulations. Realising it in practice has required a huge investment in computer technology and observing systems, such as satellites. Theoretical work since the 1960s has emphasised reasons for unpredictability, in particular chaos theory, rather than reasons for predictability. This volume redresses this balance by discussing reasons for the high predictability. These are founded in dynamical meteorological theory developed over 70 years ago, but have been reinforced by exciting recent developments in understanding why some nonlinear systems are much more stable than might be expected. In the first edition of this book, I introduced the mathematical model to be used, then described its mathematical properties, and finally applied the mathematical results to the atmosphere and ocean by relating the predictions of the model to the observed behaviour. In this second edition, I discuss how the model can be developed further to include more physical processes and show how it plays an important role in the tropics. I describe the extensive further mathematical results that have been obtained since the first edition was prepared, and describe the methods of proof more explicitly, thus showing the underlying mathematical structure more clearly and how it relates to the physics of the problem. I then show using real atmospheric data how these results can be exploited in understanding the large-scale dynamics of the atmosphere, and the workings of the complex numerical models used to simulate it.

Analysis of atmospheric or oceanic circulations directly from the governing equations of dynamics and thermodynamics is difficult because of the 'scale problem'. Theoretical results on the Navier-Stokes equations, for example, concentrate on viscous scales, which are six orders of magnitude smaller than the smallest-scale weather systems. In addition, this means that accurate solutions of the governing equations on a global scale would require nearly 30 orders of magnitude more computing power than currently available. The success of current numerical models in routinely predicting the weather for a week ahead shows that there must be a great degree of 'large-scale control'. Dynamical meteorologists have addressed this over many years by developing a hierarchy of simplified equations, which represent the behaviour of the atmosphere (or ocean) in particular regimes. These are usually obtained by identifying small parameters characteristic of the regime of interest, and deriving the simplified equations by asymptotic expansion in these parameters. Applying these equations to get rigorous mathematical results about the large scale behaviour has only been widely attempted over the last 30 years. This requires a proof that the simplified equations can be solved, and then a proof that there is a solution of the fundamental equations 'close' to the that of the simplified equations in a suitable sense. The simplified solutions can then be analysed and predictions made about atmosphere/ocean behaviour, subject to the error estimate made above. This procedure is now widely accepted, but its application is limited because of the difficulty in obtaining existence proofs and *a posteriori* error estimates. There has, however, been substantial progress in this area since the first edition of this book. More commonly, solutions of the simplified equations are obtained and used subject to the *a priori* error estimates obtained from the original asymptotic expansion.

This volume is concerned with large-scale atmosphere/ocean flows. The main focus is the atmosphere, and the desire to explain the success of large-scale weather forecasting. Most of the same principles apply to the ocean, and so this application is pointed out throughout the book. However, the examples primarily come from the atmosphere. The traditional approximations used to describe large-scale flows are based on smallness of the Rossby number, which states that large-scale flows are rotation-dominated, the Froude number, which states that the flow is strongly stratified, and the aspect ratio, which states that the atmosphere and ocean are much shallower than the horizontal scale of most weather systems and ocean circulations. These lead to the geostrophic and hydrostatic approximations, which relate the horizontal wind and the temperature to the pressure gradient. Thus, given also the equation for mass conservation, the whole of the dynamics can be determined from the pressure. The semi-geostrophic equations developed by Hoskins, following earlier work by Eliassen, give a set of simplified equations which is valid on large scales. Their asymptotic validity requires a Lagrangian form of the Rossby number to be small, which means that fluid trajectories cannot curve too sharply. This is appropriate for extra-tropical weather systems, including fronts and jet streams which are characterised by a large length-scale in one direction. They are not appropriate for smaller-scale phenomena such as tropical cyclones, which are in a different asymptotic regime. In the tropics, they predict the near-uniformity of the tropical temperature profile. Other asymptotic regimes are needed to describe tropical variability.

The analysis of these equations uses results from the theory of the Monge-Kantorovich problem, which has a long history in optimisation theory. There is thus an exciting and fundamental link between the dynamics of atmosphere/ocean flows and many other fields of study, such as economics, probability theory, magnetohydrodynamics and kinetic theory. It also exploits a 'convexity principle', due to Cullen and Purser, which requires that simplified solutions describing large-scale phenomena must be stable to parcel displacements. The convexity property is the basis of proofs of existence of the solutions to the semi-geostrophic equations.

This volume also discusses the application of the semi-geostrophic equations to real flows. The ability to simulate real data accurately on large scales is demonstrated. An important feature is that the solutions can form fronts, where the solutions are discontinuous in physical space. The formation of fronts does not destroy the predictability of the flow. Cullen and Roulstone demonstrated that an idealised quasi-periodic solution describing the growth and decay of weather systems, including fronts, remained predictable for more than 30 days. This is a case where the small-scale features are entirely slaved to larger scales. Other phenomena which can be described include the interaction of mountains with the atmospheric circulation, the inland penetration of sea-breezes and the atmospheric response to cumulus convection. In the ocean, the persistence of eddies and the outcropping of layers of constant density can be described. A similar theory for axisymmetric flow can describe the eye-wall discontinuity in hurricanes.

More generally, the flow evolution does not exhibit the systematic transfer of energy or enstrophy to small scales characteristic of two- and three-dimensional turbulence. This comes from the constraints on the flow imposed by the convexity condition. The behaviour is reminiscent of that of Hamiltonian systems of ordinary differential equations, which have no attractors and thus describe non-periodic solutions with stable statistics but which can never converge to steady states or periodic orbits. This behaviour can be applied to the real atmosphere and ocean, subject to the limitations of the semi-geostrophic approximation. Studies of the accuracy of the approximation in idealised cases suggest that the solutions are reasonably accurate on the scale of developing weather systems for about 6 days. The theory suggests that the large scale circulation of the atmosphere is non-turbulent, and that anomalous circulation patterns can persist for long periods, which explains periods of extreme weather. Smaller scale flows are turbulent but cannot be described by semi-geostrophic theory. These results thus underpin the ability of operational models to predict large-scale weather patterns a week ahead, and encourages attempts to extend this for longer periods. Thus this theory is an essential counterbalance to chaos theory, which demonstrates how predictability can be lost in nonlinear systems.

M. J. P. Cullen

Contents

Chapter 1

Introduction

Accurate extended-range weather predictions are now routinely available for up to a week ahead, and represent a major scientific achievement of the meteorological community over the last 70 years. Figure 1.1 shows the remarkable improvement in the standard of forecasts in the Northern hemisphere over the last 50 years, as measured by the root mean square error difference between forecasts of the atmospheric pressure reduced to mean sea level and the analysed observations. Five day forecasts today are as accurate as one day forecasts in the late 1960s.

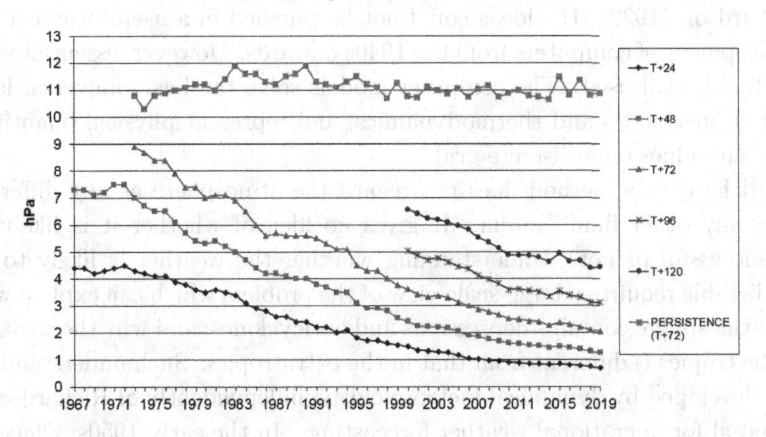

Fig. 1.1 Root mean square errors for forecasts of the atmospheric pressure reduced to mean sea level over the North Atlantic and Europe compared with analysed observations. Curves from bottom: 24, 48, 72, 96 and 120 hour forecasts. Top curve: persistence forecast. Source: Met Office. ©Crown Copyright, Met Office.

Though it is inconceivable that weather forecasting will ever reach the standard of tidal predictions, where predictions for years ahead can be readily purchased, the standard of weather forecasting indicates that the atmosphere is a highly predictable system. This runs counter to the popular conception of weather forecasting as a very empirical exercise, and to the widespread belief in 'chaos' theory and the 'butterfly effect'. Though it is undoubtedly true that there are limitations to the predictability of weather, and chaos theory can help to understand them; an important scientific question is to understand why the level of predictability is so high. This book is concerned with providing such an explanation. Realising this predictability in everyday forecasts has required a huge investment in computers and weather observations, particularly satellites. It has also involved the construction of computer models, involving hundreds of thousands of lines of code, which interpret the observations and make the predictions. This book is not concerned with describing how this is done, but only with why it is so successful.

Operational weather forecasting methods are based on the realisation that the atmosphere obeys the fundamental laws of fluid mechanics and thermodynamics; just like any other classical fluid system. It was L.F.Richardson in 1915 who made the apparently outrageous proposal that the weather could be predicted by simply writing down the fundamental equations, and solving them in a very crude discrete fashion by replacing continuously varying physical variables by values on a discrete grid, Richardson (1922). His ideas could not be pursued in a useful way till the development of computers from the 1940s onwards. However, essentially his method is still used. The computer models solve the basic universal laws of fluid mechanics and thermodynamics, and represent physical quantities by their values on a discrete grid.

Richardson's method does not regard the atmosphere as any different from any other fluid system. It gives no idea of whether it is likely to be successful or not. Understanding whether the weather is likely to be predictable requires a large-scale view of the problem which can explain why the atmosphere contains depressions and anticyclones, and why the weather in the tropics is different from that in the extratropics. Such understanding was developed by dynamical meteorologists, independently of Richardson's proposal for operational weather forecasting . In the early 1950s, Charney *et al.* (1950) developed sets of equations which were specific to the large-scale problem, and used them in the first ever numerical forecast. While, within the next 10 years, actual predictions were being made using the fundamental equations following Richardson's approach, it is the work of Charney and other dynamical meteorologists that underpins the study of why the weather is so predictable.

Theoretical work since the 1960s has emphasised reasons for unpredictability, in particular chaos theory, rather than reasons for predictability. There are two main contributors to unpredictability in addition to chaos theory. The first is that an accurate solution of the fundamental equations will be a realisation of the atmosphere at all scales down to the continuum limit, less than 1mm. Computing such a solution is many orders of magnitude beyond the power of the largest available supercomputers. The second is that the evolution of the atmosphere also depends on many physical processes which are not included in the fundamental equations. These include radiation, the effect of phase changes between water vapor, liquid water and ice, and exchanges of heat, moisture and momentum between the atmosphere and the Earth's surface. In general there are no fundamental equations that describe these processes that are as well-founded as the classical equations of motion and thermodynamics. The lack of such equations is another source of uncertainty. An example is the formation of ice crystals of different shapes in cirrus cloud, and the large difference in the interaction of crystals of different shapes with the sun's radiation.

This book, however, is concerned with the reasons for high predictability. The atmosphere contains motions on all scales; from the large-scale jet-streams associated with changes of weather type over periods of weeks to the small-scale gusts that anyone can observe when going outside on a windy day. An example is shown in the satellite picture in Figure 1.2. This shows several large-scale cloud bands associated with weather systems. These include an extended cloud band running north-south through the central Atlantic, with a separate area of cloud to the west of it, just south of 50°N. This band extends for several thousand kilometres. It is associated with a weather front, which is associated with a change of air mass. There are separate well-marked areas of cloud over the northern Baltic and over North Africa. There is also a band of cloud running east-west across central England and a less well defined area in the Mediterranean south of France. Further east in the Mediterranean there are a number of small-scale cloud features.

Figure 1.3 shows a weather chart corresponding to the same time as the picture. It shows contours of atmospheric pressure corrected to mean sea level, together with weather fronts. The weather front extending north-south in the central Atlantic is clearly marked, as is the low pressure system

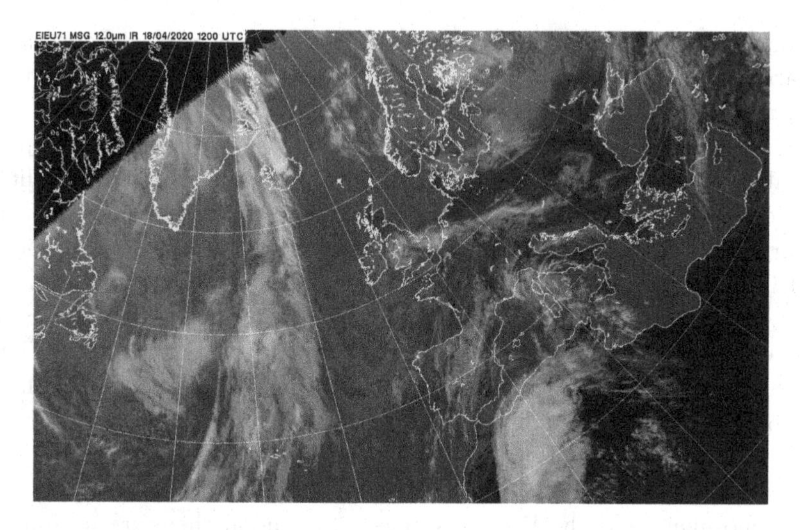

Fig. 1.2 Infra-red satellite picture for 12UTC on 18 April 2020. ©Crown Copyright, Met Office. Data: EUMETSAT.

to the west of it. The weather system in the northern Baltic is marked as is the frontal zone over England. The cloud mass over the Mediterranean does not correspond to a significant areas of low pressure, but an active area of possible thunderstorms is marked on the chart. The chart shows only the large-scale structures and gives no idea of the small-scale fluctuations that are also visible in Fig. 1.2. Figure 1.4 shows a plot of wind speed against time from an automatic wind recorder. There is a very large jump in wind speed over a few minutes. The actual change was probably faster than the resolution of the available data. In addition, there are small scale fluctuations on the shortest time-scale resolved in the plot. This illustrates that the 'large-scale' flow can have almost discontinuous behaviour as well as the ubiquitous nature of the small-scale fluctuations.

Lorenz (1963) encapsulated the existence of motions on a wide range of scales in the atmosphere into a theory of predictability. This uses the idea that each type of motion has a characteristic time-scale, and that it will only be predictable for a few multiples of that time-scale. Thus local gusts of wind will only be predictable for a few seconds. Since different scales of motion are not independent, loss of predictability at small scales will lead (the butterfly effect) to a loss of predictability at large scales. It has been realised for many years that this dynamical argument based on turbulence theory is too simplistic. In 1967, G.D.Robinson published an estimate that this would prevent the weather being forecast for more than one day ahead;

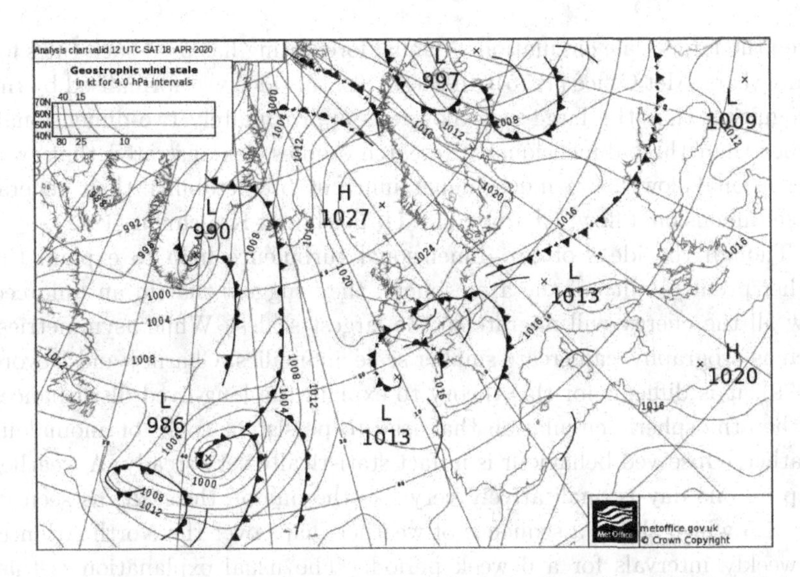

Fig. 1.3 Analysis of mean sea level pressure and weather fronts for 12 UTC on 18 April 2020. Pressure contours at 4hpa intervals. Source: Met Office. ©Crown Copyright, Met Office.

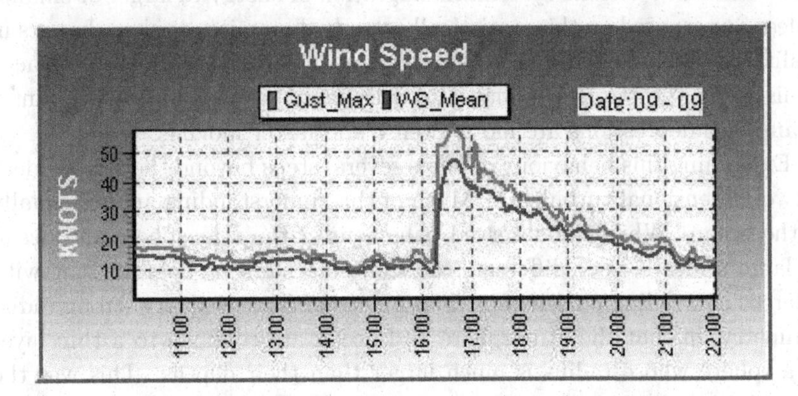

Fig. 1.4 Wind speed (knots) measured at Chichester bar, 9 September 2002. Source: CHIMET support group, Chichester Harbour Conservancy.

while it is now possible to forecast a week ahead on most occasions. The reasons for the higher predictability even then being achieved in practice were sought in the idea of 'large-scale control'. This means that the large-scale motions in the atmosphere, such as those shown in Fig. 1.3, evolve in a way that is insensitive to the small-scale details. Statistical information about the small-scale flow can be found by applying various diagnostic rules, given the large-scale circulation. Manual forecasting has used such rules for many years, Met Office (1975). In the 1970s, this idea was formalised by the recognition that the large-scale flow was approximately two-dimensional, rather than three-dimensional. Research studies demonstrated that two-dimensional flow has a much higher inherent predictability than general three-dimensional flow, Charney (1971), Leith and Kraichnan (1972).

Though the ideas of two-dimensional turbulence help to explain the higher predictability of the atmosphere, they suggest that in an unforced flow all the energy will migrate to the largest scales. While asymmetries, such as orography, can create smaller scale anomalies, Charney and Devore (1979), it is difficult for this theory to explain the long-lived disturbances to the atmospheric circulation that lead to persistent spells of anomalous weather. Observed behaviour is in fact statistically very steady. A weather map on one day is qualitatively very like the one on the next, as seen in Fig. 1.5 which shows a sequence of weather maps over the North Atlantic at weekly intervals for a 6 week period. The usual explanation is that this steady state is a balance between forcing and dissipation. However, it would better fit the facts if the natural dynamics of large-scale flows was 'non-turbulent', with no systematic migration of energy to larger or smaller scales. Superposed on this statistically steady dynamics are slow changes in qualitative behaviour from season to season. For instance, in the Northern hemisphere, the stronger equator to pole temperature gradient in winter means that depressions are more intense and faster moving.

Explaining this behaviour requires several steps beyond the simple ideas of two-dimensional turbulence. Much of this understanding applies equally to the ocean. While the observed behaviour of the atmosphere and ocean on large scales is very different, the same mechanisms operate, but with different controlling parameters. In both cases, there is a very strong radial symmetry, in that the atmosphere and ocean are confined to a thin layer on a sphere whose radius is much larger than their depths. This was the original motivation for using a two-dimensional theory. The next step is to recognise the importance of the Earth's rotation. However, the effect of the rotation is very different near the equator, where the axis of rotation is locally horizontal, and in high latitudes where it is closer to vertical. This is why the atmosphere and ocean behave in qualitatively different ways on large scales in the tropics and extratropics. In particular, an unsteady sequence of large-scale weather systems as shown in Fig. 1.5 is never seen in the tropics.

In the early days of theoretical meteorology there was little knowledge

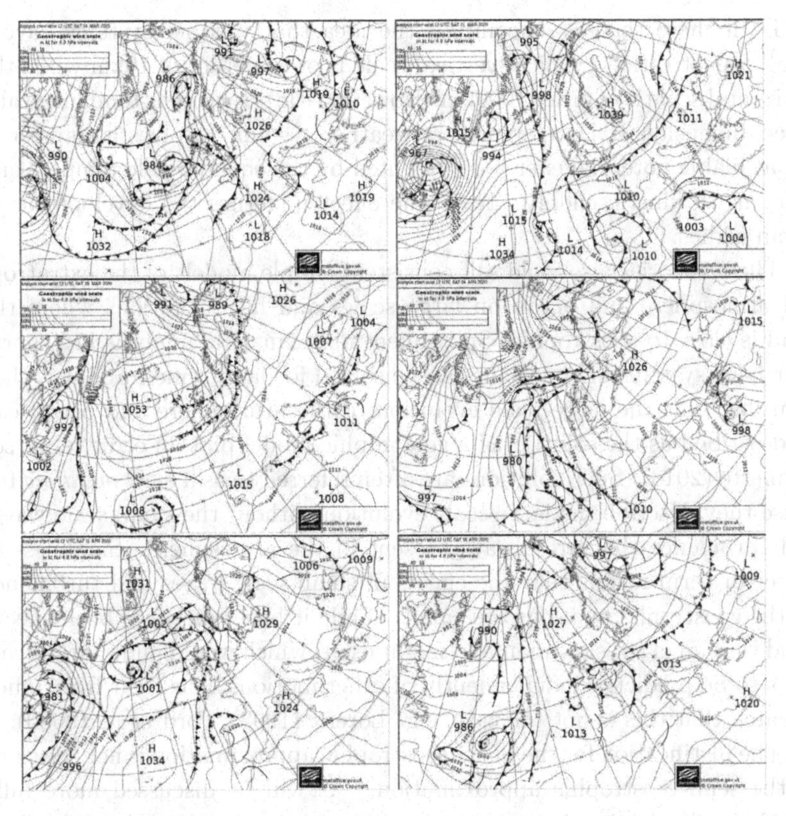

Fig. 1.5 Weather maps from 14 March to 18 April 2020 at weekly intervals. Conventions as in Fig. 1.3. Source: Met Office. ©Crown Copyright, Met Office.

of the tropics, and theories were developed for the extratropics which relied on the axis of rotation having a large vertical component. This leads to the idea of geostrophic balance, which means that the wind direction is parallel to the isobars (contours of constant pressure). This is commonly observed, and was a fundamental tool in manual weather forecasting, Met Office (1975). In the late 1940s, two theories of three-dimensional large-scale atmospheric motion were developed based on the idea that the wind was approximately geostrophic. Charney (1948) developed a simple theory of this behaviour called 'quasi-geostrophic' theory. Eliassen (1949) developed an alternative theory at about the same time which he called 'quasi-static' but is now usually called 'semi-geostrophic'. These were called 'Type 1' and 'Type 2' geostrophy by Phillips (1963). A broader perspective was given by Eliassen (1984). These theories can equally be applied to the ocean.

Both these theories encompass the idea that weather systems have a small aspect ratio, a typical vertical scale is of the order of 10km while the horizontal scale is of order 1000km. However, the structure is fundamentally three-dimensional. Development of weather systems involves a transfer of the available potential energy represented by the equator-pole temperature gradient to kinetic energy. The same effect underlies the development of ocean eddies.

These theories also yield self-consistent simple models of the extratropical atmosphere and ocean on large scales and have solutions where the wind is close to geostrophic. Subsequent mathematical analysis has shown that both systems of equations can be solved for large times, and it is thus plausible that the large-scale evolution of the atmosphere and ocean can be described as the solution of a geostrophic model plus an error term, see Vanneste (2013). Such solutions are often referred to as a *slow manifold* because they describe the large-scale evolution without the small-scale waves and turbulence that are always present in the atmosphere. Smallness of the error term means that the effects of small-scale waves and turbulence on the geostrophic motions are small. While it has sometimes been speculated that an exact slow manifold might exist, which would allow large-scale flows to be predicted independently of other motions; it is clear from much research that this is not the case, e.g. Lorenz (1992), Ford *et al.* (2000).

The justification for the quasi-geostrophic approximation is not the same as the semi-geostrophic approximation. As will be discussed more fully in this book, the quasi-geostrophic approximation involves simplifications which are not valid on scales larger than that of individual weather systems. These simplifications are extremely useful in showing that the equations can be solved and in understanding the solutions. However, they limit the application to real atmospheric and oceanic data. The semi-geostrophic approximation does not require these simplifications, so it is applicable to real atmospheric and oceanic data on large scales. However, analysing it is much more challenging. In this book, solutions of the fundamental equations are shown to be sufficiently close to the semi-geostrophic solutions on large scales that much of the high predictability of the semi-geostrophic system is inherited by the fundamental equations. This can explain the current success of weather forecasts.

The analysis of the semi-geostrophic equations exploits exciting recent developments in understanding why some nonlinear systems are much more stable than might be expected. A big advance was made by Hoskins (1975) who showed that coherent discontinuous structures, such as the weather fronts shown in Figs. 1.2 and 1.3, can be considered as part of the 'large-scale' flow and can be described by the semi-geostrophic equations. This was subsequently extended using the mathematical theory of 'optimal transport' to show that anisotropies such as fronts and jet-streams are a natural part of large-scale solutions, and that persistent anomalies in the pressure pattern, which correspond to periods of anomalous weather, are also natural. In addition, the theory shows that many observed phenomena in the atmosphere can be understood as a small-scale response to large-scale forcing in line with manual forecasting experience.

Improvements to observations of tropical meteorology show that tropical weather is fundamentally different from that in the extratropics. Much tropical weather is tied to the diurnal cycle and is the same from day to day apart from seasonal changes. The seasonal variations are related to large-scale wind patterns, in particular monsoons. Superposed on this basic climatology are various modes of variability, which are usually manifested as variations in the amount of convective precipitation. In the tropics the axis of the Earth's rotation is nearly horizontal, so the only control it exerts on the large-scale dynamics is to enforce geostrophic balance between the east-west wind and the north-south pressure gradient. Tropical cyclones are not like this, but they require some rotation to allow their initial development before they become self-sustaining. This means that they can only form more than about 10° from the equator as shown in Fig. 1.6.

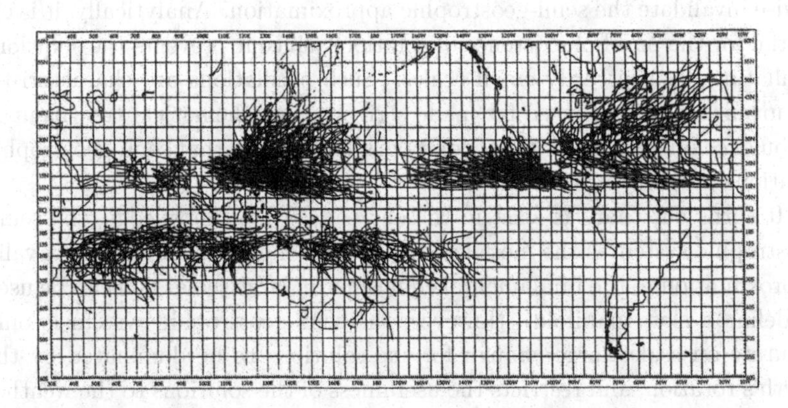

Fig. 1.6 Tracks of tropical cyclones (maximum winds > 17 ms^{-1}) for the period 1979-88 (from Neumann (1993)), Gordon *et al.* (1998). ©1998 Adrian Gordon *et al.*

This volume first sets out the fundamental equations. It then defines dimensionless parameters appropriate for large-scale weather systems, and reviews the choices of relevant asymptotic regimes. This topic is the main topic of most dynamical meteorology and geophysical fluid dynamics textbooks; and these, such as Gill (1982) and Vallis (2017), should be consulted for a more detailed and comprehensive treatment. In particular, it is shown that quasi-geostrophic theory and semi-geostrophic theory are appropriate in different regimes. Rigorous mathematical theory is required to prove that these systems can be solved for large times. Such results are necessary if it is to be true that the fundamental equations are accurately approximated by a quasi-geostrophic or semi-geostrophic solution for large times. If this is the case, then the qualitative behaviour of the geostrophic models will be inherited by the solution of the fundamental equations on large scales.

The quasi-geostrophic equations can be analysed using standard results from elliptic regularity theory. However, novel mathematical ideas are needed to prove that the semi-geostrophic system can be solved for large times. This is because the development of discontinuities means that only 'weak' solutions exist. It is possible to characterise semi-geostrophic equations as a sequence of minimum energy states. The proof that such states can be found requires the technique of 'optimal transport', originally developed in probability and statistics, Kantorovich (1942). This theory shows that a minimum energy state satisfies a convexity condition. Physically, this states that the solutions, if substituted into the complete fluid equations, would be stable to fast-growing perturbations which would otherwise invalidate the semi-geostrophic approximation. Analytically, it is essential for the proof of existence of solutions, since it prevents the solutions oscillating on arbitrarily small scales. Such oscillations are characteristic of turbulence, and have so far made a rigorous mathematical treatment of turbulence impossible. The convexity principle states that semi-geostrophic solutions are non-turbulent.

It turns out that, at least in the current state of knowledge, the semi-geostrophic system is the most general system of equations which is a valid approximation to the fundamental equations on large scales and can be used to define a slow manifold. However, the semi-geostrophic system is only accurate on scales large enough for the solutions to be dominated by the Earth's rotation, and restricts the usefulness of the solutions to the weather systems familiar from middle latitude weather maps such as Fig. 1.3 and to quasi-steady tropical circulations driven by spatially varying heat sources. Different systems of equations are required to describe tropical variability and tropical cyclones.

This book illustrates that the resulting solutions are sufficiently close to reality to describe 'traditional' extratropical synoptic meteorology. The convexity condition makes the solutions very insensitive to perturbations, which is the reason for their high predictability. Thus it is reasonable to suggest that the high predictability of semi-geostrophic solutions underpins the observed predictability of real weather. The solutions also maintain large-scale circulation anomalies for long periods, as observed, unlike two-dimensional turbulence. Thus they can explain periods of anomalous weather. In addition, they explain many well-observed smaller-scale phenomena; such as the drag on the atmosphere exerted by large-scale mountain ranges, the differing inland penetration of sea-breeze circulations at different latitudes, and the fact that tropical cyclones cannot normally be generated close to the equator. In the ocean they explain the persistence of eddies and the outcropping of density surfaces.

Chapter 2

The Governing Equations and Asymptotic Approximations to Them

2.1 Basic equations

The starting point is the traditional one, that recognises that the atmosphere and ocean can be regarded as fluid continua, and obey the basic physical laws of dynamics and thermodynamics. This was the basis of Richardson's proposal. It is remarkable how these basic equations, which were formulated many years ago, have stood the test of increasingly sophisticated computer solution and experimental verification. The study of fluid dynamics recognises that fluids exhibit a wide range of different behaviour under different circumstances. This study is facilitated by writing the equations in particular forms, and studying various approximations to them. Thus, in studying the atmosphere and ocean, it may be convenient to rewrite the basic equations in ways which are different from those used in other applications of fluid dynamics. In this book, a comprehensive description is given of the application to the atmosphere. Many of the same principles can be applied in the ocean, and these are pointed out. However, a comprehensive treatment of ocean dynamics is not attempted. The presentation of the basic equations given here is sufficient to explain and justify the large-scale theory described in the rest of the book. Much greater detail is given on the derivation of the basic equations in textbooks such as Gill (1982) and Vallis (2017). These also include a comprehensive discussion of the ocean as well as the atmosphere.

The Earth is assumed to rotate with a vector angular velocity Ω about an axis through the coordinate poles. The acceleration due to gravity and the centrifugal acceleration due to the Earth's rotation are combined. Geopotential surfaces are defined as normal to this combined acceleration. The geopotential surfaces are approximated by spherical surfaces as discussed by White *et al.* (2005). The equations are then written in spherical polar coordinates (λ, ϕ, r), with origin at the centre of the Earth. The Earth's surface is assumed to be a spherical surface with radius a with perturbations due to orography. It is defined by the equation $r = r_0(\lambda, \phi)$. The combined gravitational and centrifugal acceleration is assumed to be towards the origin, with a constant magnitude g. The atmosphere is assumed to consist of a compressible ideal gas with pressure, density and temperature p, ρ, T which are functions of position and time. It contains a mixing ratio q of water vapour. It moves with a vector velocity $\mathbf{u} = (u, v, w)$. The evolution is described by the compressible Navier-Stokes equations, the continuity equation giving mass conservation, the first law of thermodynamics and the equation of state for an ideal gas, all written in a frame of reference rotating with the Earth's angular velocity. These are

$$\begin{aligned}
\frac{\mathrm{D}\mathbf{u}}{\mathrm{D}t} + 2\Omega \times \mathbf{u} + \frac{1}{\rho}\nabla p + g\hat{\mathbf{r}} &= \nu\nabla^2\mathbf{u}, \\
\frac{\partial \rho}{\partial t} + \nabla \cdot (\rho\mathbf{u}) &= 0, \\
C_v\frac{\mathrm{D}T}{\mathrm{D}t} - \frac{RT}{\rho}\frac{\mathrm{D}\rho}{\mathrm{D}t} &= \kappa_h\nabla^2 T + S_h + LP, \\
\frac{\mathrm{D}q}{\mathrm{D}t} &= \kappa_q\nabla^2 q + S_q - P, \\
p &= \rho RT.
\end{aligned} \tag{2.1}$$

Here $\mathrm{D}/\mathrm{D}t$ is the Lagrangian time derivative $\partial/\partial t + \mathbf{u}\cdot\nabla$ and $\hat{\mathbf{r}}$ is a unit vector in the radial direction. R is the gas constant and C_v the specific heat of air at constant volume, assumed to take the same value everywhere. ν is the kinematic viscosity, also assumed constant. S_h and S_q are the total heat and moisture sources. P is the rate of conversion of water vapour to liquid water or ice, with L the associated latent heat. κ_h and κ_q are the thermal conductivity and moisture diffusivity, both assumed constant. Most of these constant coefficient assumptions are not made in operational weather forecasting and climate prediction models, which include a detailed description of atmospheric composition and cloud physics, with the necessary extra complexity in the equations. This extra sophistication is not required for the purposes of this book.

These equations form a system of seven equations for the unknowns $(\mathbf{u}, p, \rho, T, q)$. The obvious physical boundary conditions are that $\mathbf{u} = 0$ at $r = r_0$ and that $p, \rho \to 0$ as $r \to \infty$. While the no-slip condition at the

Earth's surface is standard, the issue of the correct mathematical upper boundary condition for an unbounded atmosphere is open.

In the absence of dissipation and source terms, so that the right hand side terms of the first four equations of (2.1) vanish, equations (2.1) solved in a closed time-independent region Γ with boundary conditions $\mathbf{u} \cdot \mathbf{n} = 0$, where $\mathbf{n}$ is a vector pointing outward from the boundary, conserve the energy integral

$$E = \int_\Gamma \rho \left(\frac{1}{2}(u^2 + v^2 + w^2) + C_v T + gr \right) r^2 \cos\phi \mathrm{d}\lambda \mathrm{d}\phi \mathrm{d}r. \tag{2.2}$$

The unrealistic requirement that the upper boundary be rigid can be removed by reformulating the equations in 'mass' coordinates, Wood and Staniforth (2003).

It is convenient to rewrite the first law of thermodynamics in terms of the potential temperature $\theta = T(p/p_{ref})^{-R/C_p} \equiv T/\Pi$, where C_p is the specific heat of air at constant pressure, assumed to have the same value everywhere, p_{ref} is a constant reference pressure equal to a typical pressure at the Earth's surface, and Π is the Exner pressure. This gives, noting $R = C_p - C_v$,

$$\frac{\mathrm{D}\theta}{\mathrm{D}t} = \frac{1}{C_p \Pi}(\kappa_h \nabla^2 T + S_h + LP). \tag{2.3}$$

This form of the equation is particularly useful in situations where the right hand side terms can be neglected. The momentum equations can also be rewritten using the definition of θ and the equation of state (the last equation of (2.1)) to obtain

$$\frac{\mathrm{D}\mathbf{u}}{\mathrm{D}t} + 2\Omega \times \mathbf{u} + C_p \theta \nabla \Pi + g\hat{\mathbf{r}} = \nu \nabla^2 \mathbf{u}. \tag{2.4}$$

In the absence of dissipation and source terms, equations (2.1) imply a conservation law for the *Ertel potential vorticity*

$$\mathcal{Q} = \frac{1}{\rho}(\nabla \times \mathbf{u} + 2\Omega) \cdot \nabla \theta \tag{2.5}$$

in the form

$$\frac{\mathrm{D}\mathcal{Q}}{\mathrm{D}t} = 0. \tag{2.6}$$

The main differences between the governing equations for the ocean and equations (2.1) are that compressibility effects are very weak, so that incompressible equations can be used. The time derivatives of ρ thus do not appear in the second and third equations of (2.1). There is no water vapour mixing ratio, but there is a salinity which obeys a similar conservation law. The equation of state defines the density to be a function of temperature and salinity. There is no need to introduce a potential temperature as (2.3) because the term in $\mathrm{D}\rho/\mathrm{D}t$ in the thermodynamic equation does not appear. Thus the density obeys a conservation law

$$\frac{\mathrm{D}\rho(S,T)}{\mathrm{D}t} = \text{source terms}, \tag{2.7}$$

where S is the salinity. The upper boundary condition is also different, and represents a free surface acted on by wind stress.

2.2 Scale analysis, key dimensionless parameters, and reduced equations

2.2.1 *Dimensionless equations*

This book is concerned with the large-scale flow of the atmosphere and ocean. As above, a detailed presentation is given for the atmosphere, with a brief summary of the differences required for the ocean. The first step is to identify the magnitude of the various terms in (2.1). This is aided by recognising that the acceleration due to gravity, g, is much larger than the acceleration of the air in a weather system. However, this acceleration is largely compensated by the vertical pressure gradient. Therefore define a time-independent reference state at rest, which satisfies equations (2.1), with uniform potential temperature θ_0 and with pressure p_0 and density ρ_0 depending only on the radial coordinate. This is given by an Exner pressure $\Pi_0(r)$ satisfying

$$\begin{aligned} C_p\theta_0\frac{\mathrm{d}\Pi_0}{\mathrm{d}r} + g &= 0, \\ \theta_0 &= \text{constant}, \\ \Pi &= 1 \text{ at } r = a. \end{aligned} \tag{2.8}$$

Subtract this state from (2.4). The equation becomes

$$\frac{\mathrm{D}\mathbf{u}}{\mathrm{D}t} + 2\Omega \times \mathbf{u} + C_p\theta\nabla\Pi' - g\frac{\theta'}{\theta_0}\hat{\mathbf{r}} = \nu\nabla^2\mathbf{u} \tag{2.9}$$

where $\theta' = \theta - \theta_0$ and $\Pi' = \Pi - \Pi_0$.

In the continuity equation, it is necessary to distinguish the basic state and perturbation ρ in the scale analysis, so rewrite the equation as

$$\frac{\partial \rho'}{\partial t} + \nabla \cdot ((\rho_0 + \rho')\mathbf{u}) = 0, \tag{2.10}$$

where $\rho' = \rho - \rho_0$. Similarly rewrite the equation of state from (2.1) as

$$\rho = \frac{p_{ref}\Pi^{C_v/R}}{R\theta}. \tag{2.11}$$

Applying (2.11 at $\Pi = 1$ gives $\rho_0(a) = p_{ref}/(R\theta_0)$. Subtracting the basic state from (2.11) gives

$$\rho' = \frac{p_{ref}}{R\theta}\left(\Pi^{C_v/R} - \Pi_0^{C_v/R}\right) - \frac{\rho_0\theta'}{\theta}. \tag{2.12}$$

In the ocean, the equivalent procedure would be to subtract a state with uniform density.

The next step is to non-dimensionalise equations (2.1) with (2.9). Introduce vertical and horizontal scales H and L, so that vertical and horizontal derivatives have typical magnitudes H^{-1} and L^{-1}. Introduce a time-scale τ so that time derivatives have magnitude τ^{-1}. Define scales (U, W) for the horizontal and vertical wind and $\rho_0(a)$ for the perturbation density ρ'. Write dimensionless variables with a hat, so that $u = U\hat{u}$. The Exner pressure Π' is dimensionless by definition, so $\hat{\Pi} = \Pi'$. θ_0 as already defined is used as the scale for the perturbation potential temperature θ'. The Lagrangian derivative $\frac{\mathrm{D}u}{\mathrm{D}t}$ has to be expanded in Eulerian form, and becomes

$$\frac{U}{\tau}\frac{\partial \hat{u}}{\partial \hat{t}} + \frac{U^2}{L}\left(\hat{u}\frac{\partial \hat{u}}{\partial \hat{x}} + \hat{v}\frac{\partial \hat{u}}{\partial \hat{y}}\right) + \frac{UW}{H}\hat{w}\frac{\partial \hat{u}}{\partial \hat{z}}$$

In the regimes of interest for large-scale flow, it is usual to assume that all terms in the Lagrangian time derivative scale in the same way, so that $\tau = L/U = H/W$. This will be referred to as the *Lagrangian time-scale assumption*. However, there is one particular exception discussed below, and, in addition, there are many small-scale flows, such as gravity waves, where this assumption is inappropriate.

Since the difference between vertical and horizontal scales is fundamental, the dimensionless equations are written separately for horizontal and vertical components. The subscript h indicates the horizontal component of a vector operator. The Earth's rotation vector $2\Omega = (0, 2\Omega\cos\phi, 2\Omega\sin\phi)$ where ϕ is the latitude. In some cases it will be assumed that the two horizontal scales are different, but the Lagrangian time-scale assumption will be retained.

The assumptions above mean that, for a variable χ, the horizontal derivative scales as H/L times the vertical derivative. This defines the aspect ratio A of the flow and is appropriate for wind components. This is not a realistic assumption if χ is the pressure or density, as both have a large basic state variation in the vertical. Even after subtracting the basic state (2.8), it may or may not be appropriate to assume that the horizontal and vertical derivatives are related by the aspect ratio of the flow. The same applies to the perturbation potential temperature θ'. It will be seen that in the hydrostatic regimes, which are the subject of this book, the vertical derivative of Π' is closely related to θ', so that this defines the scaling. However, making different assumptions about the ratio of derivatives of θ' leads to important different scalings. To allow for this possibility, the vertical and horizontal derivatives in the equation for θ' will be separated and scaled differently.

In this presentation of dynamical regimes, the diabatic terms on the right hand sides of (2.3) and the moisture conservation equation are omitted. These will be returned to later in the book, in particular in section 2.7.4 and in Chapter 6. Equations (2.1), using (2.3), (2.9) , (2.10) and (2.12) can then be written

$$
\begin{aligned}
&\frac{U^2}{L}\frac{\mathrm{D}(\hat{u},\hat{v})}{\mathrm{D}\hat{t}} + 2\Omega(W\hat{w}\cos\phi - U\hat{v}\sin\phi, U\hat{u}\sin\phi) \\
&\quad + \frac{C_p\theta_0}{L}(1+\hat{\theta})\nabla_h\hat{\Pi} = \nu\frac{U}{L^2}\nabla_h^2(\hat{u},\hat{v}) + \nu\frac{U}{H^2}\frac{\partial^2(\hat{u},\hat{v})}{\partial\hat{r}^2}, \\
&\frac{W^2}{H}\frac{\mathrm{D}\hat{w}}{\mathrm{D}\hat{t}} - 2\Omega U\hat{v}\cos\phi + \frac{C_p\theta_0}{H}(1+\hat{\theta})\frac{\partial\hat{\Pi}}{\partial\hat{r}} - g\hat{\theta} = \nu\nabla^2\hat{w}, \\
&\frac{\partial\hat{\rho}}{\partial\hat{t}} + \nabla\cdot\left(\left(\frac{\rho_0}{\rho_0(a)} + \hat{\rho}\right)\hat{\mathbf{u}}\right) = 0, \\
&\frac{U\theta_0}{L}\left(\frac{\mathrm{D}_h\hat{\theta}}{\mathrm{D}\hat{t}} + \alpha\hat{w}\frac{\partial\hat{\theta}}{\partial\hat{r}}\right) = 0, \\
&\hat{\rho} = \frac{\theta_0}{\theta}\left(\Pi^{C_v/R} - \Pi_0^{C_v/R}\right) - \frac{\rho_0\hat{\theta}}{\rho_0(a)(1+\hat{\theta})}.
\end{aligned}
\tag{2.13}
$$

The notation $D_h/D\hat{t}$ means $\frac{\partial}{\partial\hat{t}} + \hat{u}\frac{\partial}{\partial\hat{x}} + \hat{v}\frac{\partial}{\partial\hat{y}}$. The factor α in the fourth equation means that $\partial\hat{\theta}/\partial\hat{r}$ is scaled by $\alpha L/H$ times $\partial\hat{\theta}/\partial\hat{x}$. Physical relevance requires $\alpha \geq 1$.

2.2.2 *Asymptotic regimes and reduced equations*

Equations (2.13) include a large number of dimensionless coefficients. Special types of solution are obtained by choosing appropriate relations between them. The aim is to replace these coefficients with a single one, ϵ. The result is called a *distinguished limit*, Klein (2009). Then in the limit of small ϵ, eqs. (2.13) can be replaced by a system of reduced equations. The Lagrangian time-scale assumption is an example of this. Assuming that the reduced equations can be solved, their solutions will be a good approximation to those of (2.13) when ϵ is small. This concept is more fully discussed in Vanneste (2013). It is very important that the reduced equations can be solved for large enough times for this approximation property to be useful. While there are many ways of choosing an approximation to any given order of ϵ, only a few will lead to well-posed reduced equations.

A simple example of this procedure is the anelastic approximation (Durran (1989), Bannon (1996)). Linearise the momentum and continuity equations from (2.13), noting that the reference state has constant $\theta = \theta_0$. Combining the linear equations gives, omitting rotation and dissipation terms:

$$\frac{\partial^2 \hat{\rho}}{\partial \hat{t}^2} + \frac{\rho_0}{\rho_0(a)} \frac{C_p \theta_0}{U^2} \nabla^2 \hat{\Pi} = 0. \tag{2.14}$$

The equation of state from (2.13) implies for small $\hat{\Pi}$ and zero $\hat{\theta}$ that

$$\hat{\rho} = \frac{C_v}{R} \hat{\Pi}.$$

Inserting into (2.14) gives

$$\frac{\partial^2 \hat{\rho}}{\partial \hat{t}^2} + \frac{R C_p \theta_0}{C_v U^2} \nabla^2 \hat{\rho} = 0.$$

Standard thermodynamic relations then give

$$\frac{\partial^2 \hat{\rho}}{\partial \hat{t}^2} + \frac{\rho_0}{\rho_0(a)} \frac{c^2}{U^2} \nabla^2 \hat{\rho} = 0.$$

where the speed of sound $c = \sqrt{\gamma R \theta_0}$ at $r = a$, where γ is the ratio of specific heats C_p/C_v. Define ϵ to be the Mach number U/c, and rewrite (2.14) as

$$\frac{\partial^2 \hat{\rho}}{\partial \hat{t}^2} + \frac{\rho_0}{\rho_0(a)} \epsilon^{-2} \nabla^2 \hat{\rho} = 0. \tag{2.15}$$

This can be reconciled with the continuity equation from (2.13) if the time-scale τ is chosen to be $\epsilon U/L$ so that

$$\epsilon^{-1} \frac{\partial \hat{\rho}}{\partial \hat{t}} + \nabla \cdot \left(\left(\frac{\rho_0}{\rho_0(a)} + \hat{\rho} \right) \hat{\mathbf{u}} \right) = 0. \tag{2.16}$$

As discussed in Durran (1989), there are a number of ways to approximate (2.15) to accuracy $O(\epsilon^2)$. Some of these lead to a system of equations that fails to conserve energy. Though this is not the only requirement, it is to be expected that an approximation to (2.13) which remains accurate for a long time would need to conserve energy. Also note that the neglect of $\hat{\theta}$ in deriving the speed of sound limits the applicability of this approximation in the atmosphere significantly because the frequencies of sound waves and gravity waves overlap, see Davies *et al.* (2003). In the ocean, the continuity equation is simply $\nabla \cdot \mathbf{u} = 0$, and sound waves are not considered.

2.2.3 *The shallow atmosphere hydrostatic approximation*

Now use this procedure to derive approximations to the fundamental equations which are accurate on large horizontal scales in the atmosphere and ocean. The first step is to recognise that the atmosphere and ocean are 'shallow'. About 90% of the mass of the atmosphere is contained in the lowest 15km, which is much smaller than the radius of the earth. Thus the radial coordinate r can be replaced by a wherever it appears undifferentiated to an accuracy of 0.25%. This applies irrespective of the type of flow considered. This approximation is discussed in detail by White *et al.* (2005). In addition, depressions and anticyclones have a much larger scale than 15km, so the aspect ratio $A = H/L$ is much less than 1. This can be regarded as a characterisation of large-scale atmospheric flows. The Lagrangian time-scale assumption then requires $W/U \simeq H/L$. For convenience, write C for the dimensionless parameter $\frac{C_p\theta_0}{U^2}$. The choice of this is discussed below. The first equation of 2.13) can then be written

$$\frac{\mathrm{D}(\hat{u},\hat{v})}{\mathrm{D}\hat{t}} + Ro^{-1}(A\hat{w}\cos\phi - \hat{v}\sin\phi, \hat{u}\sin\phi) + C(1+\hat{\theta})\nabla_h\hat{\Pi} = Re^{-1}\nabla_h^2(\hat{u},\hat{v}) + (A^2 Re)^{-1}\frac{\partial^2(\hat{u},\hat{v})}{\partial\hat{r}^2}. \quad (2.17)$$

Note that two additional dimensionless parameters have had to be introduced. $Ro = U/(2\Omega L)$, the Rossby number, measures the ratio of the inertial frequency 2Ω to that associated with advection U/L. It can also be interpreted as the ratio of the Lagrangian time-scale given by $\frac{\mathrm{D}}{\mathrm{D}\hat{t}}$ to the inertial time-scale 2Ω. For a horizontal length scale L of 1000km and typical wind speeds of $10\mathrm{ms}^{-1}$, Ro is about 0.1. The Reynolds number, $Re = UL/\nu$, is extremely large, typically 10^{12}. However, near the lower boundary A is very small, so the final term in (2.17) may not be small, and is critical in representing the lower boundary condition. While considering purely internal dynamical regimes this term can be neglected. The lower boundary layer will be discussed in Chapter 6.

Now consider the vertical momentum equation, the second equation in 2.13). Using the definitions above, this becomes

$$\frac{\mathrm{D}\hat{w}}{\mathrm{D}\hat{t}} + A^{-1}\left(Ro^{-1}\hat{v}\cos\phi + CA^{-1}(1+\hat{\theta})\frac{\partial\hat{\Pi}}{\partial\hat{r}} - \frac{gL}{U^2}\hat{\theta}\right) = Re^{-1}\nabla_h^2\hat{w} + (A^2Re)^{-1}\frac{\partial^2\hat{w}}{\partial\hat{r}^2}. \tag{2.18}$$

For a fairly large horizontal length scale L of 100km and typical wind speeds of 10ms^{-1}, $\frac{gL}{U^2} \simeq 10^4$. The large value of Re allows the right hand side to be neglected. So it is necessary to choose $C = \frac{AgL}{U^2} = \frac{gH}{U^2}$ so that the third term in (2.18) balances the gravitational term. This defines the appropriate scaling for the vertical derivative of Π'. Defining $\epsilon = \frac{AU^2}{gL}$ as the ratio of the first term to the third term, (2.18) becomes to O(ϵ)

$$(1+\hat{\theta})\frac{\partial\hat{\Pi}}{\partial\hat{r}} - \hat{\theta} = 0. \tag{2.19}$$

This defines hydrostatic balance. Using the values of U and L above and an aspect ratio of 0.1 gives $\epsilon \simeq 10^{-5}$, so the hydrostatic approximation is very accurate. In dimensional terms, it is

$$C_p\theta\frac{\partial\Pi'}{\partial r} - g\frac{\theta'}{\theta_0} = 0. \tag{2.20}$$

Next, define the Brunt-Väisälä frequency N in dimensional terms as

$$N^2 = \frac{g}{\theta}\frac{\partial\theta'}{\partial r}. \tag{2.21}$$

Using scaled variables on the right hand side, noting the scaling of $\partial\theta'/\partial r$ defined after (2.13), this becomes

$$N^2 = \frac{g\alpha}{H(1+\hat{\theta})}\frac{\partial\hat{\theta}}{\partial\hat{r}}. \tag{2.22}$$

Thus write the ratio $g\alpha/H$ as N^2.

The Froude number Fr is defined as U/NH, which is a dimensionless parameter defining the ratio of the wind speed to the internal gravity wave speed. (2.22) shows that this will be smaller if α is larger. Then ϵ and C as defined above become $\epsilon = A^2\alpha Fr^2$ and $C = \alpha^{-1}Fr^{-2}$. This form of ϵ shows that the hydrostatic approximation is accurate in either the limit $A \to 0$ or $Fr \to 0$.

The resulting system of dimensionless shallow atmosphere hydrostatic equations can now be written

$$
\begin{aligned}
\frac{\mathrm{D}(\hat{u},\hat{v})}{\mathrm{D}\hat{t}} + Ro^{-1}\sin\phi(-\hat{v},\hat{u}) + \alpha^{-1}Fr^{-2}(1+\hat{\theta})\nabla_h\hat{\Pi} &= (A^2Re)^{-1}\frac{\partial^2(\hat{u},\hat{v})}{\partial\hat{r}^2},\\
(1+\hat{\theta})\frac{\partial\hat{\Pi}}{\partial\hat{r}} - \hat{\theta} &= 0,\\
\frac{\partial\hat{\rho}}{\partial\hat{t}} + \nabla\cdot\left(\left(\frac{\rho_0}{\rho_0(a)} + \hat{\rho}\right)\hat{\mathbf{u}}\right) &= 0,\\
\frac{\mathrm{D}_h\hat{\theta}}{\mathrm{D}\hat{t}} + \alpha\hat{w}\frac{\partial\hat{\theta}}{\partial\hat{r}} &= 0,\\
\hat{\rho} = \frac{\theta_0}{\theta}\left(\Pi^{C_v/R} - \Pi_0^{C_v/R}\right) &- \frac{\rho_0\hat{\theta}}{\rho_0(a)(1+\hat{\theta})}.
\end{aligned}
\tag{2.23}
$$

The shallow atmosphere hydrostatic approximation is very accurate on large scales, and is used in many weather forecasting and climate models. As such, the resulting equations are often referred to as the 'primitive' equations. These were regarded as the fundamental equations of the atmosphere before the study of and ability to predict smaller scale phenomena made the use of the fully compressible equations necessary. Given a vertical extent of 15km for the troposphere, a model with a gridlength of 15km or greater will not be able to resolve any flow with aspect ratio of order one, and so the hydrostatic approximation will be valid. Typically these equations are written in pressure coordinates, because in the original equation (2.4) with no basic state subtracted, hydrostatic balance is defined by $C_p\theta\partial\Pi/\partial r + g = 0$. This ensures that Π is a monotonic function of r. This simplifies the continuity equation and avoids the need to use the equation of state explicitly. However it makes the analysis of further approximations more difficult, so is not done at this point. In the ocean, similar arguments justify the use of the hydrostatic assumption, but the incompressibility condition means that the equations can be used as they stand and there is no need to use potential temperature as a variable or pressure as a coordinate.

2.2.4 *Properties of the shallow atmosphere hydrostatic equations*

There are two basic issues with these equations. The first is that the Coriolis term, which takes the form $2\Omega\times\mathbf{u}$ in (2.1), is no longer the curl of a vector. Thus Kelvin's circulation theorem will not hold. This reflects a loss of axial symmetry, because of the imposition of a radial symmetry enforced by the

Earth's gravity. It is also reflected in the observed behaviour, which shows that extratropical weather is qualitatively different from tropical weather. The accuracy of models making this approximation in simulating real flows shows that the circulation theorem cannot be fundamental in large-scale atmospheric or oceanic dynamics.

The second issue relates to the independence of large-scale flow from small-scale flow. In the scale analysis above, viscous effects can be neglected except at the lower boundary. It is physically clear that the no-slip constraint at the lower boundary is fundamental whatever scale is of interest. However, if viscous effects can be neglected in the interior, it requires that the equations decribing the interior large-scale flow must be solvable without including viscous effects explicitly. It may be that the solutions are 'weak', so not differentiable at every point or time, but the weak solutions must be well-defined so that they are entirely determined by large-scale information.

In recent years there have been many rigorous analyses of (2.23), mostly in the simpler ocean context and with the further Boussinesq approximation that the term $(1+\hat{\theta})$ in the first equation of (2.23) is replaced by unity. This is a very poor approximation in the atmosphere as shown by Davies *et al.* (2003). Viscosity and thermal conductivity are included. The Prandtl number $Pr = \nu/\kappa_h$. A prescribed dimensionless heat source $\hat{S}_h$ is included. Salinity is not included. The equations are then

$$
\begin{aligned}
\frac{\mathrm{D}(\hat{u},\hat{v})}{\mathrm{D}\hat{t}} + Ro^{-1}\sin\phi(-\hat{v},\hat{u}) + \alpha^{-1}Fr^{-2}\nabla_h\hat{p} &= \\
Re^{-1}\nabla_h^2(\hat{u},\hat{v}) + (A^2Re)^{-1}\frac{\partial^2(\hat{u},\hat{v})}{\partial\hat{r}^2}&, \\
\frac{\partial\hat{p}}{\partial\hat{r}} - \hat{T} = 0&, \qquad (2.24)\\
\nabla\cdot\hat{\mathbf{u}} = 0&, \\
\frac{\mathrm{D}_h\hat{T}}{\mathrm{D}\hat{t}} + \alpha\hat{w}\frac{\partial\hat{T}}{\partial\hat{r}} = (RePr)^{-1}\nabla_h^2\hat{T} + (A^2RePr)^{-1}\frac{\partial^2\hat{T}}{\partial\hat{r}^2} + \hat{S}_h&.
\end{aligned}
$$

Boundary conditions are given in Cao and Titi (2007) which are suitable for an ocean basin, these are wind-driven on the upper boundary, and free-slip and zero heat flux on the sidewalls and lower boundary.

It has been proved that the solutions can blow up if viscosity is not included, Cao *et al.* (2015). Though this does not exclude the possibility of a well-posed weak solution, no such formulation has been found to date. The difficulty can be illustrated by considering the solution of (2.24) with

T constant. This means that $\partial\hat{p}/\partial\hat{r}$ is independent of the horizontal coordinates and so $\nabla_h\hat{p}$ is independent of $\hat{r}$. If $(\hat{u},\hat{v})$ depend on $\hat{r}$, then the first equation of (2.24) is likely to have singular solutions because of intersecting fluid trajectories. There is also no means of limiting the vertical derivatives of $(\hat{u},\hat{v})$. The Brunt-Väisälä frequency (2.21) for equations (2.24) will depend on $\partial T/\partial r$ and is thus zero for constant T, so that Fr is infinite. This violates the condition for the hydrostatic approximation to be valid, which is that $\epsilon = A^2\alpha Fr^2$ is small as in the atmospheric case.

These issues suggest that (2.24) may well be solvable if viscosity is included, and that vertical viscosity might be sufficient. In Lions *et al.* (1992) it is proved that weak solutions of (2.24) exist for large times with only vertical viscosity. Uniqueness has not been established. Cao and Titi (2007) prove that unique strong solutions exist with general viscosity. They assume that initial data are given for $\hat{\mathbf{u}}$ and $\hat{T}$ which are C^∞ and satisfy the boundary conditions. The function spaces V_1 and V_2 used below express this requirement.

Theorem 2.1. *Let $\hat{S}_h \in H_1(\Gamma)$, $\hat{\mathbf{u}}_0 \in V_1$, $\hat{T}_0 \in V_2$ and $\hat{\tau} > 0$, be given. Then there exists a unique strong solution $(\hat{\mathbf{u}},\hat{T})$ of the system (2.24) on the interval $[0,\hat{\tau}]$ which depends continuously on the initial data.*

However, it is to be expected that the solutions will depend on the viscosity. In the real system, viscosity is only important on very small scales where the hydrostatic approximation is no longer valid, so the results imply a predictability limitation for hydrostatic flows.

In several papers, such as Liu and Titi (2019), the results are extended to the compressible case (2.23). However, this is only achieved for the case $\hat{\theta} = 0$ which implies $Fr = \infty$ and so is not physically relevant, as discussed above. The difficulty is that there is no theory available for the compressible Navier-Stokes equations with non-uniform $\hat{\theta}$.

2.2.5 *Classification of shallow atmosphere hydrostatic regimes*

The shallow atmosphere hydrostatic equations can describe a multiplicity of flow regimes. The next step is to identify those that are most relevant for large-scale atmospheric and oceanic flows.

There is no prognostic equation for the vertical velocity in (2.23). The hydrostatic relation (2.19) and the equation of state together give two constraints between the thermodynamic variables $\hat{\Pi}$, $\hat{\rho}$ and $\hat{\theta}$. Consistency of the separate evolution equations for $\hat{p}$ and $\hat{\theta}$ yields 'Richardson's equation', see White (2002). This is easiest derived directly from (2.1), applying the hydrostatic relation and omitting the right hand side of the thermodynamic equation as in (2.23). This gives in nondimensional form

$$\gamma\hat{p}\frac{\partial}{\partial\hat{r}}\left(\frac{\partial\hat{w}}{\partial\hat{r}}+\nabla_h\cdot\hat{\mathbf{u}}\right)=\frac{\partial\hat{p}}{\partial\hat{r}}\nabla_h\cdot\hat{\mathbf{u}}-\frac{\partial\hat{\mathbf{u}}}{\partial\hat{r}}\cdot\nabla_h\hat{p}. \tag{2.25}$$

Next, rewrite (2.23) in a simplified form to motivate the choice of regimes. The horizontal momentum equations and the thermodynamic equation can be written as

$$\begin{gathered}\frac{\partial(\hat{u},\hat{v})}{\partial\hat{t}}+Ro^{-1}\sin\phi(-\hat{v},\hat{u})+\\ \alpha^{-1}Fr^{-2}(1+\hat{\theta})\nabla_h\hat{\Pi}=(F_1,F_2),\\ \frac{\partial\hat{\theta}}{\partial\hat{t}}+\alpha\hat{w}\frac{\partial\hat{\theta}}{\partial\hat{r}}=F_3.\end{gathered} \tag{2.26}$$

where F_1, F_2 and F_3 represent all the remaining terms. The set of equations is completed by the diagnostic equations (2.19), (2.25) and the equation of state.

Next define the horizontal divergence by

$$\hat{\Delta}=\nabla_h\cdot(\hat{u},\hat{v}).$$

and the vertical component of the vorticity derived from the horizontal velocity as

$$\hat{\zeta}=\frac{1}{\cos\phi}\left(\frac{\partial\hat{v}}{\partial\lambda}-\frac{\partial}{\partial\phi}(\hat{u}\cos\phi)\right).$$

Define $\mu=\sin\phi$. Then the first two of equations (2.26) can be combined into an equation for the evolution of $\hat{\Delta}$. Transferring the terms resulting from differentiating μ and $\hat{\theta}$ to the right hand side gives

$$\frac{\partial\hat{\Delta}}{\partial\hat{t}}-Ro^{-1}\mu\hat{\zeta}+\alpha^{-1}Fr^{-2}(1+\hat{\theta})\nabla_h^2\hat{\Pi}=F_4. \tag{2.27}$$

Now calculate the second time derivative, substituting for $\partial(\hat{u},\hat{v})/\partial\hat{t}$ using the first two equations of (2.26), and again transferring some terms to the right hand side:

$$\frac{\partial^2\hat{\Delta}}{\partial\hat{t}^2}+Ro^{-2}\mu^2\hat{\Delta}+\alpha^{-1}Fr^{-2}(1+\hat{\theta})\nabla_h^2\left(\frac{\partial\hat{\Pi}}{\partial\hat{t}}\right)=F_5. \tag{2.28}$$

The next step is to differentiate with respect to $\hat{r}$, to use the hydrostatic relation (2.19) to substitute for $\partial^2\hat{\Pi}/\partial\hat{r}\partial\hat{t}$ in terms of $\partial\hat{\theta}/\partial\hat{t}$, and to use the third equation of (2.26) for $\partial\hat{\theta}/\partial\hat{t}$. This gives

$$\frac{\partial^2}{\partial\hat{t}^2}\frac{\partial\hat{\Delta}}{\partial\hat{r}} + Ro^{-2}\mu^2\frac{\partial\hat{\Delta}}{\partial\hat{r}} - Fr^{-2}\frac{\partial\hat{\theta}}{\partial\hat{r}}\nabla_h^2\hat{w} = F_6 \tag{2.29}$$

Finally, differentiate (2.29) with respect to $\hat{r}$ and use (2.25) to write $\partial\hat{w}/\partial\hat{r}$ in terms of $\hat{\Delta}$. Again transferring terms to the right hand side, this gives a second order wave equation for $\hat{\Delta}$. It takes the generic form

$$\frac{\partial^2}{\partial\hat{t}^2}\left(\frac{\partial^2\hat{\Delta}}{\partial\hat{r}^2}\right) + \mathbf{L}\hat{\Delta} = F, \tag{2.30}$$

where $\mathbf{L}$ is a linear operator. If $\mathbf{L}\hat{\Delta}$ is larger than the right hand side terms F, then equation (2.30) describes linear inertia-gravity waves. If $\mathbf{L}\hat{\Delta}$ and F are of similar magnitude, and the natural frequencies of the waves are large compared with those contained in F, then the response to the 'forcing' terms F can be expressed as the 'slow' equation

$$\mathbf{L}\hat{\Delta} = F. \tag{2.31}$$

where $\mathbf{L}$ is a positive definite linear operator. It is therefore important to identify circumstances under which $\mathbf{L}$ is large. The definition of $\mathbf{L}$ is implicit in equation (2.29). This means that the slow dynamics expressed by (2.31) are valid if either the Rossby number Ro or the Froude number Fr is small.

From now on, this book is concerned only with regimes where $\mathbf{L}$ is large. Much more detailed presentations of these arguments are given in the articles contained in Norbury and Roulstone (2002). As already noted, in the atmosphere the Rossby number is usually small for horizontal scales greater than 1000km. Weather systems of the type shown in Fig. 1.3 have a scale similar to or larger than this, so can be described by a set of equations which assume a small Rossby number. A typical value of N in the troposphere (the lowest 10km of the atmosphere) is $10^{-2}\mathrm{s}^{-1}$. The Froude number will be small for vertical scales of order 10km. Most weather systems extend over the full depth of the troposphere, so should be well described by equations assuming a small Froude number. A key parameter is the ratio of the Rossby and Froude numbers. Recall that the Rossby number was defined after (2.17) as either $U/(2\Omega L)$ or $(2\Omega)^{-1}\frac{\mathrm{D}}{\mathrm{D}t}$. It is usual to include the $\sin\phi$ factor in defining it as a function of horizontal position, so define the *Coriolis parameter* and Rossby number as

$$f = 2\Omega\sin\phi,\ Ro = U/fL,\ Ro_L = f^{-1}\frac{\mathrm{D}}{\mathrm{D}t}. \tag{2.32}$$

Ro_L is called the Lagrangian Rossby number. Define the L for which $Ro = Fr$ as the *Rossby radius of deformation* $L_R = NH/f$. Another way of expressing the distinction is whether the aspect ratio H/L is greater or less than f/N. In the troposphere this ratio is about 10^{-2}, consistent with the original assumption that it is much less than unity. If $H/L < f/N$, then $Ro < Fr$, and the flow is rotation dominated. If $H/L > f/N$ the flow is stratification dominated. This means that it is very important to identify the vertical scale of the phenomena of interest.

2.3 Various approximations to the shallow water equations

2.3.1 *The shallow water equations*

In this section the 'slow' regimes, which are those with small Ro or Fr, are illustrated using a simpler system of basic equations which is more amenable to analysis, the equations for long waves on shallow water in a rotating frame of reference. This system can be regarded as a set of equations for the vertically averaged flow of the atmosphere. It is relevant because of the assumption of small aspect ratio. Since the aim is to illustrate concepts rather than derive results directly applicable to the real system, it is sufficient to consider only the case of plane geometry.

The equations are written in non-dimensional form as

$$\begin{aligned}
\frac{\mathrm{D}\hat{u}}{\mathrm{D}\hat{t}} + \alpha^{-1}Fr^{-2}\frac{\partial \hat{h}}{\partial \hat{x}} - \mu Ro^{-1}\hat{v} &= 0,\\
\frac{\mathrm{D}\hat{v}}{\mathrm{D}\hat{t}} + \alpha^{-1}Fr^{-2}\frac{\partial \hat{h}}{\partial \hat{y}} + \mu Ro^{-1}\hat{u} &= 0,\\
\frac{\mathrm{D}\hat{h}}{\mathrm{D}\hat{t}} + (\alpha + \hat{h})\nabla \cdot \hat{\mathbf{u}} &= 0.
\end{aligned} \tag{2.33}$$

The notation follows section 2.2.5. The velocity $\hat{\mathbf{u}} = (\hat{u}, \hat{v})$ has only horizontal components. The dimensional fluid depth h has a mean value h_0. Since h_0 and $h - h_0$ may scale differently, the non-dimensional depth is written as $\alpha + \hat{h}$ with $\alpha \geq 1$. The Coriolis parameter can be a function of position, so is written as $\mu(\hat{x}, \hat{y})Ro^{-1}$. The equations are solved in a closed region $\Gamma \subset \mathcal{R}^2$ with $\hat{\mathbf{u}} \cdot \mathbf{n} = 0$ on the boundary. The equations then conserve the energy integral

$$\hat{E} = \frac{1}{2}\int_\Gamma \left((\alpha + \hat{h})(\hat{u}^2 + \hat{v}^2) + (\alpha + \hat{h})^2\right) \mathrm{d}\hat{x}\mathrm{d}\hat{y}. \tag{2.34}$$

Define the potential vorticity

$$\hat{Q} = \frac{Ro\hat{\zeta} + \mu}{(\alpha + \hat{h})} \tag{2.35}$$

where

$$\hat{\zeta} = \frac{\partial \hat{v}}{\partial \hat{x}} - \frac{\partial \hat{u}}{\partial \hat{y}}. \tag{2.36}$$

Then $\hat{Q}$ satisfies the conservation law

$$\frac{\mathrm{D}\hat{Q}}{\mathrm{D}\hat{t}} = 0. \tag{2.37}$$

2.3.2 *Key parameters*

The first step in understanding the solutions is to find analytic solutions for the special case where the equations describe small perturbations about a state of rest. Much more extensive analyses of this type can be found in Gill (1982). Note that the state $\hat{u} = \hat{v} = \hat{h} = 0$ satisfies (2.33). Then seek a general solution $\hat{u} = u'$, $\hat{v} = v'$, $\hat{h} = h'$. Assume that all primed quantities are small, so that all the terms linear in the primed quantities have to satisfy the equations on their own. Assume that the Coriolis parameter is constant, so that $\mu = 1$. The equations are solved with periodic boundary conditions in $\mathcal{R}^2$. Then

$$\begin{aligned} \frac{\partial u'}{\partial \hat{t}} + \alpha^{-1} Fr^{-2} \frac{\partial h'}{\partial \hat{x}} - Ro^{-1} v' &= 0, \\ \frac{\partial v'}{\partial \hat{t}} + \alpha^{-1} Fr^{-2} \frac{\partial h'}{\partial \hat{y}} + Ro^{-1} u' &= 0, \\ \frac{\partial h'}{\partial \hat{t}} + \alpha \left(\frac{\partial u'}{\partial \hat{x}} + \frac{\partial v'}{\partial \hat{y}} \right) &= 0. \end{aligned} \tag{2.38}$$

Seek solutions of (2.38) of the form $u' = \check{u} \exp i \left(k\hat{x} + l\hat{y} - \omega\hat{t}\right), v' = \check{v} \exp i \left(k\hat{x} + l\hat{y} - \omega\hat{t}\right), h' = \check{h} \exp i \left(k\hat{x} + l\hat{y} - \omega\hat{t}\right)$. Equations (2.38) can then be rewritten as

$$\begin{pmatrix} -i\omega & -Ro^{-1} & \alpha^{-1}Fr^{-2}ik \\ Ro^{-1} & -i\omega & \alpha^{-1}Fr^{-2}il \\ ik\alpha & il\alpha & -i\omega \end{pmatrix} \begin{pmatrix} \check{u} \\ \check{v} \\ \check{h} \end{pmatrix} = 0. \tag{2.39}$$

Equations (2.38) have the trivial solution $\check{u} = \check{v} = \check{h} = 0$. Non-trivial solutions are only possible if the determinant

$$\begin{vmatrix} -i\omega & -Ro^{-1} & \alpha^{-1}Fr^{-2}ik \\ Ro^{-1} & -i\omega & \alpha^{-1}Fr^{-2}il \\ ik\alpha & il\alpha & -i\omega \end{vmatrix} = 0. \tag{2.40}$$

This reduces to the cubic equation

$$i\omega \left(-\omega^2 + Fr^{-2}\left(k^2 + l^2\right) + Ro^{-2}\right) = 0 \tag{2.41}$$

which has roots

$$\begin{aligned}\omega &= 0\\ &= \pm\sqrt{Fr^{-2}\,(k^2+l^2)+Ro^{-2}}.\end{aligned} \tag{2.42}$$

These roots represent the eigenvalues of the matrix

$$\begin{pmatrix} 0 & -Ro^{-1} & \alpha^{-1}Fr^{-2}ik \\ Ro^{-1} & 0 & \alpha^{-1}Fr^{-2}il \\ ik\alpha & il\alpha & 0 \end{pmatrix} \tag{2.43}$$

derived from the left hand side of (2.39). The nature of the roots can be further studied by examining the associated eigenfunctions. The eigenfunction associated with $\omega = 0$ takes the form

$$\begin{pmatrix} \check{u} \\ \check{v} \\ \check{h} \end{pmatrix} = \eta \begin{pmatrix} -\alpha^{-1}Fr^{-2}il \\ \alpha^{-1}Fr^{-2}ik \\ Ro^{-1} \end{pmatrix} \tag{2.44}$$

where η is a constant. This eigenfunction is both geostrophic, satisfying

$$\alpha^{-1}Fr^{-2}\frac{\partial h'}{\partial \hat{x}} - Ro^{-1}v' = 0, \alpha^{-1}Fr^{-2}\frac{\partial h'}{\partial \hat{y}} + Ro^{-1}u' = 0 \tag{2.45}$$

and non-divergent, satisfying

$$\frac{\partial u'}{\partial \hat{x}} + \frac{\partial v'}{\partial \hat{y}} = 0. \tag{2.46}$$

This solution represents a *Rossby* wave. While at the level of this approximation, such a solution is a steady state, and thus not very interesting; it becomes a propagating wave once the effects of variable Coriolis parameter are included, as will be shown in sections 2.3.4 and 2.3.5. It forms the basis of a description of the large-scale motions studied in this book, which are observed to be close to geostrophic, and other important classes of motions which are almost non-divergent.

The other eigenfunctions, corresponding to the eigenvalues

$$\varpi = \pm\sqrt{Fr^{-2}\,(k^2+l^2)+Ro^{-2}} \tag{2.47}$$

represent inertia-gravity waves. They are neither geostrophic nor non-divergent. The eigenvalues are the shallow water equivalent of the eigenvalues of the linear operator $\mathbf{L}$ derived in equation (2.31) in section 2.2.

The inertia-gravity wave frequency ϖ is made up of two terms. The first (which describes pure gravity waves) is independent of rotation. The associated phase speed (calculated by first setting $Ro^{-1} = 0$) is Fr^{-1}, showing that the Froude number Fr is the ratio of the flow speed to the gravity wave speed, as in section 2.2.3. The second term (which describes pure inertia waves) always has a frequency Ro^{-1}, so the phase speed depends on the horizontal wavelength, being large for large wavelengths. The two terms are equal if

$$1/\sqrt{(k^2 + l^2)} \equiv \hat{L}_R = Ro/Fr. \tag{2.48}$$

$\hat{L}_R$ is the dimensionless Rossby radius of deformation for shallow water flow. $\hat{L}_R = 1$ if $Ro = Fr$.

This behaviour links directly to the three-dimensional case analysed in section 2.2.5. The case $Fr < Ro$, so that $\hat{L}_R > 1$, is the shallow water analogue of stratification dominated flow. The assumed scale of the flow (which is unity) is thus less than $\hat{L}_R$. The case $Ro < Fr$, so that $\hat{L}_R < 1$, corresponds to rotation dominated flow. The assumed scale of the flow is now greater than $\hat{L}_R$. The ratio Ro/Fr is called the Burger number.

Consider the implications in dimensional terms. Experiments in the 1950s, when the shallow water equations were used to forecast the evolution of the 500hpa pressure surface, showed that the best match to observations was obtained when h_0 was set to 2000m, corresponding to L_R =1400km, (Reiser (2000), p.65). This scale corresponds to a wavelength of about a quarter of the circumference of the Earth in middle latitudes. As shown later, the usually accepted value for L_R in the troposphere is about 1000km. The difference arises because the observed evolution depends significantly on three-dimensional effects.

An important application of this linear analysis is to extract the Rossby wave solution from general data (u', v', h'). This is called solving the 'Rossby adjustment' problem. Write $u' = \check{u}(t) \exp i(kx + ly), v' = \check{v}(t) \exp i(kx + ly), h' = \check{h}(t) \exp i(kx + ly)$. The system of equations (2.38) can be rewritten as

$$\frac{\partial}{\partial t}\begin{pmatrix} \check{u} \\ \check{v} \\ \check{h} \end{pmatrix} + \mathbf{L}\begin{pmatrix} \check{u} \\ \check{v} \\ \check{h} \end{pmatrix} = 0. \tag{2.49}$$

Let $\mathbf{E}$ be the matrix whose columns are the eigenfunctions of $\mathbf{L}$. Then a general state vector $\check{\mathbf{x}} = \begin{pmatrix} \check{u} \\ \check{v} \\ \check{h} \end{pmatrix}$ can be projected on to a basis of eigenfunctions by setting

$$\tilde{\mathbf{x}} = \mathbf{E}^{-1}\check{\mathbf{x}}. \tag{2.50}$$

The Rossby wave component $\tilde{\mathbf{x}}_0$ is isolated by setting the second and third components of $\tilde{\mathbf{x}}$ to zero. The result can be projected back to its representation in physical variables by multiplying by $\mathbf{E}$. This can be shown to give

$$\check{\mathbf{x}}_0 = \begin{pmatrix} \check{u}_0 \\ \check{v}_0 \\ \check{h}_0 \end{pmatrix} = \frac{\left((-il\check{u} + ik\check{v})\,\alpha - Ro^{-1}\check{h}\right)}{\varpi^2} \begin{pmatrix} -\alpha^{-1}Fr^{-2}il \\ \alpha^{-1}Fr^{-2}ik \\ Ro^{-1} \end{pmatrix}. \tag{2.51}$$

This amounts to determining the constant η in the definition of the eigenfunction (2.44).The numerator that appears on the right hand side of (2.51) can be written in terms of the original perturbation variables as

$$Q' = \left(-\frac{\partial u'}{\partial \hat{y}} + \frac{\partial v'}{\partial \hat{x}}\right)\alpha - Ro^{-1}h' \tag{2.52}$$

where Q' can be shown to be the linearised perturbation to the potential vorticity defined in (2.35) with $\mu = 1$. Substituting (2.52) into (2.38) gives

$$\frac{\partial Q'}{\partial \hat{t}} = 0, \tag{2.53}$$

which is the linearisation of equation (2.37). Using the definition of ϖ^2, the calculation expressed by equation (2.51) can be expressed as first calculating h_0 from

$$-Fr^{-2}\nabla^2 h_0 + Ro^{-2}h_0 = Ro^{-1}Q', \tag{2.54}$$

and then calculating u_0 and v_0 from h_0 using the geostrophic relation. This is a simple form of *potential vorticity inversion* where all the flow variables can be derived from the potential vorticity together with diagnostic relations between the variables. In the nonlinear case, the slow solutions corresponding to Rossby waves can often be written in the form of an evolution equation for the potential vorticity, together with diagnostic equations allowing the other variables to be calculated. This method was first introduced systematically by Hoskins *et al.* (1985).

Note that for length scales greater than $\hat{L}_R$ as defined in (2.48), the second term on the left hand side of (2.54) is larger than the first, and the second term on the right hand side of (2.52) is larger than the first. Thus in physical variables $h_0 \simeq h'$. In the converse case,

$$\frac{\partial v_0}{\partial x} - \frac{\partial u_0}{\partial y} \simeq \frac{\partial v'}{\partial x} - \frac{\partial u'}{\partial y}.$$

Thus the adjustment is to the depth field on large scales and to the vorticity field on small scales. A much fuller presentation is given by Vallis (2017), p.127.

2.3.3 *General equations for slow solutions*

The next step is to study a more general problem than that set out in section 2.3.2. Seek systems of equations which approximate (2.33) when the dimensionless inertia-gravity wave frequency (2.47) is greater than 1, so that either or both of Ro and Fr are small. This presentation follows McWilliams *et al.* (1999). There are many other versions of this type of analysis in the literature, for instance see Warn et al. (1995), Lynch (1989), Holm (1996) and Mohebalhojeh and Dritschel (2001).

Use the dimensionless time-scale of inertia-gravity waves ϖ^{-1} as defined in (2.47) to define a small parameter

$$\varepsilon = \frac{1}{\sqrt{Fr^{-2}(k^2+l^2)+Ro^{-2}}}. \tag{2.55}$$

This assumes that $\mu = O(1)$ in (2.33). Thus for a given dimensionless length-scale, $\varepsilon \ll 1$ if either $Fr \ll 1$ or $Ro \ll 1$. Note that ε may not be small if the dimensionless length-scale is very large and the rotation is very weak, so that μ is not O(1). This happens in the tropics, where the fast gravity wave speed associated with tropical tides does not give a high frequency. Using the condition $\varepsilon \ll 1$ yields a set of approximate equations which are valid whenever there is a clear scale separation between the assumed O(1) time-scale and the inertia-gravity wave frequency. Then rewrite equations (2.33) in terms of the vorticity (2.36) and the divergence

$$\hat{\Delta} = \frac{\partial \hat{u}}{\partial \hat{x}} + \frac{\partial \hat{v}}{\partial \hat{y}} \tag{2.56}$$

giving

$$\begin{aligned}
\frac{\partial \hat{\zeta}}{\partial \hat{t}} + u\frac{\partial(\hat{\zeta}+\mu Ro^{-1})}{\partial \hat{x}} + v\frac{\partial(\hat{\zeta}+\mu Ro^{-1})}{\partial \hat{y}} + (\hat{\zeta}+\mu Ro^{-1})\hat{\Delta} &= 0,\\
\frac{\partial \hat{\Delta}}{\partial \hat{t}} + \hat{u}\frac{\partial \hat{\Delta}}{\partial \hat{x}} + \hat{v}\frac{\partial \hat{\Delta}}{\partial \hat{y}} + \hat{\Delta}^2 - 2J(\hat{u},\hat{v}) + &\\
\alpha^{-1}Fr^{-2}\nabla^2\hat{h} - Ro^{-1}\nabla\cdot(\mu v, -\mu u) &= 0,\\
\frac{\partial \hat{h}}{\partial \hat{t}} + \hat{u}\frac{\partial \hat{h}}{\partial \hat{x}} + \hat{v}\frac{\partial \hat{h}}{\partial \hat{y}} + (\alpha+\hat{h})\hat{\Delta} &= 0,
\end{aligned} \tag{2.57}$$

where $J(\hat{u},\hat{v}) = \frac{\partial \hat{u}}{\partial \hat{x}}\frac{\partial \hat{v}}{\partial \hat{y}} - \frac{\partial \hat{u}}{\partial \hat{y}}\frac{\partial \hat{v}}{\partial \hat{x}}$. Differentiating the second of these equations with respect to time and substituting the first and third equation gives

$$\begin{aligned}\frac{\partial^2\hat{\Delta}}{\partial\hat{t}^2} + \left(-\alpha^{-1}Fr^{-2}(\alpha+\hat{h})\nabla^2\hat{\Delta} + \mu Ro^{-1}(\hat{\zeta}+\mu Ro^{-1})\hat{\Delta}\right) = \\ \alpha^{-1}Fr^{-2}\hat{\Delta}\nabla^2\hat{h} + 2\alpha^{-1}Fr^{-2}\nabla\hat{\Delta}\nabla\hat{h} + \\ \alpha^{-1}Fr^{-2}\hat{h}\nabla^2(\hat{\mathbf{u}}\cdot\nabla\hat{h}) - \mu Ro^{-1}\hat{\mathbf{u}}\cdot\nabla(\hat{\zeta}+2\mu Ro^{-1}) + \\ 2\frac{\partial}{\partial\hat{t}}J(\hat{u},\hat{v}) + \text{remainder}\end{aligned} \tag{2.58}$$

In this equation the ratio of the magnitude of the first two terms is ε^2. Equation (2.58) can then be approximated to this order by the diagnostic equation obtained by omitting the first term.

Returning to the second equation of (2.57), this approximation can be achieved by omitting the first term. The next three terms in this equation have the same magnitude as the first, so should also be omitted. The assumption that either Ro or Fr is small means that at least one of the last two terms is $O(\varepsilon^{-1})$ or greater. Consistency of the resulting equations for most cases where ε is small can be achieved by assuming that $\hat{\Delta} = \varepsilon\hat{\zeta}$. However, if μ is not constant and Ro is smaller than Fr, so that geostrophic balance holds to leading order, then $\hat{\Delta} \simeq \hat{\zeta}$. Thus the large-scale regime discussed in section 2.3.6, where variable rotation is usually important, is not included.

Next write $\hat{\mathbf{u}}_a = (\hat{u}_a, \hat{v}_a)$ for the rotational part of $(\hat{u}, \hat{v})$ satisfying

$$\begin{aligned}\frac{\partial\hat{v}_a}{\partial\hat{x}} - \frac{\partial\hat{u}_a}{\partial\hat{y}} &= \hat{\zeta}, \\ \frac{\partial\hat{u}_a}{\partial\hat{x}} + \frac{\partial\hat{v}_a}{\partial\hat{y}} &= 0.\end{aligned} \tag{2.59}$$

Write $(\hat{u}_a, \hat{v}_a)$ in terms of a stream-function $\hat{\psi}$, so that $\hat{u}_a = -\partial\hat{\psi}/\partial\hat{y}, \hat{v}_a = \partial\hat{\psi}/\partial\hat{x}$. Then the second equation of (2.57) can be approximated to $O(\varepsilon^2)$ by

$$-2J\left(\frac{\partial\hat{\psi}}{\partial\hat{x}}, \frac{\partial\hat{\psi}}{\partial\hat{y}}\right) + \alpha^{-1}Fr^{-2}\nabla^2\hat{h} - Ro^{-1}\nabla\cdot(\mu\nabla\hat{\psi}) = 0. \tag{2.60}$$

Replacing the second equation of (2.57) by (2.60) then gives a complete prognostic system. The resulting equations are called the *nonlinear balance shallow water equations.* Slightly different versions of these appear in the various references cited above. The potential vorticity equation (2.37) is not changed by this approximation. (2.60) is a nonlinear generalisation of the geostrophic condition (2.45) satisfied by linear Rossby waves. The potential vorticity equation (2.37) has already been shown to be a nonlinear generalisation of equation (2.53) which governs the evolution of linear Rossby waves.

Following McWilliams *et al.* (1999), the nonlinear balance equations can be written as

$$\begin{pmatrix} F & 1 & 0 \\ 0 & \alpha^{-1}Fr^{-2}\nabla^2 & -M \\ G & 0 & \nabla^2 \end{pmatrix} \begin{pmatrix} \hat{\chi} \\ \partial\hat{h}/\partial\hat{t} \\ \partial\hat{\psi}/\partial\hat{t} \end{pmatrix} = \begin{pmatrix} -J(\hat{\psi},\hat{h}) \\ 0 \\ -J(\hat{\psi},\hat{\zeta}+\mu Ro^{-1}) \end{pmatrix} \tag{2.61}$$

where

$$\begin{aligned} \nabla^2\hat{\chi} &= \hat{\Delta}, \\ F &= \nabla\cdot((\alpha+\hat{h})\nabla), \\ M &= \nabla\cdot(\mu Ro^{-1}\nabla) + 2\left(\frac{\partial^2\hat{\psi}}{\partial\hat{x}^2}\frac{\partial^2}{\partial\hat{y}^2} + \frac{\partial^2\hat{\psi}}{\partial\hat{y}^2}\frac{\partial^2}{\partial\hat{x}^2}\right) - 2\frac{\partial^2\hat{\psi}}{\partial\hat{x}\partial\hat{y}}\frac{\partial^2}{\partial\hat{x}\partial\hat{y}}, \\ G &= \nabla\cdot\left((\hat{\zeta}+\mu Ro^{-1})\nabla\right). \end{aligned} \tag{2.62}$$

Write (2.61) symbolically as

$$\mathbf{M}\hat{\mathbf{u}} = \mathbf{A}. \tag{2.63}$$

As discussed in McWilliams *et al.* (1999), equation (2.63) is expected to be to be solvable if $\mathbf{M}$ is elliptic or degenerate elliptic. It is unlikely to be solvable if $\mathbf{M}$ is hyperbolic. It can be shown that the necessary conditions are

(i) $\alpha+\hat{h}$ does not change sign.
(ii) $\hat{\zeta}+\mu Ro^{-1}$ does not change sign.
(iii)

$$(\hat{\zeta}+\mu Ro^{-1})^2 - \left(\frac{\partial\hat{\psi}^2}{\partial\hat{x}^2} - \frac{\partial\hat{\psi}^2}{\partial\hat{y}^2}\right)^2 - 4\left(\frac{\partial^2\hat{\psi}}{\partial\hat{x}\partial\hat{y}}\right)^2 > 0. \tag{2.64}$$

Though the first two of these conditions are ensured by potential vorticity conservation and the inherent positivity of $\alpha+\hat{h}$, condition (2.64) is not related to a constant of the motion, and spontaneous violations are to be expected. Computations, McWilliams and Yavneh (1998), show that these do indeed occur. If the initial data has small Rossby number, the velocity gradients have magnitude Ro and condition (2.64) will be satisfied. If the initial data has small Froude number but large Rossby number, (2.64) may not be satisfied. It will be seen in the following subsections that flow-dependent solvability conditions, such as these, are unavoidable for approximations to the shallow water equations valid on scales greater than $\hat{L}_R$. The difficulty caused by condition (2.64) in the case where $Fr \ll 1$, $Ro = O(1)$ is a result of trying to use the same set of simplified equations in both the asymptotic regimes $Fr \ll 1$ and $Ro \ll 1$.

For these equations to be a useful approximation, they have to be solvable for a time comparable to the time-scale of the phenomena being modelled. Condition (2.64) means that the velocity gradients have to be restricted to $O(Ro)$ in the initial data. The worst likely case is that two fluid particles initially separated by a distance $O(1)$ and with initial relative velocity $O(1)$ will come close together. This will take a time at least Ro^{-1} which by (2.55) is less than ε^{-1} and is comparable to it if Fr is not small. Thus it is likely that equation (2.63) can be solved for a time $O(\varepsilon^{-1})$ provided that $Ro = O(\varepsilon)$ in the initial data. However, this in itself is not long enough for the solutions to be useful. For instance, the condition $Ro = 0.1$, which means that the fluid velocity cannot change by more than 10ms^{-1} over a distance of 1000km in mid-latitudes, gives a solvability time of about 1 day, while the phenomena being modelled evolve over a time-scale of several days. If ε is small because Fr rather than Ro is small, the velocity gradients are not constrained and no estimate can be made in this way.

It is therefore necessary to exploit the structure of the equations to prove a better estimate. This has not been achieved for equations (2.63), though it has been achieved for the simpler equations discussed in the following sections. Several of the difficulties are discussed by McIntyre and Roulstone (2002). The outcome is that it appears not to be possible to prove the existence of a slow manifold which is accurate to more than leading order in ε and is uniformly accurate for small ε. This is consistent with the observed endemic nature of small-scale waves in the atmosphere, as illustrated in Figs. 1.2 and 1.4.

In order to find simplified models which can be solved for long enough times to be useful, the cases where $\varepsilon \ll 1$ are therefore split into the separate cases $Fr < Ro$, $Fr \ll 1$; $Fr = Ro \ll 1$ and $Ro < Fr$, $Ro \ll 1$. As shown in section 2.3.2, these cases correspond to $\hat{L}_R > 1$, $\hat{L}_R = 1$ and $\hat{L}_R < 1$ respectively. The difficulty caused by condition (2.64) in the case of small Fr is avoided by using a system of equations in this asymptotic regime which is based on a constant coefficient elliptic problem. Flow-dependent solvability conditions which require restrictions on the velocity gradients then no longer arise. Flow-dependent conditions are required for the solvability of systems of equations valid for scales larger than $\hat{L}_R$, but it is now consistent to restrict the velocity gradients in the initial data. It will be shown that, by limiting the validity of the equations to small Ro, solvability can be proved for arbitrarily large times. As noted at the start of this section, there are additional issues in the tropics where an assumption of small Ro is never uniformly valid. So this case also needs separate treatment. Each case is considered in turn in the following subsections.

2.3.4 *Slow solutions on small scales*

In this section it is assumed that $Fr \ll 1$ and $Fr < Ro$, so $\hat{L}_R > 1$. It is thus assumed that Ro=O(1), so the first two equations of (2.33) cannot be approximated and $\alpha^{-1}Fr^{-2}$ must also be O(1). The assumption $Fr \ll 1$ can only be achieved by assuming that α =O(Fr^{-2}). The third equation of (2.33) then becomes

$$\frac{\partial \hat{h}}{\partial \hat{t}} + \hat{\mathbf{u}} \cdot \nabla \hat{h} + Fr^{-2}\hat{\Delta} = 0. \tag{2.65}$$

Since the last term on the left hand side is much larger than the other two, an appropriate asymptotic approximation to (2.57) is therefore to set $\hat{\Delta} = 0$, giving

$$\begin{aligned} \frac{\partial \hat{\zeta}}{\partial \hat{t}} + \hat{u}\frac{\partial(\hat{\zeta}+\mu)}{\partial \hat{x}} + \hat{v}\frac{\partial(\hat{\zeta}+\mu)}{\partial \hat{y}} &= 0, \\ \hat{\Delta} &= 0, \\ -2J(\hat{u},\hat{v}) + \nabla^2\hat{h} - \nabla\cdot(\mu\hat{v}, -\mu\hat{u}) &= 0. \end{aligned} \tag{2.66}$$

The last equation is purely a diagnostic equation for $\hat{h}$ which does not have to be solved when advancing the system in time. It replaces (2.65) and expresses the requirement that $\partial\hat{\Delta}/\partial\hat{t} = 0$. The error in this approximation is O(Fr^2). The equations are still to be solved in a closed region $\Gamma \in \mathcal{R}^2$ with boundary conditions $\hat{\mathbf{u}} \cdot \mathbf{n} = 0$.

Equations (2.66) are the equations for two-dimensional incompressible flow in a rotating frame of reference. They conserve the energy integral

$$\int_\Gamma \left(\frac{1}{2}(\hat{u}^2 + \hat{v}^2)\right) \mathrm{d}\hat{x}\mathrm{d}\hat{y}. \tag{2.67}$$

This represents purely kinetic energy, which is consistent with the fact that the energy of disturbances in shallow water solutions with $Fr \ll 1$, $Fr < Ro$ is primarily kinetic rather than potential (Gill (1982), p.206). The role of potential vorticity is played by the absolute vorticity $\hat{\zeta}+\mu$. The first equation of (2.66) is thus the equivalent in this asymptotic regime of the equation (2.37) for conservation of potential vorticity. In this regime variations in the potential vorticity $(Ro\hat{\zeta}+\mu)/(\hat{h}+\alpha)$ defined by (2.35) will be dominated by variations in $\hat{\zeta}+\mu$ because, as noted above, $\alpha \simeq Fr^{-2}$.

An example is shown in Fig. 2.1 in dimensional variables. The mean depth of the fluid has been chosen so that $L_R = 2148$km, corresponding to a wavelength of about 13400km. A constant Coriolis parameter of $1.458 \times 10^{-4}\text{s}^{-1}$ is used. Most of the variations in h shown in Fig. 2.1 are on a smaller scale than this. The velocities, and hence the vorticity, are calculated from h by solving eq. (2.90) for $\mathbf{u}$ as discussed in section 2.3.6. This is the condition that geostrophic balance is maintained in time. It is seen that the scales of the variations in Q are much smaller than the scales of variation in h.

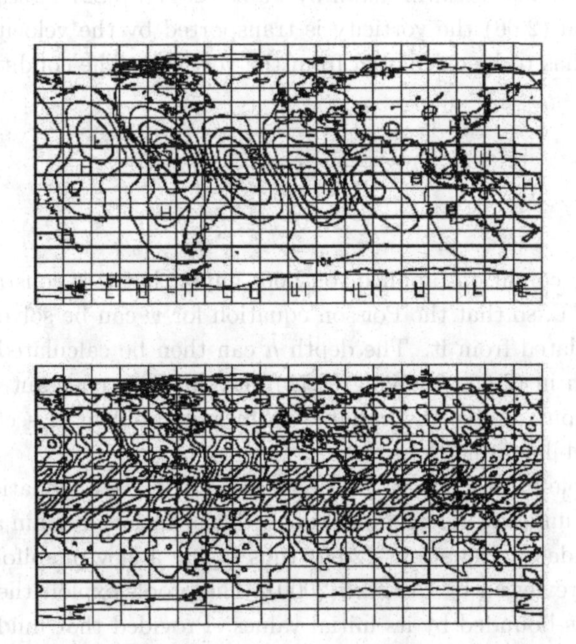

Fig. 2.1 Top: Example depth field with $gh_0 = 10^5\text{m}^2\text{s}^{-2}$. Units 100m, contour interval 80m. Bottom: Potential vorticity, units $(\text{ms})^{-1}$, contour interval 0.3×10^{-9}. From Cullen (2002). ©Royal Meteorological Society, Reading, U.K., 2002.

It is now possible to find a non-trivial Rossby wave solution. Assume that $\mu = \mu(\hat{y})$ and that $\beta = \partial\mu/\partial\hat{y}$ is a constant. This is the *beta plane* assumption. Linearising the first equation of (2.66) about a state of rest in the same manner as in section 2.3.2 gives

$$\frac{\partial \zeta'}{\partial \hat{t}} + v'\frac{\partial \mu}{\partial \hat{y}} = 0. \tag{2.68}$$

Making the same analytic substitutions as in section 2.3.2, which requires an assumption of periodic boundary conditions, gives

$$-i\omega(ikv' - ilu') + \beta v' = 0, iku' + ilv' = 0. \tag{2.69}$$

Solving this gives

$$\omega = -\frac{\beta k}{k^2 + l^2}. \tag{2.70}$$

These waves are discussed in detail by Vallis (2017), p.227 *et seq.*.

In equation (2.66) the vorticity is transported by the velocity $\hat{\mathbf{u}}$. The velocity now has to be calculated from the vorticity. The condition $\hat{\Delta} = 0$ means that

$$\hat{u} = -\frac{\partial \hat{\psi}}{\partial \hat{y}}, \hat{v} = \frac{\partial \hat{\psi}}{\partial \hat{x}}, \tag{2.71}$$
$$\nabla^2 \hat{\psi} = \hat{\zeta}.$$

The boundary conditions applied to (2.66) imply that $\hat{\psi}$ is constant on the boundaries of Γ, so that the Poisson equation for $\hat{\psi}$ can be solved and the velocity calculated from it. The depth $\hat{h}$ can then be calculated from the third equation in (2.66). Solving (2.71) only involves constant coefficient elliptic equations, unlike the variable coefficient problem of section 2.3.3. Thus no solvability conditions arise.

It is possible to prove that, given suitable initial data, equations (2.66) can be solved uniquely for all finite times. The solutions remain as smooth as the initial data. Equations (2.66) thus define a slow manifold. These theorems are reviewed by Chemin (2000). The proofs exploit the fact that the vorticity is bounded by its initial values. Provided that fluid trajectories can be shown to retain their identity, advecting the vorticity can only rearrange its values, but cannot create new ones. Fluid trajectories can be proved to retain their identity if the velocity is be smooth enough. This means that the velocity gradients, assumed bounded in the initial data, have to be controlled. Classical results quoted by Chemin (2000) state that, if $\hat{\zeta}$ is bounded above and below, and because $\hat{\Delta} = 0$, the velocity gradient norm

$$\left(\int_\Gamma \left((\nabla \hat{u})^2 + (\nabla \hat{v})^2 \right) \mathrm{d}\hat{x}\mathrm{d}\hat{y} \right)^{\frac{1}{2}} \tag{2.72}$$

is bounded uniformly in time. However, the maximum value of the velocity gradients, measured by

$$\left|\frac{\partial \hat{u}}{\partial \hat{x}}\right| + \left|\frac{\partial \hat{u}}{\partial \hat{y}}\right| + \left|\frac{\partial \hat{v}}{\partial \hat{x}}\right| + \left|\frac{\partial \hat{v}}{\partial \hat{y}}\right|, \tag{2.73}$$

cannot be determined without also knowing the vorticity gradients. The main task in the proof is thus to estimate the growth of the vorticity gradients, assuming they are bounded in the initial data.

It is also possible to prove that solutions of the shallow water equations converge to those of equation (2.66) as $Fr \to 0$. This is a special case of the result that the equations for incompressible flow are a limit of those for compressible flow, see Majda (1984). In this asymptotic regime it can be assumed that the gravity waves are governed by a linear equation with constant coefficients, giving a wave speed Fr^{-1}. The constant coefficients are essential to the proof. The techniques of Klainerman and Majda, Majda (1984), chapter 2, can then be used to derive the necessary estimates. In section 5.2 it will be seen that the qualitative behaviour exhibited by (2.66) is also exhibited by the shallow water equations (2.33) for initial data satisfying $\hat{L}_R > 1$.

2.3.5 *Quasi-geostrophic solutions*

In this section it is assumed that $Fr = Ro \ll 1$. This is the traditional scaling introduced by Charney (1948). It is derived and analysed in detail in Pedlosky (1987). The mathematical properties of the resulting quasi-geostrophic shallow water equations have been analysed by, for instance, Babin *et al.* (1999), Bourgeois and Beale (1994) and Majda (2003).

The assumption $Ro \ll 1$ means that $\alpha^{-1}Fr^{-2}$ has to be set equal to Ro^{-1} in the first two equations in (2.33). The assumption $Ro = Fr$ then implies that $\alpha = Fr^{-1}$. These equations can then be approximated to $\mathrm{O}(Ro)$ by

$$\begin{aligned} \frac{\partial \hat{h}}{\partial \hat{x}} - \mu \hat{v} &= 0, \\ \frac{\partial \hat{h}}{\partial \hat{y}} + \mu \hat{u} &= 0. \end{aligned} \tag{2.74}$$

The third equation of (2.33) can then also be approximated to $O(Ro)$ by $\hat{\Delta} = 0$.

There are now two immediate difficulties. The first is that (2.74) and $\hat{\Delta} = 0$ contain no prognostic information, the well-known geostrophic degeneracy. The second is that, if the Coriolis parameter is a function of position, the equations are contradictory because $\hat{\mathbf{u}}$ derived from (2.74) does not satisfy $\hat{\Delta} = 0$. To resolve the second issue, assume additionally that $\mu = 1 + Ro\mu(\hat{y})$. Then, to obtain prognostic information, go to the next order terms. Thus define the geostrophic wind by

$$\begin{aligned}\frac{\partial \hat{h}}{\partial \hat{x}} - \hat{v}_g &= 0,\\ \frac{\partial \hat{h}}{\partial \hat{y}} + \hat{u}_g &= 0.\end{aligned} \tag{2.75}$$

Then $\nabla \cdot \hat{\mathbf{u}}_g = 0$. The first two equations of (2.33) then imply that the ageostrophic wind $\hat{\mathbf{u}} - \hat{\mathbf{u}}_\mathbf{g} \simeq Ro\hat{\mathbf{u}}_\mathbf{g}$. The vorticity can be approximated to $O(Ro)$ by its geostrophic value, but as $\nabla \cdot \hat{\mathbf{u}}_g = 0$, the divergence is entirely ageostrophic.

Now use the vorticity- divergence form (2.57) of the shallow water equations. The first two equations are approximated by

$$\begin{aligned}\frac{\partial \hat{\zeta}_g}{\partial \hat{t}} + \hat{u}_g \frac{\partial \hat{\zeta}_g}{\partial \hat{x}} + \hat{v}_g \frac{\partial (\hat{\zeta}_g + \mu)}{\partial \hat{y}} + Ro^{-1}\hat{\Delta} &= 0,\\ \nabla^2 \hat{h} - \hat{\zeta}_g &= 0,\end{aligned} \tag{2.76}$$

and the third equation by

$$\frac{\partial \hat{h}}{\partial \hat{t}} + Ro^{-1}\hat{\Delta} = 0, \tag{2.77}$$

noting that $\hat{\mathbf{u}}_g \cdot \nabla \hat{h}$ vanishes identically.

The equations are again to be solved in a closed region $\Gamma \subset \mathcal{R}^2$ with boundary conditions $\hat{\mathbf{u}} \cdot \mathbf{n} = 0$. It is necessary to apply this condition to the geostrophic wind and divergent wind separately. The first of these conditions implies that $\hat{h}$ has to be constant along the boundary. It can then be shown that the energy integral

$$\int_\Gamma \left(\frac{1}{2}(\hat{u}_g^2 + \hat{v}_g^2) + \hat{h} \right) \mathrm{d}\hat{x}\mathrm{d}\hat{y} \tag{2.78}$$

is conserved. Equations (2.76) and (2.77) also conserve the quasi-geostrophic potential vorticity

$$\hat{Q}_g = Ro^{-1} + \mu + \hat{\zeta}_g - \hat{h}. \tag{2.79}$$

It is similar to the potential vorticity (2.52) of the linearised equations. It obeys the conservation law

$$\frac{\partial \hat{\mathcal{Q}}_g}{\partial \hat{t}} + \hat{u}_g \frac{\partial \hat{\mathcal{Q}}_g}{\partial \hat{x}} + \hat{v}_g \frac{\partial \hat{\mathcal{Q}}_g}{\partial \hat{y}} = 0. \tag{2.80}$$

$\hat{\mathcal{Q}}_g$ approximates the potential vorticity (2.35) to $\mathrm{O}(Ro)$, subject to constant scaling factors. Thus the evolution equations (2.76) and (2.77) will approximate (2.33) to $\mathrm{O}(Ro)$ provided that geostrophic balance can be maintained to $\mathrm{O}(Ro)$ as assumed in the derivation. Results to this effect are described below and in the three-dimensional case in section 2.4.5). In the latter section it is also shown how the accuracy can be improved to $\mathrm{O}(Ro^2)$.

Linearising this equation about a state of rest, as in section 2.3.4, and using (2.75) and (2.79), gives

$$\zeta' = -(k^2 + l^2)h',$$
$$\mathcal{Q}' = \zeta' - h' = -(k^2 + l^2)h' - h',$$
$$-i\omega\mathcal{Q}' = -\beta v' = -\beta ikh',$$

where $\beta = \partial\mu/\partial\hat{y}$. This leads to the dispersion relation

$$\omega = -\frac{\beta k}{(k^2 + l^2) + 1}. \tag{2.81}$$

This relation is discussed further in Vallis (2017), p. 229. Comparing with (2.70) shows an extra term in the denominator. This is related to the Rossby radius of deformation, which under the assumption $Ro = Fr$ is 1, and shows that the wave speed is reduced as the horizontal scale $\mathbf{k}^{-1}$ approaches the deformation radius.

It is easy to see that these approximations are unsuitable for large-scale flow, since the assumption $\mu = 1 + \mathrm{O}(Ro)$ will only be acceptable in a fairly small region. This led to the failure of the quasi-geostrophic equations as a basis for numerical weather forecasting as early as the 1950s, (Phillips, 2000). At that time computers were very crude, and the resolution of models was far short of that necessary to expose the small-scale limitations of quasi-geostrophic theory. However, they have proved very useful for understanding the processes governing the evolution of extratropical weather systems, where quantitative accuracy is not required, Pedlosky (1987).

The solutions of equations (2.76) and (2.77) can be found by a method analogous to that for two-dimensional incompressible flow, as described in section 2.3.4. The first step is to define an inversion procedure for calculating $\hat{h}, \hat{u}$ and $\hat{v}$ from $\mathcal{Q}_g$. Using (2.75) and (2.79) gives

$$Ro^{-1} + \mu + \nabla^2\hat{h} - \hat{h} = \hat{\mathcal{Q}}_g. \tag{2.82}$$

This is an constant coefficient elliptic equation for $\hat{h}$. It can be solved using the boundary condition that $\hat{h}$ is constant along the boundary. $\hat{\mathbf{u}}_g$ can then be calculated from (2.75). This is sufficient for the equations to be advanced in time. However, the boundary condition on $\hat{h}$ is unphysical. If normal derivative (Neumann) boundary conditions on $\hat{h}$ are used instead, these imply that the geostrophic wind parallel to the boundary is prescribed. This is also unphysical. As a result, the equations are often solved with periodic boundary conditions. The solution procedure is then analogous to that in section 2.3.4. Given bounded $\hat{Q}_g$, equation (2.82) is solved for $\hat{h}$ and $\hat{\mathbf{u}}_g$ calculated from (2.75). The solution can then be advanced in time.

If the initial data are geostrophic, and sufficiently smooth, the solutions of the shallow water equations can be proved to converge to those of (2.76) and (2.77) on a given time interval, Majda (2003), pp. 73 *et seq.* The fact that the inertia-gravity waves are governed by a linear equation with constant coefficients as in section 2.3.4 is essential. The solutions of the shallow water equations can also be analysed in this limit for initial data with large inertia-gravity waves. This involves averaging over inertia-gravity wave periods, and showing that the averaged solution satisfies (2.76) and (2.77) plus an error term which can be estimated. Such methods are described in Majda (2003), chapter 8, and Babin *et al.* (1999). Since more results of this type have been obtained for the three dimensional case, a representative set is given in section 2.4.5.

The divergence $\hat{\Delta}$ does not need to be calculated to advance the solution in time. If it is required, it can be found by differentiating the second equation of (2.76) with respect to time and eliminating the time derivatives from (2.76) and (2.77) to give the *quasi-geostrophic omega equation*:

$$\nabla^2\hat{\Delta} - \hat{\Delta} = \hat{\mathbf{u}}_{\mathbf{g}} \cdot \nabla(\hat{\zeta}_g + \mu Ro^{-1}). \tag{2.83}$$

The boundary conditions on the divergent velocity yield Neumann boundary conditions for the velocity potential $\hat{\chi}$, where $\nabla^2\hat{\chi} = \hat{\Delta}$. However, if equation (2.83) is written in terms of $\hat{\chi}$, it becomes fourth order and requires additional boundary conditions.

2.3.6 *Slow solutions on large scales*

Now consider the case $Ro \ll 1$ with $\alpha = 1$. The assumption $Ro \ll 1$ means that $\alpha^{-1}Fr^{-2}$ has to be chosen to be Ro^{-1} in (2.33), so that $Ro = Fr^2$ and $\hat{L}_R \equiv Ro/Fr < 1$. Thus the regime applies on scales larger than the deformation radius.

The first two equations of (2.33) can again be approximated by (2.74), so that (2.33) becomes:

$$\begin{aligned}\frac{\partial \hat{h}}{\partial \hat{x}} - \mu \hat{v} &= 0,\\ \frac{\partial \hat{h}}{\partial \hat{y}} + \mu \hat{u} &= 0,\\ \frac{\partial \hat{h}}{\partial \hat{t}} + \frac{\partial}{\partial \hat{x}}\left(\hat{h}\hat{u}\right) + \frac{\partial}{\partial \hat{y}}\left(\hat{h}\hat{v}\right) &= 0.\end{aligned} \tag{2.84}$$

The equations are again to be solved in a closed region $\Gamma \in \mathcal{R}^2$ with boundary conditions $\hat{\mathbf{u}} \cdot \mathbf{n} = 0$.

Substituting the first two equations into the third gives

$$\frac{\partial \hat{h}}{\partial \hat{t}} + \hat{h}\left(\frac{\partial \hat{u}}{\partial \hat{x}} + \frac{\partial \hat{v}}{\partial \hat{y}}\right) = 0. \tag{2.85}$$

These equations are the 'Type 2' equations for geostrophic flow, described by Phillips (1963).

If the Coriolis parameter is constant, so that $\mu = 1$, then $\frac{\partial \hat{u}}{\partial \hat{x}} + \frac{\partial \hat{v}}{\partial \hat{y}}$ vanishes, and the equations simply say that any choice of $\hat{h}$ can be specified, and gives a steady state. The boundary conditions on $\hat{u}$ and $\hat{v}$ imply that $\hat{h}$ is constant on the boundaries. This is usually called the 'geostrophic degeneracy' and regarded as trivial. However, there is a physical implication that anomalies in $\hat{h}$ with rapid rotation can only evolve on a timescale of Ro^{-1}, which is consistent with the observed persistence of large-scale anomalies in the atmosphere and of eddies in the ocean.

To illustrate non-trivial solutions, make the beta-plane assumption as in eq. (2.69) that $\partial\mu/\partial\hat{y} = \beta$, $\partial\mu/\partial\hat{x} = 0$, where β is constant Then

$$\frac{\partial \hat{h}}{\partial \hat{t}} - \beta \hat{h}\frac{\partial \hat{h}}{\partial \hat{x}} = 0. \tag{2.86}$$

This is 'Burger's equation', which is well-known to give discontinuous solutions in finite time for all but constant initial data. These would imply infinite values of $\hat{v}$. Though solutions of this form for $\hat{h}$ can be defined in other contexts, they are not physically relevant in the present case, so the model does not define a slow manifold. It is, nevertheless, used in ocean circulation studies since the limitations are less severe when the equivalent approximations are made in the three-dimensional case described in section 2.4.6. It is then referred to as the 'planetary geostrophic' model.

A more useful set of limiting equations for large-scale shallow water flow is obtained by seeking a more accurate approximation. As in the quasi-geostrophic case, first define the geostrophic wind:

$$\begin{aligned}\frac{\partial \hat{h}}{\partial \hat{x}} - \mu \hat{v}_g &= 0,\\ \frac{\partial \hat{h}}{\partial \hat{y}} + \mu \hat{u}_g &= 0.\end{aligned} \tag{2.87}$$

Note that the true value of the Coriolis parameter is used, rather than a constant as in equation (2.75). Next, replace the momentum $(\hat{u}, \hat{v})$ by $(\hat{u}_g, \hat{v}_g)$ in the acceleration terms of (2.33). This is called the 'geostrophic momentum' approximation and yields the 'semi-geostrophic' equations:

$$\begin{aligned}\frac{\partial \hat{u}_g}{\partial \hat{t}} + \hat{u}\frac{\partial \hat{u}_g}{\partial \hat{x}} + \hat{v}\frac{\partial \hat{u}_g}{\partial \hat{y}} + Ro^{-1}\left(\frac{\partial \hat{h}}{\partial \hat{x}} - \mu \hat{v}\right) &= 0,\\ \frac{\partial \hat{v}_g}{\partial \hat{t}} + \hat{u}\frac{\partial \hat{v}_g}{\partial \hat{x}} + \hat{v}\frac{\partial \hat{v}_g}{\partial \hat{y}} + Ro^{-1}\left(\frac{\partial \hat{h}}{\partial \hat{y}} + \mu \hat{u}\right) &= 0,\\ \frac{\partial \hat{h}}{\partial \hat{t}} + \frac{\partial}{\partial \hat{x}}\left(\hat{h}\hat{u}\right) + \frac{\partial}{\partial \hat{y}}\left(\hat{h}\hat{v}\right) &= 0.\end{aligned} \tag{2.88}$$

The equations are again solved in a region Γ with $\hat{\mathbf{u}}\cdot\mathbf{n} = \mathbf{0}$ on the boundary. They conserve the energy integral:

$$\int_\Gamma \left(\frac{1}{2}\hat{h}\left(\hat{u}_g^2 + \hat{v}_g^2\right) + \frac{1}{2}\hat{h}^2\right) \mathrm{d}\hat{x}\mathrm{d}\hat{y}. \tag{2.89}$$

Comparing (2.89) to the shallow water energy integral (2.34) shows that the kinetic energy has been approximated by its geostrophic value, but the potential energy has not been approximated. This is consistent with the fact that, for $\hat{L}_R < 1$, the energy of disturbances is primarily potential energy.

The proof of energy conservation exploits the fact that only the momentum is approximated, not the trajectory. The boundary conditions can thus be applied apply only to the trajectory, not the momentum, expressing the fact that no fluid can enter or leave Γ. No boundary conditions need to be applied to $\hat{h}$. This can be contrasted with the need to apply boundary conditions to the geostrophic wind in the quasi-geostrophic equations. As well as enabling energy conservation, this will be important in the analysis in subsequent sections.

The behaviour of equations (2.88) can be studied by first rearranging them into a single equation for the evolution of $\hat{h}$, in the manner of Schubert

(1985). The first two equations can be rewritten after multiplying by μ as

$$\mathbf{Q}\begin{pmatrix}\hat{u}\\ \hat{v}\end{pmatrix} + \frac{\partial}{\partial\hat{t}}\nabla\hat{h} = Ro^{-1}\mu\begin{pmatrix}-\frac{\partial\hat{h}}{\partial\hat{y}}\\ \frac{\partial\hat{h}}{\partial\hat{x}}\end{pmatrix},$$

$$\mathbf{Q} = \begin{pmatrix} Ro^{-1}\mu^2 + \mu\frac{\partial\hat{v}_g}{\partial\hat{x}} & \mu\frac{\partial\hat{v}_g}{\partial\hat{y}} \\ -\mu\frac{\partial\hat{u}_g}{\partial\hat{x}} & Ro^{-1}\mu^2 - \mu\frac{\partial\hat{u}_g}{\partial\hat{y}}\end{pmatrix}. \tag{2.90}$$

Use of the third equation then gives

$$\frac{\partial\hat{h}}{\partial\hat{t}} - \nabla\cdot\left(\hat{h}\mathbf{Q}^{-1}\frac{\partial}{\partial\hat{t}}\nabla\hat{h}\right) = -Ro^{-1}\nabla\cdot\left(\hat{h}\mathbf{Q}^{-1}\mu\begin{pmatrix}-\frac{\partial\hat{h}}{\partial\hat{y}}\\ \frac{\partial\hat{h}}{\partial\hat{x}}\end{pmatrix}\right). \tag{2.91}$$

This equation is elliptic if $\mathbf{Q}$ is positive definite. Note that the latitude dependence of the Coriolis parameter is included in μ, which appears as μ^2 on the diagonal. So there is no difficulty applying it to both hemispheres. It has flow-dependent coefficients, as always occurs for approximations valid for $\hat{L}_R < 1$. Applying the boundary condition $\hat{\mathbf{u}}\cdot\mathbf{n} = 0$ to the first equation of (2.90) multiplied by $\mathbf{Q}^{-1}$ yields the boundary condition

$$\mathbf{Q}^{-1}\nabla\left(\frac{\partial\hat{h}}{\partial\hat{t}}\right)\cdot\mathbf{n} = Ro^{-1}\mu\mathbf{Q}^{-1}\begin{pmatrix}-\frac{\partial\hat{h}}{\partial\hat{y}}\\ \frac{\partial\hat{h}}{\partial\hat{x}}\end{pmatrix}\cdot\mathbf{n} \tag{2.92}$$

for $\frac{\partial\hat{h}}{\partial\hat{t}}$. Provided $\mathbf{Q}$ is positive definite, this gives a Neumann-type boundary condition for equation (2.91).

In the special case of constant Coriolis parameter, so $\mu = 1$, positive definiteness of $\mathbf{Q}$ is exactly the condition for

$$\frac{1}{2}Ro^{-1}(\hat{x}^2 + \hat{y}^2) + \hat{h} \tag{2.93}$$

to be convex. This gives a very strong control over the permitted solutions for $\hat{h}$. It also shows the stabilising effect of rapid rotation since, for any given smooth $\hat{h}$, it is possible to choose Ro such that (2.93) is convex. Also in this case, Hoskins (1975) showed that equations (2.88) conserve the potential vorticity

$$\hat{\mathcal{Q}}_{sg} = \left\{Ro^{-2} + Ro^{-1}\left(\frac{\partial\hat{v}_g}{\partial\hat{x}} - \frac{\partial\hat{u}_g}{\partial\hat{y}}\right) + \frac{\partial\hat{u}_g}{\partial\hat{x}}\frac{\partial\hat{v}_g}{\partial\hat{y}} - \frac{\partial\hat{v}_g}{\partial\hat{x}}\frac{\partial\hat{u}_g}{\partial\hat{y}}\right\}/\hat{h} \tag{2.94}$$

in the sense that

$$\frac{\partial\hat{\mathcal{Q}}_{sg}}{\partial\hat{t}} + \hat{\mathbf{u}}\cdot\nabla\hat{\mathcal{Q}}_{sg} = 0. \tag{2.95}$$

The Rossby wave dispersion relation (2.81) can be obtained by linearising (2.88) about a reference state at rest, with depth 1 and $\mu = 1 + \beta\hat{y}$, and then deriving an evolution equation for the quasi-geostrophic potential vorticity (2.79). Under these conditions the semi-geostrophic equations are the same as the quasi-geostrophic equations.

It is easy to see that $\hat{Q}_{sg} = \det\mathbf{Q}$. Thus, given initial data with positive definite $\mathbf{Q}$, $\mathbf{Q}$ can be expected to remain positive definite and equation (2.91) to be solvable. Rigorous results to this effect will be given in section 3.5. If the Coriolis parameter is not constant, there is no conserved potential vorticity in the sense of (2.95) and the proof of solvability is more difficult. However, given sufficient control on the spatial derivatives of $\hat{h}$, $\mathbf{Q}$ will be positive definite as $Ro \to 0$. A rigorous argument proving solvability for short times in this case, and a formal argument extending this to large times, are given in section 4.3. Assuming the result for large times can be made rigorous, the semi-geostrophic system will define a slow manifold. This robust solvability contrasts with the nonlinear balance equations studied in section 2.3.3. However, it is achieved at the price of being accurate in a much more restricted asymptotic regime.

In order to demonstrate the accuracy of (2.88) as an approximation to (2.33), set $\alpha = 1$ and $Ro = Fr^2$ in (2.33), divide the first two equations by μRo^{-1}, and take their Lagrangian time derivative. This gives

$$Ro\frac{\mathrm{D}}{\mathrm{D}\hat{t}}\left(\mu^{-1}\begin{pmatrix}\frac{\mathrm{D}\hat{u}}{\mathrm{D}\hat{t}}\\ \frac{\mathrm{D}\hat{v}}{\mathrm{D}\hat{t}}\end{pmatrix}\right) + \begin{pmatrix}-\frac{\mathrm{D}\hat{v}}{\mathrm{D}\hat{t}}\\ \frac{\mathrm{D}\hat{u}}{\mathrm{D}\hat{t}}\end{pmatrix} + \frac{\mathrm{D}}{\mathrm{D}\hat{t}}(\mu^{-1}\nabla\hat{h}) = 0. \tag{2.96}$$

Substituting from (2.33) for the second term in (2.96), and using (2.87) for the last term gives

$$Ro\frac{\mathrm{D}}{\mathrm{D}\hat{t}}\left(\mu^{-1}\begin{pmatrix}\frac{\mathrm{D}\hat{u}}{\mathrm{D}\hat{t}}\\ \frac{\mathrm{D}\hat{v}}{\mathrm{D}\hat{t}}\end{pmatrix}\right) + Ro^{-1}\begin{pmatrix}\frac{\partial\hat{h}}{\partial\hat{y}} + \mu\hat{u}\\ -\frac{\partial\hat{h}}{\partial\hat{x}} + \mu\hat{v}\end{pmatrix} + \frac{\mathrm{D}}{\mathrm{D}\hat{t}}\begin{pmatrix}\hat{v}_g\\ -\hat{u}_g\end{pmatrix} = 0. \tag{2.97}$$

Omitting the first term in (2.97) gives exactly the first two of the semi-geostrophic equations (2.88). Then rewrite the last two terms in (2.97) as

$$Ro^{-1}\begin{pmatrix}\frac{\partial\hat{h}}{\partial\hat{y}}\\ -\frac{\partial\hat{h}}{\partial\hat{x}}\end{pmatrix} + \frac{\partial}{\partial\hat{t}}\begin{pmatrix}\hat{v}_g\\ -\hat{u}_g\end{pmatrix} + \mu^{-1}\mathbf{Q}\begin{pmatrix}\hat{u}\\ \hat{v}\end{pmatrix}, \tag{2.98}$$

where $\mathbf{Q}$ is as defined in the second equation of (2.90). Multiplying by

$\mathbf{Q}^{-1}\mu$ gives

$$\begin{pmatrix} \hat{u} \\ \hat{v} \end{pmatrix} = -Ro^{-1}\mathbf{Q}^{-1}\mu \begin{pmatrix} \frac{\partial \hat{h}}{\partial \hat{y}} \\ -\frac{\partial \hat{h}}{\partial \hat{x}} \end{pmatrix} - \mathbf{Q}^{-1}\mu \frac{\partial}{\partial \hat{t}} \begin{pmatrix} \hat{v}_g \\ -\hat{u}_g \end{pmatrix} - Ro\mathbf{Q}^{-1}\mu \frac{\mathrm{D}}{\mathrm{D}\hat{t}} \left(\mu^{-1} \begin{pmatrix} \frac{\mathrm{D}\hat{u}}{\mathrm{D}\hat{t}} \\ \frac{\mathrm{D}\hat{v}}{\mathrm{D}\hat{t}} \end{pmatrix} \right). \tag{2.99}$$

Using (2.87), this equation for $\hat{\mathbf{u}}$ is exactly the same as that derived from the first equation of the semi-geostrophic system (2.90 except for the final term. Note that $\mathbf{Q}$ is $\mathrm{O}(Ro^{-1})$. The three terms on the right hand side of (2.99) are thus of order $1, Ro$ and Ro^2 respectively. The first term is to leading order $-\mu^{-1}\begin{pmatrix} \frac{\partial \hat{h}}{\partial \hat{y}} \\ -\frac{\partial \hat{h}}{\partial \hat{x}} \end{pmatrix}$ which is $\begin{pmatrix} \hat{u}_g \\ v_g \end{pmatrix}$. The second term is a correction to geostrophy of order Ro. The final term is $\mathrm{O}(Ro^2)$ compared with the first term, so that $\hat{\mathbf{u}}$ derived from semi-geostrophic theory is a second order accurate approximation to $\hat{\mathbf{u}}$ derived from (2.33). Substituting this into the third equation of (2.33) shows that the evolution of $\hat{h}$ will also be second order accurate in Ro. As shown in the derivation of (2.91), the evolution of $\hat{h}$ determines the whole evolution.

Note that the final term, which determines the error, is a second Lagrangian time derivative. This was pointed out by Hoskins (1975). Though $\mathrm{O}(Ro^2)$, it becomes very large for the fast wave solutions of (2.33), which makes proving convergence of solutions of (2.88) to those of (2.33) very difficult, as discussed in section 2.3.5 for the quasi-geostrophic equations.

This estimate can be refined further. The estimate for $\mathbf{u}$, (2.99), only depends on the assumption that the acceleration terms in the first two equations of (2.33) are $\mathrm{O}(Ro)$ when compared with the remaining terms. Thus the Rossby number is the ratio of $\frac{\mathrm{D}}{\mathrm{D}\hat{t}}$ to the rotation rate. Therefore it is a Lagrangian Rossby number Ro_L as defined in (2.32). This only requires the horizontal scale to be large in the direction of the flow, as discussed in section 2.5.2 and illustrated in section 2.7.2. This characterisation is followed through in the next paragraph, because it is important in determining the situations where semi-geostrophic theory is applicable.

The estimate can also be extended to cover other limits than $\alpha = 1$. Starting from (2.33), equation (2.99) can be derived from it even with $\alpha > 1$ under the assumptions $Ro_L \ll 1$, $Ro_L = \alpha Fr^2$. The substitution of (2.99) into the third equation of (2.33) then gives an error $\mathrm{O}(\alpha Ro_L^2)$. Substituting for α gives the error as

$$Ro_L^3 Fr^{-2} = Ro_L(Ro_L/Fr)^2 = Ro_L Bu^2, \tag{2.100}$$

where Bu is the Burger number, as defined in section 2.3.2. This estimate is validated numerically in section 5.2.2. In the quasi-geostrophic scaling $Ro = Fr$ from section 2.3.5, the characterisation of Ro as Ro_L does not apply, and the argument leading to (2.100) gives accuracy Ro. It is also shown rigorously in section 3.6.2 that the error remains $O(Ro)$ for the incompressible case $\alpha \gg 1$, $Fr \ll Ro$.

Only limited results proving that semi-geostrophic solutions are the limit of Euler equation solutions have been proved so far. These are discussed in section 3.6. The shallow water case has not yet been included. As noted above, the difficulty is caused by the fast waves. In the semi-geostrophic case $\alpha = 1$, these can no longer be described by a constant coefficient problem because the variations of $\hat{h}$ are $O(1)$. Thus the averaging techniques discussed in sections 2.3.4 and 2.3.5 cannot be used.

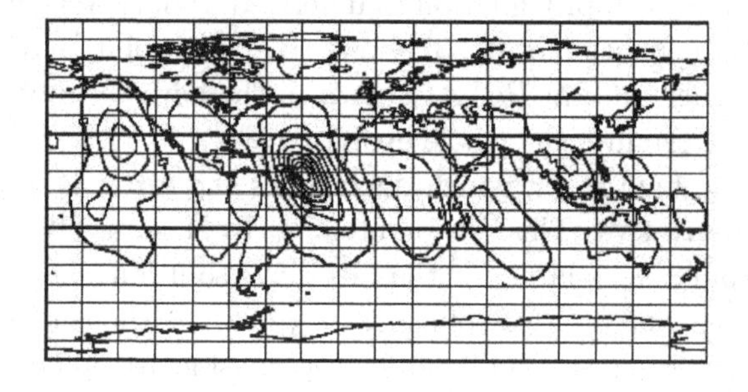

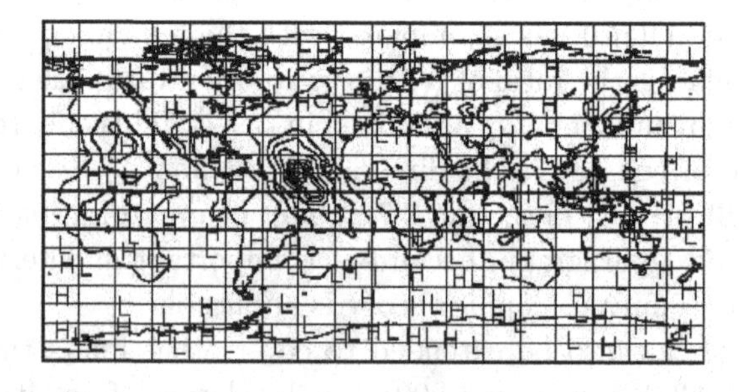

Fig. 2.2 Potential vorticities using depth field from Fig. 2.1 but with $gh_0 = 5000\text{m}^2\text{s}^{-2}$. Top: Potential vorticity, units $(\text{ms})^{-1}$, contour interval 0.8×10^{-7}. Bottom: Semi-geostrophic potential vorticity, units $10^{-13}\text{m}^{-1}\text{s}^{-2}$, contour interval 150.0. From Cullen (2002). ©Royal Meteorological Society, Reading, U.K., 2002.

The rigorous proofs that will be discussed in section 3.5 use the fact that the property $\hat{\mathcal{Q}}_{sg} = \det\mathbf{Q}$ means that the potential vorticity inversion problem involves solving a Monge-Ampère equation rather than a Poisson equation. It is well known that the solution of a Monge-Ampère equation is in general anisotropic and can readily produce singularities, which is why semi-geostrophic solutions can contain anisotropic behaviour such as fronts and jet-streams.

The semi-geostrophic potential vorticity (2.94) is approximately μRo^{-1} times the potential vorticity (2.35). Thus it is positive in both hemispheres if $\mathbf{Q}$ is positive definite. The potential vorticity (2.35) will typically change sign across the equator. To derive (2.94) from (2.35) the vorticity is approximated, but the depth dependence is not. Since, for $\hat{L}_R < 1$, variations in potential vorticity reflect variations in $\hat{h}$ rather than in $\hat{\zeta}$ it is appropriate that the depth dependence is not approximated in the definition of $\hat{\mathcal{Q}}_{sg}$. The approximation to the vorticity is the replacement of $\hat{\zeta}$ by its geostrophic value, plus an extra term. However, this extra term does not make the approximation more accurate, but actually the reverse, McIntyre and Roulstone (2002).

These points are illustrated in Fig. 2.2 which, like Fig. 2.1, is in dimensional variables and uses a constant Coriolis parameter. The same depth field is used as in Fig. 2.2, but the mean value of h is decreased to give L_R=480km, corresponding to a wavelength of about 3000km. Most of the variations in h are on larger scales than this. The potential vorticity distribution is rather similar to the h field shown in Fig. 2.1, and very different from the potential vorticity shown in Fig. 2.1. The semi-geostrophic approximation to it shown in the lower panel of Fig. 2.2 is quite accurate, though multiplied by different scaling factors.

2.3.7 *Large scale slow solutions in the tropics*

In the tropics, the dimensional Coriolis parameter is $2\Omega \sin\phi$ which goes through zero at the equator. An approximation which is uniformly valid in low Rossby number is thus not possible. Equations 2.33) still apply. Tropical solutions can be studied by choosing $\mu = \beta\hat{y}$ with β constant and the domain Γ chosen to be periodic in $\hat{x}$ and unbounded in $\hat{y}$ with a decay condition as $\hat{y} \to \pm\infty$.

The linear analysis as in section 2.3.2) can still be carried out using (2.38). The details are given in Vallis (2017), section 8.2. Choose $\alpha = Fr^{-2}$ and $Ro = 1$ so that all terms in the first two equations of 2.38) are

O(1). Seek solutions of (2.38) of the form $u' = \check{u}\nu(\hat{y})\exp i\left(k\hat{x} - \omega\hat{t}\right), v' = \check{v}\nu(\hat{y})\exp i\left(k\hat{x} - \omega\hat{t}\right), h' = \check{h}\nu(\hat{y})\exp i\left(k\hat{x} - \omega\hat{t}\right)$. Eliminating $\check{u}$ and $\check{h}$ gives an equation for the meridional structure functions ν:

$$\check{v}\left(\frac{\partial^2\nu}{\partial\hat{y}^2} - \left(-\omega^2 Fr^2 + k^2 + \frac{k\beta}{\omega} - \beta^2\hat{y}^2 Fr^2\right)\nu\right) = 0. \tag{2.101}$$

This has the trivial solution $\check{v} = 0$, corresponding to Kelvin waves as below. Non-trivial solutions correspond to eigensolutions for $\nu(\hat{y})$, giving the dispersion relation

$$\frac{1}{\beta Fr}\left(\omega^2 Fr^2 - k^2 - \frac{k\beta}{\omega}\right) = 2l + 1. \tag{2.102}$$

l is a meridional wavenumber. Associated eigenfunctions ν_l are given by

$$\nu_l(\xi) = \nu_0 H_l(\xi)\exp\left(\frac{-\xi^2}{2}\right), \tag{2.103}$$

$$\xi = (\beta Fr)^{\frac{1}{2}}\,\hat{y}.$$

H_l are Hermite polynomials so that $H_0 = 1$, $H_1 = 2\xi$, $H_2 = 4\xi^2 - 2$. The definition of ξ can be written as $\xi = \hat{y}/\hat{L}_E$, where $\hat{L}_E$ is the dimensionless equatorial deformation radius defined as $\sqrt{1/(\beta Fr)}$. These solutions describe equatorially trapped inertia-gravity waves and Rossby waves. The special case $l = 0$ describes the mixed Rossby-gravity wave.

Kelvin waves satisfy the dispersion relation $\omega^2/k^2 = Fr^{-2}$, and have meridional structure

$$\nu(\hat{y}) = \exp\left(-\frac{\beta\hat{y}^2 Fr}{2}\right) = \exp\left(-\frac{1}{2}\left(\frac{\hat{y}}{\hat{L}_E}\right)^2\right). \tag{2.104}$$

This structure applies to all the Kelvin waves.

A hierarchy of reduced models can be constructed in order to identify the large-scale behaviour. In Majda and Klein (2003) this is achieved through a formal multiscale expansion, though that procedure is not used here. In this analysis, the lowest order model is a degenerate form of the semi-geostrophic equations discussed in section 2.3.6. For the geostrophic wind defined in (2.87) to be finite with $\mu = \beta\hat{y}$, $\hat{h}$ must take the form

$$\hat{h} = 1 + c_0\hat{y}^2 + c_1(\hat{x})\hat{y}^2 + c_2(\hat{x})\hat{y}^3 + O(\hat{y}^4). \tag{2.105}$$

Then

$$\beta \hat{u}_g = -2(c_0 + c_1(\hat{x})) - 3c_2(\hat{x})\hat{y} + \mathrm{O}(\hat{y}^2), \tag{2.106}$$
$$\beta \hat{v}_g = \frac{\partial c_1}{\partial \hat{x}}\hat{y} + \frac{\partial c_2}{\partial \hat{x}}\hat{y}^2 + \mathrm{O}(\hat{y}^3).$$

As shown in section 2.3.6, semi-geostrophic solutions require $\mathbf{Q}$ as defined in (2.90) to be positive definite. This requires det$\mathbf{Q}$ to be non-negative, which gives

$$\beta^2\hat{y}^2 + \beta\hat{y}\left(\frac{\partial^2 c_1}{\partial \hat{x}^2}\hat{y} + \frac{\partial^2 c_2}{\partial \hat{x}^2}\hat{y}^2 + 3c_2 + \mathrm{O}(\hat{y})\right) - \tag{2.107}$$
$$2\left(\frac{\partial c_1}{\partial \hat{x}}\right)^2 - 3\frac{\partial c_2}{\partial \hat{x}}\frac{\partial c_1}{\partial \hat{x}}\hat{y} + 3\frac{\partial c_2}{\partial \hat{x}}\left(\frac{\partial^2 c_1}{\partial \hat{x}^2}\hat{y} + \frac{\partial^2 c_2}{\partial \hat{x}^2}\hat{y}^2\right) \geq 0.$$

This is the inertial stability condition for semi-geostrophic flow. Ensuring that this is non-negative as $\hat{y} \to 0$ requires c_1 to be constant, and thus absorbed into c_0, and $c_2 = 0$. Thus

$$\hat{h} = 1 + c_0\hat{y}^2 + \mathrm{O}(\hat{y}^4). \tag{2.108}$$

This requirement that $\hat{h}$ is almost constant near the equator is called the 'weak temperature gradient' approximation in the three-dimensional case, section 2.4.7, see Sobel *et al.* (2001). In particular, east-west variations in $\hat{h}$ are only allowed at $\mathrm{O}(\hat{y}^4)$. However (2.106) shows that a zonal geostrophic wind is allowed at the equator. While this is a dynamically trivial solution, maintaining it aganst forcing gives a non-trivial balance as discussed in section 2.4.7.

Now consider the next member of the hierarchy. Geostrophic adjustment will still operate in the north-south direction, leading to a zonal flow in geostrophic balance. However, it is never achieved in the east-west direction. With uniform rotation, a northward displacement of a particle by a distance *Ro* acted on by the Coriolis force will generate a unit zonal wind. A northward displacement from the equator by a distance $\sqrt{2/\beta}$ will also generate a unit zonal wind. However, a displacement along the equator will not generate any wind. It is plausible to suggest that the resulting signal will have an east-west scale equal to or larger than the equatorial deformation radius. The observational study by Wheeler and Kiladis (1999) suggests that most of the observed power is at frequencies less than $\beta\hat{L}_E$, which is the Coriolis frequency a distance $\hat{L}_E$ from the equator. This suggests that it is appropriate to consider a dynamical model which is specific to frequencies less than $\beta\hat{L}_E$. In order to link with semi-geostrophic dynamics outside the equatorial waveguide, the natural choice is the tropical long wave approximation, Heckley and Gill (1984), which imposes geostrophic balance of the zonal wind. This is appropriate for meridional scales comparable to or greater than $\hat{L}_E$. Majda and Klein (2003) derive these by a systematic multiscale expansion. Chan and Shepherd (2014) derive a shallow water version of these by asymptotic expansion including forcing terms.

Start from eqs. (2.33). Choose $\alpha = 1$. Then assume different length scales and thus velocity scales for the east-west and north-south directions. Thus assume velocity scales U, V such that $V = \varepsilon U$ and length scales L_x, L_y such that $L_y = \varepsilon L_x$. Assume that $\beta\hat{y}$=O(1) and $Ro = \varepsilon$. Choose $Fr = 1$. Then the acceleration term $\frac{\mathrm{D}\hat{v}}{\mathrm{D}\hat{t}}$ can be neglected to O(ε^2) in the $\hat{y}$ momentum equation. All terms in the $\hat{x}$ momentum equation become the same size.

First consider the linear problem as above. Equations (2.38) become

$$\begin{aligned}\frac{\partial \hat{u}}{\partial \hat{t}} + \frac{\partial \hat{h}}{\partial \hat{x}} - \beta\hat{y}\hat{v} &= 0,\\ \frac{\partial \hat{h}}{\partial \hat{y}} + \beta\hat{y}\hat{u} &= 0,\\ \frac{\partial \hat{h}}{\partial \hat{t}} + \left(\frac{\partial \hat{u}}{\partial \hat{x}} + \frac{\partial \hat{v}}{\partial \hat{y}}\right) &= 0.\end{aligned} \tag{2.109}$$

Carrying through the analysis as above gives the equation

$$\check{v}\left(\frac{\partial^2 \nu}{\partial \hat{y}^2} - \left(k^2 + \frac{k\beta}{\omega} - \beta^2\hat{y}^2\right)\nu\right) = 0. \tag{2.110}$$

This is a good approximation to (2.102) if $\omega^2 \ll k^2$, so the wave speed is much less than that of the gravity waves, assumed to be 1 in this scaling. The eigenvalues for the nontrivial solution can be found as before as

$$\frac{1}{\beta}\left(-k^2 - \frac{k\beta}{\omega}\right) = 2l + 1. \tag{2.111}$$

with eigenfunctions given by (2.103). As before, there is also a solution $\check{v} = 0$ corresponding to Kelvin waves which have speed ± 1 in this scaling.

Thus the solutions of (2.111) can be interpreted as Rossby wave solutions of (2.109). The gravity waves are eliminated and the mixed Rossby-gravity wave present for $l = 0$ is replaced by a Rossby wave. There is an inconsistency in the behaviour as the Rossby waves need to have $\omega^2 \ll k^2$ for the approximation to be justified, while the Kelvin wave solutions have $\omega^2 = k^2$. It arises because a filtering based on length scale is not the same as a filtering based on frequency. This exposes the issues with establishing a strict scale separation in the solutions of (2.38) in the tropics.

Observed large-scale tropical winds as illustrated in section 2.7.5 show localised anomalies which are too large to explain with linear theory. However, these are consistent with the tropical long wave approximation in that the zonal winds are close to geostrophic while the meridional winds are not. Using the same scaling as in (2.109), the nonlinear equations (2.33) become

$$\frac{\mathrm{D}\hat{u}}{\mathrm{D}\hat{t}} + \frac{\partial \hat{h}}{\partial \hat{x}} - \beta\hat{y}\hat{v} = 0,$$

$$\frac{\partial \hat{h}}{\partial \hat{y}} + \beta\hat{y}\hat{u} = 0, \tag{2.112}$$

$$\frac{\partial \hat{h}}{\partial \hat{t}} + \frac{\partial(\hat{h}\hat{u})}{\partial \hat{x}} + \frac{\partial(\hat{h}\hat{v})}{\partial \hat{y}} = 0.$$

In principle, these can be solved by defining

$$Y = \frac{1}{2}\beta\hat{y}^2 - \hat{u}. \tag{2.113}$$

There is no evolution equation for $\hat{v}$, so define $\mathrm{D}^*/\mathrm{D}\hat{t} \equiv \partial/\partial\hat{t} + \hat{v}\partial/\partial\hat{y}$. Then the first equation of (2.112) implies

$$\frac{\mathrm{D}^*Y}{\mathrm{D}\hat{t}} + \hat{u}\frac{\partial Y}{\partial \hat{x}} - \frac{\partial \hat{h}}{\partial \hat{x}} = 0. \tag{2.114}$$

$\hat{v}$ has to be diagnosed by using the time derivative of the second equation of (2.112), giving

$$-\frac{\partial^2}{\partial \hat{y}^2}(\hat{h}\hat{v}) + \beta\hat{y}\hat{v}\frac{\partial Y}{\partial \hat{y}} = \frac{\partial^2(\hat{h}\hat{u})}{\partial\hat{x}\partial\hat{y}} - \beta\hat{y}\left(\hat{u}\frac{\partial Y}{\partial \hat{x}} - \frac{\partial \hat{h}}{\partial \hat{x}}\right). \tag{2.115}$$

This is equivalent to the diagnostic equation found by Chan and Shepherd. It is elliptic and thus can be solved subject to the inertial stability condition $\hat{y}\frac{\partial Y}{\partial \hat{y}} > 0$, so that Y has to be a monotone increasing function of $\hat{y}$ north of the equator and vice versa. At the equator $\hat{h}$ has to be continuous and $\partial\hat{h}/\partial\hat{y} = \mathrm{O}(\hat{y})$ as required by (2.108). It is possible to solve this equation in an equatorial strip with $\hat{v}$ given on the northern and southern boundaries. These values of $\hat{v}$ could then be used as boundary conditions for the semi-geostrophic solution (2.91) in an extratropical region, thus giving a coupled solution.

The multiscale approach of Majda and Klein (2003) would define Y using a semi-geostrophic solution (2.106) subject to (2.108) and derive a quasi-linear form of (2.115) using Y from the semi-geostrophic solution, which then gives an elliptic equation to be solved for each $\hat{x}$. The inertial stability condition above is guaranteed by the positive definiteness of (2.90). Since this Y has to be very smooth in the longitudinal direction, it is likely that it can be proved that the variation of $\hat{v}$ in $\hat{x}$ would be no greater than that of the right hand side. The equations should then be well-posed. It is not clear whether the fully nonlinear equations (2.112) are well-posed. Computed solutions of the nonlinear equations are given in Chan and Shepherd (2014) which agree closely with those of (2.33).

2.4 Various approximations to the three-dimensional hydrostatic Boussinesq equations

2.4.1 *The hydrostatic Boussinesq equations*

This section illustrates the 'slow' regimes discussed in section 2.3 in three dimensions. For simplicity, this is done using the incompressible Boussinesq version of the shallow atmosphere hydrostatic equations, (2.24), which are appropriate for the ocean, but can be applied to the atmosphere with limited accuracy. In order to allow the same equations to be used for the atmosphere and the ocean, the variables have different meanings in the two cases. In the ocean, minus $\hat{\theta}$ is the density, which is a function of temperature and salinity as in (2.7), $\hat{\varphi}$ is the pressure variable and the vertical coordinate $\hat{z}$ is depth. In the atmosphere, $\hat{\theta}$ is the potential temperature, $\hat{\varphi}$ is a perturbation geopotential as discussed below, and the vertical coordinate $\hat{z}$ is a function of pressure as defined below. Remove all the dissipation and source terms from (2.24) and again write $\mu = \sin\phi$. Then with the changes of notation listed above, the equations become

$$\begin{aligned} \frac{\mathrm{D}(\hat{u},\hat{v})}{\mathrm{D}\hat{t}} + Ro^{-1}\mu(-\hat{v},\hat{u}) + \alpha^{-1}Fr^{-2}\nabla_h\hat{\varphi} &= 0, \\ \frac{\partial\hat{\varphi}}{\partial\hat{z}} - \hat{\theta} &= 0, \\ \nabla\cdot\hat{\mathbf{u}} &= 0, \\ \frac{\mathrm{D}_h\hat{\theta}}{\mathrm{D}\hat{t}} + \alpha\hat{w}\frac{\partial\hat{\theta}}{\partial\hat{z}} &= 0. \end{aligned} \tag{2.116}$$

McWilliams and Gent (1980) show how to apply these equations to the atmosphere. First describe the procedure using dimensional variables. As in section 2.2.3, note that in the original equation (2.4) with no basic state subtracted, hydrostatic balance is defined by $C_p\theta\partial\Pi/\partial r + g = 0$. This ensures that Π is a monotonic function of r so that the equations can be rewritten with pressure as a vertical coordinate. Next, following Hoskins and Bretherton (1972), define

$$z = \left(1 - \left(\frac{p}{p_{ref}}\right)^{\frac{R}{C_p}}\right) H_s \tag{2.117}$$

where the scale height $H_s = \frac{C_p p_{ref}}{R\rho_0(a)g}$, p_{ref} is the reference pressure used in section 2.2.1, and $\rho_0(a)$ and θ_0 are as used to define the reference state (2.8) and to derive (2.12). z is a function of pressure, defined as the height at which a given pressure is reached in the reference atmosphere defined by (2.8). Then Hoskins and Bretherton (1972) show that the hydrostatic relation becomes $\partial\varphi/\partial z = g\theta/\theta_0$.

Now return to dimensionless variables. Set $\hat{z} = z/H_s$, $\hat{\theta} = \theta/\theta_0$ and $\hat{\varphi} = \varphi/(gH_s)$. It is shown by Hoskins that the continuity equation reduces to $\nabla\cdot(\xi(\hat{z})\hat{\mathbf{u}}) = 0$, where $\xi(\hat{z})$ is a function depending on the reference state. The Boussinesq approximation is to set $\xi(\hat{z}) = 1$, which assumes that fluid particles do not change their pressure very fast following the motion. The limitations of this assumption in the atmosphere are discussed in Davies *et al.* (2003).

Equations (2.116) are solved in a region $\Gamma \in \mathbb{R}^3$ with boundary conditions $\hat{\mathbf{u}} \cdot \mathbf{n} = 0$. In the atmosphere, this implies that the upper and lower boundary conditions are applied on constant pressure surfaces. In the ocean it excludes surface effects. These features, together with the Boussinesq approximation, are idealisations which will not be accurate at large scales but which aid analysis and are useful for illustrative purposes. Quantitative justification is given by Hoskins (1982).

Equations (2.116) conserve the energy integral

$$E = \int_\Gamma \left(\frac{1}{2}(\hat{u}^2 + \hat{v}^2) - \hat{\theta}\hat{z}\right) \mathrm{d}\hat{x}\mathrm{d}\hat{y}\mathrm{d}\hat{z}. \tag{2.118}$$

They also conserve the potential vorticity

$$\hat{\mathcal{Q}} = (\hat{\Xi} + (0, 0, \mu Ro^{-1})) \cdot \nabla\hat{\theta} \tag{2.119}$$

in the sense that $\mathrm{D}\hat{\mathcal{Q}}/\mathrm{D}\hat{t} = 0$. $\hat{\Xi}$ is defined by

$$\hat{\Xi} = \left(-\frac{\partial\hat{v}}{\partial\hat{z}}, \frac{\partial\hat{u}}{\partial\hat{z}}, \frac{\partial\hat{v}}{\partial\hat{x}} - \frac{\partial\hat{u}}{\partial\hat{y}}\right). \tag{2.120}$$

This equation is derived in White (2002).

2.4.2 *Key parameters*

In order to identify the key asymptotic regimes, follow the same procedure as in section 2.3.2. Much more extensive linear analysis is given in Gill (1982). Linear analyses of (2.116) are given by Davies *et al.* (2003) and compared with that of the anelastic and fully compressible equations. This illustrates that the vertical structure of the normal modes is quite different. In the fully compressible system the modes are sinusoidal in height with an exponential scaling factor. In the Boussinesq case there is no exponential scaling factor.

Linearise (2.116) about a state of rest given by

$$\begin{aligned} \hat{u} = \hat{v} = \hat{w} = 0, \\ \hat{\theta} = \hat{z}, \\ \frac{\partial \hat{\varphi}}{\partial \hat{z}} - \hat{\theta} = 0. \end{aligned} \tag{2.121}$$

Seek a general solution $u = u', v = v', w = w', \theta' = \hat{z} + \theta', \varphi = \hat{\varphi} + \varphi'$. Temporarily assume periodic boundary conditions on $u', v', w', \theta', \varphi'$ in all three directions, and assume that the Coriolis parameter is constant so that $\mu = 1$. Assume that u' takes the form $\breve{u} \exp^{i(kx+ly+mz-\omega t)}$, with similar definitions for $\breve{v}, \breve{w}, \breve{\theta}$ and $\breve{\varphi}$. Following the same method as in section 2.3.2 shows that the condition for non-trivial solutions is that

$$\begin{aligned} \omega = & \quad 0, \\ = & \pm \tfrac{1}{m}\sqrt{Fr^{-2}(k^2+l^2) + Ro^{-2}m^2}. \end{aligned} \tag{2.122}$$

As in the shallow water case, section 2.3.2, the solution $\omega = 0$ corresponds to a Rossby wave. The other solutions correspond to inertia-gravity waves. The inertia-gravity wave frequency is

$$\varpi = \sqrt{Fr^{-2}m^{-2}(k^2+l^2) + Ro^{-2}}. \tag{2.123}$$

ϖ is made up of two terms, as in the shallow water case (2.47). The first describes pure gravity waves. This term can tend to zero if $(k^2+l^2)/m^2 \to 0$, which implies that the aspect ratio tends to zero. This would be inconsistent with the original assumption that the dimensionless aspect ratio of the flow is O(1). The second term in ϖ is the same as in the shallow water case, implying that the inertia-gravity wave frequency is always greater than Ro^{-1}. The two terms are equal if

$$1/\sqrt{(k^2+l^2)} = Ro(mFr)^{-1} \equiv \hat{L}_R \tag{2.124}$$

where $\hat{L}_R$ is the Rossby radius of deformation which is equal to Ro/Fr as in section 2.3.2. As discussed at the end of section 2.2.5, the condition $\hat{L} = \hat{L}_R$

can also be written in dimensional terms as that the aspect ratio $H/L = f/N$, recalling that $Fr = U/NH$ where the Brunt-Väisälä frequency N is defined in (2.22). A typical value of this ratio is 0.01 in the troposphere, which is similar to the observed aspect ratio of weather systems. Flows with frequency less than ϖ and aspect ratios greater than f/N are stratification dominated, flows with frequency less than ϖ and aspect ratios less than f/N are rotation dominated.

2.4.3 *General equations for slow solutions*

As in section 2.3.3, consider the asymptotic regimes where the horizontal flow speed is smaller than the horizontal component of the inertia-gravity wave speed. Using (2.123), define

$$\varepsilon = 1/\sqrt{Fr^{-2}m^{-2} + Ro^{-2}(k^2 + l^2)^{-1}}. \quad (2.125)$$

Equations (2.116) can be rewritten in terms of the vertical component of the vorticity, $\hat{\zeta} = \partial\hat{v}/\partial\hat{x} - \partial\hat{u}/\partial\hat{y}$, and the horizontal divergence, $\hat{\Delta} = \partial\hat{u}/\partial\hat{x} + \partial\hat{v}/\partial\hat{y}$, in the manner of (2.57). The horizontal momentum equations become

$$\frac{\partial\hat{\zeta}}{\partial\hat{t}} + \nabla_h \cdot (\hat{\mathbf{u}} \cdot \nabla\hat{v} + \mu Ro^{-1}\hat{u}, -\hat{\mathbf{u}} \cdot \nabla\hat{u} + \mu Ro^{-1}\hat{v}) = 0, \quad (2.126)$$

$$\frac{\partial\hat{\Delta}}{\partial\hat{t}} + \hat{u}\frac{\partial\hat{\Delta}}{\partial\hat{x}} + \hat{v}\frac{\partial\hat{\Delta}}{\partial\hat{y}} + \hat{\Delta}^2 - 2J(\hat{u},\hat{v}) +$$

$$\frac{\partial\hat{w}}{\partial\hat{x}}\frac{\partial\hat{u}}{\partial\hat{z}} + \frac{\partial\hat{w}}{\partial\hat{y}}\frac{\partial\hat{v}}{\partial\hat{z}} + \alpha^{-1}Fr^{-2}\nabla_h^2\hat{\varphi} - Ro^{-1}\nabla_h \cdot (\mu\hat{v}, -\mu\hat{u}) = 0,$$

where $\nabla_h \equiv (\partial/\partial\hat{x}, \partial/\partial\hat{y})$ and $\nabla_h^2 \equiv \nabla_h \cdot \nabla_h$. Write $(\hat{u},\hat{v},0) \equiv \hat{\mathbf{u}}_h$. As before, differentiate the second equation with respect to time and substitute from the first equation. This gives

$$\frac{\partial^2\hat{\Delta}}{\partial\hat{t}^2} + \mu Ro^{-1}(\hat{\zeta} + \mu Ro^{-1})\hat{\Delta} + \alpha^{-1}Fr^{-2}\nabla_h^2\frac{\partial\hat{\varphi}}{\partial\hat{t}} =$$

$$2\alpha^{-1}Fr^{-2}\hat{\Delta}\nabla_h^2\hat{\varphi} - 2\mu Ro^{-1}\hat{\Delta}\hat{\zeta} \quad (2.127)$$

$$-\mu Ro^{-1}\hat{\mathbf{u}} \cdot \nabla(\hat{\zeta} + \mu Ro^{-1}) + 2\frac{\partial}{\partial\hat{t}}J(\hat{u},\hat{v}) + \text{remainder}.$$

Then differentiate with respect to $\hat{z}$ and substitute from the third, fourth and fifth equations of (2.116). This gives, after including additional terms in 'remainder',

$$-\frac{\partial^2}{\partial \hat{t}^2}\frac{\partial^2 \hat{w}}{\partial \hat{z}^2} - \mu Ro^{-1}\frac{\partial}{\partial \hat{z}}\left((\hat{\zeta} + \mu Ro^{-1})\frac{\partial \hat{w}}{\partial \hat{z}}\right) - Fr^{-2}\nabla_h^2\left(\hat{w}\frac{\partial \hat{\theta}}{\partial \hat{z}}\right) = \alpha^{-1}Fr^{-2}\nabla_h^2\left(\mathbf{u}_h \cdot \nabla_h \hat{\theta}\right)$$

$$+\frac{\partial}{\partial \hat{z}}\left(-\mu Ro^{-1}\hat{\mathbf{u}} \cdot \nabla(\hat{\zeta} + \mu Ro^{-1}) + 2\frac{\partial}{\partial \hat{t}}J(\hat{u}, \hat{v})\right) + \text{new remainder.}$$

The linearisation of this equation about the basic state (2.121) is

$$\frac{\partial^2}{\partial \hat{t}^2}\frac{\partial^2 \hat{w}}{\partial \hat{z}^2} + \mu^2 Ro^{-2}\frac{\partial^2 \hat{w}}{\partial \hat{z}^2} + Fr^{-2}\nabla_h^2 \hat{w} = 0. \tag{2.128}$$

If μ is constant, this equation describes inertia-gravity waves with the frequency ϖ calculated in (2.123).

Thus, as in section 2.3.3, equation (2.128) can be approximated to $O(\varepsilon^2)$ by the diagnostic equation obtained by omitting the first term. Again as in section 2.3.3, a consistent approximation to (2.116) for most cases where $\varepsilon \ll 1$ can be obtained by assuming that $\hat{\Delta} \simeq \varepsilon\hat{\zeta}$, with the reservation that if μ is not constant, this is inconsistent with geostrophic balance.

Introduce a stream-function $\hat{\psi}$ defined after (2.59). Then the second equation of (2.126) can be approximated to $O(\varepsilon^2)$ by

$$-2J\left(\frac{\partial \hat{\psi}}{\partial \hat{x}}, \frac{\partial \hat{\psi}}{\partial \hat{y}}\right) + \alpha^{-1}Fr^{-2}\nabla_h^2\hat{\varphi} - \nabla_h \cdot (\mu Ro^{-1}\nabla_h \hat{\psi}) = 0. \tag{2.129}$$

Differentiate the fourth equation of (2.116) with respect to $\hat{z}$ and use the third equation to give

$$\frac{\partial}{\partial \hat{t}}\left(\frac{\partial^2 \hat{\varphi}}{\partial \hat{z}^2}\right) + \nabla_\alpha \cdot \left(\frac{\partial^2 \hat{\varphi}}{\partial \hat{z}^2}\hat{\mathbf{u}}\right) + \nabla_\alpha \cdot \left(\frac{\partial \hat{\varphi}}{\partial \hat{z}}\frac{\partial \hat{\mathbf{u}}}{\partial \hat{z}}\right) = 0, \tag{2.130}$$

where $\nabla_\alpha = (\frac{\partial}{\partial \hat{x}}, \frac{\partial}{\partial \hat{y}}, \alpha\frac{\partial}{\partial \hat{z}})\cdot$. Note that the fourth equation of (2.116) is not approximated in general because, in the term $\alpha\hat{w}\partial\hat{\theta}/\partial\hat{z}$, $\hat{w}$ will be $O(\varepsilon)$ and $\alpha \geq 1$, so the product may not be small.

The system of equations comprising the first equation of (2.126), equation (2.130) and the time derivative of equation (2.129) are essentially the equations used in McWilliams *et al.* (1999). In that paper, isentropic coordinates, (i.e. using potential temperature as a coordinate), were used. That considerably simplifies the analysis because the Lagrangian time derivative only involves the 'horizontal' winds. Write $\hat{\chi}$ for the velocity potential, so that $\hat{\Delta} = \nabla_h^2 \hat{\chi}$. Then the equations in McWilliams *et al.* (1999), applied to the present case, can be written

$$\begin{pmatrix} F\,\partial^2/\partial\hat{z}^2 & 0 & \\ 0 & \nabla_h^2 & -M \\ G & 0 & \nabla_h^2 \end{pmatrix} \begin{pmatrix} \hat{\chi} \\ \partial\hat{\varphi}/\partial\hat{t} \\ \partial\hat{\psi}/\partial\hat{t} \end{pmatrix} = \tag{2.131}$$
$$\begin{pmatrix} -J(\hat{\psi}, \partial^2\hat{\varphi}/\partial\hat{z}^2) - J(\partial\hat{\psi}/\partial\hat{z}, \partial\hat{\varphi}/\partial\hat{z}) \\ 0 \\ -J(\hat{\psi}, \hat{\zeta} + \mu Ro^{-1}) \end{pmatrix},$$

where

$$\begin{aligned} \nabla_h^2 \hat{\chi} &= \hat{\Delta}, \\ F &= \nabla_\alpha \cdot \left(\frac{\partial^2 \hat{\varphi}}{\partial \hat{z}^2} \nabla_h \right) + \nabla_\alpha \cdot \left(\frac{\partial \hat{\varphi}}{\partial \hat{z}} \frac{\partial}{\partial \hat{z}} \nabla_h \right), \\ M &= \nabla_h \cdot (\mu Ro^{-1} \nabla_h) + 2 \left(\frac{\partial^2 \hat{\psi}}{\partial \hat{x}^2} \frac{\partial^2}{\partial \hat{y}^2} + \frac{\partial^2 \hat{\psi}}{\partial \hat{y}^2} \frac{\partial^2}{\partial \hat{x}^2} \right) - 2 \frac{\partial^2 \hat{\psi}}{\partial \hat{x} \partial \hat{y}} \frac{\partial^2}{\partial \hat{x} \partial \hat{y}}, \\ G &= \nabla_h \cdot \left((\hat{\zeta} + \mu Ro^{-1}) \nabla_h \right) + \nabla_h \left(\frac{\partial \hat{\psi}}{\partial \hat{z}} \right) \cdot \nabla_h \int_0^{\hat{z}} \mathrm{d}\hat{z}', \end{aligned} \tag{2.132}$$

noting that $\nabla_h \int_0^{\hat{z}} = -\hat{w}$.

Write (2.131) symbolically as

$$\mathbf{Lu} = \mathbf{A}. \tag{2.133}$$

Solvability of this equation depends on the same conditions as those written out in section 2.3.3. Condition (i) is more restrictive in that it implies that $\partial\hat{\theta}/\partial\hat{z}$ does not change sign. It can be shown that the product of conditions (i) and (ii) requires that the potential vorticity does not change sign, so that spontaneous violations are only possible if both terms change sign simultaneously. Condition (iii) is again liable to spontaneous violations as it is not a constant of the motion. The same discussion as in section 2.3.3 applies to estimating the time for which condition (iii) will be satisfied, given assumptions on the initial data.

It can be inferred that it is not possible to define a slow manifold by a single set of equations for all cases where $\varepsilon \ll 1$. Therefore seek limit solutions in the cases $Fr < Ro$, $Fr = Ro$ and $Ro < Fr$ separately. In dimensional terms, these correspond to aspect ratios H/L greater than, equal to, or less than f/N respectively.

2.4.4 *Slow solutions with large aspect ratio*

This case corresponds to $\varepsilon \ll 1$ with $Fr < Ro$. The limiting behaviour is extensively analysed in Majda (2003) and only a few key results are given here. In the shallow water case described in section 2.3.4, it was shown that the limiting solution was given by the solution of the equations for two-dimensional incompressible flow. In the present case it was shown in Embid and Majda (1998) that the limiting solution is given by two-dimensional incompressible flow for each $\hat{z}$. Laboratory experiments demonstrating this were shown by Fincham *et al.* (1996).

In a similar way to the shallow water case, the assumption that $Fr \ll 1$ but $Ro = O(1)$ requires that $\alpha = Fr^2$, so that the horizontal gradients of $\hat{\theta}$ are $O(Fr^2)$ of the vertical gradients. The fourth equation of (2.116) can then only be satisfied if $\hat{w} \simeq Fr^2 \hat{\mathbf{u}}_h$ so that the third equation of (2.116) implies that $\hat{\Delta} \simeq Fr^2 \hat{\zeta}$ as discussed in section 2.3.3. In addition, the vertical advection terms in the first two equations of (2.126) are $O(Fr^2)$. In the vorticity equation, the first equation of (2.126), the assumption $Ro = O(1)$ means that the term $\mu Ro^{-1}\hat{\Delta}$ is $O(Fr^2)$. Therefore the first equation of (2.126) can be further approximated by

$$\frac{\partial \hat{\zeta}}{\partial \hat{t}} - \frac{\partial \hat{\psi}}{\partial \hat{y}} \frac{\partial (\hat{\zeta} + \mu Ro^{-1})}{\partial \hat{x}} + \frac{\partial \hat{\psi}}{\partial \hat{x}} \frac{\partial (\hat{\zeta} + \mu Ro^{-1})}{\partial \hat{y}} = 0, \tag{2.134}$$

where $\hat{\psi}$ is the stream-function defined after (2.59). Equations (2.134) and (2.129) are exactly the equations for two-dimensional incompressible flow (2.66). Their analytic properties are thus as summarised in section 2.3.4. $\hat{\varphi}$ is determined diagnostically from (2.129). The fourth equation of (2.116) is no longer used, but is replaced by a calculation of $\hat{\theta}$ from $\hat{\varphi}$ using the second equation of (2.116).

The conservation of the potential vorticity, (2.119), is replaced by the conservation of $(\hat{\zeta} + \mu Ro^{-1})$ for each $\hat{z}$. The simplification of the thermodynamic contribution to the potential vorticity is expected because potential vorticity variations are similar to vorticity variations if the aspect ratio is large, exactly analogously to the shallow water case.

The weakness of this system is that there is no control over the vertical scale. Since each level can evolve independently, if there is any initial variation of $\hat{\mathbf{u}}$ in $\hat{z}$ it will magnify in time. Such behaviour is illustrated by exact solutions in Majda (2003) and in the experiments of Fincham *et al.* (1996). This means that the vertical scale of the flow will reduce in time. If this happens, the calculation of $\hat{\theta}$ from $\hat{\varphi}$ may give unrealistic values, which is inconsistent with the fourth equation of (2.116) which states that values of $\hat{\theta}$ have to be bounded by their initial values. Thus the system of equations cannot be expected to be solvable for large times and cannot form a slow manifold. In addition, the reduction in vertical scale will lead to a violation of the assumpton of large aspect ratio used to justify this approximation to (2.116).

Return to dimensional variables, so that $Fr = U/NH$ as in section 2.2.5. Then, as discussed by Majda (2003), the vertical scale H can be controlled by viscosity, in which case the small value of Fr can be maintained. It may also be controlled by rotation, in which case the reduction in aspect ratio is arrested when $H/L \simeq f/N$, and the quasi-geostrophic limit equations described in section 2.4.5 become appropriate.

2.4.5 *Quasi-geostrophic solutions*

In this section it is assumed that $\varepsilon = Ro = Fr \ll 1$. Then a consistent scaling of the momentum equations requires $\alpha = Fr^{-1}$ as in section 2.3.5.

As in the shallow water case, section 2.3.5, the lowest order solution contains no prognostic information and is self-contradictory if the Coriolis parameter is a function of position so that $\mu \neq 1$. Therefore, as in that section, assume that $\mu = 1 + Ro\mu(\hat{y})$. Then define the geostrophic wind by

$$\begin{aligned} \frac{\partial \hat{\varphi}}{\partial \hat{x}} - \hat{v}_g &= 0, \\ \frac{\partial \hat{\varphi}}{\partial \hat{y}} + \hat{u}_g &= 0. \end{aligned} \tag{2.135}$$

Then $\nabla_h \cdot \hat{\mathbf{u}}_g = 0$ and so the vertical velocity $\hat{w}$ is entirely ageostrophic. The first equation of (2.116) then implies that the ageostrophic horizontal wind $(\hat{u}, \hat{v}, 0) - (\hat{u}_g, \hat{v}_g, 0)$ is of order ε. The vertical component of the vorticity $\hat{\zeta}$ can thus be approximated to $\mathrm{O}(\varepsilon)$ by its geostrophic value $\hat{\zeta}_g$. Equations (2.129) and the first equation of (2.126) are further approximated by

$$\nabla_h^2 \hat{\varphi} - \hat{\zeta}_g = 0, \tag{2.136}$$

and

$$\frac{\partial \hat{\zeta}_g}{\partial \hat{t}} + \hat{u}_g \frac{\partial \hat{\zeta}_g}{\partial \hat{x}} + \hat{v}_g \frac{\partial (\hat{\zeta}_g + \mu)}{\partial \hat{y}} + Ro^{-1} \hat{\Delta} = 0. \tag{2.137}$$

Assume additionally that $\frac{\partial \hat{\theta}}{\partial \hat{z}} = 1 + \mathrm{O}(\varepsilon)$. Since $\hat{w}$ is order $\varepsilon(\hat{u}, \hat{v})$, the fourth equation of (2.116) can be approximated to $\mathrm{O}(\varepsilon)$ by

$$\frac{\partial \hat{\theta}}{\partial \hat{t}} + \hat{u}_g \frac{\partial \hat{\theta}}{\partial \hat{x}} + \hat{v}_g \frac{\partial \hat{\theta}}{\partial \hat{y}} + \alpha \hat{w} = 0. \tag{2.138}$$

The system of equations is completed by the second and third equations of (2.116).

Once again the equations are to be solved in a closed region Γ. As in the shallow water case, in order to achieve energy conservation both $\hat{\mathbf{u}}_{\mathbf{g}} \cdot \mathbf{n}$ and $\hat{\mathbf{u}} \cdot \mathbf{n}$ are required to vanish on the boundary of Γ. The former condition implies that $\hat{\varphi}$ is constant along the intersection of the boundary of Γ with each constant $\hat{z}$ surface. The conserved energy integral is

$$E = \int_{\Gamma} \left(\frac{1}{2}(\hat{u}_g^2 + \hat{v}_g^2) - \hat{\theta}\hat{z} \right) \mathrm{d}\hat{x}\mathrm{d}\hat{y}\mathrm{d}\hat{z}. \tag{2.139}$$

The quasi-geostrophic potential vorticity,

$$\hat{Q}_g = Ro^{-1} + (\hat{\zeta}_g + \mu) + \frac{\partial \hat{\theta}}{\partial \hat{z}} \tag{2.140}$$

satisfies the conservation law

$$\frac{\partial \hat{Q}_g}{\partial \hat{t}} + \hat{\mathbf{u}}_{\mathbf{g}} \cdot \nabla \hat{Q}_g = 0. \tag{2.141}$$

Note that both the vorticity and the thermodynamic contributions to the potential vorticity (2.119) have been approximated. The solution of equations (2.136), (2.137) and (2.138) can therefore be obtained by transporting $\hat{Q}_g$ with velocity $\hat{\mathbf{u}}_{\mathbf{g}}$ and calculating the other fields from $\hat{Q}_g$. The potential vorticity inversion relation obtained from (2.140) is

$$\hat{Q}_g = Ro^{-1} + (\nabla_h^2 \hat{\varphi} + \mu) + \frac{\partial^2 \hat{\varphi}}{\partial \hat{z}^2}. \tag{2.142}$$

This is a constant coefficient Poisson equation for φ, so no solvability issues arise. It requires boundary conditions. In addition to those imposed by energy conservation requirements, it is normal to specify Neumann boundary conditions on constant $\hat{z}$ surfaces as in Hoskins *et al.* (1985). This is equivalent to specifying $\hat{\theta}$. While the solution of (2.142) for $\hat{\varphi}$ means that good regularity estimates can be obtained in the interior of Γ, it is an open question whether the regularity of $\hat{\theta}$ can be maintained along the boundary. Most results have therefore been proved with $\hat{\theta}$ required to be a constant on the parts of the boundary which are surfaces of constant $\hat{z}$. However, the more general boundary condition leads to 'surface quasi-geostrophic dynamics', which have been widely studied, see Held *et al.* (1995).

Linearising (2.142) about a state of rest (2.121) with $\mu = \beta\hat{y}$, and using the same notation as in section 2.4.2 gives

$$\begin{aligned} \zeta' &= -(k^2 + l^2)\varphi', \\ Q' = \zeta' + \frac{\partial^2 \varphi'}{\partial \hat{z}^2} &= -(k^2 + l^2 + m^2)\varphi', \\ -i\omega Q' &= -\beta v' = -\beta i k \varphi'. \end{aligned}$$

This leads to the dispersion relation

$$\omega = -\frac{\beta k}{(k^2 + l^2 + m^2)}. \tag{2.143}$$

This relation is discussed further in Vallis (2017), p. 229. Comparing with (2.81) shows that the vertical squared wavenumber replaces the mean depth from the shallow water case. As discussed in section 2.4.2, the Rossby radius of deformation from the shallow water case becomes an aspect ratio. This will be 1 under the assumption $Ro = Fr$, which is when the vertical and horizontal wavenumbers are equal.

In order to show that these equations define a slow manifold, it has to be shown that they can be solved for large times. The necessary results quoted are from Bourgeois and Beale (1994). As stated in that paper, the regularity assumptions are probably far from optimal. The quasi-geostrophic system (2.136), (2.137), (2.138) and the last two equations of (2.116) are solved in a rectangular box with periodic boundary conditions with period 1 in $\hat{x}$ and $\hat{y}$ and rigid boundaries at $\hat{z} = 0$ and $\hat{z} = 1$. Define the region of integration Γ as $B \times (0, 1)$, where $B = \left((-\frac{1}{2}, \frac{1}{2}) \times (-\frac{1}{2}, \frac{1}{2})\right)$. Write Γ_z for the horizontal cross-section $B \times \{\hat{z}\}$. Since the solution is unaltered by adding an arbitrary constant to the geopotential φ, set $\int_\Gamma \hat{\varphi}(\hat{x}, \hat{y}, \hat{z}) \mathrm{d}\hat{x}\mathrm{d}\hat{y}\mathrm{d}\hat{z} = 0$. Assume that $\hat{\theta}$ is constant on $\hat{z} = 0$ and $\hat{z} = 1$. Since $\hat{w} = 0$ on $\hat{z} = 0, 1$, equation (2.138) ensures that $\hat{\theta}$ will remain constant there if it is constant initially. μ is again given by the beta-plane approximation $\mu = 1 + \beta\hat{y}$. Define the perturbation potential vorticity to be $\hat{\mathcal{Z}} = \hat{\mathcal{Q}} - Ro^{-1} - \mu$. This gives the system QGS of Bourgeois and Beale (1994), p. 1030, subject to changes in notation.

In order to set out their results, some function spaces have to be defined. Only brief definitions are given here. More detailed background can be found in the textbook of Adams (1975). Here H^s denotes the Sobolev space of horizontally periodic functions on Γ with square integrable generalised derivatives up to order s. C denotes the space of continuous functions. The L^p norm of a function $\hat{u}$ on Γ is

$$\left(\int_\Gamma \hat{u}^p \mathrm{d}\hat{x}\mathrm{d}\hat{y}\mathrm{d}\hat{z} \right)^{\frac{1}{p}}. \tag{2.144}$$

If $p = \infty$, the L^p norm is the maximum value of $|\hat{u}|$. The notation $|\hat{u}|_n$ indicates the sum of the L^2 norms of all partial derivatives of total orders up to n. If $\hat{u}(\hat{t}, \cdot)$ is a function of $\hat{x}, \hat{y}, \hat{z}$ and $\hat{t}$, the notation $|\hat{u}|_{n,\hat{T}}$ represents $\sup_{0 \le \hat{t} \le \hat{T}} |\hat{u}(\hat{t}, \cdot)|_n$

The first result is

Theorem 2.2. *(QGS short time existence). If the initial perturbation potential vorticity* $\hat{\mathcal{Z}}(0,\hat{x},\hat{y},\hat{z})$ *is in* $H^s(\Gamma)$ *for some* $s > 3$*, with* $|\hat{\mathcal{Z}}_g(0,\cdot)| \leq M$*, then there exists a time* $\hat{T}^* > 0$ *and a solution* $\hat{\mathcal{Z}}$ *in* $C([0,\hat{T}^*)]; H^s(\Gamma))$ *to QGS, where* $\hat{T}^*$ *depends only on* M, Γ, α *and* β*. The perturbation potential vorticity satisfies the estimate* $|\hat{\mathcal{Z}}|_{s,T^*} \leq 2M$*.*

The next result is that

Theorem 2.3. *(QGS global existence). If* $\hat{\mathcal{Z}}(0,\cdot)$ *is in* $H^s(\Gamma)$ *for some* $s \geq 3$*, then given any time* $\hat{T} > 0$ *there exists a solution* $\hat{\mathcal{Z}}(\hat{t},\cdot) \in C([0,\hat{T}]; H^s(\Gamma))$ *to QGS.*

The latter result means that the quasi-geostrophic equations, with isentropic upper and lower boundary conditions, can be integrated for arbitrarily long times. They therefore form a slow manifold.

The next step is to estimate the distance of the exact solution from the slow manifold. It would be natural to use equations (2.116) as the exact equations. However, the hydrostatic approximation included in these equations makes them impossible to solve unless visocisity is included, as shown in section 2.2.4. However, the latter is unphysical. Instead, the hydrostatic approximation is removed, while retaining the Boussinesq approximation. The improved behaviour is demonstrated by Lions *et al.* (1992).

The quasi-geostrophic equations can then be derived rigorously as giving the limiting behaviour of the Boussinesq equations by exploiting the uniformity of the basic state given by the assumptions on $\partial\hat{\theta}/\partial\hat{z}$ and on μ. The linearised analysis as in section 2.4.2 extended to the Boussinesq equations gives inertia-gravity waves like (2.123) with a modified dispersion relation. If the solutions are averaged over these 'fast' waves, the limiting behaviour can be shown to correspond to the quasi-geostrophic equations above. Such results are given by Babin *et al.* (1999), Babin *et al.* (2002), Embid and Majda (1998) and Majda (2003). Then the convergence results above can be fed back to provide rigorous control on the behaviour of the Boussinesq equations. In Bourgeois and Beale (1994) a quantitative convergence result is obtained for short times, but the constant in the estimate depends on $\exp(\hat{t}/\varepsilon)$ so is of little use as $\varepsilon \to 0$.

It is also shown in Bourgeois and Beale (1994) that the quasi-geostrophic system can be expanded to higher order in ε and still be solved for large times. The method is first to solve QGS, and then to solve a similar equation for the next order correction. The result is

Theorem 2.4. *Given time $\hat{T} > 0$, a solution $\hat{\mathbf{U}}_g$ of QGS in $C([0,\hat{T}]); H^{s+1}(\Gamma))$ and initial data $\hat{\mathbf{U}}_1$ in $H^s(\Gamma)$, for some $s \geq 3$, there exists a unique solution $\hat{\mathbf{U}}_1$ in $C([0,\hat{T}]); H^s(\Gamma))$ for the first order correction to the geostrophic wind and potential temperature.*

The effect of this is that, after solving QGS, an auxiliary problem can be solved to correct the solution to higher accuracy. In Muraki et al. (1999) a similar procedure is followed, and shown by computations to agree much more closely with numerical solutions of (2.116). The result means that the QGS solutions have to be qualitatively correct over the entire time interval, so that they can be refined by a subsequent correction step. Thus the feedback of the correction term on the original QGS solution can be ignored, and does not prevent convergence of the Boussinesq equations to the corrected QGS at an $O(\varepsilon^2)$ rate over the entire time interval. Thus there is potentially very high predictability in this regime.

The main limitation on the applicability of the quasi-geostrophic results is that the constant coefficient assumptions in equations (2.135) and (2.138) are not valid on large scales, thus explaining the failure of quasi-geostrophic models in operational weather forecasting trials in the 1950s. Another limitation is the choice of boundary conditions. However, the constant coefficient assumption and simple boundary conditions enable extensive studies of the analytic behaviour, as has been illustrated. They also enable analytic solutions which have been very important in understanding extratropical weather systems, see Pedlosky (1987), Gill (1982) and Vallis (2017), chapter 12.

2.4.6 *Slow solutions with small aspect ratio*

In this section assume that $\alpha = 1$, so that a consistent scaling of the momentum equations requires $\varepsilon = Ro \ll 1$, $Ro = Fr^2$. In dimensional terms this corresponds to assuming an aspect ratio less than f/N.

The assumption $Ro \ll 1$ means that the first two equations of (2.116) can be approximated to $O(\varepsilon)$ by the geostrophic relations

$$\begin{aligned} \frac{\partial \hat{\varphi}}{\partial \hat{x}} - \mu \hat{v} &= 0, \\ \frac{\partial \hat{\varphi}}{\partial \hat{y}} + \mu \hat{u} &= 0. \end{aligned} \tag{2.145}$$

All terms in the remaining equations in (2.116) are $O(1)$ so no further approximations can be made. The resulting equations are called the planetary geostrophic equations, or the Type 2 geostrophic equations of Phillips

(1963). Unlike the shallow water case analysed in section 2.3.6, the equations now have non-trivial solutions for $\mu = 1$ since (2.145) does not imply that $\hat{\mathbf{u}} \cdot \nabla\hat{\theta} = 0$. The equations conserve the energy integral

$$\int_\Gamma -\hat{\theta}\hat{z}\mathrm{d}\hat{x}\mathrm{d}\hat{y}\mathrm{d}\hat{z}, \tag{2.146}$$

provided that $\hat{w} = 0$ on those parts of the boundary which are not parallel to the $\hat{z}$ axis. In contrast to the case of large aspect ratio, the energy is entirely potential energy. However, equations (2.145) and the third equation of (2.116) may not be compatible with this boundary condition unless $\mu = 1$, in which case $\hat{w} = 0$ everywhere. The conserved potential vorticity is $\mu\partial\hat{\theta}/\partial\hat{z}$, which retains the temperature dependence of the exact potential vorticity (2.119) but loses the velocity dependence.

The difficulty with the boundary conditions also makes the equations impossible to solve in a sufficiently general context to be useful. If the potential vorticity is known, $\partial^2\hat{\varphi}/\partial\hat{z}^2$ can be determined from the second equation of (2.116). It would then be possible to solve for $\hat{\varphi}$ if $\hat{\theta}$ was given on all parts of the boundary not parallel to the $\hat{z}$ axis. If Γ is rectangular, with boundaries including $\hat{z} = 0, 1$, then if $\mu = 1$ and $\hat{\theta}$ is constant on $\hat{z} = 0, 1$ initially, it will remain so and the equations can be solved. However the assumption of large horizontal scale means that μ should be regarded as a function of position, in which case no compatible boundary conditions are available.

As in the shallow water case, the semi-geostrophic approximation gives a more useful and mathematically well-posed system. The geostrophic wind is first defined by

$$\begin{aligned} \frac{\partial\hat{\varphi}}{\partial\hat{x}} &= \mu\hat{v}_g, \\ \frac{\partial\hat{\varphi}}{\partial\hat{y}} &= -\mu\hat{u}_g. \end{aligned} \tag{2.147}$$

The momentum is then replaced by its geostrophic value in (2.116), while the remaining equations are not approximated. The resulting system is

$$
\begin{aligned}
\frac{\mathrm{D}\hat{u}_g}{\mathrm{D}\hat{t}} + Ro^{-1}\left(\frac{\partial\hat{\varphi}}{\partial\hat{x}} - \mu\hat{v}\right) &= 0,\\
\frac{\mathrm{D}\hat{v}_g}{\mathrm{D}\hat{t}} + Ro^{-1}\left(\frac{\partial\hat{\varphi}}{\partial\hat{y}} + \mu\hat{u}\right) &= 0,\\
\frac{\mathrm{D}\hat{\theta}}{\mathrm{D}\hat{t}} &= 0,\\
\frac{\partial\hat{\varphi}}{\partial\hat{z}} - \hat{\theta} &= 0,\\
\nabla\cdot\hat{\mathbf{u}} &= 0.
\end{aligned} \tag{2.148}
$$

These equations were introduced and analysed by Hoskins (1975). Given the boundary conditions $\hat{\mathbf{u}}\cdot\mathbf{n} = 0$ on the boundary of Γ, they conserve the energy integral

$$
E_g = \int_\Gamma \left(\frac{1}{2}(\hat{u}_g^2 + \hat{v}_g^2) - \hat{\theta}\hat{z}\right) \mathrm{d}\hat{x}\mathrm{d}\hat{y}\mathrm{d}\hat{z}. \tag{2.149}
$$

The kinetic energy has been approximated by its geostrophic value, but the potential energy is not approximated.

As in section 2.3.6, the behaviour of equations (2.148) can be studied by first rearranging them into a single equation for the evolution of $\hat{\varphi}$, in the manner of Schubert (1985). The first three equations can be rewritten using (2.147) and the hydrostatic relation from (2.148) as

$$
\mathbf{Q}\begin{pmatrix}\hat{u}\\ \hat{v}\\ \hat{w}\end{pmatrix} + \frac{\partial}{\partial\hat{t}}\nabla\hat{\varphi} = Ro^{-1}\mu\begin{pmatrix}-\frac{\partial\hat{\varphi}}{\partial\hat{y}}\\ \frac{\partial\hat{\varphi}}{\partial\hat{x}}\\ 0\end{pmatrix}, \tag{2.150}
$$

where

$$
\mathbf{Q} = \begin{pmatrix}
Ro^{-1}\mu^2 + \mu\frac{\partial\hat{v}_g}{\partial\hat{x}} & \mu\frac{\partial\hat{v}_g}{\partial\hat{y}} & \mu\frac{\partial\hat{v}_g}{\partial\hat{z}}\\
-\mu\frac{\partial\hat{u}_g}{\partial\hat{x}} & Ro^{-1}\mu^2 - \mu\frac{\partial\hat{u}_g}{\partial\hat{y}} & -\mu\frac{\partial\hat{u}_g}{\partial\hat{z}}\\
\frac{\partial\hat{\theta}}{\partial\hat{x}} & \frac{\partial\hat{\theta}}{\partial\hat{y}} & \frac{\partial\hat{\theta}}{\partial\hat{z}}
\end{pmatrix}. \tag{2.151}
$$

Using the fifth equation from (2.148) then gives

$$
\nabla\cdot\left(\mathbf{Q}^{-1}\frac{\partial}{\partial\hat{t}}\nabla\hat{\varphi}\right) = Ro^{-1}\nabla\cdot\left(\mathbf{Q}^{-1}\mu\begin{pmatrix}-\frac{\partial\hat{\varphi}}{\partial\hat{y}}\\ \frac{\partial\hat{\varphi}}{\partial\hat{x}}\\ 0\end{pmatrix}\right). \tag{2.152}
$$

This equation is elliptic if $\mathbf{Q}$ is positive definite. This implies both inertial and static stability and is discussed much further in section 3.3. Boundary conditions similar to (2.92) can be derived which are consistent with the requirement $\hat{\mathbf{u}} \cdot \mathbf{n} = 0$.

In the case when $\mu = 1$, the equations also conserve the potential vorticity which is given by the determinant of $\mathbf{Q}$. The vorticity dependence of the true potential vorticity has been approximated. The temperature dependence is the same as that for the potential vorticity (2.119) derived from (2.116). However, if $\mu \neq 1$ there is no invariant of this kind.

Now demonstrate that the solution of (2.148) approximates the solution of the shallow atmosphere hydrostatic Boussinesq equations (2.116) to $O(Ro^2)$. The same procedure is followed as that used in section 2.3.6 to derive (2.99). Set $\alpha = 1$ and $Ro = Fr^2$. Divide the first two equations of (2.116) by μRo^{-1}, and take their Lagrangian time derivative. This gives

$$Ro\frac{\mathrm{D}}{\mathrm{D}\hat{t}}\left(\mu^{-1}\begin{pmatrix}\frac{\mathrm{D}\hat{u}}{\mathrm{D}\hat{t}}\\ \frac{\mathrm{D}\hat{v}}{\mathrm{D}\hat{t}}\end{pmatrix}\right)+\begin{pmatrix}-\frac{\mathrm{D}\hat{v}}{\mathrm{D}\hat{t}}\\ \frac{\mathrm{D}\hat{u}}{\mathrm{D}\hat{t}}\end{pmatrix}+\frac{\mathrm{D}}{\mathrm{D}\hat{t}}(\mu^{-1}\nabla\hat{\varphi})=0. \tag{2.153}$$

Substituting from (2.116) for the second term in (2.153), and using (2.147) for the last term gives

$$Ro\frac{\mathrm{D}}{\mathrm{D}\hat{t}}\left(\mu^{-1}\begin{pmatrix}\frac{\mathrm{D}\hat{u}}{\mathrm{D}\hat{t}}\\ \frac{\mathrm{D}\hat{v}}{\mathrm{D}\hat{t}}\end{pmatrix}\right)+Ro^{-1}\begin{pmatrix}\frac{\partial\hat{\varphi}}{\partial\hat{y}}+\mu\hat{u}\\ -\frac{\partial\hat{\varphi}}{\partial\hat{x}}+\mu\hat{v}\end{pmatrix}+\frac{\mathrm{D}}{\mathrm{D}\hat{t}}\begin{pmatrix}\hat{v}_g\\ -\hat{u}_g\end{pmatrix}=0. \tag{2.154}$$

Omitting the first term in (2.154) gives exactly the first two of the semi-geostrophic equations (2.148). Then rewrite the last two terms in (2.154) together with the fifth equation of (2.116) as

$$Ro^{-1}\begin{pmatrix}\frac{\partial\hat{\varphi}}{\partial\hat{y}}\\ -\frac{\partial\hat{\varphi}}{\partial\hat{x}}\\ 0\end{pmatrix}+\frac{\partial}{\partial\hat{t}}\begin{pmatrix}\hat{v}_g\\ -\hat{u}_g\\ \mu^{-1}\hat{\theta}\end{pmatrix}+\mu^{-1}\mathbf{Q}\begin{pmatrix}\hat{u}\\ \hat{v}\\ \hat{w}\end{pmatrix}, \tag{2.155}$$

where $\mathbf{Q}$ is defined by (2.151). Multiplying by $\mathbf{Q}^{-1}\mu$ and including in (2.154) gives

$$\begin{pmatrix}\hat{u}\\ \hat{v}\\ \hat{w}\end{pmatrix}=-Ro^{-1}\mathbf{Q}^{-1}\mu\begin{pmatrix}\frac{\partial\hat{\varphi}}{\partial\hat{y}}\\ -\frac{\partial\hat{\varphi}}{\partial\hat{x}}\\ 0\end{pmatrix}-\mathbf{Q}^{-1}\mu\frac{\partial}{\partial\hat{t}}\begin{pmatrix}\hat{v}_g\\ -\hat{u}_g\\ \mu^{-1}\hat{\theta}\end{pmatrix}- \tag{2.156}$$
$$Ro\mathbf{Q}^{-1}\mu\frac{\mathrm{D}}{\mathrm{D}\hat{t}}\left(\mu^{-1}\begin{pmatrix}\frac{\mathrm{D}\hat{u}}{\mathrm{D}\hat{t}}\\ \frac{\mathrm{D}\hat{v}}{\mathrm{D}\hat{t}}\\ 0\end{pmatrix}\right).$$

As with (2.99), the Rossby number that appears in (2.156) is the Lagrangian Rossby number Ro_L defined in (2.32). The first term on the right hand side is approximately $\hat{\mathbf{u}}_g$ with a correction of $\mathrm{O}(Ro_L)$, the second term is $\mathrm{O}(Ro_L)$ and the final term is $\mathrm{O}(Ro_L^2)$. The first two terms are the value of $\hat{\mathbf{u}}$ calculated from the semi-geostrophic equations 2.148. Thus the error in $\hat{\mathbf{u}}$ is $\mathrm{O}(Ro_L^2)$. The evolution equations 2.148) can be written as the single equation (2.152) for $\hat{\varphi}$. The error in the evolution of $\hat{\varphi}$ will be the same as the error in $\hat{\mathbf{u}}$ which is $\mathrm{O}(Ro_L^2)$. Thus the error in the solution of (2.148) as an approximation to the solution of (2.116) will also be of this order. As with the derivation of (2.100), the error is a second Lagrangian time derivative, as noted by Hoskins (1975), which will be large for fast waves and make a convergence proof difficult.

As in section 2.3.6, the estimate can be extended to cover other limits than $\alpha = 1$. Returning to (2.116), allow $\alpha > 1$ and set $Ro = \alpha Fr^2$. Then (2.154) can still be derived. The substitution of (2.156) into the fourth equation of (2.116) now gives the overall error as $\mathrm{O}(\alpha Ro_L^2)$. Substituting for α gives the error as

$$Ro_L^3 Fr^{-2} = Ro_L (Ro_L/Fr)^2 = Ro_L Bu^2, \tag{2.157}$$

where Bu is the Burger number. As in the shallow water case (2.100), this gives an error $\mathrm{O}(Ro)$ in the quasi-geostrophic case $Ro = Fr$.

The estimate for the case $\alpha = 1$ is exploited in a rigorous convergence result described in section 3.6.4. However, proving convergence for large times requires assuming the existence of a class of solutions of (2.116) where the assumption $\frac{\mathrm{D}^2}{\mathrm{D}\hat{t}^2} = \mathrm{O}(1)$ is valid, which may well not be possible. This is why convergence of the quasi-geostrophic equations to the Boussinesq equations discussed in section 2.3.5 uses a time-averaging technique. However, that cannot be applied without the restrictive assumptions made in quasi-geostrophic theory.

As noted in section 2.4.5, a rigorous convergence proof would need to use the non-hydrostatic version of (2.116) to ensure the existence of a solution without viscosity. If the non-hydrostatic term $\frac{\mathrm{D}\hat{w}}{\mathrm{D}\hat{t}}$ is included in the third equation of (2.116), the derivation of (2.156) is unchanged. However, the third equation of (2.116) is used when deriving the evolution equation for $\hat{\varphi}$ (2.152). There would now be an extra term which is proportional to $\frac{\mathrm{D}^2\hat{w}}{\mathrm{D}\hat{t}^2}$ contributing to the error in (2.148).

The geostrophic momentum approximation used to derive (2.148) is a Lagrangian approximation in the sense that the neglected term is $\mathrm{D}(\hat{\mathbf{u}} - \hat{\mathbf{u}}_g)/\mathrm{D}\hat{t}$, but the $\mathrm{D}/\mathrm{D}\hat{t}$ operator has not been approximated. This is equivalent to requiring the rate of change of wind direction following a fluid trajectory to be much less than Ro^{-1}, but the rate of change of the magnitude of the wind is not restricted. In addition, because the thermodynamic terms in the equations have not been approximated, the equations are also asymptotically correct in the limit $Fr \to 0$, even if there is no rotation. This is because the limiting solution of (2.116) as $Fr \to 0$ with $\mu = 0$ is a state of rest in hydrostatic balance where the fluid has been rearranged so that $\hat{\theta}$ is monotonically increasing with $\hat{z}$. This state is determined only by consistency with the last three equations of (2.116), none of which have been approximated. The semi-geostrophic equations, however, do not describe the flow accurately in this limit, the appropriate equations are those discussed in section 2.4.4. The correctness of this limit is important in applying semi-geostrophic theory globally, since it allows the correct limiting rest state to be obtained at the equator. This is a useful leading order solution in the tropics as will be discussed in section 2.4.7

2.4.7 *Large scale slow solutions in the tropics*

The hierarchical approach of section 2.3.7 is followed. Equations (2.116) still apply. Tropical solutions can again be studied by choosing $\mu = \beta\hat{y}$ and the domain Γ chosen to be periodic in $\hat{x}$ and unbounded in $\hat{y}$ with a decay condition as $\hat{y} \to \pm\infty$.

The linear analysis in section 2.4.2 was carried out by linearising (2.116) about the basic state (2.121). In the constant rotation case this leads to the dispersion relation (2.122), while the shallow water analysis for constant rotation leads to (2.42). Comparing these shows that they are equivalent if Fr^{-2} in (2.42) is replaced by $(mFr)^{-2}$ to give (2.122). Thus each vertical mode of the three-dimensional problem is equivalent to a shallow water solution with an appropriate mean depth. The same argument can be applied to the tropical problem. As in section 2.3.7, choose $\alpha = Fr^{-2}$ and $Ro = 1$. Then the equation (2.101) for the meridional structure functions ν becomes

$$\check{v}\left(\frac{\partial^2 \nu}{\partial \hat{y}^2} - \left(-\omega^2 m^2 Fr^2 + k^2 + \frac{k\beta}{\omega} - \beta^2 m^2 Fr^2 \hat{y}^2\right)\nu\right) = 0. \qquad (2.158)$$

The same wave characteristics apply as in 2.3.7 with Fr replaced by mFr.

Now consider the hierarchy of solutions discussed in section 2.3.7. The semi-geostrophic solution requires the matrix $\mathbf{Q}$ defined in (2.151) to be positive definite. The argument leading to (2.108) in section 2.3.7 now gives

$$\hat{\varphi} = 1 + c_0\hat{y}^2 + \mathrm{O}(\hat{y}^4). \tag{2.159}$$

Applying the hydrostatic relation gives a similar expression for $\hat{\theta}$ which defines the weak temperature gradient approximation.. This was introduced by Sobel *et al.* (2001) to define the lowest order response to forcing in the tropics. Suppose a forcing term $\hat{S}_h$ is included in the $\hat{\theta}$ equation. Assume that the horizontal derivatives of $\hat{\theta}$ are zero, which is an extreme form of the condition (2.159). This equation and the continuity equation then become

$$\frac{\partial \hat{\theta}}{\partial \hat{t}} + Fr^{-2}\hat{w}\frac{\partial \hat{\theta}}{\partial \hat{z}} = \hat{S}_h, \tag{2.160}$$
$$\nabla \cdot \hat{\mathbf{u}} = 0.$$

Enforce the condition that $\frac{\partial \hat{\theta}}{\partial \hat{t}}$ is a function of $\hat{z}$ only. Then if, as is typical, $\hat{S}_h$ depends on $\hat{x}$ and $\hat{y}$ as well, the solution is

$$\frac{\partial \hat{\theta}}{\partial \hat{t}} = \overline{\hat{S}_h}^{xy}, \tag{2.161}$$
$$Fr^{-2}\hat{w}\frac{\partial \hat{\theta}}{\partial \hat{z}} = \hat{S}_h - \overline{\hat{S}_h}^{xy}.$$

$(\hat{u}, \hat{v})$ are then determined by the continuity equation. This shows that localised heating is spread uniformly over the region by the ageostrophic circulation, which is the basic mechanism behind the Hadley and Walker circulations, Beare and Cullen (2019).

It will be shown in section 6.1.4 that the inclusion of boundary layer friction allows the strong constraint (2.159) to be relaxed within the boundary layer. This means that the horizontal pressure gradient will be very small above the boundary layer, but horizontal temperature gradients in the boundary layer will create horizontal pressure gradients with the opposite sign in the boundary layer.

Now consider the tropical long wave regime as in section 2.3.7. The linear analysis is essentially the same, with mFr replacing Fr. The nonlinear equations are

$$\frac{\mathrm{D}\hat{u}}{\mathrm{D}\hat{t}} + \frac{\partial \hat{\varphi}}{\partial \hat{x}} - \mu\hat{v} = 0,$$
$$\frac{\partial \hat{\varphi}}{\partial \hat{y}} + \mu\hat{u} = 0,$$
$$\frac{\mathrm{D}_h\hat{\theta}}{\mathrm{D}\hat{t}} + Fr^{-2}\hat{w}\frac{\partial \hat{\theta}}{\partial z} = 0, \tag{2.162}$$
$$\frac{\partial \hat{\varphi}}{\partial \hat{z}} - \hat{\theta} = 0,$$
$$\nabla \cdot \hat{\mathbf{u}} = 0.$$

These can be rewritten in the manner of (2.150) as

$$\mathbf{Q}\begin{pmatrix}\hat{v}\\ \hat{w}\end{pmatrix}+\frac{\partial}{\partial \hat{t}}\begin{pmatrix}\frac{\partial \hat{\varphi}}{\partial \hat{y}}\\ \frac{\partial \hat{\varphi}}{\partial \hat{z}}\end{pmatrix}=\begin{pmatrix}\mu\frac{\partial \hat{\varphi}}{\partial \hat{x}}+\mu\hat{u}\frac{\partial \hat{u}}{\partial x}\\ -\hat{u}\frac{\partial \hat{\theta}}{\partial x}\end{pmatrix}, \tag{2.163}$$

where

$$\mathbf{Q}=\begin{pmatrix}(\mu^2-\mu\frac{\partial \hat{u}}{\partial \hat{y}}) & -\mu\frac{\partial \hat{u}}{\partial \hat{z}}\\ \frac{\partial \hat{\theta}}{\partial \hat{y}} & Fr^{-2}\frac{\partial \hat{\theta}}{\partial \hat{z}}\end{pmatrix}. \tag{2.164}$$

Defining Y as in (2.113) gives

$$\mathbf{Q}=\begin{pmatrix}\mu\frac{\partial Y}{\partial \hat{y}} & \mu\frac{\partial Y}{\partial \hat{z}}\\ \frac{\partial \hat{\theta}}{\partial \hat{y}} & Fr^{-2}\frac{\partial \hat{\theta}}{\partial \hat{z}}\end{pmatrix}. \tag{2.165}$$

Solvability would require $\mathbf{Q}$ to be positive definite, as with (2.150). This is consistent with the condition (2.115) for the shallow water version of these equations to be solvable. Boundary conditions consistent with the requirement $(\hat{v},\hat{w})\cdot\mathbf{n}=0$ can be derived in a similar way to (2.92). Alternatively, $\hat{v}$ could be prescribed on the boundary of an equatorial strip, as in the shallow water case. This would allow coupling to a semi-geostrophic solution in the extratropics.

These equations have a rather similar structure to the non-axisymmetric vortex equations of Craig (1991). In the axisymmetric case, these can be solved as discussed in section 4.4. Initial data consistent with the tropical long wave assumption would have small values of the derivatives with respect to $\hat{x}$ on the right hand side of (2.163). Solvability would require $\mathbf{Q}$ to be positive definite, as with (2.150). Assume $\mathbf{Q}$ is strictly positive definite in the initial data. However, solving them by inverting $\mathbf{Q}$ will only maintain small $\hat{x}$ derivatives for a short time while $\mathbf{Q}$ remains strictly positive definite. Thus it is unlikely that this system can define a slow manifold.

There is a greater chance that the equation (2.115) required to solve the long wave equations in the shallow water case can be solved for long times since it is a scalar elliptic equation. Nearly all the theoretical work in this area thus uses shallow water equations. The observational study of Wheeler and Kiladis (1999) shows that a significant part of the observed variability can be explained by shallow water solutions with appropriate choices of Fr. They identify two modes in particular, one corresponding to a vertical mode with a full wavelength in the troposphere, and another which represents bulk variability of the tropospheric layer. In dimensional terms, a typical tropospheric depth of 16km and Brunt-Väisälä frequency of 10^{-2} imply a gravity wave speed for the internal tropospheric mode of about 25ms^{-1} and for the full tropospheric layer of about 50ms^{-1}, as found by Wheeler and Kiladis (1999). Given the horizontal uniformity of the tropical region, as expressed by (2.159), and the prevalence of moist convection which will generate a strong upscale cascade in the vertical to the scale of the tropospheric depth, it is not surprising that these shallow water solutions are useful. Some theoretical work has been done on two-layer tropical shallow water models, for instance Stechmann *et al.* (2008), though no rigorous results governing long term behaviour are available.

In dimensional terms, the equatorial deformation radii L_E associated with these gravity wave speeds are about 1100km and 1550km, or 10° and 14° latitude. These are the latitudes at the edge of the equatorial waveguide where tropical long wave dynamics gives way to semi-geostrophic dynamics. This is illustrated in section 2.7.5. In principle, solutions of the two sets of equations can be matched at an east-west boundary because the normal velocity on such a boundary can be used as a boundary condition for both sets of equations. In the ocean, the internal gravity wave speed will be much smaller, so L_E will also be smaller.

The control over the vertical scale exerted by convection also makes the layered two-dimensional vortex models described in section 2.4.4 relevant. The Froude numbers for the two vertical modes discussed above would be about 0.4 and 0.2 for a typical wind speed of 10ms^{-1}, which is small enough for the approximation to be useful. It would be necessary to include the convective mass flux to maintain the vertical coherence of a vortex with horizontal winds that vary in the vertical. In the stratosphere there is no convection to control the vertical scale, and so existence of robust nonlinear large-scale solutions is problematic.

2.5 Large scale solutions of the fully compressible equations in spherical geometry

2.5.1 *Choice of semi-geostrophic scaling*

Now consider the fully compressible equations in spherical geometry and height coordinates r. The original dimensional variables are used. The appropriate regime for studying the large-scale flow of the atmosphere is the small aspect ratio regime described in section 2.4.6. This is because the condition $H/L < f/N$ is satisfed in the troposphere for the typical values $f \simeq 10^{-4}s^{-1}, N \simeq 10^{-2}s^{-1}, H = 10\text{km}$ noted in section 2.2.5 if $L > 1000\text{km}$. Thus the semi-geostrophic equations are used. As noted in section 2.4.6, they have the advantage that only the geostrophic and hydrostatic approximations are made. The additional assumptions made in the quasi-geostrophic equations derived in section 2.4.5 are not required. In particular, the full Coriolis parameter is used wherever it appears, the static stability $\partial\theta/\partial r$ is not replaced by a reference state value, and the natural boundary conditions of no normal flow are all that are required. These additional assumptions prevented the quasi-geostrophic system being a useful tool for operational weather forecasting. Note, however, that the no-normal flow condition assumes that the upper boundary is a rigid surface, as commonly assumed in atmospheric models, while a decay condition as $r \to \infty$ would be more appropriate, as introduced in section 2.1. It can be shown that the energy conservation property still holds if a decay condition is used.

It has also been shown that the semi-geostrophic equations can be solved for sufficiently large times to be able to describe the behavior of weather systems because the solvability condition described in section 2.4.6 can be maintained, unlike the equivalent condition for the more general models discussed in section 2.4.3. These results are described in the rest of the book. These benefits are obtained at the price of restricted validity. However, observed large-scale flows satisfy the conditions for the semi-geostrophic approximation to be valid as will be demonstrated in section 2.7.4.

As discussed in section 2.2.5, the assumption of small aspect ratio implies rotation-dominated flow, so that the time-scale of the flow must be larger than f^{-1} which is $(2\Omega \sin\phi)^{-1}$. It is important to note at the outset that this time-scale is quite long, it corresponds to a period of 12 hours at the poles and becomes much greater as the equator is approached. Thus the semi-geostrophic model is a rather coarse approximation to the real flow, all the small-scale detail visible in Fig. 1.2 will be absent.

2.5.2 *Derivation and properties of the equations*

The starting point is the shallow atmosphere hydrostatic equation system (2.23), which is very accurate on large horizontal scales. This incorporates the subtraction of a reference state as in (2.8). The source terms in (2.1) are included, together with the vertical part of the dissipation terms which are important near the lower boundary. Write this system in dimensional variables and spherical polar coordinates (r, λ, ϕ):

$$\begin{aligned}
\frac{\mathrm{D}u}{\mathrm{D}t} - \frac{uv\tan\phi}{a} - fv + \frac{C_p\theta}{a\cos\phi}\frac{\partial\Pi'}{\partial\lambda} &= \nu\frac{\partial^2 u}{\partial r^2},\\
\frac{\mathrm{D}v}{\mathrm{D}t} + \frac{u^2\tan\phi}{a} + fu + \frac{C_p\theta}{a}\frac{\partial\Pi'}{\partial\phi} &= \nu\frac{\partial^2 v}{\partial r^2},\\
\frac{\mathrm{D}\theta}{\mathrm{D}t} &= \frac{1}{C_p\Pi}\left(\kappa_h\frac{\partial^2 T}{\partial r^2} + S_h + LP\right),\\
\frac{\mathrm{D}q}{\mathrm{D}t} &= \kappa_q\frac{\partial^2 q}{\partial r^2} + S_q - P,\\
C_p\theta\frac{\partial\Pi'}{\partial r} - g\frac{\theta'}{\theta_0} &= 0,\\
p &= \rho RT,\\
\gamma\frac{\partial}{\partial r}\left\{p\left(\frac{\partial w}{\partial r} + \nabla_r\cdot\mathbf{u} - \frac{1}{C_pT}(\kappa_h\nabla^2 T + S_h + LP)\right)\right\} &=\\
\frac{\partial p}{\partial r}\nabla_r\cdot\mathbf{u} &- \frac{\partial\mathbf{u}}{\partial r}\cdot\nabla_r p.
\end{aligned} \tag{2.166}$$

f is the Coriolis parameter $2\Omega\sin\phi$ defined in (2.32).

As in section 2.4.6, the semi-geostrophic equations are derived by making the geostrophic momentum approximation in (2.166). First define the geostrophic wind (u_g, v_g) by

$$\begin{aligned}
-fv_g + \frac{C_p\theta}{a\cos\phi}\frac{\partial\Pi'}{\partial\lambda} &= 0,\\
fu_g + \frac{C_p\theta}{a}\frac{\partial\Pi'}{\partial\phi} &= 0.
\end{aligned} \tag{2.167}$$

If the Lagrangian Rossby number is small, so that $\frac{D}{Dt} \ll f$, the horizontal wind will be close to its geostrophic value.

Next replace the horizontal momentum by its geostrophic value. Thus the first two equations of (2.166) are replaced by

$$\begin{aligned}
\frac{\mathrm{D}u_g}{\mathrm{D}t} - \frac{uv_g\tan\phi}{a} - fv + \frac{C_p\theta}{a\cos\phi}\frac{\partial\Pi'}{\partial\lambda} &= \nu\frac{\partial^2 u}{\partial r^2},\\
\frac{\mathrm{D}v_g}{\mathrm{D}t} + \frac{uu_g\tan\phi}{a} + fu + \frac{C_p\theta}{a}\frac{\partial\Pi'}{\partial\phi} &= \nu\frac{\partial^2 v}{\partial r^2}.
\end{aligned} \tag{2.168}$$

The remaining equations in (2.166) are not approximated. In the absence of dissipation and source terms, and with the boundary conditions $\mathbf{u}\cdot\mathbf{n} = 0$ on the boundary of Γ, the semi-geostrophic equations can be shown to conserve the energy integral

$$E_g = \int_\Gamma \rho\left(\frac{1}{2}(u_g^2 + v_g^2) + C_vT + gr\right)a^2\cos\phi\mathrm{d}\lambda\mathrm{d}\phi\mathrm{d}r. \tag{2.169}$$

The shallow atmosphere semi-geostrophic equations as derived here do not have an exact potential vorticity conservation law of the form (2.6). As discussed in section 2.2.4, this is because the radial symmetry used to derive the shallow atmosphere hydrostatic approximation is a stronger constraint than the axial symmetry about the rotation axis. This is consistent with the fact that the semi-geostrophic approximation is not uniformly valid over the sphere, and as a result the qualitative behaviour of the atmosphere on large scales is different in the tropics and extratropics.

In order to represent real data accurately, it is necessary to include a lower boundary layer so that the boundary condition $\mathbf{u} = 0$ at $r = a$ can be satisfied. This is achieved by retaining the friction terms in the first two equations of (2.166. Adding these to (2.167) gives the *geotriptic* wind $\mathbf{u}_e$:

$$\begin{aligned} -fv_e + \frac{C_p\theta}{a\cos\phi}\frac{\partial\Pi'}{\partial\lambda} &= \nu\frac{\partial^2 u_e}{\partial r^2}, \\ fu_e + \frac{C_p\theta}{a}\frac{\partial\Pi'}{\partial\phi} &= \nu\frac{\partial^2 v_e}{\partial r^2}. \end{aligned} \tag{2.170}$$

Adding the friction terms to (2.168) gives the semi-geotriptic equations:

$$\begin{aligned} \frac{\mathrm{D}u_c}{\mathrm{D}t} - \frac{uv_c\tan\phi}{a} - fv + \frac{C_p\theta}{a\cos\phi}\frac{\partial\Pi'}{\partial\lambda} &= \nu\frac{\partial^2(2u_c - u)}{\partial r^2}, \\ \frac{\mathrm{D}v_e}{\mathrm{D}t} + \frac{uu_e\tan\phi}{a} + fu + \frac{C_p\theta}{a}\frac{\partial\Pi'}{\partial\phi} &= \nu\frac{\partial^2(2v_e - v)}{\partial r^2}. \end{aligned} \tag{2.171}$$

This form of the friction terms is required to ensure that the evolution equations dissipate energy with the given boundary conditions. This is discussed more fully in section 6.1 and Beare and Cullen (2010). The friction terms result in a stabilisation of the equations, so that the matrix $\mathbf{Q}$ defined in eq. (2.151) for the semi-geostrophic equations has an extra term on the diagonal which makes $\mathbf{Q}$ more positive definite and thus relaxes the inertial stability condition. This is demonstrated in section 6.1.4.

Much of the rest of this book is taken up with analysing the semi-geostrophic equations. Some key points are noted based on the preceding sections:

(i) The only approximations made are to the horizontal momentum. As noted in sections 2.3.6 and 2.4.6, this ensures large-scale validity.

(ii) It appears strange that only the horizontal momentum is approximated, not the trajectory. This choice retains energetic consistency, which is essential in proving that the equations can be solved for large times.

(iii) The solutions will only make sense if the horizontal pressure gradient $\nabla_h \Pi'$ goes to zero at the equator, as discussed in sections 2.3.7 and 2.4.7. Otherwise the geostrophic wind cannot be defined. The existence of solutions which satisfy this constraint will be demonstrated in section 4.3.

(iv) Only long-term mean solutions can be described in the tropics. Tropical variability is excluded, but can be described by the tropical long wave equations discussed in sections 2.3.7 and 2.4.7.

(v) The assumption made is that $D\mathbf{u}_h/Dt \ll f\hat{\mathbf{r}} \times \mathbf{u}_h$, where $\mathbf{u}_h = (u, v, 0)$. This states that the Lagrangian Rossby number Ro_L is small. It only controls the component of $D\mathbf{u}_h/Dt$ normal to $\mathbf{u}_h$. It thus requires the rate of change of horizontal wind direction to be much less than f but does not restrict the rate of change of magnitude.

The remaining chapters of this book analyse the semi-geostrophic system, showing that it is well-posed and can thus define a slow manifold. They also demonstrate that the equations have solutions which contain much of the physics of observed large-scale flows and give important but limited information in the tropics.

2.6 Illustrations of asymptotic behaviour from observations

2.6.1 *Atmosphere and ocean spectra*

The asymptotic behaviour of 'slow' solutions in three dimensions was described in section 2.4. In section 2.4.4, the limit solution is two-dimensional incompressible flow at each z, which will self-destruct because the requirement of small Fr cannot be maintained. In section 2.4.5 the solution is quasi-geostrophic flow with a conserved energy and potential vorticity. These have different characteristic horizontal and vertical scales. It is shown in Salmon (1998) that this double conservation property implies a cascade of energy to large horizontal and vertical scales, and of potential enstrophy to small horizontal and vertical scales. There is a characteristic -3 slope in the energy spectrum. This regime is called 'geostrophic turbulence', following Charney (1971). In section 2.4.6 the solution is semi-geostrophic flow. Energy is conserved and there is a potential vorticity invariant if the rotation rate is constant. Hwever, the energy is primarily available potential energy determined by the pressure and temperature, and the potential vorticity is largely determined by the static stability. While these have different vertical scales, they have the same horizontal scale. Thus there will be no systematic horizontal energy or enstrophy cascade and no preferred slope in the horizontal energy spectrum. There will still be a vertical upscale energy cascade. In the troposphere, this will be arrested by the inhomogeneities at the boundary layer top and the tropopause.

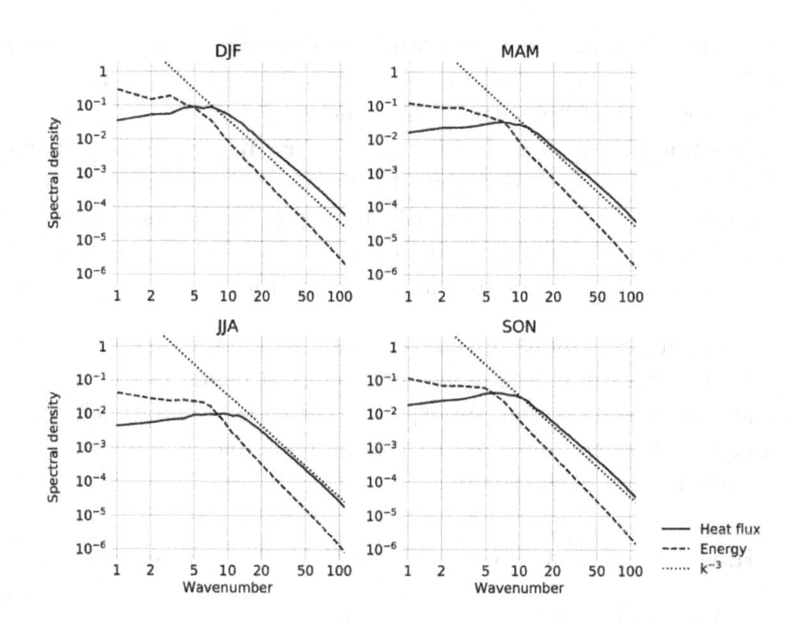

Fig. 2.3 Dashed line: Zonal energy spectra, average of 300hpa, 500hpa and 700hpa averaged over 40°N and 50°N from 5 years of ERA5 data. Solid line: Zonal spectrum of meridional heat flux, averaged in the same way. Dotted line: reference -3 slope. Source: Climate Dynamics Group, Met Office. ©Crown Copyright, Met Office.

Figs. 2.3 and 2.4 show the observed behaviour in the extratropics. Fig 2.3 is derived from ERA5 reanalysis data. There is sufficient satellite data available for there to be high confidence in these analyses at large scales. Fig 2.4, from Nastrom and Gage (1985), is derived from observations along aircraft tracks and thus does not see the largest scales. The smaller scales may well be more reliable than the reanalysis data because of the limited analysis resolution.

Fig. 2.3 shows an almost flat energy spectrum for wavenumbers up to 5. There is then a fairly sharp transition to a spectral slope somewhat greater than -3 which is achieved by wavenumber 10. The transition wavenumber is different for different seasons and is at a smaller wavenumber in autumn and winter. Fig. 2.4 shows the spectra for wind components and potential

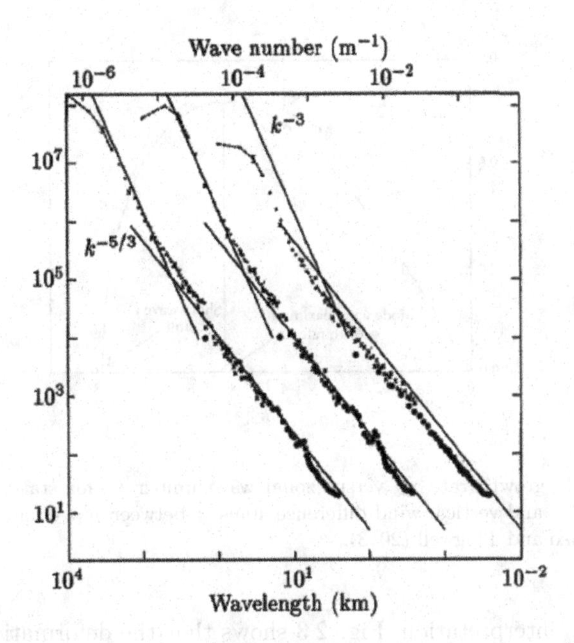

Fig. 2.4 Variance power spectra of wind and potential temperature near the tropopause from aircraft data. The spectra for meridional wind and temperature are shifted one and two decades to the right. Solid lines with slopes -3 and -$\frac{5}{3}$ given for reference. From Nastrom and Gage (1985). ©American Meteorological Society. Used with permission.

temperature. In comparing with Fig. 2.3, note that wavenumber 10 in Fig. 2.3 corresponds to a wavelength of 3000km at the latitude illustrated. The transition from a flat to a -3 spectrum can just be seen at the largest scales. The -3 spectrum then extends down to a wavelength of about 500km and then shallows to a -$\frac{5}{3}$ spectrum which extends to the smallest observed wavelength of 1km.

Fig. 2.3 also shows the spectrum of the meridional heat transport $v'T'$ averaged over the troposphere, where v', T' are departures from the zonal mean meridional wind and temperature respectively. The classical Eady growth rate derived from linear theory, e.g. Vallis (2017), p. 360, and illustrated in Fig. 2.5, has a peak at the Rossby radius of deformation. It shows similar characteristics to the observed heat transport, though the short wave cutoff is much sharper in linear theory than the observations. This suggests that the deformation radius corresponds to the peak in the meridional heat transport spectrum shown in Fig. 2.3. The average scale of the heat transport can then be calculated to be about 50% larger than the deformation radius.

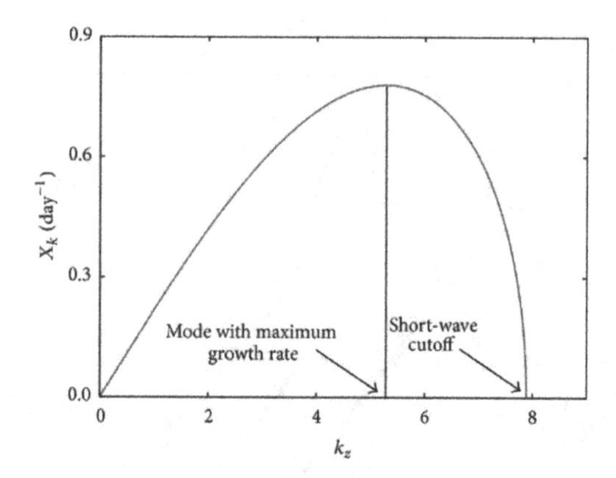

Fig. 2.5 Eady growth rate χ_k versus zonal wavenumber k_z for static stability $2 \times 10^{-6}\text{m}^2\text{Pa}^{-2}\text{s}^{-2}$ and vertical wind difference 40ms^{-1} between $p = 0$ and $p = 1000$hpa. From Soldatenko and Tingwell (2013).

With this interpretation, Fig. 2.3 shows that the deformation radius is about wavenumber 7 in all seasons except the summer when there is a broad peak in the heat transport from wavenumbers 6-11. This corresponds to a wavelength of 4000km at the latitude illustrated and thus a length-scale of 650km. This is somewhat less than that given by NH/f as discussed in section 2.4.2 where H is the tropopause depth, which would give a value of 1000km. The change in slope of the energy spectrum occurs at wavenumber 5 in winter and autumn, and 7 in spring and summer. Thus the geostrophic turbulence regime can extend to somewhat larger scales than the deformation radius. This is consistent with the gradual inhibition of the upscale energy cascade once the scale is larger than the deformation radius.

The results are therefore consistent with the theoretical expectations above. The geostrophic turbulence regime extends from wavelengths of 4000-5500km, depending on season, down to 500km. On horizontal scales larger than the deformation radius, the restriction on vertical scale means that the aspect ratio will be less than f/N and semi-geostrophic dynamics will become applicable. The nearly flat energy spectrum in this regime is consistent with the expected lack of an upscale horizontal energy cascade. On small horizontal scales, the transition to a $-\frac{5}{3}$ spectrum at wavelengths of about 500km suggests that 'slow' solutions no longer dominate at these scales as discussed in Waite and Bartello (2006).

Fig. 2.6 gives an example of the behaviour of the upper ocean in the

Gulf Stream region, taken from Callies and Ferrari (2013). That paper identifies three regimes, geostrophic eddies on a horizontal scale greater than 100km, which is the radius of deformation, sub-mesoscale turbulence on scales between 10 and 100km, and internal waves on scales below 10km. The spectrum becomes flat in the large-scale regime. A -3 slope is observed in the submesoscale turbulence regime, and a -2 slope in the internal wave regime. This behaviour is analogous to the equivalent regimes in the atmosphere. In addition, Ragone and Badin (2016) show that models of eddies using semi-geostrophic rather than quasi-geostrophic theory exhibit anisotropic structure, which reduces the spectral slope to between -2 and -3 at the bottom of the sub-mesoscale range. This is because of the use of the Monge-Ampère equation rather than an isotropic Poisson equation in calculating the solutions. This shallowing is seen in Fig. 2.6 for wavelengths less than 20km.

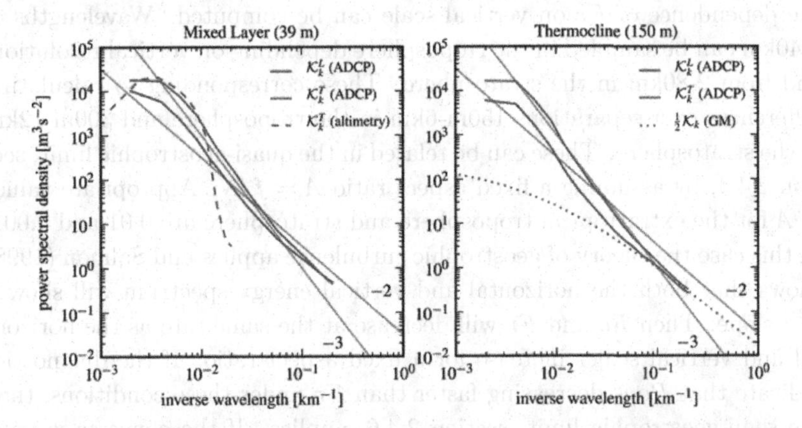

Fig. 2.6 Gulf Stream region wavenumber spectra of longitudinal and transverse kinetic energies K_k^L and K_k^T in the (left) mixed layer and (right) thermocline from (solid lines) ship track observations and (dashed line) altimeter observations. For other details see Callies and Ferrari (2013) from where the figure is taken. ©American Meteorological Society. Used with permission.

2.6.2 *Identification of different regimes in the atmosphere*

Now consider the atmospheric case in more detail. The spectra in Figs. 2.7-2.10 are derived from the U.K. Met Office Unified Model (UM) at 40km horizontal resolution and various vertical resolutions with grid spacings between 180m and 720m in the upper troposphere. The labels L62, L123 and L242 in the figure captions refer to successive doublings of the number of vertical levels. The methodology is described in Cullen (2017). The aim is to identify the different regimes occuring at various horizontal and vertical scales, and to study the differences between the troposphere and stratosphere and between the tropics and the extratropics.

The analyses in sections 2.4.4-2.4.6 show the importance of the ratio of the Rossby number Ro to the Froude number Fr when taking the asymptotic limit. Diagnostics from the model can be used to show the dependence of Ro on horizontal scale, where Ro is calculated as the ratio of the vertical component of the relative vorticity to the Coriolis parameter and the vorticity is calculated from the velocity differences over increasing horizontal distances. This allows wavelengths of 250-3000km to be sampled, which correspond to calculating differences over separations 40-500km. In order to allow a meaningful estimate in the tropics, the value of the Coriolis parameter at 5° latitude is used for latitudes within 5° of the equator. Similarly, the dependence of Fr on vertical scale can be computed. Wavelengths of 1-40km can be sampled in the troposphere depending on vertical resolution, and from 2-80km in the stratosphere. These corresponding to calculating differences over separations 150m-6km in the troposphere and 300m-12km in the stratosphere. These can be related in the quasi-geostrophic limit, section 2.4.4, by assuming a fixed aspect ratio $A = f/N$. Appropriate values of A for the extratropical troposphere and stratosphere are 0.01 and 0.001. In this case the theory of geostrophic turbulence applies and Salmon (1998) shows that both the horizontal and vertical energy spectrum will show a k^{-3} slope. Then Ro and Fr will decrease at the same rate as the horizontal and vertical scales increase for a fixed aspect ratio. If the diagnostics indicate that Ro is decreasing faster than Fr under these conditions, then the semi-geostrophic limit, section 2.4.6, applies. If the converse is true, the analysis of section 2.4.4 applies, which predicts a vertical scale collapse. In the latter case Fr would then start to increase, so this situation should not occur. If it does, it is likely that inadequate vertical resolution of the model is preventing the scale collapse.

Fig. 2.7 compares the tropospheric Rossby number Ro plotted as a function of horizontal wavelength for the extratropical southern hemisphere (south of 20°S) and the tropics (20°N to 20°S). It is shown in Cullen (2017) that if the kinetic energy spectrum $E(k)$ is proportional to k^{-m}, Ro calculated from the velocity gradients will then be approximately proportional to k^n where $n = (3-m)/2$. The case $n = 0$ would then correspond to a k^{-3} kinetic energy spectrum as observed in the atmosphere on wavelengths from 500 to 4000 km. The case $m = \frac{5}{3}$ observed at smaller scales gives $n = \frac{2}{3}$

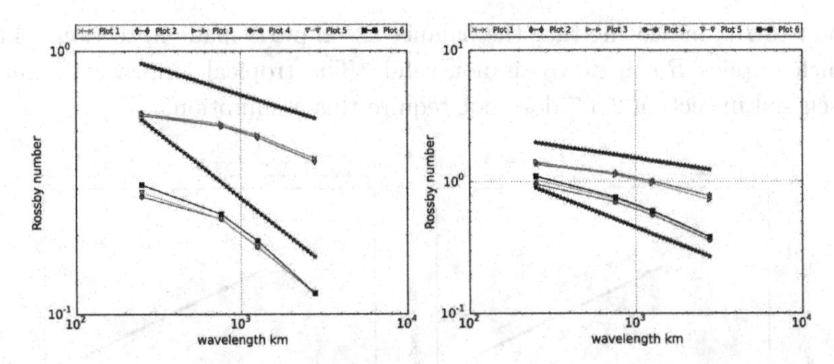

Fig. 2.7 Loglog plots against wavelength (km) of tropospheric Rossby number based on rotational wind for (a) the Southern hemisphere, and (b) the tropics. Plot 1: L62, plot 2: L123, plot 3: L242, and based on divergent wind, plot 4: L62, plot 5: L123, and plot 6: L242. Reference lines with stars give power law dependencies of $k^{0.2}$ and $k^{0.5}$. From Cullen (2017). ©Crown Copyright, Met Office.

and the case $m = 1$ gives $n = 1$. Note that this correspondence is only approximate. For instance, if Ro is independent of horizontal wavelength, so $n = 0$, the relative vorticity at the coarsest scale used is the same as that at the finest scale, so there is no small scale vorticity at all. However, a velocity field with a k^{-3} energy spectrum will still have some vorticity at small scales.

Fig. 2.7 shows the scale dependence of Ro and also of a similar quantity Ro_D which is the ratio of the horizontal divergence to the Coriolis parameter. In the extratropics, Ro has a $k^{0.2}$ dependence at large scales, suggesting an energy spectrum of $k^{-2.6}$ which is not far from k^{-3}. At wavelengths less than 500km, $Ro \simeq k^{0.1}$ which implies a steeper energy spectrum. In Fig. 2.3, the total energy spectrum deduced from reanalyses was steeper than a k^{-3} dependence at wavelengths greater than 300km. Smaller scale observations such as those shown in Fig 2.4 show a shallower $-\frac{5}{3}$ kinetic energy spectrum. It is likely that the steeper spectra seen in data from numerical models result from the numerical smoothing required to get satisfactory results with limited resolution. This seems to happen when the wavelength is less than 12 gridlengths of the model. Ro_D has a dependence $k^{0.5}$ at large scales, which implies a divergent kinetic energy spectrum with a slope of k^{-2}. The shallower slope is consistent with other studies such as Waite and Bartello (2006). The spectrum becomes steeper at small scales, again indicating numerical smoothing.

The results in the tropics show that both Ro and Ro_D have similar scale dependence to the extratropics. The difference is that Ro_D has values much closer to Ro, indicating that the balance assumption made in section 2.4.3, which implies $Ro \simeq \varepsilon Ro_D$, is not valid. The tropical long wave regime discussed in section 2.4.7 does not require this assumption.

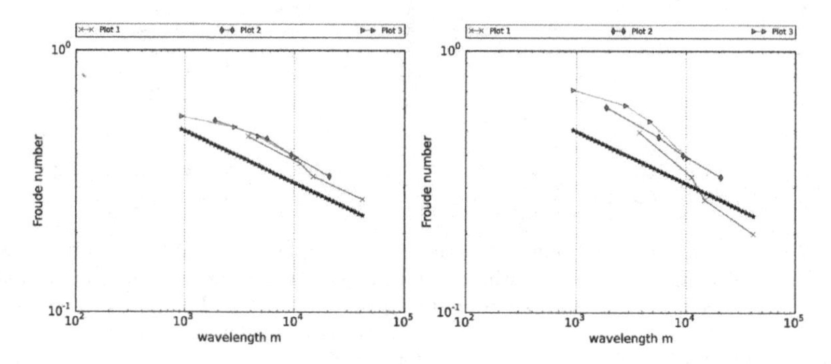

Fig. 2.8 Loglog plots against wavelength (m) of tropospheric Froude number for (a) the Southern hemisphere and (b) the tropics. Plot 1: L62, plot 2: L123, and plot 3: L242. The reference line with stars gives a power law dependency of $k^{0.2}$. From Cullen (2017). ©Crown Copyright, Met Office.

Now consider the vertical spectrum. Fig. 2.8 shows Fr for the tropics and extratropics as a function of vertical scale. As noted above, compare the horizontal dependence of Ro with the vertical dependence of Fr assuming an aspect ratio of 0.01 in the extratropics. Comparing Figs. 2.7 with 2.8 shows that Fr behaves the same way as Ro, indicating the presence of the quasi-geostrophic regime, and $Fr \simeq Ro$ at all equivalent scales. In the tropics, since $Ro = O(1)$ at all scales measured, the asymptotic behaviour as $Ro, Fr \to 0$ together does not apply. The data shows that $Fr \simeq k^{0.5}$ at large vertical scales, so that the limit discussed in section 2.4.4 applies. At small vertical scales $Fr \simeq k^{0.2}$ indicating a steeper spectrum and the effect of numerical smoothing, so the vertical scale collapse is arrested. In section 2.4.7, it was pointed out that convection can also arrest the vertical scale collapse in the tropics. Given that the spectrum of Fr steepens at a vertical wavelength of about 5km, this could be playing a part.

Now consider the stratosphere. Fig. 2.9 shows Ro and Ro_D for the stratosphere in the extratropics and tropics. In the extratropics, Ro is independent of scale, suggesting a very steep spectrum. It has a value of about 0.3, so the asymptotic theory is relevant. $Ro_D \simeq k^{0.5}$ at large scales, but at small scales $Ro \simeq Ro_D$ for the highest vertical resolution (plot 6 in the figure). At large scales $Ro > Ro_D$, so that the vorticity is larger than the divergence. This is consistent with a 'slow' solution as discussed in

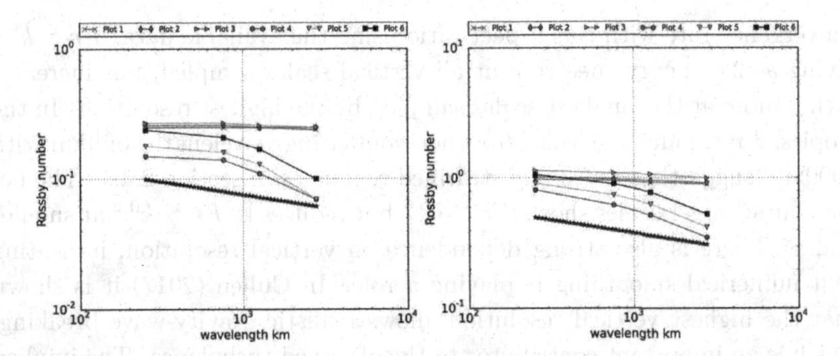

Fig. 2.9 Loglog plots against wavelength (km) of stratospheric Rossby number based on rotational wind for (a) the Southern hemisphere and (b) the tropics. Plot 1: L62, plot 2: L123, plot 3: L242, and based on divergent wind, plot 4: L62, plot 5: L123, and plot 6: L242. The reference line with stars gives a power law dependency of $k^{0.2}$. From Cullen (2017). ©Crown Copyright, Met Office.

section 2.4.3. The scale dependence and the relative magnitude of Ro and Ro_D are similar in the tropics. However, both Ro and Ro_D are much larger in the tropics and are typically O(1) rather than O(0.3) so the asymptotic theory is not relevant.

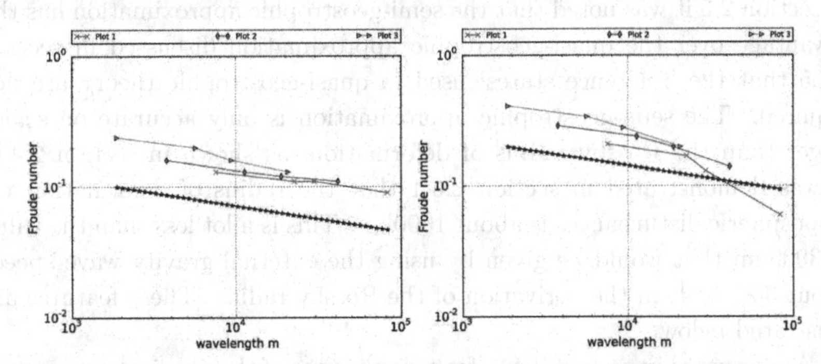

Fig. 2.10 Loglog plots against wavelength (m) of stratospheric Froude number for (a) the Southern hemisphere and (b) the tropics. Plot 1: L62, plot 2: L123, and plot 3: L242. The reference line with stars gives a power law dependency of $k^{0.2}$. From Cullen (2017). ©Crown Copyright, Met Office.

Fig. 2.10 shows Fr for the stratosphere. In the extratropics, Fr should be compared with Ro for an aspect ratio of 0.001. Thus Fr for a vertical wavelength of 3km has to be compared with Ro at a wavelength of 3000km. These are only just attainable with the data available, but show that Fr and Ro are both about 0.3. It is not possible to compare the asymptotic convergence rate with this aspect ratio using the available data. $Fr \simeq k^{0.2}$ giving a -2.6 energy spectrum at all vertical scales sampled, but increases rather more at the smallest scales sampled by the highest resolution. In the tropics, Fr is much less than Ro when comparing wavelengths of 3km with 3000km, suggesting a strongly stratified regime as in section 2.4.4. The behaviour at larger scales shows $Fr \simeq k^{0.5}$ but reduces to $Fr \simeq k^{0.2}$ at smaller scales. There is also strong dependence on vertical resolution, indicating that numerical smoothing is playing a role. In Cullen (2017) it is shown that the highest vertical resolution allows realistic gravity-wave breaking, which is an important contributor to the observed turbulence. The implied energy spectrum is shallower, k^{-2}, at large vertical scales which is close to the $k^{-\frac{5}{3}}$ expected for gravity waves and turbulence, e.g. Skamarock *et al.* (2014).

2.7 The appropriateness of the semi-geostrophic approximation for large-scale flows in the atmosphere

2.7.1 *Implications of the atmospheric basic state*

In section 2.5 it was noted that the semi-geostrophic approximation has the advantage over the quasi-geostrophic approximation discussed in section 2.4.5 that the 'reference states' used in quasi-geostrophic theory are not required. The semi-geostrophic approximation is only accurate on scales larger than the Rossby radius of deformation, as shown in section 2.4.6. It was demonstrated in section 2.6.1 that the radius of deformation for tropospheric disturbances is about 1000km. This is a lot less than the value of 3000km that would be given by using the external gravity wave speed, about 300ms^{-1}, in the derivation of the Rossby radius. These features are illustrated below.

As discussed in section 2.4.2, for a given value of the physical parameters f/N, the radius of deformation can only be defined for a given vertical scale H. The reason that the radius of deformation is less than might be expected is shown in Fig. 2.11. This shows that the tropospheric jet-stream reaches a maximum velocity at the tropopause. The vertical scale associated with the zonal wind structure is thus the tropopause depth, rather than the depth of the atmosphere.

The zonal wind structure is linked to the zonal mean temperature structure by geostrophic and hydrostatic balance. The zonal mean temperature cross-section is also shown in Fig. 2.11. The maximum velocity of the

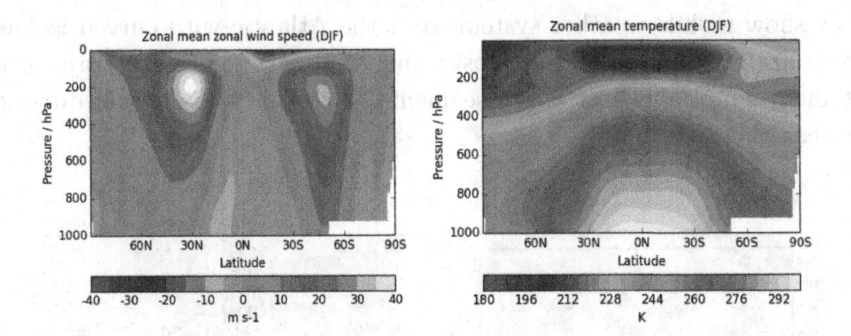

Fig. 2.11 Left: zonal mean winds and right: zonal mean temperatures for Northern hemisphere winter from ERA data. Source: Met Office. ©Crown Copyright, Met Office.

jet-stream at the tropopause requires a reversal of the equator-pole temperature gradient in the stratosphere. This occurs primarily because of the effects of ozone.

The temperature structure shown in Fig. 2.11 also illustrates why the quasi-geostrophic assumption of small departures from a global reference temperature profile in the vertical is not appropriate for large-scale flows. While the zonal mean static stability $\frac{\partial\theta}{\partial z}$ does not actually vary much in the troposphere, the height of the tropopause and the value of θ at the surface do vary a lot with latitude.

2.7.2 *Curvature of trajectories*

It is important to recognise that the condition that the Lagrangian Rossby number Ro_L, as defined in section 2.5.2, is small is less restrictive than the condition that the Rossby number $Ro = U/fL$, as defined in section 2.2.3 and used in section 2.6.2, is small because it only restricts the spatial derivatives of $\mathbf{u}_h$ in the direction of the trajectory. Semi-geostrophic theory can therefore be used for (almost) straight flows, possibly with small scales in the direction normal to the wind direction. The Rossby number based on the cross-wind scale would then be large. A jet-stream in the atmosphere is an example of this case. Using a mid-latitude value of $f{=}10^{-4}\text{s}^{-1}$, a value $Ro_L = 0.1$ means that a trajectory cannot change its direction by more than an angle of $\pi/4$ in 24 hours. Figure 2.12 shows some typical trajectories inferred from the Met Office Unified Model (UM) and used in the Met Office atmospheric dispersion model (NAME). The surface pressure maps on the right are for 3 days before and just before the arrival time.

They show mobile weather systems over the Atlantic with curved isobars, which are streamlines of the geostrophic wind. However the trajectories are made up of almost straight segments, with sharp changes of direction between them.

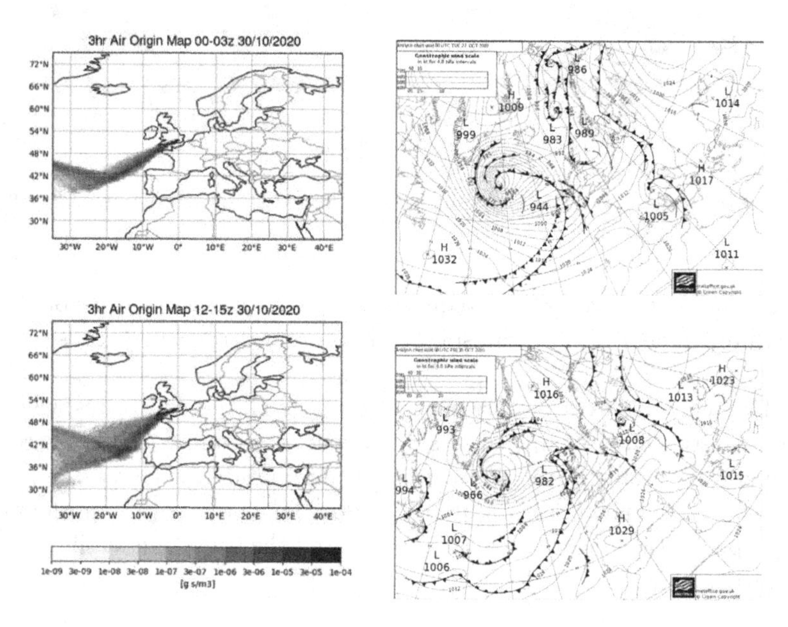

Fig. 2.12 Model trajectories obtained from the Met Office NAME model. Trajectories arriving in London at 00UTC and 12UTC on 30 October 2020 are shown. The output shows the source of the air back over 7 days from the arrival time. The spread reflects the diffusive processes included in the NAME model and the averaging over a 3hr arrival time window. Surface pressure maps for 00UTC on 27 and 30 October are shown in the right hand column. Source: Atmospheric Dispersion and Air Quality Group, Met Office. ©Crown Copyright, Met Office.

2.7.3 *Validation of the inertial stability condition*

In section 2.3.7 it was shown that h has to satisfy the condition (2.108) near the equator to be an admissible solution of the semi-geostrophic shallow water equations on the sphere. The term $c_0\hat{y}^2$ in this equation implies a zonally uniform geostrophic wind. So write (2.108) in dimensional form as

$$h = h_0 + \frac{a}{g}\Omega\left(U_0\phi^2 + \frac{1}{2}U_1(\lambda)\phi^4\right) + O(\phi^5). \tag{2.172}$$

Since the north-south component v_g of the geostrophic wind is given by $v_g = \frac{g}{2\Omega a \sin\phi\cos\phi}\frac{\partial h}{\partial \lambda}$, it follows that

$$
\begin{aligned}
\frac{gh}{a} &\simeq \Omega U_0 \phi^2 + \frac{1}{2}\Omega U_1(\lambda)\phi^4 + \mathrm{O}(\phi^5) \\
u_g &\simeq U_0 + U_1(\lambda)\phi^2 + \mathrm{O}(\phi^3), \\
v_g &\simeq \frac{1}{4\cos\phi}\frac{\mathrm{d}U_1}{\mathrm{d}\lambda}\phi^3 + \mathrm{O}(\phi^4).
\end{aligned} \tag{2.173}
$$

The condition for inertial stability that $\mathbf{Q}$ as defined in (2.90) is positive definite requires in particular that

$$
2\Omega\sin\phi \geq \frac{1}{a}\frac{\partial u_g}{\partial \phi}, \tag{2.174}
$$

which gives $U_1 \leq \Omega a$, and the amplitude of the $\mathrm{O}(\phi^4)$ term in h as less than $\frac{1}{2g}\Omega^2 a^2$. At latitude 6°, this means the variation in h is restricted to about 1m. (2.174) is also the inertial stability condition required for the tropical long wave equations to be solvable derived after eq. (2.115) in section 2.3.7.

A verification of (2.174) against observations was carried out by Veitch and Mawson (1993). Aircraft data was used from flights in the region 30°N to 30°S. (2.174) could only be tested by direct calculation from the data in cases of parallel flow. Since most aircraft cross the tropics in a north-south direction, these flights allow verification of (2.174). Only 5 cases out of 121 showed any violations of (2.174) using wind data 80km apart. This is consistent with the validity of the semi-geostrophic model and the tropical long wave model. It does not distinguish between them. The lack of east-west aircraft tracks meant that values of $\frac{\partial v_g}{\partial \lambda}$ could not be estimated. That would distinguish between the two models. Large values would invalidate both of them.

2.7.4 *Demonstration of the accuracy of semi-geostrophic theory for large-scale flows*

In this subsection, the ability of the semi-geostrophic model to reproduce the large-scale aspects of a comprehensive atmospheric model are demonstrated, following Cullen (2018). In the example shown, the Met Office UM is run wih a 40km horizontal grid and 70 vertical levels. The data come from an operational analysis, followed by a 5 day forecast to remove unrealistic features resulting from the analysis procedure.

It is necessary to prepare the data before applying semi-geostrophic theory to it. The theory calculates the winds and temperatures from the

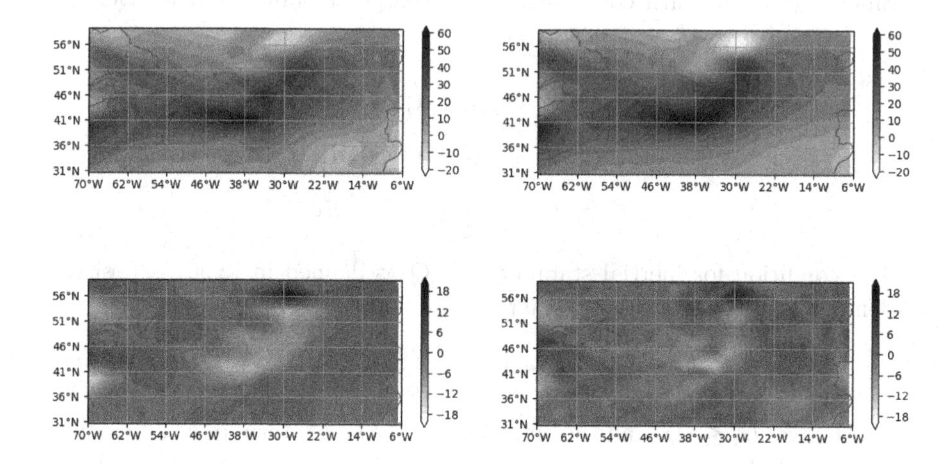

Fig. 2.13 Zonal component of wind at 4000 m, units ms^{-1}: top left, direct UM output; top right: geotriptic wind calculated from (2.170); bottom left: ageotriptic wind calculated using (2.171) and (2.166); and bottom right: ageotriptic wind from UM output. Source: Met Office. ©Crown Copyright, Met Office.

pressure, so that it is necessary that the pressure data used is consistent with the theory. This is achieved by reducing the horizontal resolution of the data to 160km, thus eliminating gravity waves and other small-scale features which are represented in the UM. Further filtering is applied in the tropics, because semi-geostrophic pressure fields have to take the form (2.159) in order to satisfy the inertial stability condition. The diagnostic procedure then uses the UM pressure field to solve a generalisation of eq. (2.152) to calculate $\frac{\partial p}{\partial t}$ and then deduces the total wind $\mathbf{u}$ from a generalisation of (2.150). The dynamics in this equation are extended to the UM equations, which are fully compressible, use terrain-following coordinates to represent orography and incorporate spherical geometry. The right hand side of (2.150) is generalised to

$$\mathbf{Q}\begin{pmatrix}\hat{u}\\ \hat{v}\\ \hat{w}\end{pmatrix} + \frac{\partial}{\partial \hat{t}}\nabla\hat{\varphi} = Ro^{-1}\mu\begin{pmatrix}-\frac{\partial\hat{\varphi}}{\partial\hat{y}} - \hat{S}_y\\ \frac{\partial\hat{\varphi}}{\partial\hat{x}} + \hat{S}_x\\ \hat{S}_h\end{pmatrix}, \tag{2.175}$$

where $\hat{S}_x, \hat{S}_y, \hat{S}_h$ represent the dimensionless forcing terms from other physical effects, such as radiation or phase changes of moisture, which were included in the original equations (2.1). The resulting circulation $\hat{\mathbf{u}}$ thus represents the the circulation $\hat{\mathbf{u}}$ needed to maintain geostrophic and hydrostatic balance against forcing. The response to the forcing will be amplified if det$\mathbf{Q}$ is small, which corresponds to small potential vorticity. This will typically represent weak inertial or static stability. This greatly aids the understanding of solutions of the UM, as will be demonstrated in chapter 6.

The solution procedure only works if the matrix $\mathbf{Q}$ defined in (2.151) and generalised to the UM equations, is positive definite. This is not guaranteed for UM data, and so some regularisation has to be applied. A boundary layer has to be included, so the equations solved in Cullen (2018) are the semi-geotriptic equations (2.171) defined in section 2.5.2. This greatly expands the applicability of the model in the tropics by relaxing the inertial stability condition, as shown in section 6.1.4. It also improves the numerical conditioning of the solution procedure. This extension is not very significant in the example shown but very important in the tropics, as shown in section 6.1.6.

As shown in section 2.4.6, semi-geostrophic theory is accurate to $O(\varepsilon^2)$ where $\varepsilon = Ro = Fr^2$. The results shown in Fig. 2.7 show that in the extratropics Ro is about 0.4 for wavelengths of 3000km, thus a horizontal scale of 500km. However, as noted in section 2.5.2, the definition of the Rossby number is different from that used in section 2.6.2 and is likely to give smaller values, particularly for straight trajectories. The accuracy of semi-geotriptic theory in the boundary layer is discussed in section 6.1.5.

Fig. 2.13 shows a test of the theory over the extratropical Atlantic Ocean, where the theory should work best. The UM zonal wind at an altitude of 4km is illustrated. The geotriptic wind is a good approximation to the model wind, but is somewhat stronger. This is because of the cyclonic curvature, which makes the geostrophic wind an overestimate of the true wind. For the area and level illustrated, the r.m.s. UM zonal wind is 18ms^{-1} and the r.m.s. difference between the UM wind and the geotriptic wind is 3.1ms^{-1}. In the third panel the ageotriptic wind derived from solving a generalisation of (2.150) is shown. This is compared with the ageotriptic wind obtained by subtracting the computed geotriptic wind from the UM wind. There is a reasonable correspondence, the correlation between the two estimates of ageotriptic wind is 0.64. The r.m.s. difference between the two estimates is 2.6ms^{-1}.

2.7.5 *The transition from the semi-geostrophic regime to the tropical long wave regime*

In this subsection, data from the Met Office UM is used to illustrate the transition between the extratropical semi-geostrophic regime described in section 2.4.6 and the tropical long wave regime described in section 2.4.7. The data used are a two-week average of UM integrations in order to focus on the large-scale behaviour. As in section 2.7.4, the UM data are interpolated to a 160km grid to remove gravity waves and allow the calculation of the geostrophic wind from eq. (2.167) to be meaningful. However, the tropical filtering required to enforce inertial stability is not used, as the tropical long-wave regime does not require this in the east-west direction. In order to allow (2.167) to be solved very near the equator, the value of the Coriolis parameter is replaced by its value at 5° in the appropriate hemisphere. In interpreting the results, it is important to remember that the geostrophic wind is simply a function of pressure.

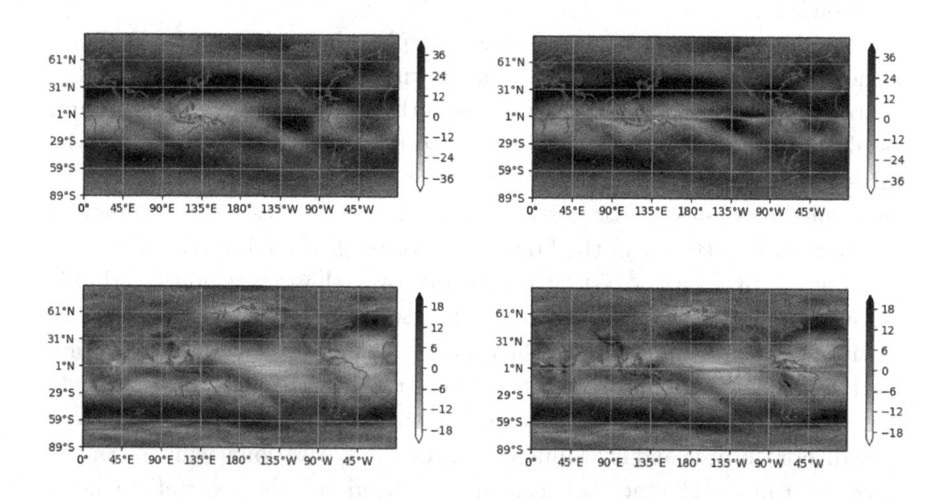

Fig. 2.14 Zonal component of winds, units ms^{-1}: top left, direct UM output at 14km altitude; top right: geostrophic wind calculated from (2.167) at same altitude; bottom left: direct UM output at 1400 m altitude; bottom right: geostrophic wind calculated from (2.167) at same altitude. Source: Met Office. ©Crown Copyright, Met Office.

In the tropical long wave equations, the zonal component of the wind is still geostrophic. The solution of the equations requires the inertial stability condition as discussed after (2.165) which implies that (2.159) is satisfied for each x. There is no control over variations in x, but validity of the equations requires this to be small. Fig. 2.14 compares the UM winds in the upper and lower troposphere with the geostrophic winds calculated from the UM pressure. At the upper level, the geostrophic wind is very accurate except at the equator where the regularisation has been applied. At the lower level some differences are also visible over Africa and west of South America where there are high mountains. Also, note that in the deep tropics the horizontal structure of the winds is very different at the two levels shown. This is consistent with the idea of there being two dominant vertical modes, as discussed in section 2.4.7 and suggested by the analysis of Wheeler and Kiladis (1999).

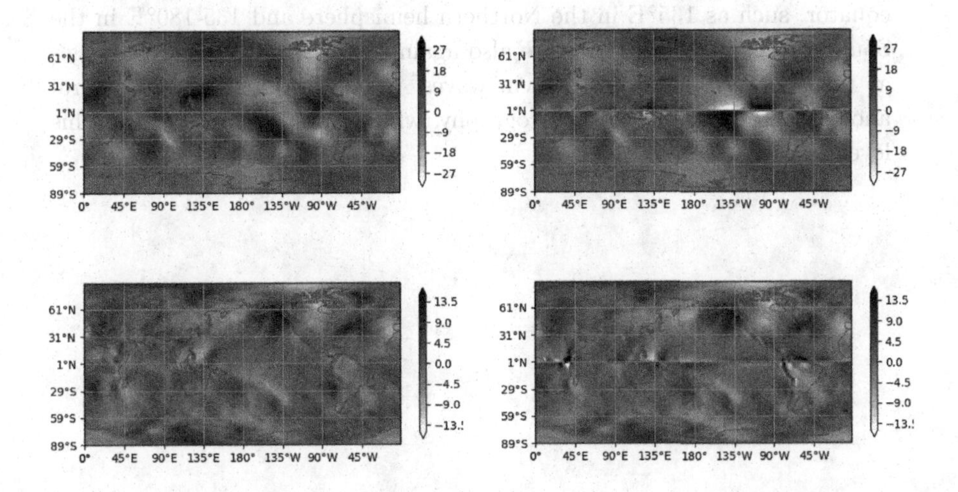

Fig. 2.15 Meridional component of winds, units ms^{-1}: top left, direct UM output at 14km altitude; top right: geostrophic wind calculated from (2.167) at same altitude; bottom left: direct UM output at 1400 m altitude; bottom right: geostrophic wind calculated from (2.167) at same altitude. Source: Met Office. ©Crown Copyright, Met Office.

Fig. 2.15 makes the same comparison for the meridional wind. While both wind components should be close to geostrophic in the semi-geostrophic regime, the meridional wind will not be geostrophic in the tropical long wave regime. Plots of this component will thus show the transition, which should occur at the equatorial deformation radius discussed in section 2.4.7. At the upper level, the geostrophic wind accurately approximates the UM down to 30° latitude. This is rather larger than the estimates of the equatorial deformation radius given in section 2.4.7. At some longitudes, such as 135-180°W, it remains accurate much closer to the equator. At the equator there are several places where it changes sign across the equator. This indicates that there is a coherent pressure perturbation at the equator which varies with longitude, probably a Kelvin wave. In the UM picture, there is no obvious discontinuity at the equatorial deformation radius. Thus the two regimes must be closely coupled, as discussed in section 2.3.7. In low latitudes the meridional wind has a number of clear features crossing the equator with a NW-SE orientation. These features usually lose their identity beyond 30° latitude. At the lower level, as with the zonal wind, the pattern is very different from the upper level. There are several places where the geostrophic approximation remains accurate very close to the equator, such as 135°E in the Northern hemisphere and 135-180°E in the Southern hemisphere. There are also again places where the sign changes at the equator, suggesting Kelvin waves. In addition, there are several anomalous regions related to orography, which would be expected at this level.

Chapter 3

Solution of the Semi-geostrophic Equations in Plane Geometry

3.1 The solution as a sequence of minimum energy states

3.1.1 *The evolution equation for the geopotential*

In this chapter it is shown that, in the special case of constant Coriolis parameter f, the semi-geostrophic equations are well-posed, in the sense that they can be solved for arbitrarily large times given physically reasonable boundary conditions. The assumption of constant f is, of course, not compatible with the study of large-scale atmospheric flow. In sections 4.2 and 4.3 it is shown how the results can be extended to variable f. However, explicit solutions are much easier to construct with constant f, so this case is useful in understanding the physical nature of the solutions. For the same reasons, the three-dimensional Boussinesq formulation in pressure-related coordinates derived in section 2.4.1 is used. The Boussinesq assumption is withdrawn in section 4.1. In this section and section 3.2, equations (2.147) and (2.148) in dimensional form are used, with f assumed constant.

The semi-geostrophic equations are then an approximation to the shallow atmosphere hydrostatic incompressible Boussinesq Euler equations, stated in non-dimensional form in (2.116). These equations are solved with rigid wall boundary conditions. In dimensional form, they are

$$\begin{aligned}
\frac{\mathrm{D}u}{\mathrm{D}t} + \frac{\partial \varphi}{\partial x} - fv &= 0, \\
\frac{\mathrm{D}v}{\mathrm{D}t} + \frac{\partial \varphi}{\partial y} + fu &= 0, \\
\frac{\partial \varphi}{\partial z} - \frac{g\theta}{\theta_0} &= 0, \\
\frac{\mathrm{D}\theta}{\mathrm{D}t} &= 0, \\
\nabla \cdot \mathbf{u} &= 0.
\end{aligned} \tag{3.1}$$

The semi-geostrophic approximation to (3.1) is

$$\begin{aligned}
\frac{\mathrm{D}u_g}{\mathrm{D}t} + \frac{\partial \varphi}{\partial x} - fv &= 0, \\
\frac{\mathrm{D}v_g}{\mathrm{D}t} + \frac{\partial \varphi}{\partial y} + fu &= 0, \\
\frac{\mathrm{D}\theta}{\mathrm{D}t} &= 0, \\
\nabla \varphi &= \left(fv_g, -fu_g, \frac{g\theta}{\theta_0} \right), \\
\nabla \cdot \mathbf{u} &= 0.
\end{aligned} \tag{3.2}$$

The combination of incompressible equations with rigid wall boundary conditions in three dimensions is unrealistic as a description of the large-scale flow of the atmosphere. In section 3.3 the shallow water semi-geostrophic equations derived in section 2.3.6 are therefore studied. The free surface boundary condition allows a correct description of height-independent large-scale atmospheric flow. This is extended to a free surface formulation of three-dimensional flow appropriate to the ocean in section 3.3.5 and to the atmosphere in section 3.3.6.

Following Schubert (1985), equations (3.2) can be written in the form:

$$\mathbf{Q} \begin{pmatrix} u \\ v \\ w \end{pmatrix} + \frac{\partial}{\partial t} \nabla \varphi = \begin{pmatrix} f^2 u_g \\ f^2 v_g \\ 0 \end{pmatrix},$$

$$\mathbf{Q} = \begin{pmatrix} f^2 + f\frac{\partial v_g}{\partial x} & f\frac{\partial v_g}{\partial y} & f\frac{\partial v_g}{\partial z} \\ -f\frac{\partial u_g}{\partial x} & f^2 - f\frac{\partial u_g}{\partial y} & -f\frac{\partial u_g}{\partial z} \\ \frac{g}{\theta_0}\frac{\partial \theta}{\partial x} & \frac{g}{\theta_0}\frac{\partial \theta}{\partial y} & \frac{g}{\theta_0}\frac{\partial \theta}{\partial z} \end{pmatrix}. \tag{3.3}$$

Use of the fifth equation of (3.2) then gives

$$\nabla \cdot \mathbf{Q}^{-1} \frac{\partial}{\partial t} \nabla \varphi = \nabla \cdot \mathbf{Q}^{-1} \begin{pmatrix} f^2 u_g \\ f^2 v_g \\ 0 \end{pmatrix}. \tag{3.4}$$

In principle, equation (3.4) can be solved for $\partial\varphi/\partial t$. This requires boundary conditions on $\partial\varphi/\partial t$. Apply the boundary conditions $\mathbf{u}\cdot\mathbf{n} = 0$ to $\mathbf{Q}^{-1}$ times the first equation of (3.3). This gives the boundary condition for (3.4) as

$$\mathbf{Q}^{-1}\frac{\partial}{\partial t}\nabla\varphi\cdot\mathbf{n} = \mathbf{Q}^{-1}\begin{pmatrix} f^2 u_g \\ f^2 v_g \\ 0 \end{pmatrix}\cdot\mathbf{n} \tag{3.5}$$

on the boundary of Γ. If $\mathbf{Q}$ is positive definite and $(\mathbf{Q}\mathbf{n})\cdot\mathbf{n} > 0$, then equation (3.4) with (3.5) becomes an elliptic equation to be solved for $\partial\varphi/\partial t$ with a Neumann type boundary condition, and is expected to be solvable. It was stated in section 2.4.6 that det $\mathbf{Q}$ is a constant of the motion. Thus, if det $\mathbf{Q}> 0$ throughout Γ at $t = 0$ it will remain so for all t. However, det $\mathbf{Q}> 0$ is not a sufficient condition for $\mathbf{Q}$ to be positive definite, so there is some work to be done. It is possible to write $\mathbf{Q}$ as a function of φ as below:

$$\mathbf{Q} = \begin{pmatrix} f^2 + \frac{\partial^2\varphi}{\partial x^2} & \frac{\partial^2\varphi}{\partial x\partial y} & \frac{\partial^2\varphi}{\partial x\partial z} \\ \frac{\partial^2\varphi}{\partial y\partial x} & f^2 + \frac{\partial^2\varphi}{\partial y^2} & \frac{\partial^2\varphi}{\partial y\partial z} \\ \frac{\partial^2\varphi}{\partial z\partial x} & \frac{\partial^2\varphi}{\partial z\partial y} & \frac{\partial^2\varphi}{\partial z^2} \end{pmatrix}. \tag{3.6}$$

It can be seen that $\mathbf{Q}$ is the Hessian matrix of $\varphi + \frac{1}{2}f^2(x^2 + y^2)$.

Equations (3.3) can be interpreted physically as stating that $\mathbf{u}$ is determined by the requirements of maintaining geostrophic and hydrostatic balance. In a sense, the equations describe a continuous geostrophic adjustment problem which is a nonlinear generalisation of the classical Rossby adjustment problem discussed in section 2.2, equation (2.51).

3.1.2 *Solutions as minimum energy states*

In this section it is shown that a geostrophic and hydrostatic state can be characterised as a stationary point of the energy with respect to Lagrangian parcel displacements, neglecting pressure perturbations resulting from the displacements. The physical significance of the class of variations used will be discussed below. More details of these arguments are given in Shutts and Cullen (1987), section 3.

Start with a state of the fluid with an associated vector field $(\tilde{u}, \tilde{v})$ and scalar field $\tilde{\theta}$. Associate with this state an energy given by the dimensional formula analogous to (2.149)

$$\tilde{E} = \int_\Gamma \left(\frac{1}{2}(\tilde{u}^2 + \tilde{v}^2) - g\tilde{\theta}z/\theta_0\right) \mathrm{d}x\mathrm{d}y\mathrm{d}z. \tag{3.7}$$

This is a functional of $\tilde{u}, \tilde{v}$ and $\tilde{\theta}$, regarded as functions of position over Γ, which has the following property:

Theorem 3.1. *The conditions for the energy $\tilde{E}$ to be stationary with respect to variations $\Xi = (\xi, \eta, \chi)$ of particle positions satisfying continuity $\delta(\mathrm{d}x\mathrm{d}y\mathrm{d}z) = 0$ via*

$$\nabla \cdot \Xi = 0 \tag{3.8}$$

in Γ and

$$\delta\tilde{u} = f\eta,\ \delta\tilde{v} = -f\xi,\ \delta\tilde{\theta} = 0, \tag{3.9}$$

together with $\Xi \cdot \mathbf{n} = 0$ on the boundary of Γ, are that

$$(f\tilde{v}, -f\tilde{u}, g\tilde{\theta}/\theta_0) = \nabla\tilde{\varphi} \tag{3.10}$$

for some scalar $\tilde{\varphi}$.

Proof Write

$$\begin{aligned} \delta\tilde{E} &= \int_\Gamma \left(\tilde{u}\delta\tilde{u} + \tilde{v}\delta\tilde{v} - gz\delta\tilde{\theta}/\theta_0 - g\tilde{\theta}\chi/\theta_0\right) \mathrm{d}x\mathrm{d}y\mathrm{d}z, \\ &= \int_\Gamma \left(f\tilde{u}\eta - f\tilde{v}\xi - g\tilde{\theta}\chi/\theta_0\right) \mathrm{d}x\mathrm{d}y\mathrm{d}z, \\ &= \int_\Gamma \left(-\Xi \cdot (f\tilde{v}, -f\tilde{u}, g\tilde{\theta}/\theta_0\right) \mathrm{d}x\mathrm{d}y\mathrm{d}z. \end{aligned} \tag{3.11}$$

For this to vanish for any Ξ satisfying (3.8) and the boundary conditions, (3.10) must be satisfied. □

Equation (3.10) means that $(\tilde{u}, \tilde{v}, \tilde{\theta})$ represents a state in geostrophic and hydrostatic balance, with geopotential $\tilde{\varphi}$. The next step is to show that if the stationary point is a minimum, the matrix $\mathbf{Q}$ calculated by replacing (u_g, v_g, θ) by $(\tilde{u}, \tilde{v}, \tilde{\theta})$ in (3.3) is positive definite.

Theorem 3.2. *The condition for $\tilde{E}$ to be minimised with respect to the class of variations defined in Theorem 3.1 is that the matrix*

$$\mathbf{Q} = \begin{pmatrix} f^2 + f\frac{\partial\tilde{v}}{\partial x} & f\frac{\partial\tilde{v}}{\partial y} & f\frac{\partial\tilde{v}}{\partial z} \\ -f\frac{\partial\tilde{u}}{\partial x} & f^2 - f\frac{\partial\tilde{u}}{\partial y} & -f\frac{\partial\tilde{u}}{\partial z} \\ \frac{g}{\theta_0}\frac{\partial\tilde{\theta}}{\partial x} & \frac{g}{\theta_0}\frac{\partial\tilde{\theta}}{\partial y} & \frac{g}{\theta_0}\frac{\partial\tilde{\theta}}{\partial z} \end{pmatrix} \tag{3.12}$$

is positive definite.

Proof A minimum is also a stationary point, so characterise the stationary point as satisfying $(\tilde{u}, \tilde{v}, \tilde{\theta}) = (u_g, v_g, \theta_g)$ with

$$(fv_g, -fu_g, g\theta_g/\theta_0) = \nabla\varphi_g. \tag{3.13}$$

Using (3.11), write

$$\delta\tilde{E} = \int_\Gamma \left(\Xi \cdot (-f(\tilde{v} - v_g), f(\tilde{u} - u_g), -g(\tilde{\theta} - \theta_g)/\theta_0\right) dxdydz. \quad (3.14)$$

Then, taking a second variation

$$\delta^2\tilde{E} = \int_\Gamma \left(\delta\left(f\Xi \cdot (-(\tilde{v} - v_g), (\tilde{u} - u_g), -g(\tilde{\theta} - \theta_g)/(f\theta_0)\right)\right) dxdydz, \quad (3.15)$$

and since $\tilde{\mathbf{u}} = \mathbf{u}_g$ when $\delta\tilde{E} = 0$, this reduces to

$$\int_\Gamma \left((f\xi, f\eta) \cdot (-\delta(\tilde{v} - v_g), \delta(\tilde{u} - u_g)) - g\chi\delta(\tilde{\theta} - \theta_g)/\theta_0\right) dxdydz. \quad (3.16)$$

Substituting for $\delta\tilde{\mathbf{u}}$ and $\delta\tilde{\theta}$ from (3.9) gives

$$\int_\Gamma ((f\xi, f\eta) \cdot ((f\xi, f\eta) + \delta(v_g, -u_g)) + g\chi\delta\theta_g/\theta_0)\, dxdydz. \quad (3.17)$$

The Lagrangian nature of the displacement means that $\delta(u_g, v_g) = \Xi \cdot \nabla(u_g, v_g)$ and $\delta(g\theta_g/\theta_0) = \Xi \cdot \nabla(g\theta_g/\theta_0)$, so (3.17) becomes

$$\int_\Gamma \Xi \cdot ((f^2\xi, f^2\eta, 0) + (f\xi, f\eta, \chi) \cdot \nabla(v_g, -u_g, \theta_g/\theta_0))\, dxdydz. \quad (3.18)$$

Therefore the result is

$$\delta^2\tilde{E} = \int_\Gamma (\Xi \cdot \mathbf{Q} \cdot \Xi)\, dxdydz \quad (3.19)$$

where $\mathbf{Q}$ is as defined in (3.12). Thus positive definiteness of $\mathbf{Q}$ is equivalent to positive definiteness of $\delta^2\tilde{E}$, which is the condition for E to be minimised. □

Note that the proofs of Theorems 3.1 and 3.2 have been written in a way that is equally valid when f is a function of position. This will be exploited in section 4.3.1. Since the main physical applicability of semi-geostrophic theory is to the case where f is a function of position, it is important that the energy minimisation property carries over to this case.

The requirement that $(v_g, -u_g, \theta_g)$ are the gradient of φ_g is necessary for $(v_g, -u_g, \theta_g)$ to be a solution of equations (3.2). However, this only means that the energy has to be stationary with respect to variations (3.8) and (3.9) while solvability of (3.4) requires that the energy is minimised. This is an additional constraint, which does not form part of equation (3.2).

3.1.3 *Physical meaning of the energy minimisation*

This section considers the physical meaning of the energy minimisation condition. For this, the non-hydrostatic version of the shallow atmosphere Boussinesq incompressible Euler equations (3.1) are used, from which the semi-geostrophic equations (3.2) were derived. These are

$$\begin{aligned}\frac{\mathrm{D}(u,v)}{\mathrm{D}t} + f(-v,u) + \nabla_h\varphi &= 0,\\ \frac{\mathrm{D}w}{\mathrm{D}t} + \frac{\partial\varphi}{\partial z} - \frac{g\theta}{\theta_0} &= 0,\\ \nabla\cdot\mathbf{u} &= 0,\\ \frac{\mathrm{D}\theta}{\mathrm{D}t} &= 0.\end{aligned} \tag{3.20}$$

Following Shutts and Cullen (1987), consider the conditions for a flow to be stable to perturbations consisting of small Lagrangian displacements. Such an analysis assumes that the time-scale over which the perturbations evolve is much shorter than the time-scale over which the basic state evolves. This is most straightforward if the displacements are applied to a steady state basic flow. Examples are straight or circular flows.

Consider a straight flow in the y direction. Give an impulsive velocity to a fluid parcel and study the ensuing motion under the condition that the perturbation pressure is zero (Godson (1950), Van Meighem (1952)). This is referred to as *parcel stability analysis*, Emanuel (1983). This assumption is discussed below. The resulting motion is assumed to be independent of y and so it is assumed that the parcel extends infinitely far in the y direction. The equations for the undisturbed straight flow $(\overline{v},\overline{\theta},\overline{\varphi})$ are derived from equations (3.20) as

$$\begin{aligned}-f\overline{v} + \frac{\partial\overline{\varphi}}{\partial x} &= 0,\\ \frac{\partial\overline{\varphi}}{\partial z} - g\frac{\overline{\theta}}{\theta_0} &= 0.\end{aligned} \tag{3.21}$$

Let the displacement be $\Xi = (\xi, 0, \chi)$, so that $u = \dot{\xi}, w = \dot{\chi}$. Then v and θ for the displaced parcel evolve according to (3.20) as

$$\begin{aligned}\frac{\mathrm{D}v}{\mathrm{D}t} + f\dot{\xi} \equiv \frac{\mathrm{D}(v+f\xi)}{\mathrm{D}t} &= 0,\\ \frac{\mathrm{D}\theta}{\mathrm{D}t} &= 0,\end{aligned} \tag{3.22}$$

and the third equation of (3.20) implies that, for small displacements

$$\frac{\partial \dot{\xi}}{\partial x} + \frac{\partial \dot{\chi}}{\partial z} = 0. \tag{3.23}$$

The first equation of (3.22) means that the value of v following a parcel is equal to $\overline{v} - f\xi$ and it is assumed that φ is not changed by the displacement. Using these facts in (3.20) gives the equations for the evolution of the displacement as

$$\begin{aligned} \ddot{\xi} + f^2\xi + \Xi \cdot \nabla\left(\frac{\partial \overline{\varphi}}{\partial x}\right) &= 0, \\ \ddot{\chi} + \Xi \cdot \nabla\left(\frac{\partial \overline{\varphi}}{\partial z}\right) &= 0. \end{aligned} \tag{3.24}$$

These equations can be rewritten as

$$\begin{pmatrix} \ddot{\xi} \\ \ddot{\chi} \end{pmatrix} + \mathbf{Q}\begin{pmatrix} \xi \\ \chi \end{pmatrix} = 0,$$

$$\mathbf{Q} = \begin{pmatrix} f^2 + \frac{\partial^2 \overline{\varphi}}{\partial x^2} & \frac{\partial^2 \overline{\varphi}}{\partial x \partial z} \\ \frac{\partial^2 \overline{\varphi}}{\partial z \partial x} & \frac{\partial^2 \overline{\varphi}}{\partial z^2} \end{pmatrix}. \tag{3.25}$$

Using the fourth equation from (3.2) shows that the matrix $\mathbf{Q}$ that appears in (3.25) is a two-dimensional version of the matrix $\mathbf{Q}$ appearing in (3.6). Thus, for this case, the condition for parcel stability is that $\mathbf{Q}$ is positive definite, which is exactly the condition for (3.4) to be elliptic and therefore solvable.

It was shown by Hoskins (1975) that the three-dimensional semi-geostrophic equations (3.2) are exactly those for the straight flow case but with the flow at an arbitrary direction to the coordinate axes. Thus the three-dimensional matrix $\mathbf{Q}$ given in (3.6) governs the parcel stability of a straight flow in an arbitrary direction to the coordinate axes.

The conditions under which the perturbation pressure can be neglected are discussed fully in Shutts and Cullen (1987). It is first assumed that the response to the perturbation is hydrostatic. As discussed in section 2.2, this will be true in regimes where semi-geostrophic theory is relevant. The perturbation pressure $\delta\varphi$ is then generated by the perturbation $\delta\theta$ to the potential temperature. If the displacement has a vertical scale H, then $\delta\varphi$ will be of order $gH\delta\theta/\theta_0 = H^2N^2$, where N is the Brunt-Väisälä frequency. Thus the perturbation to the horizontal pressure gradient will be of order N^2H^2/L, where L is the horizontal scale of the parcel. The parcel stability analysis assumes that this is small compared to the change in fv given by (3.22) which is of order f^2L. Thus the condition is that the aspect ratio of the displacement is much less than f/N. This is the same condition as the flow itself has to satisfy for semi-geostrophic theory to be accurate, as discussed in section 2.4.6. Since displacements can be chosen to satisfy the condition for the analysis to be valid, the parcel stability condition is necessary for stability; but as other displacements could also be chosen, it is not sufficient.

It was shown in section 2.5 that semi-geostrophic theory was accurate if the Lagrangian Rossby number (2.32) was small. This condition requires $\frac{\mathrm{D}}{\mathrm{D}t} \ll f$, so that for displacements of small aspect ratio as required above, (3.25) shows that there will be a separation between the time-scales of the perturbation and the time-scale on which the basic state evolves. The parcel stability condition for straight flow in the y direction is equivalent to the symmetric stability condition of Bennetts and Hoskins (1979). For purely horizontal motion in the y direction, it reduces to the inertial stability condition given by the positivity of the corresponding diagonal element of $\mathbf{Q}$:

$$f^2 + \frac{\partial^2 \varphi}{\partial x^2} \geq 0. \tag{3.26}$$

For purely vertical motion it reduces to the static stability condition $\partial^2\varphi/\partial z^2 \geq 0$, which is equivalent to $\partial\theta/\partial z \geq 0$. A different form of the inertial stability condition applies to axisymmetric flows, as shown in section 4.4. For two-dimensional horizontal flows, the condition that $\mathbf{Q}$ is positive definite is a form of inertial stability condition, but is only physically relevant under the conditions discussed above. Other forms of the condition are discussed by McWilliams *et al.* (1999).

The analysis above leads to the *stability principle* which is imposed on solutions of the semi-geostrophic equations (3.2).

Definition 3.1. An *admissible* solution of the semi-geostrophic equations (3.2) on a region Γ is one that is characterised by a function $\varphi(\mathbf{x}, t)$ whose evolution satisfies (3.4) in a suitable sense and where the matrix $\mathbf{Q}$ calculated from φ using (3.6) is positive definite at each t.

In the rest of the chapter it is shown that existence of such solutions can be proved under appropriate assumptions. The results do not preclude the existence of additional solutions not satisfying the condition on $\mathbf{Q}$. However, such solutions are ignored as unphysical. This is analogous to the rejection of solutions of the equations of gas dynamics containing expansion shocks on the grounds that these are entropy reducing and so unphysical.

3.2 Solution as a mass transport problem

3.2.1 *Solution by change of variables*

The analysis in section 3.1 suggests that robust solvability of equations (3.2) is plausible since it depends on the positive definiteness of $\mathbf{Q}$ and det$\mathbf{Q}$ is a constant of the motion. Rigorous arguments to this effect are developed in this and subsequent sections. The extra rigour is non-trivial, since positivity of det$\mathbf{Q}$ is not sufficient for positive definiteness of $\mathbf{Q}$. In addition, it turns out that the solutions can be discontinuous in space. This invalidates the derivations leading to equation (3.4) but can be reconciled with the derivation of (3.2).

The first step is to use the change of variables introduced by Hoskins (1975). Set

$$X = f^{-1}v_g + x,\ Y = -f^{-1}u_g + y,\ Z = g\theta/(f^2\theta_0). \tag{3.27}$$

Defining P by

$$P = \frac{1}{2}(x^2 + y^2) + f^{-2}\varphi, \tag{3.28}$$

then

$$(X, Y, Z) = \nabla P. \tag{3.29}$$

Equations (3.2) can then be written as

$$\begin{aligned} \frac{\mathrm{D}X}{\mathrm{D}t} &= u_g, \\ \frac{\mathrm{D}Y}{\mathrm{D}t} &= v_g, \\ \frac{\mathrm{D}Z}{\mathrm{D}t} &= 0, \\ \nabla \cdot \mathbf{u} &= 0. \end{aligned} \tag{3.30}$$

Equations (3.29) and (3.30) are a system of seven equations for the unknowns (X, Y, Z, P, u, v, w). It was noted by Hoskins that this change of variables made the trajectory $\mathbf{u}$ implicit, while the geostrophic wind appears explicitly on the right hand side of (3.30). As noted after (3.6), the defintion (3.28) means that $\mathbf{Q}$ is the Hessian matrix of f^2P. Thus positive definiteness of $\mathbf{Q}$ is equivalent to convexity of P.

Now describe the energy minimisation property derived in section 3.1.2 in these variables. Associate with each particle a vector field $(\tilde{X}, \tilde{Y}, \tilde{Z})$, and that given this field the energy of the system is defined by

$$\tilde{E} = \int_{\Gamma} f^2 \left(\frac{1}{2} \left((x - \tilde{X})^2 + (y - \tilde{Y})^2 \right) - z\tilde{Z} \right) \mathrm{d}x\mathrm{d}y\mathrm{d}z. \tag{3.31}$$

Theorems 3.1 and 3.2 can then be rewritten as follows:

Theorem 3.3. *The conditions for the energy $\tilde{E}$ defined by (3.31) to be stationary with respect to variations Ξ of particle positions satisfying continuity $\delta(\mathrm{d}x\mathrm{d}y\mathrm{d}z) = 0$ via*

$$\nabla \cdot \Xi = 0 \tag{3.32}$$

in Γ and

$$\delta\tilde{X} = \delta\tilde{Y} = \delta\tilde{Z} = 0, \tag{3.33}$$

together with $\Xi \cdot \mathbf{n} = 0$ on the boundary of Γ, are that

$$(\tilde{X}, \tilde{Y}, \tilde{Z}) = \nabla\tilde{P} \tag{3.34}$$

for some scalar $\tilde{P}$. The condition for $\tilde{E}$ to be minimised with respect to this class of variations is that $\tilde{P}$ is convex.

Proof The proof of the first statement is identical to the proof of Theorem 3.1 with the substitutions suggested by (3.27), namely $\tilde{X} = f^{-1}\tilde{v} + \tilde{x}$, $\tilde{Y} = -f^{-1}\tilde{u} + \tilde{y}$, $\tilde{Z} = g\tilde{\theta}/(f^2\theta_0)$, where $\tilde{x}, \tilde{y}$ are the coordinates of the particle positions. Theorem 3.2 and the substitution $\tilde{P} = \frac{1}{2}(x^2 + y^2) + f^{-2}\tilde{\varphi}$ show that the Hessian of $\tilde{P}$ is positive definite. Positive definiteness of the Hessian implies convexity. □

Now update Definition 3.1 to apply to this form of the equations:

Definition 3.2. An *admissible* solution of the semi-geostrophic equations (3.29), (3.30) on a region Γ is one that is characterised by a convex function $P(\mathbf{x}, t)$ whose gradient $(X(\mathbf{x}, t), Y(\mathbf{x}, t), Z(\mathbf{x}, t))$ satisfies (3.30) in a suitable sense.

This definition is called the *convexity* principle.

The class of variations used in Theorem 3.3 can be described as *rearrangements* of $\tilde{X}$, $\tilde{Y}$ and $\tilde{Z}$ viewed as functions of (x, y, z). The theory of rearrangements is reviewed in this context by Douglas (2002). Formal definitions will be given in section 3.5.2. If the energy minimisation problem of Theorem 3.3 can be solved uniquely, the solution of (3.29) and (3.30) can be viewed as constructing a sequence of energy minimising states, with X, Y and Z evolving in time according to (3.30). The velocity (u, v, w) defines a trajectory which performs the rearrangement required to maintain the energy minimisation property. If the velocity is smooth enough, the incompressibility condition in equations (3.30) means that the trajectory takes the form of a *measure-preserving mapping* from positions of fluid particles at a given time to the positions at a later time. This means that the volume of any subset of the fluid is conserved in time, though it may become highly distorted. It will be shown in sections 3.5.2 and 3.5.4 that solutions can indeed be defined in this way.

3.2.2 *The equations in dual variables*

In this section the definitions of dependent and independent variables are interchanged, so that instead of $\mathbf{X} \equiv (X, Y, Z)$ being a function of (x, y, z), $\mathbf{x} \equiv (x, y, z)$ is regarded as a function of (X, Y, Z). The right hand sides of the first three equations of (3.30) now define a 'velocity' in (X, Y, Z) space. This interpretation was noted by Hoskins. However, he did not transform the z coordinate. His *geostrophic* coordinates are (X, Y, Z_g) with $Z_g = z$. The relation between his choice and the present one is discussed by Chynoweth and Sewell (1989). Rewrite equations (3.29) and (3.30) as a set of equations using (X, Y, Z) as independent variables. noting that equations (3.27) give $u_g = f(y - Y), v_g = f(X - x)$. Write the first three equations of (3.30) as

$$\frac{\mathrm{D}X}{\mathrm{D}t} = U, \quad \frac{\mathrm{D}Y}{\mathrm{D}t} = V, \quad \frac{\mathrm{D}Z}{\mathrm{D}t} = W, \tag{3.35}$$

where

$$(U, V, W) = (f(y - Y), f(X - x), 0). \tag{3.36}$$

Consider $\mathbf{U} \equiv (U, V, W)$ also to be a function of (X, Y, Z), then (3.35) means that $\mathbf{U}$ defines a velocity in (X, Y, Z) space. In order to calculate the velocity, it is necessary to calculate (x, y, z) as a function of (X, Y, Z). To do this, define

$$\begin{aligned} R(X, Y, Z) = {} & x(X, Y, Z)X + y(X, Y, Z)Y + z(X, Y, Z)Z \\ & - P(x(X, Y, Z), y(X, Y, Z), z(X, Y, Z)). \end{aligned} \tag{3.37}$$

Then

$$\frac{\partial R}{\partial X} = \frac{\partial x}{\partial X}X + x + \frac{\partial y}{\partial X}Y + \frac{\partial z}{\partial X}Z - \frac{\partial P}{\partial x}\frac{\partial x}{\partial X} - \frac{\partial P}{\partial y}\frac{\partial y}{\partial X} - \frac{\partial P}{\partial z}\frac{\partial z}{\partial X}. \tag{3.38}$$

Using (3.29) then gives

$$\frac{\partial R}{\partial X} = \frac{\partial x}{\partial X}X + x + \frac{\partial y}{\partial X}Y + \frac{\partial z}{\partial X}Z - X\frac{\partial x}{\partial X} - Y\frac{\partial y}{\partial X} - Z\frac{\partial z}{\partial X} = x. \quad (3.39)$$

Similar calculations for the Y and Z components give

$$\nabla R = (x, y, z). \quad (3.40)$$

This gives a characterisation of (x, y, z) as a function of (X, Y, Z), but to solve the equations it is necessary to calculate R.

The first step in doing this is to calculate $\nabla \cdot \mathbf{U}$. (3.36) gives

$$\begin{aligned} \nabla \cdot \mathbf{U} &= \frac{\partial f(y-Y)}{\partial X} + \frac{\partial f(X-x)}{\partial Y}, \\ &= f\frac{\partial^2 R}{\partial Y \partial X} - f\frac{\partial^2 R}{\partial X \partial Y} = 0. \end{aligned} \quad (3.41)$$

The next step is to note that, according to the convexity principle (Definition 3.2), it is necessary to find solutions with P convex. Equation (3.37) is exactly the statement that P and R are *Legendre transforms.* (3.37) is called the *duality relation.* A full discussion of this transformation and its use in the theory of convex functions is given in Rockafellar (1970) and Sewell (2002). In particular, the Legendre transform of a convex function is also convex, so that R is a convex function of (X, Y, Z). The effect of the convexity of P is that X is a monotone function of x, Y of y and Z of z, and similarly x is a monotone function of X, y of Y and z of Z. The coordinate transformation between (x, y, z) and (X, Y, Z) therefore makes sense.

Next define a *potential density* in (X, Y, Z) coordinates. This concept was introduced by Schubert and Magnusdottir (1994). Since the equations being used in physical space are incompressible, this is defined as the volume in physical space associated with a given volume in (X, Y, Z) space. Therefore the potential density σ is given by

$$\sigma \equiv \frac{\partial(x, y, z)}{\partial(X, Y, Z)} = \det \begin{pmatrix} \frac{\partial^2 R}{\partial X^2} & \frac{\partial^2 R}{\partial X \partial Y} & \frac{\partial^2 R}{\partial X \partial Z} \\ \frac{\partial^2 R}{\partial Y \partial X} & \frac{\partial^2 R}{\partial Y^2} & \frac{\partial^2 R}{\partial Y \partial Z} \\ \frac{\partial^2 R}{\partial Z \partial X} & \frac{\partial^2 R}{\partial Z \partial Y} & \frac{\partial^2 R}{\partial Z^2} \end{pmatrix}. \quad (3.42)$$

For a given $\sigma(X, Y, Z)$, equation (3.42) is a Monge-Ampère equation for R. The appropriate boundary condition to use is that $(x, y, z) = \nabla R \in \Gamma$. This expresses the fact that all the fluid has to stay within Γ. This form of boundary condition is standard in the theory of Monge-Ampère equations, and is called the 'second boundary value problem', Pogorelov (1964). There are a number of proofs that the Monge-Ampère equation can be solved with this boundary condition. Some of these results are described and used in sections 3.4 and 3.5. The definition of σ and the boundary condition means that the integral of σ over all (X, Y, Z) in $\mathbb{R}^3$ must be equal to the volume of Γ.

For any choice of velocity $\mathbf{U}(X, Y, Z)$, conservation of volume in physical space means that

$$\frac{\partial \sigma}{\partial t} + \nabla \cdot (\sigma \mathbf{U}) = 0. \tag{3.43}$$

In the present case, equation (3.41) means that

$$\frac{\partial \sigma}{\partial t} + \mathbf{U} \cdot \nabla \sigma = 0. \tag{3.44}$$

The full system (3.29) and (3.30) can then be rewritten as a set of equations in *dual variables*:

$$\begin{aligned} \frac{\partial \sigma}{\partial t} + \mathbf{U} \cdot \nabla \sigma &= 0, \\ \det(\text{Hess } R) &= \sigma, \\ (x, y, z) &= \nabla R, \\ (U, V, W) &= (f(y - Y), f(X - x), 0). \end{aligned} \tag{3.45}$$

This is a set of eight equations for the unknowns $(U, V, W, x, y, z, R, \sigma)$. Given initial data for σ, R can be found from the second equation, (x, y, z) from the third, and (U, V, W) from the fourth. The first equation can then be used to advance σ in time. Solvability requires that the integral of σ at $t = 0$ is equal to the volume of Γ. The only boundary conditions required are that $(x, y, z) = \nabla R \in \Gamma$. These are compatible with the condition $\mathbf{u} \cdot \mathbf{n} = 0$ stated with equation (3.2) and thus with the conservation of energy. No boundary conditions have to be imposed on the pressure or geostrophic wind. However, the problem is a free boundary problem in (X, Y, Z) space. If σ at $t = 0$ is only non-zero over a finite subset $\Sigma(0)$ of (X, Y, Z) space, the region $\Sigma(t)$ over which it is non-zero at later times will vary and can only be determined after the equations have been solved.

Some concepts of formal analysis are needed to discuss the solutions further. Brief definitions are given here. More detailed background can be found in the textbook by Halmos (1950). The *support* of a function σ on $\mathbb{R}^3$ is the set where it is non-zero. It is called *compact* if it is closed and bounded. It is called a *probability measure* if it is non-negative and integrates to 1. Since the total mass in the problem is usually fixed, this definition can be satisfied by normalising the total mass to 1. The *measure* of a set is a positive function which measures the size of the set in some sense, for instance area, volume or mass. In particular, the standard Lebesgue measure (essentially area on $\mathbb{R}^2$, volume on $\mathbb{R}^3$) is denoted by $\mathcal{L}$.

The support of σ in (X, Y, Z) space is considered to be a subset Σ of $\mathbb{R}^3$. The definition (3.42) of the potential density σ can be written as $\sigma \mathrm{d}X \mathrm{d}Y \mathrm{d}Z = \mathrm{d}x \mathrm{d}y \mathrm{d}z$.

Given a probability measure $\sigma(\mathbf{X})$, define the measure of a set A as $\sigma(A) = \int_A \sigma \mathrm{d}\mathbf{X}$. Given a mapping $\mathbf{s}$ from $\mathbb{R}^3$ to Γ, it is then natural to define the potential density associated with $\mathbf{s}$ in terms of the ratio of the measure of a set to its image under $\mathbf{s}$ as follows:

Definition 3.3. A mapping $\mathbf{s} : \mathbb{R}^m \to \mathbb{R}^n$ *pushes forward* a probability measure μ on $\mathbb{R}^m$ to another probability measure ν on $\mathbb{R}^n$ if, for any set B in $\mathbb{R}^n$ with measure $\nu(B)$, $\mathbf{s}^{-1}(\mathbf{x}) \in \mathbb{R}^m$ is calculated for every point $\mathbf{x}$ in B, then the total measure of these points in $\mathbb{R}^m$ is $\mu(B)$. This property is written as $\mathbf{s}_{\#}\mu = \nu$.

This definition is closely associated with the ideas of rearrangements and measure-preserving mappings introduced in section 3.2.1. The potential density σ (3.42) associated with a mapping $\mathbf{s}$ from $\mathbb{R}^3$ to Γ satisfies

$$\mathbf{s}_{\#}\sigma = \mathcal{L}_\Gamma, \tag{3.46}$$

where $\mathcal{L}_\Gamma$ is the Lebesgue measure on Γ.

Rewrite Theorem 3.3 in terms of these mappings. Let S be the set of all mappings $\mathbf{s}$ from $\mathbb{R}^3$ to Γ satisfying (3.46) with potential density σ. Given a mapping $\mathbf{s} \in S$, define the energy integral

$$E_s = \int_{\mathbb{R}^3} f^2 \left(\frac{1}{2} \left((\tilde{x} - X)^2 + (\tilde{y} - Y)^2 \right) - \tilde{z} Z \right) \sigma \mathrm{d}X \mathrm{d}Y \mathrm{d}Z, \tag{3.47}$$

where $(\tilde{x}, \tilde{y}, \tilde{z}) = \mathbf{s}(X, Y, Z)$. Theorem 3.3 then becomes

Theorem 3.4. *Given $\sigma : \mathbb{R}^3 \to \mathbb{R}^+$, with $\int_{\mathbb{R}^3} \sigma \mathrm{d}X \mathrm{d}Y \mathrm{d}Z = \mathcal{L}(\Gamma)$, the condition for the energy E_s to be minimised for maps $\mathbf{s} \in S$ is that*

$$\mathbf{s}(X, Y, Z) = \nabla \tilde{R}, \tag{3.48}$$

where $\tilde{R}$ is convex. Such a minimising map, if it exists, is called an optimal map and written as $\mathbf{t} : \mathbb{R}^3 \to \Gamma$.

Proof First note that the variations defined in the theorem are the same as the variations used in Theorem 3.3. Given a map $\mathbf{s} \in S$, identify fluid particles with fixed positions $\mathbf{X}$ in $\mathbb{R}^3$. Any variations in $\mathbf{s}$ generate variations Ξ of the position of the particles in Γ. (3.46) means that these variations will not be able to change the volume that the particles occupy within Γ, and all particles have to remain inside Γ. Therefore Ξ must satisfy (3.32) and (3.33). Conversely, variations satisfying (3.32) and (3.33) clearly generate a map $\mathbf{s} \in S$. Theorem 3.3 then gives the condition for a stationary point to be $\mathbf{X}(x, y, z) = \nabla P(x, y, z)$ and for a minimiser that P is convex. Equations (3.37) and (3.40) are then used to define $\tilde{R}$ with $\tilde{R}$ convex. $\mathbf{s}(\mathbf{X})$ is then defined to be equal to $\nabla \tilde{R}$. □

A rigorous version of this theorem due to Brenier (1991), which proves that there is a unique minimiser under appropriate assumptions, is given in section 3.5. The problem as stated in this theorem is an example of a *mass transport problem.* Given a mapping between two spaces, the energy (3.47) is regarded as a 'cost'. This has to be minimised subject to the constraint that the mapping pushes forward a given probability measure μ to another probability measure ν. In the present case, the potential density σ defined in (3.42) is one probability measure and the constraint (3.46) states that $\mathbf{s}$ pushes σ forward to Lebesgue measure on Γ. This type of problem was first stated by Monge (1781) in a military context, but has been found since in many other contexts, particularly probability and statistics. A review of applications and analysis of these problems is given in the book by Villani (2003). A direct proof that such a minimiser exists is not practicable because of the nonlinearity of the constraint (3.46). The successful proof was achieved by Kantorovich (1942), who showed that the problem can be shown to be equivalent to a linear programming problem. This reformulation is discussed in the next subsection.

Finally Definition 3.2 is updated to apply to the equations in dual variables.

Definition 3.4. An admissible solution of the semi-geostrophic equations (3.45) in dual variables is one that is characterised by a convex function $R(\mathbf{X}, t)$, where $\det(\mathrm{Hess}(R))$ satisfies (3.43) in a suitable sense, with $\mathbf{U}$ given by (3.45).

3.2.3 *Solution using optimal transport*

In this section some important properties of semi-geostrophic solutions are derived, exploiting the Legendre duality defined by equation (3.37). These are discussed much more fully in Purser and Cullen (1987), and in more general contexts in Sewell (2002). The reformulation of the energy minimisation problem as a linear programming problem developed by Kantorovich (1942) is also described. This is often referred to as the *dual problem* and is essential in proving that the dual variable formulation of the semi-geostrophic equations can be solved.

First illustrate the meaning of the Legendre transform. It exploits the fact that a convex surface P can be regarded as the intersection of its tangent planes. Regard the surface as being given by the equation

$$p = P(x, y, z) \tag{3.49}$$

in $\mathbb{R}^4$ with coordinates (x, y, z, p). The convention, following Pogorelov (1964), is that if P is a convex function of (x, y, z), then the surface defined by equation (3.49) is called convex in the direction $p < 0$. Since $(X, Y, Z) = \nabla P$, the equation of a typical tangent plane can be written as

$$p = -R + xX + yY + zZ \tag{3.50}$$

where R is a constant. This is consistent with the definition of R in equation (3.37), so $-R(X, Y, Z)$ can be interpreted as the point where the tangent plane to p with gradient (X, Y, Z) intersects the hyperplane $(x, y, z) = 0$. This is illustrated for a one-dimensional case in Fig. 3.1.

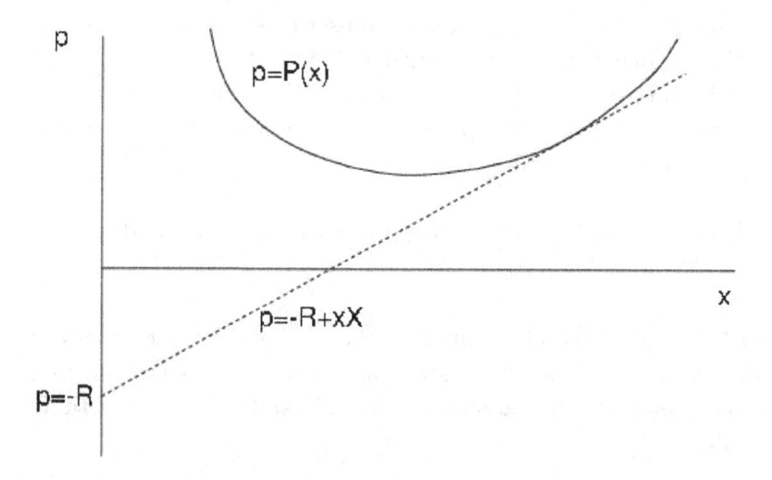

Fig. 3.1 The construction of the Legendre transform $R(X)$ of a convex surface $P(x)$.

Next write down the definition of a convex surface as the intersection of its tangent planes. This means that P takes the largest value of p at each (x, y, z) associated with any tangent plane, so that

$$P(x, y, z) = \sup_{(X,Y,Z)}(-R(X, Y, Z) + xX + yY + zZ). \tag{3.51}$$

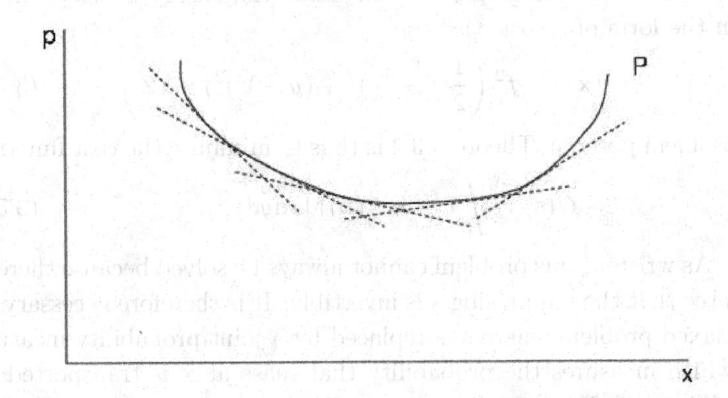

Fig. 3.2 The convex surface P as an intersection of its tangent planes.

This is illustrated in Fig. 3.2.

Equation (3.51) implies that, for any (x, y, z) and (X, Y, Z),

$$P(x,y,z) + R(X,Y,Z) \geq xX + yY + zZ. \qquad (3.52)$$

Now use equation (3.28) to replace P by φ. Similarly, $\Psi(X,Y,Z)$ can be defined by

$$\Psi = f^2\left(\frac{1}{2}(X^2+Y^2) - R\right). \qquad (3.53)$$

Using equation (3.36) gives

$$(U,V,W) = f^{-1}\left(-\frac{\partial \Psi}{\partial Y}, \frac{\partial \Psi}{\partial X}, 0\right). \qquad (3.54)$$

Thus Ψ is a stream-function for the flow in (X,Y,Z) coordinates in the same sense that φ is a stream-function for the geostrophic wind in (x,y,z) coordinates. Using (3.53) and (3.28) in (3.52) gives

$$-\varphi(x,y,z) + \Psi(X,Y,Z) \leq \frac{1}{2}f^2\left((x-X)^2 + (y-Y)^2 - zZ\right). \qquad (3.55)$$

The right hand side of this equation is now recognisable as the energy density which is the integrand of the energy (3.47). Equation (3.55) is fundamental to proving that the minimiser characterised in Theorem 3.4 exists.

Next exploit this identification by making some further definitions. Define a functional called the *cost function* depending on $\mathbf{x}$ and $\mathbf{X}$ equal to the energy integral. Assume, as in section 3.2.2, that $\mathbf{X} = \mathbf{s}^{-1}(\mathbf{x})$, and that the image of Γ under the map $\mathbf{s}^{-1}$ is Σ. Then the energy density can be written in the form of a cost c as

$$c(\mathbf{x}, \mathbf{s}^{-1}(\mathbf{x})) = f^2 \left(\frac{1}{2}(x - X)^2 + (y - Y)^2) - zZ \right). \tag{3.56}$$

The problem posed in Theorem 3.4 is thus to minimise the cost function

$$C(\mathbf{s}) = \int_\Gamma c(\mathbf{x}, \mathbf{s}^{-1}(\mathbf{x}))\mathrm{d}x\mathrm{d}y\mathrm{d}z, \tag{3.57}$$

for $\mathbf{s} \in S$. As written, this problem cannot always be solved because there is no guarantee that the minimising $\mathbf{s}$ is invertible. It is therefore necessary to solve a relaxed problem where $\mathbf{s}$ is replaced by a joint probability measure $\gamma(\mathbf{x}, \mathbf{X})$ which measures the probability that mass at $\mathbf{x}$ is transported to $\mathbf{X}$. Integrating $\gamma(\mathbf{x}, \mathbf{X})$ over all $\mathbf{x}$ gives $\sigma(\mathbf{X})$ and integrating $\gamma(\mathbf{x}, \mathbf{X})$ over all $\mathbf{X}$ gives Lebesgue measure on Γ. This condition is stated as that the *marginals* of γ are $\mathcal{L}$ and σ.

Where there is an invertible map $\mathbf{s}$, this means that

$$\gamma(\mathbf{x}, \mathbf{X}) = \sigma(\mathbf{X})\delta(\mathbf{x} - \mathbf{s}(\mathbf{X})). \tag{3.58}$$

Then the relaxed version of (3.57) is to minimise

$$C(\gamma) = \int_{\Gamma \times \Sigma} c(\mathbf{x}, \mathbf{X})\gamma(\mathbf{x}, \mathbf{X})\mathrm{d}\mathbf{x}\mathrm{d}\mathbf{X}, \tag{3.59}$$

over all $\gamma \in \mathbb{P}$ where $\mathbb{P}$ is the set of joint probability measures γ such that $\gamma(A \times \Sigma) = \mathcal{L}^3(A)$ for all $A \subset \Gamma$ and $\gamma(\Gamma \times B) = \sigma(B)$ for all $B \subset \Sigma$. This is called a *Monge-Kantorovich* problem.

This is too hard to solve directly, so Kantorovich (1942) posed a dual problem and proved

Theorem 3.5. *Given potentials* $\varphi(\mathbf{x}), \Psi(\mathbf{X})$, *define*

$$J(\varphi, \Psi) = \int_\Gamma -\varphi \mathrm{d}\mathbf{x} + \int_\Sigma \Psi\sigma \mathrm{d}\mathbf{X}, \tag{3.60}$$

$$I(\gamma) = \int_{\Gamma \times \Sigma} c(\mathbf{x}, \mathbf{X})\gamma(\mathbf{x}, \mathbf{X})\mathrm{d}\mathbf{x}\mathrm{d}\mathbf{X}.$$

Let Φ_C *be the set of* φ, Ψ *such that*

$$-\varphi(\mathbf{x}) + \Psi(\mathbf{X}) \leq c(\mathbf{x}, \mathbf{X}). \tag{3.61}$$

This is exactly the relation (3.55). Then

$$\inf_{\mathbb{P}(\mathcal{L}^3, \sigma)} I(\gamma) = \sup_{\Phi_C} J(\varphi, \Psi). \tag{3.62}$$

This shows the equivalence of the minimisation of $C(\mathbf{s})$ as defined in (3.57) and the maximisation of Φ_C under the constraint (3.61). The latter is a linear programming problem. It can be solved under certain technical conditions as in Theorem 3.15 in section 3.5.2 and Villani (2003), Theorem 1.3.

At the maximiser, (3.61) gives

$$\Psi(\mathbf{X}) - \varphi(\mathbf{x}) = c(\mathbf{X}, \mathbf{x}). \tag{3.63}$$

This is a form of duality relation which is another way of viewing Legendre duality as expressed in (3.51). It is discussed in more detail in Sewell (2002), p.162.

Differentiating (3.63) with respect to $\mathbf{x}$ and using the definition (3.56) of c gives

$$-\nabla\varphi(\mathbf{x}) = f^2\left((x - X), (y - Y), -Z\right). \tag{3.64}$$

and differentiating (3.63) with respect to $\mathbf{X}$ gives

$$\nabla\Psi(\mathbf{X}) = -f^2\left((x - X), (y - Y), -z\right). \tag{3.65}$$

Equation (3.64), using (3.27), gives

$$\nabla\varphi = (fv_g, -fu_g, g\theta/\theta_0),$$

which are the geostrophic and hydrostatic relations from (3.2). Equation (3.65), using (3.45), gives

$$f(U, V) = \left(-\frac{\partial\Psi}{\partial Y}, \frac{\partial\Psi}{\partial X}\right).$$

This identifies Ψ as the stream-function in (X, Y, Z) coordinates as in (3.54).

The convexity property of the minimiser can be deduced by defining the *c-transform* of a function Ψ on Σ as

$$\Psi^c(\mathbf{x}) = \inf_{\mathbf{X}\in\Sigma}\{c(\mathbf{X}, \mathbf{x}) - \Psi(\mathbf{X})\}. \tag{3.66}$$

Apply the equivalent definition to a function φ on Γ. Then

$$-\varphi^c(\mathbf{X}) = \inf_{\mathbf{x}\in\Gamma}\{c(\mathbf{X}, \mathbf{x}) + \varphi(\mathbf{x})\}. \tag{3.67}$$

At the solution of the dual problem in Theorem 3.5, both inequalities (3.66) and (3.67) will become equalities, so that (3.63) means that $\Psi^c = -\varphi$ and $-\varphi^c = \Psi$. Together these imply that $(\Psi^c)^c = \Psi$ and $(\varphi^c)^c = \varphi$. Then φ and Ψ are called *c-convex.*

In addition, these relations mean that $-\varphi$ and Ψ satisfy the respective inequalities (3.66) and (3.67). In the present case, with c defined by (3.56), Brenier (1991) proved under a non-degeneracy condition that a minimiser exists and is attained by a map $\mathbf{t}$ which is invertible. Since the cost (3.56) is the energy density, this corresponds to energy minimisation. Then, using (3.28) and (3.53) to replace φ and Ψ by P and R, (3.67) and (3.66) become

$$P(\mathbf{x}) = \sup_{\mathbf{X}\in\Sigma}\{\mathbf{x}\cdot\mathbf{X} - R(\mathbf{X})\}. \tag{3.68}$$

and

$$R(\mathbf{X}) = \sup_{\mathbf{x}\in\Gamma}\{\mathbf{x}\cdot\mathbf{X} - P(\mathbf{x})\}. \tag{3.69}$$

Note that (3.68) is exactly the same as the formula (3.51) for constructing a convex surface. Since at the solution of the dual problem, the suprema in (3.68) and (3.69) become equalities, the duality relation(3.37) is recovered, together with the convexity of P and R discussed in section 3.2.2. In particular, P and R are Legendre transforms.

This formulation covers more general cost functions than (3.56). Thus c-convexity can be used in contexts where the idea of a convex function cannot be applied, such as on the sphere in section 4.3.3 and where the cost c is more complicated, as in the compressible case treated in section 4.1. The convexity property is critical in proving that the equations can be integrated over a long time period, beacuse it prevents small-scale oscillations developing and allows sequences of approximate solutions to converge to a reasonable limit. Where there is no convex function, the c-convexity defined above can achieve the same results. This is essential in proving the rigorous theorems set out in section 3.5.

This derivation also illustrates the symmetry between the equations defined with (x, y, z) as independent variables and those using (X, Y, Z) as independent variables. This was exploited by Purser and Cullen (1987) to derive solutions for two different physical problems from any one solution. One aspect of this is that the potential density conservation law (3.44) can be inverted to give a conservation law for a dimensional form of the potential vorticity, which is the determinant of the matrix $\mathbf{Q}$ defined by (2.151). Thus the potential vorticity is the mass in (X, Y, Z) space associated with a region in physical space. This relates to the idea of potential vorticity 'substance' introduced by Haynes and McIntyre (1990). This has to be satisfied by the physical space equations (3.29) and (3.30). This takes the form

$$\mathcal{Q}_{sg} \equiv \frac{\partial(X, Y, Z)}{\partial(x, y, z)}, \tag{3.70}$$

$$\frac{\partial \mathcal{Q}_{sg}}{\partial t} + \mathbf{u}\cdot\nabla\mathcal{Q}_{sg} = 0.$$

This conservation law was derived by Hoskins (1975) directly from (3.2). Together with incompressibility, it implies mass conservation in (X, Y, Z) space as in (3.43).

The symmetry is, however, lost if the equations are solved in a closed region Γ. As noted in the previous subsection, if σ has compact support $\Sigma(t)$, the problem in (X, Y, Z) coordinates is a free boundary problem. The symmetry could be recovered if periodic boundary conditions were used in all three directions, but this does not correspond to a physically useful case.

An important property of the Legendre transform is that it maps the boundary of a convex set to the boundary of another convex set, as stated in the following theorem from Cullen and Purser (1984):

Theorem 3.6. *Given a domain Σ in (X, Y, Z) and a convex domain Γ in (x, y, z), and a Legendre transform between them generated by a convex function $P(x, y, z)$, all points on the convex hull of Σ correspond to points on the boundary of Γ.*

Proof Let $A = (X_A, Y_A, Z_A)$ be a point on the convex hull of Σ whose image (x_A, y_A, z_A) is strictly inside Γ. Let (a, b, c) be the outward unit normal to the convex hull of Σ at A. Then convexity of the convex hull means that no point with coordinates of the form

$$
\begin{aligned}
&(X_A, Y_A, Z_A) + (X', Y', Z'), \\
&(X', Y', Z') \cdot (a, b, c) > 0,
\end{aligned}
\tag{3.71}
$$

is in Σ, and therefore no such point has an image in Γ. However, by convexity of P, $\nabla P\left((x_A, y_A, z_A) + \alpha(a, b, c)\right) \cdot (a, b, c)$ is an increasing function of α. Since for some α the point $(x_A, y_A, z_A) + \alpha(a, b, c)$ lies inside Γ, this implies that it must have a gradient of the form given in equation (3.71), which is a contradiction. Thus any point on the boundary of Σ must map to a point on the boundary of Γ, as illustrated in Fig. 3.3. □

Rigorous results proving this and more general statements, due to Caffarelli, are described in section 3.5.2.

If $\Sigma(t)$ is not convex, then points on the boundary of Σ can map to the interior of Γ. An example is shown in Figure 3.4. This creates a situation where two separate points, marked A and B, on the boundary of the convex hull of Σ can map onto adjacent points of the boundary of Γ. The intermediate point C, on the boundary of Σ, but not on the convex hull, maps to the interior of Γ. Now consider the integral of the potential vorticity Q defined in equation (3.70) over Γ. If the integral is taken round the boundary of Γ, it will include the portion of the convex hull of Σ within

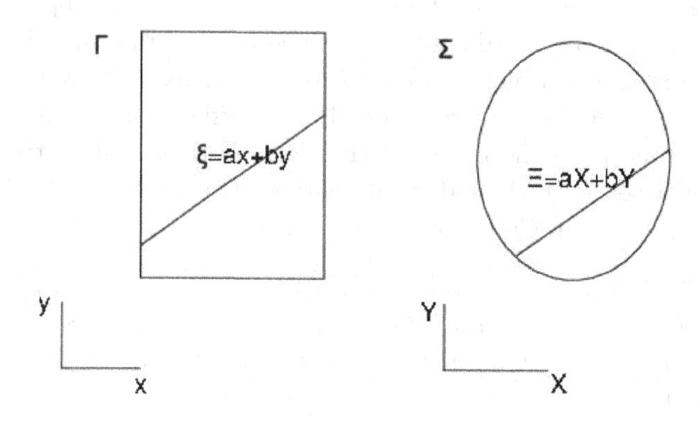

Fig. 3.3 The convex hull of Σ must map to the boundary of Γ.

the dashed line connecting A and B in Fig. 3.4. Thus the integral will be greater than the ratio of the volume of Σ to the volume of Γ. As time evolves, equation (3.41) shows that the volume of Σ is conserved. However, the shape of Σ will vary in time, and the volume of extra parts of (X, Y, Z) space included in the integral of $\mathcal{Q}$ may vary. Thus the integrated potential vorticity is not conserved, even though the Lagrangian conservation law (3.70) is satisfied. In effect, potential vorticity from the boundary has been sucked into the fluid, an interpretation first made by F.P.Bretherton. The physical interpretation of this is discussed in section 3.4.2 in the context of frontogenesis. Examples are given in sections 3.4.2 and 5.3.4.

3.3 The shallow water semi-geostrophic equations

3.3.1 *Solutions as minimum energy states*

This section describes the solutions of the shallow water semi-geostrophic equations. The same steps are followed as in sections 3.1 and 3.2, but only the main results are summarised.

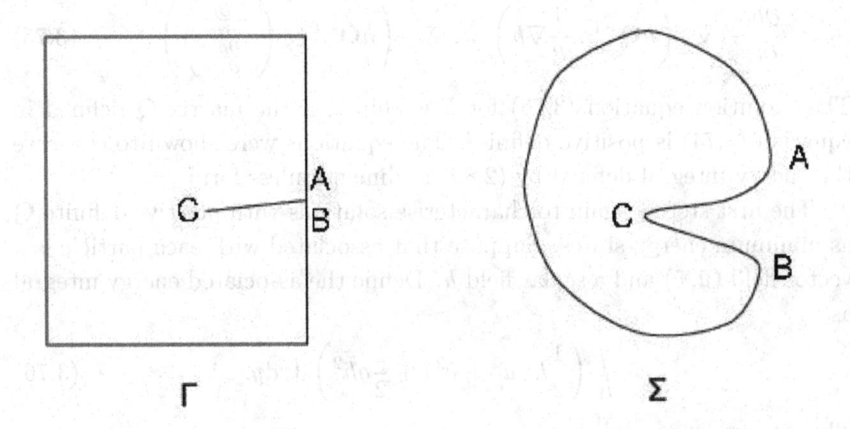

Fig. 3.4 Mapping a non-convex Σ to Γ. The points A, B and C indicate corresponding points in Γ and Σ.

The governing equations (2.87) and (2.88) in dimensional form are

$$g\frac{\partial h}{\partial x} - fv_g = 0, \qquad (3.72)$$
$$g\frac{\partial h}{\partial y} + fu_g = 0.$$

$$\frac{\partial u_g}{\partial t} + u\frac{\partial u_g}{\partial x} + v\frac{\partial u_g}{\partial y} + g\frac{\partial h}{\partial x} - fv = 0,$$
$$\frac{\partial v_g}{\partial t} + u\frac{\partial v_g}{\partial x} + v\frac{\partial v_g}{\partial y} + g\frac{\partial h}{\partial y} + fu = 0, \qquad (3.73)$$
$$\frac{\partial h}{\partial t} + \frac{\partial}{\partial x}(hu) + \frac{\partial}{\partial y}(hv) = 0.$$

In this subsection it will again be assumed that the Coriolis parameter f is constant. Analysis of the more physically relevant case where f is a function of position will be carried out in section 4.3. The rewritten equations (2.90) and (2.91), which are convenient for analysis, now become

$$\mathbf{Q}\begin{pmatrix} u \\ v \end{pmatrix} + g\frac{\partial}{\partial t}\nabla h = fg\begin{pmatrix} -\frac{\partial h}{\partial y} \\ \frac{\partial h}{\partial x} \end{pmatrix},$$
$$\mathbf{Q} = \begin{pmatrix} f^2 + f\frac{\partial v_g}{\partial x} & f\frac{\partial v_g}{\partial y} \\ -f\frac{\partial u_g}{\partial x} & f^2 - f\frac{\partial u_g}{\partial y} \end{pmatrix}. \qquad (3.74)$$

$$\frac{\partial h}{\partial t} - \nabla \cdot \left(h\mathbf{Q}^{-1} g \frac{\partial}{\partial t} \nabla h \right) = -\nabla \cdot \left(h\mathbf{Q}^{-1} f g \begin{pmatrix} -\frac{\partial h}{\partial y} \\ \frac{\partial h}{\partial x} \end{pmatrix} \right). \tag{3.75}$$

The evolution equation (3.75) for h is elliptic if the matrix $\mathbf{Q}$ defined in equation (3.74) is positive definite. The equations were shown to conserve the energy integral defined by (2.89) in dimensionless form.

The first step is again to characterise solutions with positive definite $\mathbf{Q}$ as minimum energy states. Suppose that associated with each particle is a vector field $(\tilde{u}, \tilde{v})$ and a scalar field $\tilde{h}$. Define the associated energy integral as

$$\int_\Gamma \left(\frac{1}{2} \tilde{h} \left(\tilde{u}^2 + \tilde{v}^2 \right) + \frac{1}{2} g \tilde{h}^2 \right) dx dy. \tag{3.76}$$

Then it can be proved that

Theorem 3.7. *The conditions for the energy E (3.76) to be stationary with respect to variations $\Xi = (\xi, \eta)$ of particle positions satisfying continuity $\delta(\tilde{h} dx dy) = 0$ via*

$$\delta \tilde{h} = -\tilde{h} \nabla \cdot \Xi \tag{3.77}$$

in Γ and

$$\delta \tilde{u} = f\eta, \ \delta \tilde{v} = -f\xi, \tag{3.78}$$

together with $\Xi \cdot \mathbf{n} = 0$ on the boundary of Γ, are that

$$(f\tilde{v}, -f\tilde{u}) = \nabla \tilde{h}. \tag{3.79}$$

The condition for the stationary point to be a minimum is that the matrix

$$\mathbf{Q} = \begin{pmatrix} f^2 + f\frac{\partial \tilde{v}}{\partial x} & f\frac{\partial \tilde{v}}{\partial y} \\ -f\frac{\partial \tilde{u}}{\partial x} & f^2 - f\frac{\partial \tilde{u}}{\partial y} \end{pmatrix}, \tag{3.80}$$

is positive definite.

Proof Write δ for a change following the particles, and ∂ for a change at a fixed position in space. The change to the energy resulting from the variation is then

$$\int_\Gamma \left(\tilde{h} \left(\tilde{u} \delta \tilde{u} + \tilde{v} \delta \tilde{v} \right) + g \tilde{h} \partial \tilde{h} \right) dx dy, \tag{3.81}$$

recalling that $\delta(\tilde{h} dx dy) = 0$. (3.77) can be rewritten in Eulerian form as

$$\partial \tilde{h} + \nabla \cdot (\tilde{h} \Xi) = 0. \tag{3.82}$$

Using this, and integrating by parts, gives

$$\int_\Gamma \left(\tilde{u}\delta\tilde{u} + \tilde{v}\delta\tilde{v} + g\Xi\cdot\nabla\tilde{h}\right)\tilde{h}dxdy. \tag{3.83}$$

Using (3.78) then gives

$$\int_\Gamma \Xi\cdot\left(-f\tilde{v} + g\frac{\partial\tilde{h}}{\partial x}, f\tilde{u} + g\frac{\partial\tilde{h}}{\partial y}\right)\tilde{h}dxdy. \tag{3.84}$$

Therefore the condition for E to be stationary is (3.79) as required. At a stationary point, set $(\tilde{u},\tilde{v},\tilde{h}) = (u_g, v_g, h_g)$, with

$$(fv_g, -fu_g) = g\nabla h_g. \tag{3.85}$$

Now take a second variation

$$\delta^2 E = \int_\Gamma \delta\left(f\Xi\cdot\left(-\tilde{v} + f^{-1}g\frac{\partial\tilde{h}}{\partial x}, \tilde{u} + f^{-1}g\frac{\partial\tilde{h}}{\partial y}\right)\right)\tilde{h}dxdy. \tag{3.86}$$

Since (3.79) holds at a stationary point, this reduces to

$$\int_\Gamma f\Xi\cdot\delta\left(-\tilde{v} + f^{-1}g\frac{\partial\tilde{h}}{\partial x}, \tilde{u} + f^{-1}g\frac{\partial\tilde{h}}{\partial y}\right)\tilde{h}dxdy. \tag{3.87}$$

Using (3.77) to substitute for $\delta\tilde{\mathbf{u}}$ gives

$$\int_\Gamma f\Xi\cdot\left((f\xi, f\eta) + \delta(f^{-1}g\nabla\tilde{h})\right)\tilde{h}dxdy. \tag{3.88}$$

Writing $\delta(f^{-1}g\nabla\tilde{h}) = \partial(f^{-1}g\nabla\tilde{h}) + \Xi\cdot\nabla(f^{-1}g\nabla\tilde{h})$, using (3.82) for $\partial\tilde{h}$, and using (3.85) gives $\delta(f^{-1}g\nabla\tilde{h}) = -f^{-1}g\nabla\nabla\cdot(\tilde{h}\Xi) + \Xi\cdot\nabla(\tilde{v}_g, -\tilde{u}_g)$. Substituting this into equation (3.88) and integrating by parts then gives

$$\int_\Gamma f\Xi\cdot((f\xi, f\eta) + \Xi\cdot(\tilde{v}_g, -\tilde{u}_g))\,\tilde{h} + g(\nabla\cdot(\tilde{h}\Xi))^2 dxdy. \tag{3.89}$$

The second term is positive definite. The energy will therefore be minimised if the first term is positive definite. It can be seen that it takes the form $\Xi\cdot\mathbf{Q}\cdot\Xi$ where $\mathbf{Q}$ is given by (3.80). Thus the energy will be minimised if $\mathbf{Q}$ is positive definite. □

Note that the variations used in Theorem 3.7 assume the perturbations to the depth $\tilde{h}$ do not affect $\delta\tilde{u}$ and $\delta\tilde{v}$. Recall the definition, (2.48), of the Rossby radius of deformation L_R. In dimensional terms, this is $\sqrt{gH}/f$ where H is the depth. The assumption above is justified if the scale of the variations is greater than L_R, because the change to $g\nabla\tilde{h} \simeq g\nabla H\nabla\cdot\Xi$ will be of order $gH\Xi/L^2$ and the change to $f\tilde{u}$ is of order $f^2\eta$. The ratio of these is less than L_R/L. Thus, as in the three-dimensional case, the energy minimisation principle is valid under the same conditions as the semi-geostrophic approximation. The form of (3.89) suggests that positive definiteness of $\mathbf{Q}$ might not be necessary for the stationary point to be a minimiser, only sufficient. However, it is shown in section 3.3.4 that this condition is actually necessary as well. Now state the convexity principle under which solutions of the shallow water semi-geostrophic equations (3.72) and (3.73) are sought.

Definition 3.5. An admissible solution of the semi-geostrophic shallow water equations (3.72) and (3.73) on a region Γ is one that is characterised by a depth $h(\mathbf{x}, t)$ whose evolution satisfies (3.75) in a suitable sense and where the matrix $\mathbf{Q}$ calculated from h using (3.72) and (3.73) is positive definite at each t.

Note that this definition implies that $gh + \frac{1}{2}f^2(x^2 + y^2)$ is convex.

Theorem 3.7 and Definition 3.5 both also hold if f is a function of position. This is because f does not have to be varied in a Lagrangian sense in the proof of Theorem 3.7. This will be exploited in section 4.3.

3.3.2 *Solution by change of variables*

Introduce the same new variables as in the three-dimensional case (3.27). Then (3.73) can be rewritten as

$$\begin{aligned}\frac{\mathrm{D}X}{\mathrm{D}t} &= u_g,\\ \frac{\mathrm{D}Y}{\mathrm{D}t} &= v_g,\\ \frac{\partial h}{\partial t} + \nabla \cdot (h\mathbf{u}) &= 0.\end{aligned} \tag{3.90}$$

Equation (3.72) becomes

$$\begin{aligned}(X, Y) &= \nabla P,\\ P &= \frac{1}{2}(x^2 + y^2) + f^{-2}gh.\end{aligned} \tag{3.91}$$

Now describe the energy minimisation property in these variables. Suppose that associated with each particle is a vector field $(\tilde{X}, \tilde{Y})$ and a scalar field $\tilde{h}$. Given these fields, the energy of the system is defined by

$$E = \int_\Gamma \left(\frac{1}{2}hf^2\left((x - \tilde{X})^2 + (y - \tilde{Y})^2\right) + \frac{1}{2}g\tilde{h}^2\right) dxdy. \tag{3.92}$$

Theorem 3.7 can then be rewritten as follows

Theorem 3.8. *The conditions for the energy E to be stationary with respect to variations Ξ of particle positions satisfying continuity $\delta(\tilde{h}dxdy) = 0$ via*

$$\delta\tilde{h} = -\tilde{h}\nabla \cdot \Xi \tag{3.93}$$

in Γ and

$$\delta\tilde{X} = \delta\tilde{Y} = 0, \tag{3.94}$$

together with $\Xi \cdot \mathbf{n} = 0$ on the boundary of Γ, are that

$$(\tilde{X}, \tilde{Y}) = \nabla\tilde{P}, \tag{3.95}$$

where $\tilde{P}$ is given by (3.91) with $h = \tilde{h}$. The condition for E to be minimised with respect to this class of variations is that $\tilde{P}$ is convex.

Proof The proof of the first statement is identical to the proof of the first part of Theorem 3.7 with the substitutions $\tilde{X} = f^{-1}\tilde{v} + \tilde{x}$, $\tilde{Y} = -f^{-1}\tilde{u} + \tilde{y}$, where $\tilde{x}$, $\tilde{y}$ are coordinates of particle positions. Use equation (3.85) to define values X_g, Y_g and h_g at the stationary point. The second statement then follows from the second part of Theorem 3.7, the definition (3.91) of $\tilde{P}$, and the fact that, in this case, the matrix $\mathbf{Q}$ defined in (3.74) can be rewritten in terms of h_g as

$$\mathbf{Q} = \begin{pmatrix} f^2 + g\frac{\partial^2 h_g}{\partial x^2} & g\frac{\partial^2 h_g}{\partial x \partial y} \\ g\frac{\partial^2 h_g}{\partial y \partial x} & f^2 + g\frac{\partial h_g}{\partial y^2} \end{pmatrix}. \square \tag{3.96}$$

Admissible solutions of the equations are still defined by Definition 3.5.

3.3.3 *The equations in dual variables*

In this subsection the definitions of dependent and independent variables are exchanged, as in section 3.2.3. The same manipulations can be carried out as in the three-dimensional case. In particular, equation (3.37) becomes

$$R(X,Y) = x(X,Y)X + y(X,Y)Y - P(x(X,Y), y(X,Y)). \tag{3.97}$$

From this it can be shown that $\nabla R = (x, y)$. Now define the potential density in (X, Y) coordinates as the mass in physical space associated with a given region of (X, Y) space.

$$\sigma \equiv \frac{h\partial(x,y)}{\partial(X,Y)} = \det\left\{ h \begin{pmatrix} \frac{\partial^2 R}{\partial X^2} & \frac{\partial^2 R}{\partial X \partial Y} \\ \frac{\partial^2 R}{\partial Y \partial X} & \frac{\partial^2 R}{\partial Y^2} \end{pmatrix} \right\}. \tag{3.98}$$

Assume σ is non-zero in a set $\Sigma \subset \mathbb{R}^2$. This definition implies that

$$\int_{\Sigma} \sigma \mathrm{d}X\mathrm{d}Y = \int_{\Gamma} h\mathrm{d}x\mathrm{d}y. \tag{3.99}$$

Both sides of this equation can be normalised to 1, so that σ and h are probability measures. It can again be shown that $\nabla \cdot \mathbf{U} = 0$ where $\mathbf{U} = (f(y-Y), f(X-x))$. Then rewrite (3.90) and (3.91) as a set of equations in dual variables:

$$\begin{aligned}
\frac{\partial \sigma}{\partial t} + \mathbf{U} \cdot \nabla \sigma &= 0, \\
(U, V) = (f(y-Y), f(X-x)&), \\
h\det(\text{Hess } R) &= \sigma, \\
(x, y) &= \nabla R, \\
\frac{1}{2}(x^2 + y^2) + f^{-2}gh + R &= xX + yY.
\end{aligned} \tag{3.100}$$

Now rewrite Theorem 3.8 in dual variables as in Theorem 3.4. The definition of the potential density, (3.98), can be rewritten as $\sigma \mathrm{d}X\mathrm{d}Y = h\mathrm{d}x\mathrm{d}y$. Then, using Definition 3.3,

$$\mathbf{s}_{\#}\sigma = h. \tag{3.101}$$

Since the problem is now being solved in (X, Y) coordinates, σ is determined by the first equation of (3.100) and h has to be found by constructing a map from (X, Y) to (x, y) and applying (3.98). Since h is unknown, this problem is not a mass transport problem.

Thus let S be the set of all mappings $\mathbf{s}$ from Σ to Γ. Any such mapping generates a depth h by (3.98). Associate with $\mathbf{s}$ and σ the energy E defined by

$$E = \int_{\Sigma} \frac{1}{2} f^2 |\mathbf{X} - \mathbf{s}(\mathbf{X})|^2 \sigma \mathrm{d}X\mathrm{d}Y + \int_{\Gamma} \frac{1}{2} gh^2 \mathrm{d}x\mathrm{d}y, \tag{3.102}$$

where $\mathbf{X} = (X, Y)$. Theorem 3.8 then becomes

Theorem 3.9. *Given $\sigma : \Sigma \to \mathbb{R}^+$, suppose that the energy E defined in (3.102) is minimised over maps $\mathbf{s} \in S$ by a map $\mathbf{t}$. Let h be the depth calculated by using $\mathbf{t}$ to evaluate (3.98). Then*

$$\mathbf{t}(\mathbf{X}) = \nabla R, \tag{3.103}$$

where R is convex and satisfies the last equation of (3.100). Such a minimising map, if it exists, is called an optimal map.

Proof As in the proof of Theorem 3.4, the result follows from identifying condition (3.101) with the variation $\delta(h\mathrm{d}x\mathrm{d}y) = 0$ used in Theorem 3.7. Then use Theorem 3.7 itself and, using the definition of R given in equation (3.97), the result follows. $\square$

3.3.4 *Solution using optimal transport*

Now describe the proof that such a map exists. The rigorous proof is given in Cullen and Gangbo (2001). Suppose that there is an h which minimises (3.102) for a given σ. Then the associated minimising map $\mathbf{t}$ satisfies (3.101). To characterise $\mathbf{t}$, define a set of maps $S_h \subset S$ such that if $\mathbf{s} \in S_h$, $\mathbf{s}_{\#}\sigma = h$. The second term in (3.102) depends only on h and not on the choice of $\mathbf{s} \in S_h$. Thus the minimising map $\mathbf{t}$ must minimise the cost function

$$\int_\Sigma c(\mathbf{x}, \mathbf{X}))\sigma \mathrm{d}X\mathrm{d}Y = \int_\Sigma f^2\left(\frac{1}{2}(x-X)^2 + (y-Y)^2)\right)\sigma \mathrm{d}X\mathrm{d}Y \quad (3.104)$$

over maps $\mathbf{S} \in S_h$. Finding an optimal map for this cost is a two-dimensional version of the problem solved in section 3.2.3. The minimising value of the cost function is called the *Wasserstein distance* between the probability measures σ and h, and written as $W_2(\sigma, h)$. This concept will be used again in section 3.5.3. The minimiser will be characterised, as in (3.64), by a potential φ such that

$$-\nabla\varphi(\mathbf{x}) = f^2((x-X), (y-Y)). \quad (3.105)$$

The energy E defined in (3.102) can now be written as

$$E = W_2(\sigma, h) + \int_\Gamma \frac{1}{2} gh^2 \mathrm{d}x\mathrm{d}y \quad (3.106)$$

Note that this now only depends on h for a given σ. Now seek to minimise E subject to the constraint $\mathbf{s}_{\#}\sigma = h$ given in (3.101).

It is shown in Cullen and Gangbo (2001) that the functional $W_2(\sigma, h)$ is weakly semi-continuous as a function of h. Since the second term in (3.106) is strictly convex in h, then E is strictly convex in h and therefore a unique minimiser exists. This will satisfy the requirement that E is minimised with respect to perturbations of h with σ fixed. The variations of h used in Theorem 3.8 satisfy $\delta(h\mathrm{d}x\mathrm{d}y) = 0$, which is consistent with the constancy of σ. Thus Theorem 3.8 characterises the minimising h, so in particular $\mathbf{X} = \nabla P$, with P convex and h is given from P by (3.91). Comparing with (3.105) shows that $\varphi = gh$, and so h will have the convexity properties deduced in section 3.2.3.

There is no reason why the fluid should occupy the whole of Γ. If the support Σ of σ is such that $x - X$ and $y - Y$ are large for $(x, y) \in \Gamma$, then the solution will have large $|\nabla h|$ for all $(x, y) \in \Gamma$. If the total available mass $\int_\Sigma \sigma \mathrm{d}X\mathrm{d}Y$ is small, then this cannot be achieved if the fluid fills the whole of Γ, as illustrated in Fig. 3.5.

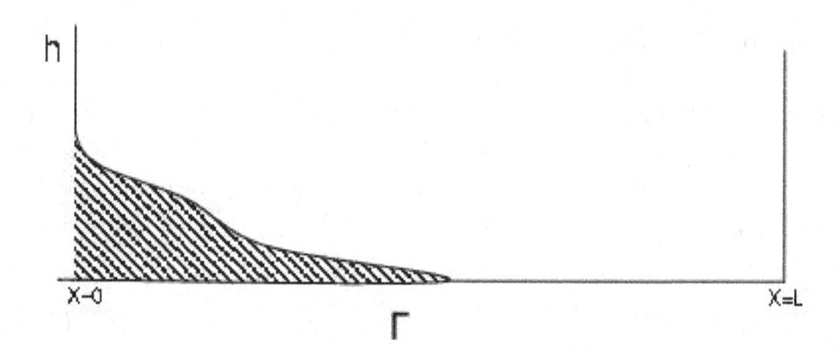

Fig. 3.5 A typical cross-section of the solution of the shallow water model showing that the water need not fill the whole of Γ.

It is not possible to determine whether the fluid fills Γ until the whole problem is solved. It is possible in principle for the fluid to fill Γ at $t = 0$ and then fail to fill it at a later time. The physical implications of this process, called 'outcropping', are discussed in section 3.4.3.

3.3.5 *The Boussinesq semi-geostrophic equations with a free surface*

This section considers the three-dimensional Boussinesq semi-geostrophic equations (3.2) with a free surface boundary condition replacing the rigid boundaries used in section 3.1.1. This would be an upper boundary condition in a model for the large-scale behaviour of the ocean. It is a lower boundary condition when applied to the atmosphere, as it then allows for variations of surface pressure. As noted in section 2.4.1, the variables have different meanings when applied to the atmosphere and ocean. In order to avoid unnecessary changes of notation, the following presentation retains the atmospheric notation. There is an extra evolution equation for the surface z, which implies using (2.117) an evolution equation for the surface pressure. This is similar to the third equation in (3.73). This problem was initially solved by Cullen *et al.* (2015b). Their result was extended by Cheng (2017) and Cullen *et al.* (2019) as discussed in section 3.5.

Equations (3.2) are solved in a domain $\Gamma_h \subset \mathbb{R}^3$ where

$$\Gamma_h = \Gamma \times [h(x, y, t), H_s]; \Gamma \subset \mathbb{R}^2, \tag{3.107}$$

where H_s is the scale height where the pressure is zero as in (2.117). The equations are supplemented by an evolution equation for h,

$$\frac{\partial h}{\partial t} + u\frac{\partial h}{\partial x} + v\frac{\partial h}{\partial y} = 0, \tag{3.108}$$

and the additional boundary condition that the geopotential $\varphi = 0$ at $z = h$. These equations make the non-trivial assumption that h is a single-valued function of (x, y). An important feature of the analysis is that a solution of the equations can be found so that this is the case.

The energy of this system for a notional state $(\tilde{u}, \tilde{v}, \tilde{\theta}, \tilde{h})$ is

$$\tilde{E} = \int_{\Gamma_{\tilde{h}}} \left(\frac{1}{2} (\tilde{u}^2 + \tilde{v}^2) - g\tilde{\theta}z/\theta_0 \right) \mathrm{d}x\mathrm{d}y\mathrm{d}z. \tag{3.109}$$

First generalise the identification of geostrophic and hydrostatic states with stationary points of the energy to the free surface case.

Theorem 3.10. *The conditions for the energy $\tilde{E}$ (3.109) to be stationary with respect to variations $\Xi = (\xi, \eta, \chi)$ of particle positions satisfying continuity $\delta(\tilde{h}\mathrm{d}x\mathrm{d}y\mathrm{d}z)$ via*

$$\nabla \cdot \Xi = 0, \quad \int_{\Gamma} \chi(x, y, \tilde{h})\mathrm{d}x\mathrm{d}y = 0 \tag{3.110}$$

in $\Gamma_{\tilde{h}}$ and

$$\delta\tilde{u} = f\eta,\ \delta\tilde{v} = -f\xi,\ \delta\tilde{\theta} = 0, \tag{3.111}$$

together with $\Xi \cdot \mathbf{n} = 0$ on the boundaries of $\Gamma_{\tilde{h}}$ other than $z = \tilde{h}$, and $\delta\tilde{h} = \chi$ are that

$$(f\tilde{v}, -f\tilde{u}, g\tilde{\theta}/\theta_0) = \nabla\tilde{\varphi}, \tag{3.112}$$

for some scalar $\tilde{\varphi}$ with $\tilde{\varphi} = 0$ at $z = \tilde{h}(x, y)$. Furthermore, $\tilde{h}$ satisfies

$$\begin{aligned} &f\tilde{v} + \frac{g\tilde{\theta}}{\theta_0}\frac{\partial\tilde{h}}{\partial x}, \\ &-f\tilde{u} + \frac{g\tilde{\theta}}{\theta_0}\frac{\partial\tilde{h}}{\partial y}. \end{aligned} \tag{3.113}$$

If the variations do not alter $\tilde{h}$, then the condition for the stationary point to be a minimum is that the matrix $\mathbf{Q}$ defined in 3.12 is positive definite. If the variations alter $\tilde{h}$, then any minimiser of the energy has a single-valued $\tilde{h}$.

Proof Write δ for a change following the particles, and ∂ for a change at a fixed position in space. The change to the energy resulting from the variation is then given by (3.11).

Integrating by parts gives the condition for E to be stationary as (3.112) together with $\varphi = 0$ at $z = \tilde{h}(x,y)$, as required. Note that the assumption that $\tilde{h}$ is not varied means that $\chi = 0$ at $z = \tilde{h}(x,y)$. The condition $\varphi = 0$ at $z = \tilde{h}(x,y)$ together with the relations $(f\tilde{v}, -f\tilde{u}) = \left(\frac{\partial\tilde{\varphi}}{\partial x}, \frac{\partial\tilde{\varphi}}{\partial y}\right)$ give (3.113).

The proof that the stationary point is a minimiser when $\tilde{h}$ is not varied, is exactly the same as the proof of Theorem 3.1, and gives the condition (3.12). Suppose that there is a minimising state with $\tilde{h}$ multivalued. Since the variations conserve mass, it is possible to rearrange this state so that $\tilde{h}$ is single-valued by purely vertical displacements which conserve $\tilde{\theta}$. These increase z and so will reduce the total energy defined by (3.109) since the last term, the potential energy, is reduced. Thus the state with multivalued $\tilde{h}$ cannot be a minimiser. □.

The factor $g\tilde{\theta}/\theta_0$ in (3.113) makes the characterisation of a minimum energy state much harder. In the shallow water case, section 3.3.1, which can be obtained by assuming $\tilde{\theta} = \theta_0$, $\tilde{h}$ satisfies a convexity condition given after Definition 3.5. This is no longer the case. The proofs in Cullen *et al.* (2015b), Cheng (2017) and Cullen *et al.* (2019) are all slightly different. The description here is an informal presentation which is closest to Cheng (2017), though the latter was written for the ocean case. It is necessary to start by defining dual variables as in (3.27):

$$X = f^{-1}v_g + x,\ Y = -f^{-1}u_g + y,\ Z = g\theta/(f^2\theta_0). \tag{3.114}$$

Then P is defined by $P = \frac{1}{2}(x^2+y^2)+f^{-2}\varphi$ as in (3.28) so that $(X,Y,Z) = \nabla P$ as in (3.29). Define R as in (3.37) so that $(x,y,z) = \nabla R$. The condition $\varphi = 0$ at $z = h$ means that $P = \frac{1}{2}(x^2+y^2)$ at $z = h$. In addition, control of the condition (3.113) requires the assumption that

$$\frac{g\theta}{f^2\theta_0} = \frac{\partial P}{\partial z} = Z \in \left[\frac{1}{\eta}, \eta\right] \tag{3.115}$$

for some $\eta > 0$.

Now define admissible solutions for this case, as in Definition 3.2.

Definition 3.6. An *admissible* solution of the semi-geostrophic equations (3.29), (3.30) on a region Γ_h defined by 3.107), where $h(x,y,t)$ is a single-valued positive bounded function at each t which satisfies (3.108) in a suitable sense, is one that is characterised by a convex function $P(\mathbf{x}, t)$ whose gradient $(X(t), Y(t), Z(t))$ satisfies (3.30) in a suitable sense, satisfies $P = \frac{1}{2}(x^2 + y^2)$ at $z = h(x, y)$ and satisfies (3.115).

Define σ as the potential density in (X, Y, Z) coordinates as in section 3.2.2 and ν_h as the Lebesgue measure $\mathcal{L}$ restricted to Γ_h. Assume for the present that the support of σ is a compact set Σ in $\mathbb{R}^3$. Then seek a map $\mathbf{s}$ from Σ to Γ_h which minimises (3.109) and satisfies

$$\mathbf{s}_{\#}\sigma = \nu_h. \tag{3.116}$$

Note that as h is unknown, so is ν_h. It is assumed that $h \leq H_s$, is continuous and that $\int_\Gamma h = 1$. It is also assumed that the minimising P satisfies the conditions above.

As in section 3.2.3, the minimiser may not be achieved by an invertible map, so the problem has to be solved in terms of a joint probability measure $\gamma(\Gamma_h, \Sigma)$ with marginals ν_h and σ. Let G be the set of such measures. The mass transport problem is then to minimise

$$C(\gamma) = \int_{\Gamma_h \times \Sigma} c(\mathbf{x}, \mathbf{X})\gamma(\mathbf{x}, \mathbf{X})\mathrm{d}\mathbf{x}\mathrm{d}\mathbf{X}, \tag{3.117}$$

where

$$c(\mathbf{x}, \mathbf{X}) = f^2\left(\frac{1}{2}((x - X)^2 + (y - Y)^2) - zZ\right), \tag{3.118}$$

for all $\gamma \in G$

As in the fixed boundary case, section 3.2.3, this can only be solved using the Kantorovich dual problem, which is (3.60) in the fixed boundary case. In the present case, it is easier to use the convex potentials P and R rather than φ and Ψ. Then the problem is to maximise

$$J(P, R) = \sup_{h \leq H_s} \int_{\Gamma_h} \left(\frac{1}{2}(x^2 + y^2) - P(\mathbf{x})\right) \mathrm{d}\mathbf{x} + \int_\Sigma \left(\frac{1}{2}(X^2 + Y^2) - R(\mathbf{X})\right) \sigma \mathrm{d}\mathbf{X} \tag{3.119}$$

over all P, R such that

$$P(\mathbf{x}) + R(\mathbf{X}) \geq \mathbf{x} \cdot \mathbf{X}, \tag{3.120}$$

for all $\mathbf{x} \in \Gamma_h$ and $\mathbf{X} \in \Sigma$. The supremum in the first term on the right hand side of (3.119) eliminates states with multivalued h as shown below.

Cheng (2017) shows that a unique maximiser can be found under suitable technical conditions. Slightly different results are proved by Cullen *et al.* (2015b) and Cullen *et al.* (2019). These results are stated in section 3.5. The difficult part is to control the unknown fluid depth h. The first step is to show that, because Σ is compact, there is a lower bound H on h. The next step is to consider the functional

$$\Pi_P(x,y,s) = \int_s^{H_s} \left(\frac{1}{2}(x^2+y^2) - P(x,y,z)\right) \mathrm{d}z. \tag{3.121}$$

Note that it has been assumed that $\frac{\partial P}{\partial z} \geq \frac{1}{\eta}$ and convexity of P implies that $\frac{\partial P}{\partial z}$ increases with z. Then the function $s \to \Pi_P(x,y,s)$ is uniformly concave for each (x,y), so it achieves a unique maximum s^*. Then define $h_P(x,y)$ to be the unique $s^* \in [0,H)$. This can be proved to give the required supremum of the first term on the right hand side of (3.119). This eliminates states with a multivalued h.

3.3.6 *The fully compressible semi-geostrophic equations in pressure coordinates with a free surface*

This section considers the fully compressible semi-geostrophic equations (2.166)-(2.168) in dimensional form. The hydrostatic relation, (2.8), applied to the full state of the fluid is

$$C_p\theta\frac{\partial \Pi}{\partial r} + g = 0. \tag{3.122}$$

This means that Π, which is a monotonic function of pressure, can be used as a vertical coordinate instead of physical height z. In practice the pressure p is used. However, the Boussinesq approximation made in section 2.2.4 is not made. This model is the most suitable form of the semi-geostrophic equations for the atmosphere, apart from the constant rotation assumption. The rigid upper boundary assumed in equations (3.2) is not realistic, though often used in numerical models.

The equations are then

$$\begin{aligned}
\frac{\mathrm{D}u_g}{\mathrm{D}t} + \frac{\partial \varphi}{\partial x} - fv &= 0, \\
\frac{\mathrm{D}v_g}{\mathrm{D}t} + \frac{\partial \varphi}{\partial y} + fu &= 0, \\
\frac{\mathrm{D}\theta}{\mathrm{D}t} &= 0, \\
f(u_g, v_g) &= \left(-\frac{\partial \varphi}{\partial y}, \frac{\partial \varphi}{\partial x}\right), \\
\frac{\partial \varphi}{\partial p} + \frac{RT}{p} &= 0, \\
\nabla \cdot \mathbf{u} &= 0.
\end{aligned} \tag{3.123}$$

R is the gas constant as in section 2.1. The coordinates are (x, y, p) in a region $\Gamma_p = \Gamma \times [0, p_s(x, y, t)]$ where $\Gamma \subset \mathbb{R}^2$. The velocity components are (u, v, ω). The boundary conditions are that $(u, v) \cdot \mathbf{n} = 0$ on the boundary of Γ, $\omega = 0$ on $p = 0$ and

$$\frac{\mathrm{D}p_s}{\mathrm{D}t} = \omega, \tag{3.124}$$

and $\varphi = 0$ at $p = p_s$.

The energy takes the form

$$E = \int_{\Gamma_p} \left(\frac{1}{2}(u_g^2 + v_g^2) + C_p T\right) \mathrm{d}x\mathrm{d}y\mathrm{d}p, \tag{3.125}$$

noting that $T = \theta\Pi$ as defined after eq. (2.2). The solution procedure is essentially the same as in section 3.3.5, as described in Cullen *et al.* (2015b). It is necessary to ensure that p_s is single-valued if it represents an energy minimiser by an argument similar to that for the single-valuedness of h. Introduce the dual variables

$$X = f^{-1}v_g + x,\ Y = -f^{-1}u_g + y,\ Z = -\frac{C_p \theta}{f^2 p_{ref}^{(R/C_p)}}. \tag{3.126}$$

The energy (3.125) can be rewritten as

$$E = \int_{\Gamma_p} f^2 \left(\frac{1}{2}((x - X)^2 + (y - Y)^2) - p^{(R/C_p)} Z\right) \mathrm{d}x\mathrm{d}y\mathrm{d}p. \tag{3.127}$$

This takes the same form as (3.118). The rest of the argument follows that in section 3.3.5.

3.4 Discrete solutions of the semi-geostrophic equations

3.4.1 *The discrete problem*

In this section it is shown that explicit solutions of the incompressible Boussinesq semi-geostrophic equations (3.29) and (3.30) can be constructed for the special case of piecewise constant data. This construction was developed independently by Cullen and Purser (1984). It was subsequently discovered that this method is an example of a general method of constructing convex (hyper)surfaces with faces of given area or volume developed by Alexandrov and described by Pogorelov (1964). The method also applies to the shallow water case, equations (3.90) and (3.91). Cullen and Purser (1989) showed how it could be related to the general solution procedure using optimal transport described in the preceding sections.

Consider the case described in Theorem 3.3, so that a vector field $(\tilde{X}, \tilde{Y}, \tilde{Z})$ is given for each (x, y, z). Assume this data is piecewise constant so that Γ is divided into n segments with volumes $\{\sigma_i\}$. On each of these segments set $\tilde{X} = X_i, \tilde{Y} = Y_i, \tilde{Z} = Z_i$. The data is thus as illustrated for a two-dimensional example in Figure 3.6.

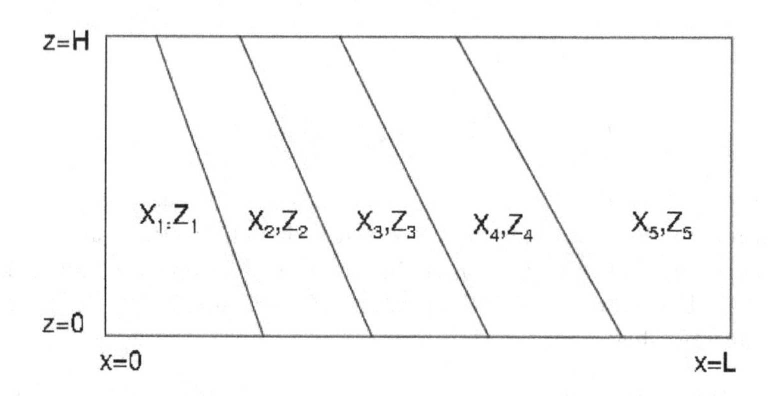

Fig. 3.6 Example data for (X_i, Z_i) as functions of (x, z) in Γ.

The problem is then to minimise the energy integral (3.31) by choosing an appropriate set of n segments with the volumes σ_i. According to Theorem 3.3, the minimiser will satisfy (3.29), so that $(\tilde{X}, \tilde{Y}, \tilde{Z})$ have to form the gradient of a convex function $\tilde{P}$. Since $(\tilde{X}, \tilde{Y}, \tilde{Z})$ take the discrete values (X_i, Y_i, Z_i), this means that $\tilde{P}$ is a polyhedral hypersurface, with n hyperfaces with gradients (X_i, Y_i, Z_i) and each hyperface having a projection onto Γ with volume σ_i. The result of applying this construction to the data shown in Fig. 3.6 is shown in Fig. 3.7.

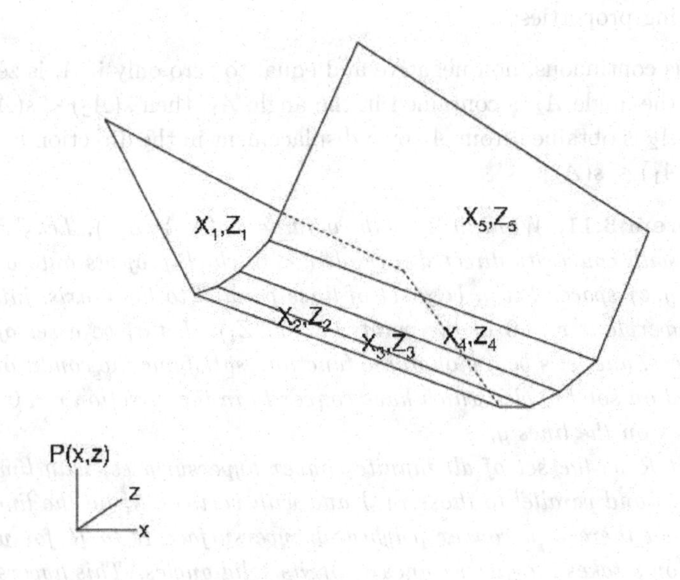

Fig. 3.7 The surface $\tilde{P}$ constructed by rearranging the data shown in Fig. 3.6 to give a convex polyhedron with faces with specified area.

The desired solution can thus be constructed if it can be proved that it is possible to construct a convex hypersurface on Γ with n hyperfaces of specified gradient and specified volume. The proof that this is possible is given in a theorem due to Alexandrov, independently rediscovered by R. J. Purser, Cullen and Purser (1984). Under the Legendre transform, the hyperfaces become vertices and their volumes become the solid angles associated with those vertices, see Sewell (2002). Therefore a hypersurface in (X, Y, Z, r) space has to be constructed, given specified vertices (X_i, Y_i, Z_i), and solid angles associated with the vertices which project onto regions of Γ with specified volumes σ_i. The theorem and proof given here follow Pogorelov (1964), Chapter 2, Theorem 2, translated into the notation used here and written for the three-dimensional case.

First define the *limit angle* of the hypersurface. Given an infinite hypersurface R which is not a prism, draw from some point A above the hypersurface all rays which do not intersect the hypersurface. Their directions fill some solid angle which is called the limit angle of the hypersurface. It is defined to within a parallel displacement which depends on the choice of A.

Next, let s be a monotonic function defined on solid angles A_i with the following properties:

(i) s is continuous, non-negative and equal to zero only if A_i is zero.
(ii) If the angle A_2 is contained in the angle A_1, then $s(A_2) < s(A_1)$.
(iii) If A_2 is obtained from A_1 by a displacement in the direction $r < 0$, then $s(A_1) \leq s(A_2)$.

Theorem 3.11. *Work in $\mathbb{R}^4$ with coordinates (X, Y, Z, r). Let $\mathcal{G}$ be a solid angle with convexity directed towards $r < 0$ which projects onto a region Γ of (x, y, z) space. Let g_i be a set of lines parallel to the r-axis, intersecting the hyperplane $r = 0$ at the points (X_i, Y_i, Z_i). Let σ_i be a set of positive numbers, and let s be a monotonic function, satisfying the conditions above, defined on solid angles which have convexity in the direction $r < 0$ and with vertices on the lines g_i.*

Let $\mathcal{R}$ be the set of all infinite convex hypersurfaces, with limit angles equal to and parallel to those in $\mathcal{G}$ and with vertices A_i on the lines g_i.

Then there is a convex polyhedral hypersurface R in $\mathcal{R}$ for which the function s takes the given values σ_i on its solid angles. This hypersurface is unique, except that an arbitrary displacement in the r direction is permitted unless s is strictly increasing under a displacement in the direction $r < 0$.

Proof Let $R_0 \in \mathcal{R}$ be a hypersurface with a single vertex A_1 with r coordinate r_1 with solid angle $\mathcal{G}$, so all the vertices $A_i : 2 \leq i \leq n$ are degenerate, with zero solid angles. Then, trivially,

$$s(A_i) \leq \sigma_i : \; i \geq 2. \tag{3.128}$$

Let R_s be an arbitrary hypersurface in $\mathcal{R}$ for whose solid angles the condition (3.128) is satisfied and with the vertex A_1 having r coordinate r_1. The set $\mathcal{R}_s$ of such hypersurfaces is not empty, as the hypersurface R_0 belongs to it. It is closed because of the assumption that s is monotonic. Assume that the r coordinate of the ith vertex is r_{is}. Set $\varsigma(s) = \sum_{i=1}^{i=n} r_{is}$. Because of the fixing of r_1, the function ς is bounded on the set $\mathcal{R}_s$, and therefore attains its absolute minimum.

Let R_s^* be a hypersurface for which ς attains its minimum. Then show that, for this hypersurface, $s(A_i) = \sigma_i$ for all i. Suppose not, and that for some j, $s(A_j) < \sigma_j$. Displace this vertex a small distance δ in the direction $r < 0$ to give A'_j. Let $\mathcal{G}_s$ be the limit angle of the hypersurface R^*_S. Denote by R'_s the hypersurface with the maximum values of r for each (X, Y, Z) which contains the points $A_1, A_2..A'_j, ...A_n$ and the angle $\mathcal{G}_s$. This hypersurface belongs to $\mathcal{R}$. If δ is sufficiently small, it also belongs to $\mathcal{R}_s$. The remaining solid angles can only become smaller, and their values of s will still satisfy (3.128). However, $\varsigma(s')$ is clearly less than $\varsigma(s^*)$, contradicting the assumption that S^* is a minimiser of ς. This establishes existence of the required hypersurface.

Now consider uniqueness. Suppose there are two hypersurfaces R_1 and R_2 whose vertices correspond by projection along the r axis and that s takes the same values on the corresponding vertices. If the hypersurfaces do not coincide, the r coordinates of their vertices r_{1i}, r_{2i} will differ for some i. Without loss of generality, let $\delta > 0$ be the maximum of $r_{1i} - r_{2i}$. Displace the hypersurface R_2 a distance δ in the direction $r > 0$. After this displacement

$$r_{1i} - r_{2i} \leq 0. \tag{3.129}$$

Let V_1 and V_2 be the set of vertices of R_1 and R_2 respectively which coincide after this displacement. Call two vertices *adjacent* if they belong to the same edge. Then it can be claimed that a vertex of R_1 which is adjacent to a vertex in V_1 also belongs to V_1.

To show this, let A_1 be a vertex of R_1 which belongs to V_1. Suppose that there is a vertex of R_1 adjacent to A_1 which is not in V_1. Then the inequality (3.129) is strict, and the solid angle at A_2 is contained in the interior of a solid angle of R_1 at the vertex A_1. Since s takes the same value at A_1 and A_2, this contradicts the monotonicity of s.

Since all vertices which are adjacent to vertices in V_1 belong to V_1, all vertices of R_1 belong to V_1. Therefore R_1 and R_2 coincide after some displacement along the r-axis. If s strictly increases when the angle is displaced in the direction $r < 0$, then R_1 and R_2 have to coincide, since otherwise s would have to take on different values on the vertices situated on the same lines g_i but with different r coordinates. □

Theorem 3.11 can be translated into a theorem on the existence of a convex polyhedral hypersurface on Γ with n hyperfaces of specified gradient and specified volume.

Theorem 3.12. *Work in $\mathbb{R}^4$ with coordinates (x, y, z, p). Let $\mathcal{P}$ be the set of convex polyhedral hypersurfaces with convexity directed towards $p < 0$ defined on a region Γ of (x, y, z) space. Let (X_i, Y_i, Z_i) be a set of gradients of* *the hyperfaces of the hypersurfaces. Let σ_i be a set of positive numbers, and let s be a monotonic function defined on hyperfaces A_i with the following properties:*

(i) s is continuous, non-negative and equal to zero only if A_i has zero volume.
(ii) If the hyperface $A_2 \subset A_1$, then $s(A_2) < s(A_1)$.
(iii) If A_2 is obtained from A_1 by a displacement in the direction $p > 0$, then $s(A_1) \leq s(A_2)$.

Then there is a convex polyhedral hypersurface P in $\mathcal{P}$ for which the function s takes the given values σ_i on its hyperfaces. This hypersurface is unique, except that an arbitrary displacement in the p direction is permitted unless s is strictly increasing under a displacement in the direction $p > 0$.

Proof To start the construction, let $P_0 \in \mathbb{R}$ be a hypersurface with a single hyperface A_1 defined by the equation $p = xX_1 + yY_1 + zZ_1 + p_1$. Then, trivially,

$$s(A_i) \leq \sigma_i \; i \geq 2. \tag{3.130}$$

The remainder of the proof is simply a rewrite of the proof of Theorem 3.11.□

Theorem 3.12 can be interpreted as the solution of a mass transport problem, as discussed in section 3.2.2. This is done by defining the potential density σ to be a sum of Dirac masses

$$\sigma(X, Y, Z) = \sum_{i=1}^{i=n} \sigma_i \delta(X_i, Y_i, Z_i). \tag{3.131}$$

The theorem then asserts the existence of a unique map $\mathbf{t} : \mathbb{R}^3 \to \Gamma$ which pushes forward the measure defined in (3.131) to Lebesgue measure on Γ. Thus each of the points (X_i, Y_i, Z_i) is mapped to a hyperface A_i. The mapping is not defined at other points in $\mathbb{R}^3$, as the measure σ is zero at these points. The mapping generates a convex P, and therefore represents an energy minimiser in the sense of Theorem 3.4. The energy integral that is minimised is

$$E = \sum_{i=1}^{i=n} \int_{A_i} f^2 \left(\frac{1}{2} \left((x - X_i)^2 + (y - Y_i)^2 \right) - zZ_i \right) \mathrm{d}x\mathrm{d}y\mathrm{d}z. \tag{3.132}$$

The theorem proves the existence and uniqueness of an energy minimiser for this special choice of σ. It can also be interpreted in terms of an energy minimising rearrangement of a function which takes the values (X_i, Y_i, Z_i) on sets of volume σ_i, see Douglas (2002), section 5.

In applying this result to equations (3.29) and (3.30), make the choice that σ_i is the volume of the ith hyperface projected onto the hyperplane $x = y = z = 0$. It is then necessary to ensure that the total of the σ_i is equal to the volume of Γ. The actual construction is best started by choosing a P^* which has n hyperfaces with the required gradients and non-zero areas. This is achieved by choosing the 'Voronoi' solution. The simplest form of this is $P^* = \frac{1}{2}(x^2 + y^2 + z^2)$. The hyperplanes are then defined as

$$p = xX_i + yY_i + zZ_i - \frac{1}{2}(X_i^2 + Y_i^2 + Z_i^2). \tag{3.133}$$

The intersection of these hyperplanes will give a convex surface as shown in Fig. 3.2. The volumes of each hyperface can be calculated and compared with the required value σ_i. The solution can then be found by iteration on the coordinates p_i where the hyperplanes intersect $x = y = z = 0$. Convergence is guaranteed by Theorem 3.12. A more general form of (3.133) and an efficient implementation of this algorithm, due to R. J. Purser, are described in section 5.1.2.

The application of Theorem 3.12 to the shallow water case is discussed in section 3.4.3. Convergence of solutions with a finite set of σ_i to continuous solutions has been proved by Cullen *et al.* (2007c) as described in section 3.5.3.

3.4.2 *Example: frontogenesis*

In section 3.2.3 it was shown, in Fig. 3.4, that it might be possible for two points in (X, Y, Z) space a finite distance apart to be mapped to adjacent points in (x, y, z) space. An example of this is provided by the simple frontogenesis problem introduced by Hoskins and Bretherton (1972). This is set in a two-dimensional cross-section which is embedded in a three-dimensional deformation field. The cross-section thus shrinks in time. The equations are, using the formulation (3.28)-(3.30):

$$\begin{aligned}
P(x,z) &= \frac{1}{2}x^2 + f^{-2}\varphi, \\
(X,Z) &= \nabla P, \\
\frac{\mathrm{D}X}{\mathrm{D}t} &= -\alpha X, \\
\frac{\mathrm{D}Z}{\mathrm{D}t} &= 0, \\
\frac{\partial u}{\partial x} + \frac{\partial w}{\partial z} &= -\alpha.
\end{aligned} \tag{3.134}$$

The final equation of (3.134) implies that the area of a fluid cross-section shrinks at a rate α. The equations are to be solved in the time-dependent domain $\Gamma(t) = [-L\exp(-\alpha t), L\exp(-\alpha t)] \times [0,H]$.

Choose piecewise constant initial data, in the manner of Fig. 3.6. $\Gamma(0)$ is divided into elements with areas σ_i on each of which $(X,Z) = (X_i, Z_i)$. Then equations (3.134) give

$$\begin{aligned}
X_i(t) &= X_i(0)\exp(-\alpha t), \\
Z_i(t) &= Z_i(0), \\
\sigma_i(t) &= \sigma_i(0)\exp(-\alpha t).
\end{aligned}$$

The solution can now be constructed using Theorem 3.12.

An example is shown using the initial data given in Fig. 3.8, taken from Cullen and Purser (1984). X is chosen so that $X = X(Z)$ is constant along lines of constant Z. This corresponds to the 'zero potential vorticity' data used by Hoskins and Bretherton (1972). The relation $(X,Z) = \nabla P$ means that the slope of the contours has to be given by

$$\frac{\mathrm{d}z}{\mathrm{d}x} = \frac{\mathrm{d}X}{\mathrm{d}Z}. \tag{3.135}$$

Since this slope is constant along Z contours, these contours have to be straight lines.

The data used are shown in Fig. 3.8, plotted on a 200×20 grid. The change of slope at the upper and lower boundaries is because of the extrapolation of the data from the interior grid points. X is given implicitly by

$$\frac{x - X(x,z)}{z - \frac{1}{2}H} = \frac{-5}{1 + [5X(x,z)/L]^2}\frac{L}{H}, \tag{3.136}$$

and $Z = \left(1 + \tan^{-1}[5X(x,z)]\right)H$.

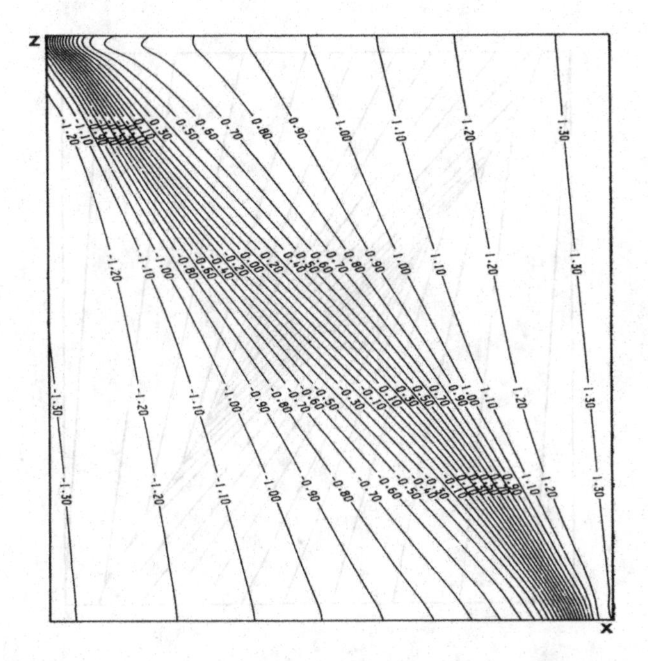

Fig. 3.8 Initial Z contours for the solution of equations (3.134) with $L = H = 1$. From Cullen and Purser (1984). ©Crown Copyright, Met Office.

As time evolves, the third and fourth equations of (3.134) show that X becomes smaller as a function of Z. Then (3.135) shows that the slopes of the contours becomes shallower. The boundary conditions require the domain to shrink in the x-direction while staying the same size in the z-direction. The combination of the two effects means that contours collide. For the data defined in (3.136), this happens at time $\alpha t = 1.5$.

This can be most convincingly demonstrated by using piecewise constant data of the type illustrated in Fig. 3.6. X and Z are constant on each element, and X_i, Z_i and σ_i evolve in time according to (3.135). The solution as a function of (x, z) is then found by constructing the surface P as described in the previous section. The solution at time $\alpha t = 2.5$ is shown in Fig. 3.9. Contours of Z have been forced away from the boundary, because this is the only way the required slope can be maintained.

The physical meaning of this solution is that air initially in contact with the boundaries is forced away from it. This is what happens when an 'occluded front' forms in the atmosphere and warm air initially in contact with the ground is lifted away from it. In Fig. 1.3 the fronts marked with a combination of semi-circular and triangular symbols are occluded fronts. It

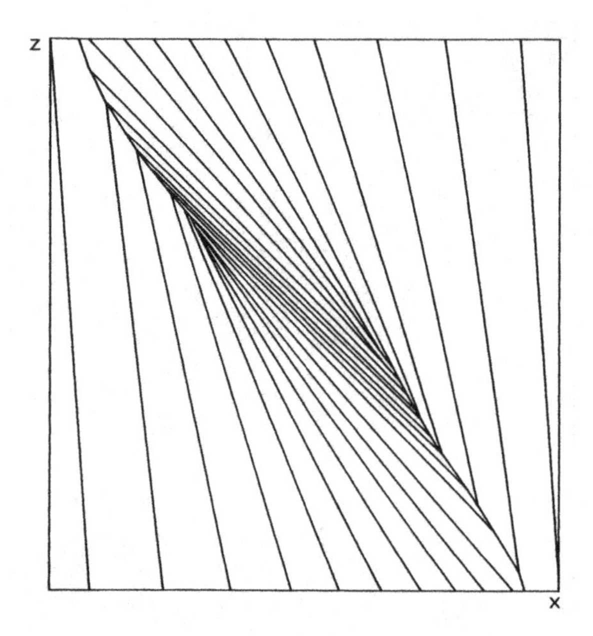

Fig. 3.9 Solution of (3.134) at time $\alpha t = 2.5$, using piecewise constant data of the type shown in Fig. 3.6. The x scale has been expanded so that the domain appears to be the same size as at $t = 0$. From Cullen and Purser (1984). ©Crown Copyright, Met Office.

also suggests that the strongest frontal discontinuities will be at the upper and lower boundaries. The rigid upper boundary used in these calculations is not physically realistic, so the conclusion only applies to the lower boundary. In the real atmosphere, it is certainly true that frontal discontinuities are strongest near the surface. However, their detailed structure is governed by other approximations to the Navier-Stokes equations which take into account, in particular, the effects of frictional drag. Fig. 1.4 is an example of a gust front, governed by a different type of dynamics, which is triggered in the presence of an air-mass discontinuity created in the manner described here.

These solutions have been derived purely from the Lagrangian form of equations (3.134), in which the last equation is written in the form

$$\frac{DV}{Dt} = -\alpha V \tag{3.137}$$

where V is the specific volume. While it is clear that the solution by construction of a convex polyhedron is consistent with the Lagrangian equations for piecewise constant data, which is extended to general data in section 3.5.4, it is not clear whether the solution makes sense as a solution of the Eulerian form of the equations. In particular, as noted in section 3.2.3, the integrated potential vorticity over Γ will not be conserved. Fig. 3.10 shows a solution of the Eulerian semi-geostrophic equations with the initial data shown in Fig. 3.8 at the same time as the solution shown in Fig. 3.9. Conventional finite difference methods are used. Artificial viscosity has to be used to allow the discontinuity to be captured. Apart from local smoothing, the solutions agree quite well. This is gratifying, since finite difference methods are based on the existence of smooth solutions to the governing Eulerian equations, which is certainly not the case here. In the more challenging case of the Eady model discussed in section 5.3, it will be shown that the disagreement is small initially but grows with time.

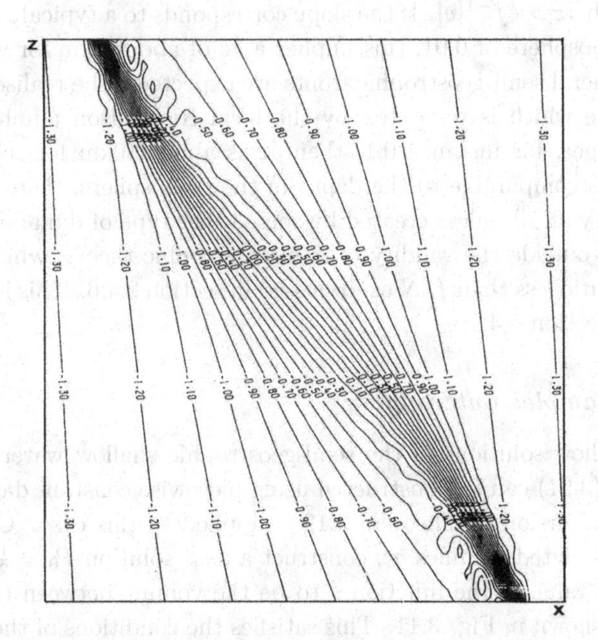

Fig. 3.10 A finite difference solution of equations (3.134) after a time $\alpha t = 2.5$. The x scale has been expanded so that the domain appears to be the same size as at $t = 0$. From Cullen and Purser (1984). ©Crown Copyright, Met Office.

The important feature of the solutions as an approximation to a true solution of the Navier-Stokes equations is that there is no discontinuity in a Lagrangian sense. The evolution of X and Z following fluid parcels is continuous in time. The requirement that no fluid trajectory changes direction on a time-scale comparable to f^{-1} will certainly be met if the deformation rate $\alpha \ll f$. The main restriction on the validity of the solutions comes from the assumption of two-dimensionality. This is discussed by Hoskins and Bretherton (1972). While a straight flow $v(x,z)$ in the y direction with a discontinuity in v in the (x,z) plane is an exact solution of the inviscid governing equations, it will be unstable to three-dimensional perturbations if the Richardson number,

$$Ri = \frac{g}{\theta_0}\frac{\partial \theta}{\partial z} / \left(\frac{\partial v}{\partial z}\right)^2, \tag{3.138}$$

is less than 1/4. The slope of the discontinuity is given by (3.135) to be $f[v]\theta_0/g[\theta]$ where $[v]$ and $[\theta]$ are the jumps in v and θ across the discontinuity. Then for a slope ς, the condition $Ri > 1/4$ implies that the jumps occur over a depth $\delta z > \varsigma f^{-1}[v]$. If the slope corresponds to a typical aspect ratio for the troposphere of 0.01, this implies a δz of about 1km for v=10ms^{-1}. Thus in general semi-geostrophic fronts are expected to be realised as shear zones whose width is restricted by the local Richardson number. If the slope is larger, for instance 0.1, then δz is about 10km for $[v]$=10ms^{-1}. Since this is comparable to the depth of the troposphere, there will be no discontinuity at all unless created by some other type of dynamics. Such a situation is outside the validity of semi-geostrophic theory, which requires an aspect ratio less than f/N as discussed in section 2.4.6. This is discussed further in section 5.4.2.

3.4.3 *Example: outcropping*

Now show how solutions of the semi-geostrophic shallow water equations (3.90) and (3.91) can be constructed using piecewise constant data. A two-dimensional version of Theorem 3.12 is applied to this case. Given faces A_i with associated volumes σ_i, construct a base solution $P_0 = \frac{1}{2}(x^2 + y^2)$. Choose the value of the function s to be the volume between the face A_i and P_0, as shown in Fig. 3.11. This satisfies the conditions of the theorem. Start the construction by choosing r_i so that the intersection of all the tangent planes with $\Gamma \times [0, \infty)$ lie below the surface P_0, giving $s_i = 0$ for all i, trivially less than σ_i. The rest of the proof follows that of Theorem 3.11.

Next illustrate solutions to the equations obtained using a conventional Eulerian finite difference method similar to that described in section 5.1.3 by Cloke and Cullen (1994). The model represents the evolution of a wind-driven ocean circulation. Equations (3.73) are rewritten as conservation

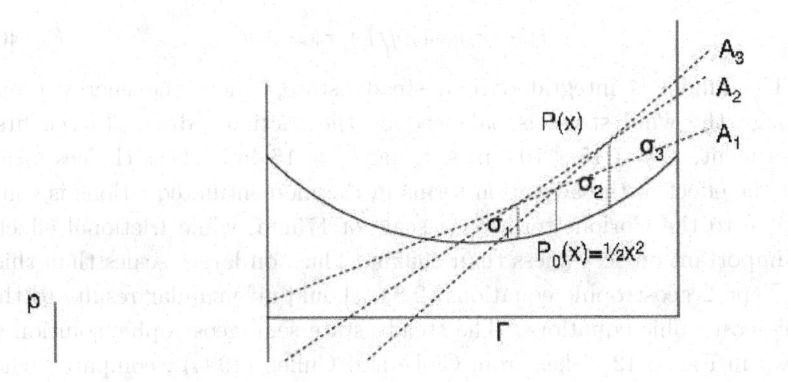

Fig. 3.11 Construction of the depth function for the solution of (3.90) and (3.91) in a one-dimensional cross-section, illustrating the notation used in the text.

laws for the momentum and mass. A wind stress (τ_x, τ_y) and frictional drag $-\epsilon(u_g, v_g)$ are applied to the momentum. This gives

$$\begin{aligned} h\frac{\partial u_g}{\partial t} + hu\frac{\partial u_g}{\partial x} + hv\frac{\partial u_g}{\partial y} + fh(v_g - v) &= \tau_x - \epsilon u_g, \\ h\frac{\partial v_g}{\partial t} + hu\frac{\partial v_g}{\partial x} + hv\frac{\partial v_g}{\partial y} - fh(u_g - u) &= \tau_y - \epsilon v_g, \\ \frac{\partial h}{\partial t} + \frac{\partial}{\partial x}(hu) + \frac{\partial}{\partial y}(hv) &= 0. \end{aligned} \tag{3.139}$$

The equations are solved by a semi-implicit finite difference method. The mass fluxes (hu, hv) are solved for iteratively to enforce the conditions (3.72). The only boundary conditions used are that $\mathbf{u} \cdot \mathbf{n} = 0$ on the walls of the basin. Values of derivatives of u_g and v_g near the boundaries are calculated using one-sided differencing. If values of h become zero in the true solution during the integration period, the finite difference method may predict negative values of h. This is prevented by using a flux limiting procedure in the discretisation of the third equation of (3.139). The use of the mass fluxes rather than the velocities as implicit variables makes it easier to do this.

The problem is solved on a beta plane, so $f = f_0 + \beta y$. It is solved in a square domain with side 10000km. The constants are set as $f_0 = 10^{-4}\text{s}^{-1}$, $\beta = 0.7 \times 10^{-11}\text{m}^{-1}\text{s}^{-1}$, $g = 10\text{ms}^{-2}$, $\epsilon = 3.3 \times 10^{-5}\text{ms}^{-1}$. The initial data has constant h. The wind stress is chosen to act only in the

x-direction, so that

$$\tau_x = \tau_0 \cos(\pi y/L), \tau_y = 0. \tag{3.140}$$

The model is integrated to a steady state, where the energy input through the wind stress is balanced by the frictional drag. In the first experiment, $\tau_0 = 1.15 \times 10^{-4}\text{m}^2\text{s}^{-1}$ and $h_0 = 16.4$m. These choices mean that the effect of the advection terms in the momentum equations is comparable to the Coriolis term on a scale of 175km, while frictional effects are important on scales less than 200km. Thus, on larger scales than this, the Type 2 geostrophic equations (2.84) should give similar results to the semi-geostrophic equations. The steady state semi-geostrophic solution is shown in Fig. 3.12, taken from Cloke and Cullen (1994) , compared with an analytic calculation of the solution due to Parsons (1969). The solutions show a classic western boundary current. Because of the small mean depth, this can only be sustained if the fluid fails to fill the domain, giving an outcrop in the north-western corner. The solutions show qualitatively similar outcrop lines. The differences are due to the further simplifications made in the analytic model. Cloke and Cullen (1994) demonstrate that the semi-geostrophic solution is almost identical to a solution of (2.84) with the same forcing and drag obtained by Bogue *et al.* (1986).

In the second experiment, $\tau_0 = 9.2 \times 10^{-4}\text{m}^2\text{s}^{-1}$ and $h_0 = 32.8$m.The other parameters are unaltered. The effect of the advection terms is now important on scales less than 350km, but the effect of the friction is only important on scales less than 100km. It can be shown that the solution of equations (2.84) and the analytic solution are unaltered from the previous experiment, but the solution of equations (3.139) is now different, as shown in Fig. 3.13. The western boundary current overshoots its previous position, due to the effect of the advection terms in the momentum equation.

The importance of this model is in the explanation of the separation of the Gulf Stream from the coast of the United States. It is believed that inertial effects as shown in Fig. 3.13 are important in this process. The outcropping process is also important in the modelling of layers of constant density in the ocean, which can reach the surface at some points. The depth of the layer then becomes zero as a function of horizontal position. A model which can treat this effect is very useful.

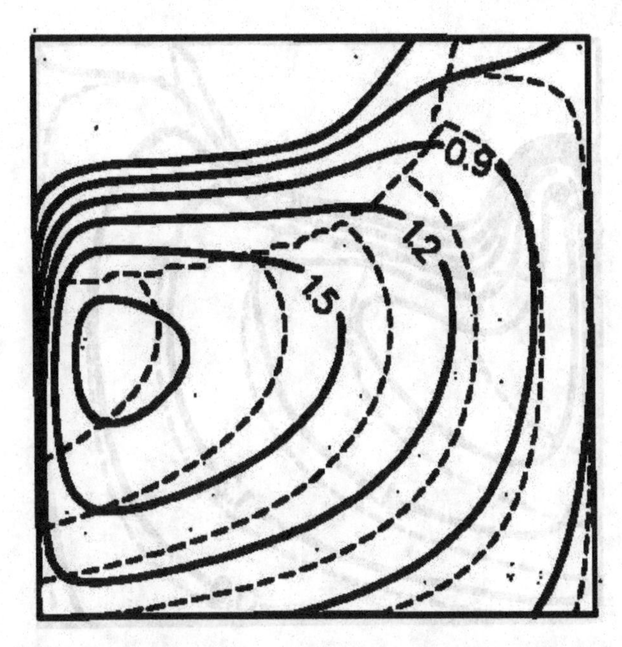

Fig. 3.12 Solid lines: Steady state solutions of equations (3.139) with parameters as given in the text. Dashed lines: Analytic solution. From Cloke and Cullen (1994).

3.5 Rigorous results on existence of solutions

3.5.1 *Summary of the solution procedure*

In sections 3.2 and 3.3 the solutions of the semi-geostrophic equations in four cases, three-dimensional incompressible Boussinesq with and without a free upper boundary, three-dimensional compressible with a free upper boundary, and shallow water have been analysed in plane geometry with constant rotation. In all cases, solutions were identified with time-evolving minimum energy states of the fluid, but it was assumed that such states could actually be found. Much rigorous mathematics has been done to show that this is possible, based on the theory of optimal transport. This theory is described by Villani (2003) and Villani (2008). The minimising state is characterised by a potential P satisfying a convexity property.

Finding a time-evolving minimum energy state not only requires proving the existence of a minimiser, but finding a fluid trajectory that evolves the solution forward in time. The *polar factorisation theorem* of Brenier (1991) achieves this. However, as shown in eqs. (3.4) and (3.75) for the

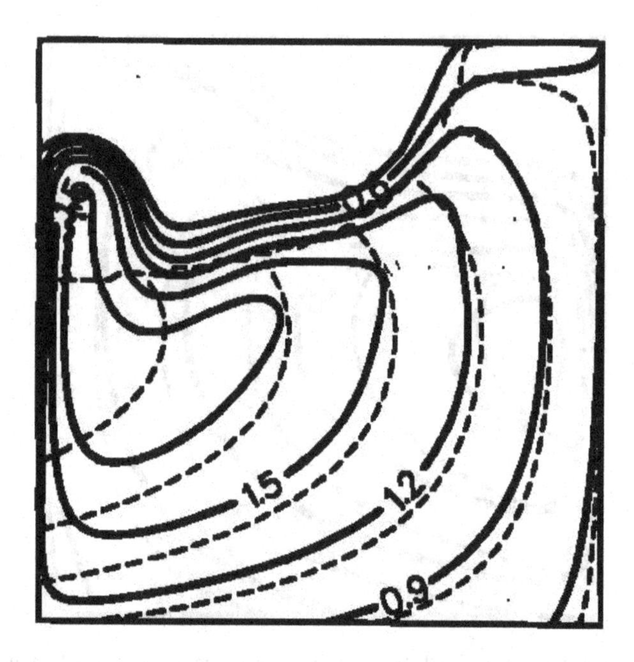

Fig. 3.13 Solid lines: Steady state solution of equations (3.139) using the second set of parameters defined in the text. Dashed lines: Analytic solution as in Fig. 3.12. From Cloke and Cullen (1994).

three-dimensional incompressible Boussinesq and shallow water cases respectively, the problem can be solved if the evolution of P in time can be found. The fluid trajectory can then be found by post-processing. This is a big advantage when proving long-time existence of the solutions, as the convexity of P makes it easier to prove convergence of approximate solutions, while the fluid trajectory may not even be deterministic in some physically relevant cases.

The applicability of optimal transport is most obvious when the equations are written in Lagrangian form in the dual variables introduced in sections 3.2.1 and 3.3.2. This gives equations (3.30) and (3.90) for the two problems above. Both systems consist of evolution equations for a vector of dual variables $\mathbf{X}$ following fluid particles. Each particle represents a given volume in physical space. The associated velocity is unknown and has to be determined by finding a mapping of the particles to physical space that minimises the energy and preserves the volume associated with each particle. In particular, all particles have to remain in the domain Γ, which enforces the prescribed boundary conditions. This can be written as an optimal transport problem. In the free boundary and shallow water cases, the depth of the fluid is also unknown and has to be determined by the energy minimisation. Analysis of the Eulerian form of the equations shows that this procedure would give a solution of the equations even if only a stationary point of the energy rather than a minimiser was required. However, the requirement that the energy is minimised is needed to make the solution well-posed. This extra constraint is not part of the Eulerian equations and has to be imposed in addition. A physical justification is found that suggests that this constraint is necessary if the semi-geostrophic equations are to be a good approximation to the Euler equations as the solutions evolve in time.

The optimal transport problem will find a mapping of the fluid particles to physical space that minimises a cost function. It can be shown, as described below, that this will always give some form of convexity condition. In the three-dimensional Boussinesq problem, it was shown that choosing the cost function to be the energy gave the required form of the constraints. In the free boundary and shallow water problems, the need to predict the fluid depth means that optimal transport only forms part of the solution procedure.

This mapping is then used to construct the right hand sides of the equations for $\frac{D\mathbf{X}}{Dt}$. The resulting evolution equations will not make sense if a finite mass of fluid has the same values of $\mathbf{X}$. This degeneracy of the Lagrangian description of a fluid is well known. In such cases the evolution equations can only be solved in a very weak sense. However, the discrete scheme described in section 3.4.1 uses an approximation which is degenerate, but the discrete evolution equations can nevertheless be solved. This raises the question of the convergence of the discrete scheme. Even when this degeneracy does not occur, the solution $\mathbf{X}$ may be discontinuous in physical space, as illustrated in section 3.4.2. In such cases the evolution equations may only make sense in Lagrangian form. It is therefore of interest to identify when the solution is smooth, and can be regarded as a solution of the Eulerian equations.

In the following subsections the rigorous results that have been obtained at the time of writing are summarised. The results are explained in terms of the framework above. No proofs are given, as these can be found in the original references.

3.5.2 *Solutions of the mass transport problem*

The condition required to make the solution procedures work is the existence of a unique energy minimiser in the sense of Theorem 3.3. Informally, that can be stated as the existence of a unique rearrangement of a vector-valued function $\mathbf{X} = (X, Y, Z)$ as the gradient of a convex function. The polar factorisation theorem of Brenier (1991) achieves this.

Before stating the theorem, it is necessary to define the term *rearrangement* properly. For general applicability, the notations μ and ν are used for arbitrary measures, not necessarily the Lebesgue measure. The statement $\mathbf{X} = \nabla P \circ \mathbf{s}$ means that $\mathbf{X}$ and ∇P will be rearrangements of each other if $\mathbf{s}$ is measure preserving, where a rearrangement is defined by Douglas (2002) as follows:

Definition 3.7. Let Γ be a bounded set in $\mathbb{R}^3$ with measure μ, and let $F(\mathbf{x}), G(\mathbf{x}) : \Gamma \to \mathbb{R}^d$ be integrable functions. Then F and G are *rearrangements* if

$$\mu\left(\{\mathbf{x} : F(\mathbf{x}) \in B\}\right) = \mu\left(\{\mathbf{x} : G(\mathbf{x}) \in B\}\right), \tag{3.141}$$

for every Borel subset B of $\mathbb{R}^3$. A *Borel* set is any set that can be formed from open (or closed) sets through the operations of countable union, countable intersection, and relative complement.

A simple one-dimensional example is shown in Fig. 3.14.

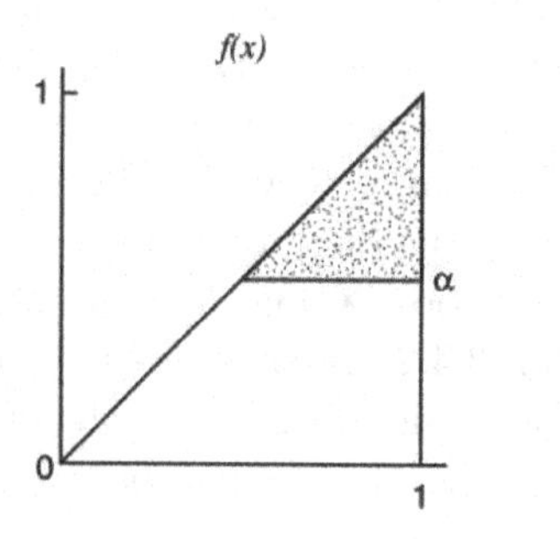

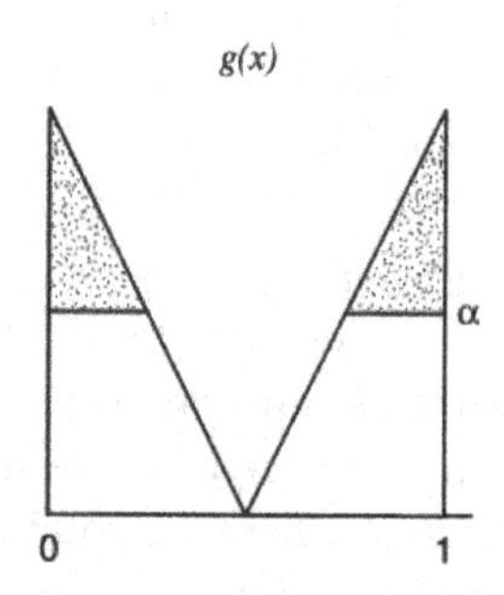

Fig. 3.14 Two scalar functions $F(x)$ and $G(x)$ on the unit interval which are rearrangements of each other. From Cullen and Douglas (2003). ©Royal Meteorological Society, Reading, U.K., 2003.

There is a close link between this definition and Definition 3.3. The statement that F and G are rearrangements according to Definition 3.7 is equivalent to saying that, if $F_{\#}\mu = \nu$, then $G_{\#}\mu = \nu$, where ν is a measure on $\mathbb{R}^d$. Thus the set $S = \left\{\mathbf{s} : \mathbb{R}^3 \to \Gamma,\ \mathbf{s}_{\#}\sigma = \mathcal{L}\right\}$ used in Theorem

3.4 is equivalent to stating that $\mathbf{X}(\mathbf{s}(\mathbf{X}))$ is a rearrangement of a given function $\mathbf{X}(\mathbf{x})$ which takes values in a set $B \subset \mathbb{R}^3$ on a set of given measure $\int_B \sigma \mathrm{d}X\mathrm{d}Y\mathrm{d}Z$ in Γ.

A measure-preserving mapping is defined by

Definition 3.8. Given a measure μ on bounded subsets of $\mathbb{R}^3$, a measure-preserving mapping $\mathbf{s} : \mathbb{R}^3 \to \mathbb{R}^3$ satisfies

$$\mu(\{\mathbf{x} : \mathbf{s}(\mathbf{x}) \in B\}) = \mu(B) \tag{3.142}$$

for all measurable sets B.

Brenier's theorem uses the function spaces introduced in section 2.4.5 and, in addition, the Sobolev space

$$W^{1,p}(\Gamma) = \{u \in L^p(\Gamma), \nabla u \in L^p(\Gamma : \mathbb{R}^3)\}, \tag{3.143}$$

where u is a scalar function and $p \in [1, \infty)$. Write $\mathcal{L}(A)$ for the Lebesgue measure of a set $A \subset \mathbb{R}^3$. As stated in Brenier (1991) it is also necessary that the dual variable $\mathbf{X}$ is in a bounded subset $\Sigma \subset \mathbb{R}^3$.

Brenier's theorem is stated in slightly simplified form using the notation of section 3.2.3. It is in two parts and states:

Theorem 3.13. *Given a map $\mathbf{X} = (X, Y, Z) \in L^p$ from Γ to Σ, there is a unique rearrangement $\mathbf{X}^* = \nabla P$ where $P \in W^{1,p}(\Gamma : \Sigma)$, P is convex and $\nabla P \in L^p(\Gamma : \Sigma)$, $p \in [1, \infty)$. The mapping $\mathbf{X} \to \mathbf{X}^*$ is continuous.*

Theorem 3.14. *Given a map $\mathbf{X} = (X, Y, Z) \in L^p(\Gamma : \Sigma)$, such that $\mathcal{L}(\mathbf{X}^{-1}(E)) = 0$ for each set $E \subset \Sigma$ for which $\mathcal{L}(E) = 0$, there is a unique pair $(\mathbf{X}^*, \mathbf{s})$ such that*

(i) $\mathbf{X}^*$ *has the properties in Theorem 3.13.*
(ii) $\mathbf{s} : \Gamma \to \Gamma$ *is the unique measure-preserving that maximises*

$$\int_\Gamma \mathbf{X} \cdot \mathbf{s}(\mathbf{x}) \mathrm{d}\mathbf{x}. \tag{3.144}$$

(iii) $\mathbf{X} = \mathbf{X}^* \circ \mathbf{s}$.
(iv) The factorisation is unique up to treatment of sets of zero measure and continuously dependent on the data.

Item (iii) is called the *polar factorisation* of $\mathbf{X}$.

Theorem 3.13 states that there is a unique rearrangement of a vector-valued function (X, Y, Z) as the gradient of a convex function under quite general conditions, including the finite dimensional case treated in section

3.4 where volumes in Γ are mapped onto single points in Σ. This form of the result establishes the existence of a unique energy minimiser in the sense of Theorem 3.3. The gradient of a convex function is the natural multi-dimensional generalisation of a monotonically increasing scalar-valued function, since, as discussed in section 3.2.3, each component of the gradient is an increasing function of the corresponding coordinate.

Theorem 3.14 states that $\mathbf{X} = (X, Y, Z)$ can be uniquely written as the polar factorisation $\mathbf{X}^* \circ \mathbf{s}$ where $\mathbf{s}$ is a measure-preserving mapping as defined above. The restrictions on this second result, called the 'non-degeneracy condition', exclude the finite dimensional case treated in section 3.4. The degenerate case corresponds to infinite values of the potential density σ defined in section 3.2.2. The non-degeneracy condition can be expressed as the requirement that σ is *absolutely continuous* with respect to the Lebesgue measure on Σ. This is written as $\sigma << \mathcal{L}$

In the physical problem, it is quite possible that the image under (X, Y, Z) of a volume of finite size in Γ has zero measure in Σ. Using (3.27), this will be true, for instance, for a volume with uniform potential temperature θ. The non-degeneracy condition required by Theorem 3.14 will not then be satisfied. However, Theorem 3.13 covers this case with $p = 1$. Integrability of (X, Y, Z), so $\mathbf{X} \in L^1(\Gamma)$, corresponds essentially to finite energy so will be satisfied in the physical problem. Therefore Theorem 3.13 will apply under all physically relevant conditions.

The cost (3.144) with $n = 3$ is equivalent to the energy minimised in Theorem 3.4 provided that the integral $\int_\Sigma \frac{1}{2} f^2 \left(z^2 + Z^2\right) \sigma \mathrm{d}X \mathrm{d}Y \mathrm{d}Z$ is preserved for all maps in the set $S = \{\mathbf{s} : \Sigma \to \Gamma,\ \mathbf{s}_{\#}\sigma = \mathcal{L}\}$ defined above. This is certainly true if $\mathbf{s}$ is a smooth map, and is true if $\mathbf{s}$ is the optimal map because of its monotonicity property.

Theorem 3.13 solves the Monge-Kantorovich problem set out in section 3.2.3. The desired solution is obtained by minimising the relaxed constraint (3.50). In order to show that this can be done, it is necessary to show that this is equivalent to solving the Kantorovich dual problem by maximising (3.60). However, as stated in section 3.2.3, the solution may correspond to a joint probability measure rather than a map. In that case, the solution would not describe a rearrangement in the sense of Definition 3.7. Brenier therefore proves in Theorem 3.14 that, under the non-degeneracy condition, the maximiser corresponds to a deterministic map $\mathbf{s}$ rather than just a joint probability measure γ.

A general result on the solution of the Monge-Kantorovich problem is contained in Villani (2008), p.70, Theorem 5.10. This uses the term *Polish space* to mean a complete separable metric space, and the term *c-cyclically monotone* to mean for a cost function $c(x, y)$ that for each $m \geq 1$ and any choices $(x_1, ..., x_m)$ and $(y_1, ..., y_m)$

$$\sum_{i=1}^{m} c(x_i, y_i) \leq \sum_{i=1}^{m} c(x_i, y_{i-1}), \tag{3.145}$$

where $y_0 = y_m$. These terms are described by Villani (2003), p.86.

Theorem 3.15. *Given Polish spaces (X, μ) and (Y, ν), where μ, ν are probability measures on these spaces, let $c : X \times Y \to \mathbb{R} \cup \{+\infty\}$ be a lower semicontinuous cost function, such that*

$$\forall (x, y) \in X \times Y; c(x, y) \geq a(x) + b(y)$$

for some real-valued upper semicontinuous functions $a \in L^1(\mu)$ and $b \in L^1(\nu)$. Then

(i) There is duality

$$\begin{aligned} &\min_{\pi \in \Pi(\mu,\nu)} \textstyle\int_{X \times Y} c(x, y) \mathrm{d}\pi \\ &= \sup_{(\psi,\phi) \in C_b(X) \times C_b(Y): \phi - \psi \leq c} \left(\textstyle\int_Y \phi(y) \mathrm{d}\nu(y) - \int_X \psi(x) \mathrm{d}\mu(x)\right) \\ &= \sup_{(\psi,\phi) \in L^1(\mu) \times L^1(\nu): \phi - \psi \leq c} \left(\textstyle\int_Y \phi(y) \mathrm{d}\nu(y) - \int_X \psi(x) \mathrm{d}\mu(x)\right) \\ &= \sup_{\psi \in L^1(\mu)} \left(\textstyle\int_Y \psi^c(y) \mathrm{d}\nu(y) - \int_X \psi(x) \mathrm{d}\mu(x)\right) \\ &= \sup_{\phi \in L^1(\nu)} \left(\textstyle\int_Y \phi(y) \mathrm{d}\nu(y) - \int_X \phi^c(x) \mathrm{d}\mu(x)\right). \end{aligned} \tag{3.146}$$

(ii) If c is real-valued and the optimal cost $C(\mu, \nu) = \inf_{\pi \in \Pi(\mu,\nu)} \int c \mathrm{d}\pi$ is finite, then there is a measurable $c-$cyclically monotone set $\Gamma \subset X \times Y$ (closed if a, b and c are continuous) such that for any $\pi \in \Pi(\mu, \nu)$ the following five statements are equivalent:

(a) π is optimal;

(b) π is $c-$cyclically monotone;

(c) There is a $c-$convex ψ such that $\pi-$almost surely, $\psi^c(y) - \psi(x) = c(x, y)$;

(d) There exist $\psi : X \to \mathbb{R} \cup \{+\infty\}$ and $\phi : Y \to \mathbb{R} \cup \{-\infty\}$ such that $\phi(y) - \psi(x) \leq c(x, y)$ for all (x, y), with equality $\pi-$almost surely;

(e) π is concentrated on Γ.

(iii) If c is real-valued, $C(\mu, \nu) < +\infty$, and there is the pointwise upper bound

$$c(x, y) \leq c_X(x) + c_Y(y), \; (c_X, c_Y) \in L^1(\mu) \times L^1(\nu), \tag{3.147}$$

then both the primal and dual Kantorovich problems have solutions, so

$$\begin{aligned} &\min_{\pi\in\Pi(\mu,\nu)} \int_{X\times Y} c(x,y)\mathrm{d}\pi \qquad (3.148)\\ &= \max_{(\psi,\phi)\in L^1(\mu)\times L^1(\nu):\phi-\psi\leq c} \left(\int_Y \phi(y)\mathrm{d}\nu(y) - \int_X \psi(x)\mathrm{d}\mu(x)\right)\\ &= \sup_{\psi\in L^1(\mu)} \left(\int_Y \psi^c(y)\mathrm{d}\nu(y) - \int_X \psi(x)\mathrm{d}\mu(x)\right). \end{aligned}$$

If in addition a, b and c are continuous, then there is a closed $c-$cyclically monotone set $\Gamma \subset X \times Y$, such that for any $\pi \subset \Pi(\mu,\nu)$ and for any $c-$convex $\psi \in L^1(\mu)$,

(a) π is optimal in the Kantorovich problem if and only if $\pi(\Gamma) = 1$,
(b) ψ is optimal in the dual Kantorovich problem if and only if $\Gamma \subset \partial_c(\psi)$.

Theorem 3.15 only establishes the existence of a joint probability measure $\pi(\mu,\nu)$ that solves the problem, thus proving Theorem 3.13. In order to prove Theorem 3.14, it is necessary to identify when this corresponds to a deterministic map. Villani (2008), chapter 9, discusses this issue. Conditions can only be found for specific classes of cost function. An example for the quadratic cost $c(x,y) = (x-y)^2$ is given below (Theorem 9.4 in Villani (2008)). This covers the requirements of Theorem 3.14.

Theorem 3.16. *Let $c(x,y) = |x-y|^2$ in $\mathbb{R}^n$. Let μ,ν be two probability measures on $\mathbb{R}^n$ such that $\int |x|^2\mathrm{d}\mu(x) + \int |y|^2\mathrm{d}\nu(y) < +\infty$ and μ does not give mass to sets of dimension at most $n-1$. (This is true in particular if $\mu << \mathcal{L}$.) Then there is a unique optimal coupling (x,y) of μ and ν; it is deterministic, and characterized, among all couplings of (μ,ν), by the existence of a lower semicontinuous convex function ψ such that $y = \nabla\psi(x)$ almost surely. In other words, there is a unique optimal transference π; it is a Monge transport plan, and it is characterized by the existence of a lower semi-continuous convex function ψ whose subdifferential contains the support of π.*

Further versions and generalisations of this theorem are given in Gangbo and McCann (1996) as well as in Villani (2008).

Further progress beyond Theorem 3.14 and 3.16 in establishing conditions where there is a deterministic coupling and thus a unique polar factorisation was made by Burton and Douglas (1998). Call **X** *countably degenerate* if the non-degeneracy condition can be satisfied by removing countably many level sets. The piecewise constant data discussed in section 3.4 thus generates a map which is countably degenerate. Then Burton and Douglas (1998) prove

Theorem 3.17. *If* $\mathbf{X}$ *on* Γ *is countably degenerate, then it has a polar factorisation* $\mathbf{X} = \mathbf{X}^* \circ \mathbf{s}$, *where* $\mathbf{s} : \Gamma \to \Gamma$ *is measure preserving. Alternatively, if* $\mathbf{X}^*$ *is almost injective, then the polar factorisation exists and is unique.*

This extension allows the finite dimensional case of section 3.4 to be covered. It does not, however, cover all cases of physical interest since any restriction on the existence of sets of constant potential temperature is unphysical.

Since $\mathbf{X}$ evolves continuously in time according to equations (3.30) or (3.90), it can be hoped that the function P whose existence is guaranteed by the polar factorisation will also evolve smoothly in time. The trajectory generated by the incompressible velocity (u, v, w) in physical space, concealed in the D/Dt operators in equations (3.30), is given by the measure-preserving mapping $\mathbf{s}$ as discussed in section 3.2.1. However, Theorem 3.17, which relaxes the non-degeneracy condition, is not sufficient to guarantee the existence of a measure-preserving mapping, so that the trajectory in physical space required to maintain the energy minimisation property may not be achievable by a smooth velocity field. Even if a measure-preserving mapping exists, it may not be unique. This situation is addressed in sections 3.5.3 and 3.5.4.

Another important extension is to the periodic case. Section 3.4.2 demonstrated the formation of discontinuities in $\mathbf{X}$ at the boundaries of Γ, but it was conjectured that no discontinuities could be generated in the interior. In order to find regularity estimates which do not see the effect of the boundaries, choose $\Gamma = \mathbb{T}^3$, which represents a three-dimensional plane periodic domain with periodicity L in all directions. Though only the case where the horizontal coordinates are periodic is physically relevant, use of the three-dimensional periodic problem allows the behaviour of the interior flow to be studied independently of the boundaries. The following theorem is given by Loeper (2005), based on earlier work by Cordero-Erausquin (1999).

Theorem 3.18. *Given a map* $\mathbf{X} = (X, Y, Z)$ *from* $\mathbb{T}^3$ *to* $\mathbb{T}^3$ *that does not collapse sets of positive measure into sets of measure zero,* $\mathbf{X}$ *can be written as* $\nabla P \circ \mathbf{s}$, *where* $\mathbf{s} : \mathbb{T}^3 \to \mathbb{T}^3$ *is measure-preserving,* P *is convex and* $P - \frac{1}{2}(x^2 + y^2 + z^2)$ *is periodic. The factorisation is unique up to treatment of sets of zero measure and continuously dependent on the data.*

If R is the Legendre transform of P, then the periodic boundary conditions mean that

$$|\nabla R(X,Y,Z) - (X,Y,Z)| \leq \frac{1}{2}\sqrt{3}L. \tag{3.149}$$

This shows that the velocity $\mathbf{U} = f(\frac{\partial R}{\partial Y} - Y, X - \frac{\partial R}{\partial X})$ is bounded uniformly in time. This is not true in the non-periodic case. Such bounds are needed to obtain long-time existence of solutions of the evolution equations, as discussed in section 3.5.3.

In Theorem 3.6 it was shown that points on the boundary of Γ are mapped to points on the boundary of the convex hull of the support Σ of σ. This result suggests that there can be discontinuities in (X, Y, Z) as a function of (x, y, z) on the boundary of Γ if Σ is non-convex, as illustrated in Fig. 3.4. The example shown in section 3.4.2 shows that these can penetrate into the interior of Γ. It was suggested in Cullen and Purser (1984) that interior discontinuities were not possible if the potential vorticity $\mathcal{Q}$ given by (3.70) is non-zero and bounded. In the finite-dimensional example of section 3.4.1, the potential density σ consists of Dirac masses and $\mathcal{Q}$ is zero almost everywhere, so the solution is 'full' of discontinuities.

These questions are addressed by the regularity theory due to Caffarelli. Some further definitions are needed to explain the results.

Definition 3.9. A weak solution in the sense of Alexandrov to the Monge-Ampère equation (3.42) is a potential $R : \Sigma \to \mathbb{R}$ such that for every Borel set B in Σ with Lebesgue measure $\mathcal{L}(B)$, the Lebesgue measure of $\nabla R(B)$ in Γ makes sense and is equal to $\sigma\mathcal{L}(B)$.

Definition 3.10. The space C^α : $0 < \alpha < 1$ contains all continuous functions $F(\mathbf{X})$ on a set $B \subset \mathbb{R}^3$ such that there exists a constant c for which $|F(\mathbf{X}) - F(\mathbf{Y})| \leq c|\mathbf{X} - \mathbf{Y}|^\alpha$ for all $\mathbf{X}, \mathbf{Y}$ in B.

In addition, in an generalisation of (3.143), the space $W^{2,p}$ is the space of all functions whose second derivatives are integrable according to the L^p norm.

The main results as applied to the cases treated here are as follows. In Caffarelli (1992a) it is proved that if Γ is convex and $\log(\sigma)$ is bounded, then the Monge-Ampère equation (3.42) for R is satisfied in the weak sense due to Alexandrov, and ∇R is a C^α map. Further, if σ is continuous, then R is $W^{2,p}$ for every $p < \infty$ and if σ is C^α, then R is $C^{2,\alpha}$ in the interior of Σ. The restriction $p < \infty$ means that the second derivatives can be locally infinite, so equation (3.42) may not have a solution in the classical sense.

In Caffarelli (1992b), it is proved that if both Σ and Γ are convex, then both ∇P and ∇R satisfy estimates of the type $|\nabla P(\mathbf{x}) - \nabla P(\mathbf{y})| \geq c|\mathbf{x} - \mathbf{y}|^M$. Since ∇P and ∇R are inverses of each other, this means that both are continuous in this sense up to the boundary. This is consistent with Theorem 3.6. If the problem is posed with periodic boundary conditions, this suggests that no discontinuities in (X, Y, Z) can be present. If the problem is not given with periodic boundary conditions, then there is no reason why Σ, which evolves as a result of equations (3.45), should remain convex and discontinuities can form. In the atmosphere there are no lateral boundaries, but there is a rigid lower boundary. Discontinuities extending from the lower boundary, as illustrated in section 3.4.2, can be expected. The result that the formation of discontinuities in ∇P by the semi-geostrophic equations requires the presence of boundaries was conjectured by Hoskins and Bretherton (1972).

The following theorem is proved in Caffarelli (1996).

Theorem 3.19. *If Γ and Σ are convex, and σ is strictly positive and bounded, then there exist a pair of strictly convex $C^{1,\alpha}$ functions P, R on Γ and Σ respectively such that*

(i) ∇P *maps* Γ *onto* Σ *and* ∇R *maps* Σ *onto* Γ.
(ii) ∇P *and* ∇R *are inverses of each other.*
(iii) $\det \mathrm{Hess} R = \sigma$ *in the Alexandrov sense.*
(iv) If σ is C^α, then P, R are strongly convex, i.e. their Hessians are strictly positive and $C^{2,\alpha}$.

This result shows that, if σ is bounded at $t = 0$ and strictly positive on a convex set $\Sigma(0) \subset \mathbb{R}^3$, then if σ is also C^α, both P and R will have second derivatives everywhere, and thus (X, Y, Z) will be continuously differentiable as a function of (x, y, z) everywhere provided that σ remains C^α.

The necessity for the condition that σ is C^α was demonstrated by Wang (1995). He constructed specific examples to show that, if σ is not continuous, then R is not $W^{2,p}$ for any p. He also included the example of the function

$$R(X, Y) = X^2/\log|\log(X^2 + Y^2)| + Y^2 \log|\log(X^2 + Y^2)|. \qquad (3.150)$$

This function is strictly convex, and the potential density calculated from it is continuous and positive. However, it does not have bounded second derivatives at the origin.

It is also shown in Jian and Wang (2007) that, if the modulus of continuity of σ, defined by $\omega(r) = \sup(|\sigma(\mathbf{X}) - \sigma(\mathbf{Y})|) : |\mathbf{X} - \mathbf{Y}| < r$, satisfies

$$\int_0^1 \frac{\omega(r)}{r} < \infty, \tag{3.151}$$

then R is locally twice differentiable.

If this were the best result obtainable, the only way to prove that σ remains C^α if it is C^α at $t = 0$ is to control $\omega(r)$. This requires use of the evolution equations, and so no such result has yet been proved except for short times, as will be discussed in the next section. However, improved regularity estimates have been made by De Philippis and Figalli (2012) which show that P and R are in $W^{2,1}_{loc}$ under suitable conditions.

The key estimate of De Philippis and Figalli (2012) forms part of the following theorems, which cover the two-dimensional periodic case and the three-dimensional case for a bounded domain respectively. These are stated as Theorem 2.2 in Ambrosio *et al.* (2012) and Theorem 3.1 in Ambrosio *et al.* (2014). The result has been slightly simplified for the present application and the notation has been changed for consistency with Theorems 3.18 and 3.19 above.

Theorem 3.20. *Let P be as found in Theorem 3.18 but in $\mathbb{T}^2$ rather than $\mathbb{T}^3$. Assume without loss of generality $P(0) = 0$. Assume σ as used in Theorem 3.19 satisfies $\sigma << \mathcal{L}(\mathbb{T}^2)$ and there are constants λ, Λ such that $0 < \lambda \leq \sigma \leq \Lambda < \infty$. Then:*

(i) $P \in C^{1,\beta}(\mathbb{T}^2)$ for some $\beta(\lambda, \Lambda) \in (0, 1)$ and there exists a constant $C(\lambda, \Lambda)$ such that

$$\|P\|_{C^{1,\beta}} \leq C.$$

(ii) $P \in W^{2,1}(\mathbb{T}^2)$, and for any k there exists a constant $C(\lambda, \Lambda, k)$ such that

$$\int_{\mathbb{T}^2} |\nabla^2 P| \log_+^k |\nabla^2 P| d\mathbf{x} \leq C.$$

(iii) If $\sigma \in C^{k,\alpha}(\mathbb{T}^2)$ for some k and $\alpha \in (0, 1)$, then $P \in C^{k+2,\alpha}(\mathbb{T}^2)$ and there exists a constant $C(\lambda, \Lambda, \|\sigma\|_{C^{k,\alpha}})$ such that

$$\|P\|_{C^{k+2,\alpha}} \leq C.$$

Moreover, there exist two positive constants (c_1, c_2) depending only on λ, Λ and $\|\sigma\|_{C^{0,\alpha}}$ such that

$$c_1 Id \leq \nabla^2 P(\mathbf{x}) \leq c_2 Id, \forall \mathbf{x} \in \mathbb{T}^2.$$

The conditions on σ in the first two parts of the theorem are expected to be robust during the time evolution as (3.44) shows that σ is conserved along fluid trajectories. Item (i) of the result means that no discontinuities in ∇P can form, as noted above. Item (ii) means that weak Eulerian solutions can be found, as shown in Theorem 3.40 below.

The three-dimensional case considers a bounded convex domain Γ in $\mathbb{R}^3$ with Lebesgue measure, and a 'dual' domain Σ with measure $\sigma(\mathbf{X})$ where Σ is an open subset of $\mathbb{R}^3$ and with $\int_\Sigma \sigma \mathrm{d}\mathbf{X} = 1$. Note that $\int_\Sigma \sigma^2$ is not assumed to be bounded. Given that Γ is bounded, the result cannot be true if σ is compactly supported in $\mathbb{R}^3$, or else the discontinuous solutions illustrated in section 3.4 would not be possible. The assumptions in Ambrosio *et al.* (2014) exclude the compactly supported case. The theorem in that paper includes the proof of existence of an optimal map under these assumptions, as well as the $W^{2,1}_{loc}$ regularity of P. As will be seen in section 3.5.4, this allows the existence of weak Eulerian solutions in physical space. When σ is non-zero for very large values of $\mathbf{X}$, it implies very large values of the state variables in physical space, which is unrealistic. Thus the achievement of $W^{2,1}_{loc}$ regularity for a bounded domain Γ is not physically relevant. The result for the periodic problem in Theorem 3.20 is more important

Theorem 3.21. *Assume that for any compact set $K \subset \Sigma$ there exist $\lambda, \Lambda(K)$ such that*

$$0 < \lambda \leq \sigma \leq \Lambda.$$

Then

(i) There exists a unique optimal transport map, namely a unique (up to an additive constant) convex function $R(\mathbf{X}) : \Sigma \to \mathbb{R}$ such that $\nabla R_{\#}\sigma = \mathcal{L}(\Gamma)$. R is a strictly convex Alexandrov solution of

$$\det \nabla^2 R(\mathbf{X}) = \sigma(\mathbf{X}).$$

(ii) $R \in W^{2,1}_{loc}(\Sigma) \cap C^{1,\beta}_{loc}(\Sigma)$. More precisely, if Σ_0 is an open set strictly contained in Σ and $0 < \lambda \leq \sigma \leq \Lambda < \infty$ in Σ_0, then for any k there exist constants $C_1(k, \Sigma_0, \Gamma, \lambda, \Lambda)$, $\beta(\lambda, \Lambda)$ and $C_2(\Sigma_0, \Gamma, \lambda, \Lambda)$ such that

$$\int_{\Sigma_0} |\nabla^2 R| \log^k_+ |\nabla^2 R| \mathrm{d}\mathbf{X} \leq C_1,$$

and

$$\|R\|_{C^{1,\beta}(\Sigma_0)} \leq C_2.$$

(iii) Assume also that Σ, Γ are bounded and uniformly convex, with boundaries $\partial\Sigma, \partial\Gamma \in C^{2,1}$ and that $\sigma \in C^{1,1}(\Sigma)$ and $\lambda \le \sigma \le \Lambda$ in Σ. Then

$$R \in C^{3,\alpha}(\Sigma) \cap C^{2,\alpha}(\bar{\Sigma}) \ \forall \alpha \in (0,1),$$

and there exists a constant $C(\alpha, \Sigma, \Gamma, \lambda, \Lambda, \|\sigma\|_{C^{1,1}})$ such that

$$\|R\|_{C^{3,\alpha}(\Sigma)} \le C \text{ and } \|R\|_{C^{2,\alpha}(\bar{\Sigma})} \le C.$$

Moreover, there exist positive constants c_1, c_2, κ depending only on $(\lambda, \Lambda, \|\sigma\|_{C^{0,\alpha}})$ such that

$$c_1 Id \le \nabla^R(\mathbf{X}) \le c_2 Id, \ \forall \mathbf{X} \in \Sigma$$

and

$$\nabla R(\mathbf{X}) \cdot \mathbf{X} \ge \kappa, \ \forall \mathbf{X} \in \partial\Sigma.$$

An alternative approach is given by O'Neill (2020). C^2 regularity for the solution of the Monge-Ampère equation (3.42) for the periodic case can only be achieved by restricting the variation of σ because of the example of Wang (1995) above. The theorem as stated below applies to the Dirichlet problem for a bounded convex set $\Gamma_0 \subset \mathbb{R}^3$, but can be extended to the periodic case by embedding the periodic box Γ in a much larger convex set Γ_0 to which the theorem applies. This formalises the discussion above, which argues that the interior behavour of the solutions can be studied by using periodic geometry.

Theorem 3.22. *Let $R \in C^2(\Gamma_0) \cap C(\bar{\Gamma}_0)$ be a strictly convex solution of (3.42) with Dirichlet boundary conditions in a convex set Γ_0 such that $B_{n^{-1}}(0) \subset \Gamma_0 \subset B_1(0)$ with right hand side σ satisfying*

$$0 < 1 - \epsilon \le \sigma \le 1 + \epsilon,$$

where $\epsilon \le \epsilon_0 < 1/2$ is a positive constant which depends only on n, and

$$\|\sigma\|_{C^{1/2}} := \left(\int_0^1 \frac{(\omega_\sigma^{1/2}(r))}{r} \mathrm{d}r \right)^2 \equiv \left\| \frac{\omega_\sigma(r)}{r^2} \right\|_{L^{1/2}([0,1])} < \infty, \qquad (3.152)$$

where $\omega(r)$ is the modulus of continuity introduced before (3.151) and (3.152) contains the definition of $\mathcal{C}^{1/2}$. Then, when restricted on $\Gamma' \subset\subset \Gamma_0$ such that $B_{r_1} \subset \Gamma' \subset B_{r_2}$, $n^{-1} < r_1 < r_2 < 1$, every element of the Hessian matrix in (3.42) has modulus of continuity bounded as

$$\omega_{D^2R}(d) \le c_0 k \left(k^{1/2}(r_1^2 - r_2^2)d + \int_0^{k^2 d} \frac{\omega_\sigma(r)}{r} \mathrm{d}r + k^{3/2} d \int_d^k \frac{\omega_\sigma(r)}{r^2} \mathrm{d}r \right), \qquad (3.153)$$

with $0 < d < 1$, $k := c_1 \exp\left(c_2 \|\sigma\|_{C^{1/2}}^{1/2}\right)$, and constants c_0, c_1, c_2 depending only on n, ϵ_0, Γ_0 and $dist(\Gamma', \partial\Gamma_0)$.

The significance of this result is that it allows the existence of classical solutions of the semi-geostrophic equations with periodic boundary conditions to be proved for a finite time T which increases as $\epsilon \to 0$, as will be shown in the following sections. This is consistent with the numerical results which will be shown in section 5.2 and also aids the proof that the solutions of the Euler equations converge to those of the semi-geostrophic system which is discused in section 3.6.

3.5.3 *Existence of semi-geostrophic solutions in dual variables*

Now turn to the solution of the evolution equations. It is easiest to start with the equations in dual variables. These are given in (3.45) for the three-dimensional incompressible Boussinesq case, and in (3.100) for the shallow water case. The free surface equations described in sections 3.3.5 and 3.3.6 can also be described this way. In principle, these can be solved by choosing initial data σ satisfying the conditions of Theorem 3.4, solving the Monge-Ampère equation for R, and then calculating $\mathbf{U}$. The velocity $\mathbf{U}$ is used to solve the transport equation for σ. Since $\mathbf{U}$ is non-divergent, provided the velocity $\mathbf{U}$ is smooth enough for the transport equation to make sense, values of σ will be conserved along trajectories and the initial bounds on σ will be maintained uniformly in time. The difficulty is that boundedness of σ is not enough to make $\mathbf{U}$ differentiable, as discussed in the previous section. The main task in showing that (3.45) or (3.100) can be solved is to circumvent this difficulty.

It has been known for some time, Diperna and Lions (1989), that the transport equation could be solved if $\mathbf{U}$ and $\nabla\cdot\mathbf{U}$ were in the space L^p, $p \in [1,\infty]$ and the solutions were unique if $\mathbf{U} \in W^{1,1}_{loc}$, i.e. it has integrable first derivatives, and $\nabla\cdot\mathbf{U}$ is bounded. However, the discussion above shows that this is only true in the present case if $\sigma \in [\frac{1}{C}, C]$ for some $C > 0$. This result was extended to vector fields $\mathbf{U}$ of bounded variation (BV) and bounded divergence by Ambrosio (2004). Since $\mathbf{U}$ is generated from the gradient of a convex function, and the gradient of a convex function cannot oscillate, it is BV. In the constant rotation cases treated in this chapter, $\nabla\cdot\mathbf{U} = 0$. This means that a solution of the semi-geostrophic equations in dual variables should be obtainable under quite general conditions.

The method of solution for dual variables described above is quite general, and was formalised by Ambrosio and Gangbo (2008) for a wide class of equations, including the semi-geostrophic equations in dual variables. Some definitions are needed to explain the results. The notation of Theorem 3.16 is used. Thus let μ, ν be two probability measures on $\mathbb{R}^n$ such that $\int |x|^2 d\mu(x) + \int |y|^2 d\nu(y) < +\infty$. Define a cost function $c(x, y) : \mathbb{R}^n \times \mathbb{R}^n \to \mathbb{R}$.

As in section 3.3.4, the minimising value of the cost is called the *Wasserstein distance* between the measures μ and ν, associated with the cost function c. This can be very general, as in Theorem 3.15. In Ambrosio and Gangbo (2008) it is chosen to be the quadratic cost used in Theorem 3.16, so that the Wasserstein distance between μ and ν is

$$W_2^2(\mu, \nu) = \min_{\pi \in \Pi_{\mu,\nu}} \left(\int_{\mathbb{R}^n \times \mathbb{R}^n} |x - y|^2 d\pi(x, y) \right). \tag{3.154}$$

Π is the set of joint probability measures on $\mathbb{R}^n \times \mathbb{R}^n$ with μ and ν as marginals. This concept was introduced in section 3.2.3 after eq. (3.57). Write $\mathcal{P}_2(\mathbb{R}^n)$ for the set of probability measures with finite second moments on $\mathbb{R}^n$ with W_2^2 as a metric and $\mathcal{P}_2^a$ the subset of $\mathcal{P}_2$ which are absolutely continuous with respect to Lebesgue measure. This geometry is described in Ambrosio *et al.* (2008).

Next define a potential $H : \mathcal{P}_2 \to \mathbb{R}$ (called a Hamiltonian by Ambrosio and Gangbo (2008)). It is defined on a subset $D(H)$ of $\mathcal{P}_2$. Write $\partial H(\mu)$ for the derivative of H with respect to μ. This may be multivalued, so write $\nabla H(\mu)$ for the element of $\partial H(\mu)$ with minimal $L^2(\mu; \mathbb{R}^n)$ norm. The generic problem is then, given an initial measure $\bar{\mu} \in \mathcal{P}_2(\mathbb{R}^n)$, find a path $t \to \mu_t \in \mathcal{P}_2(\mathbb{R}^n)$ such that

$$\begin{aligned} \frac{d\mu_t}{dt} + \nabla \cdot (J \nabla H(\mu_t) \mu_t) = 0, \; t \in (0, T) \\ \mu_0 = \bar{\mu}, \end{aligned} \tag{3.155}$$

where $\|H(\mu_t)\|_{L^2(\mu_t)} \in L^1(0, T)$.

J is a $n \times n$ symplectic matrix, such as $\begin{pmatrix} 0 & 1 \\ -1 & 0 \end{pmatrix}$.

Next a method of interpolation between measures is needed. This would consist of interpolation along the optimal map that connects μ_0 to μ_1, if it exists. The definition has to work if there is no deterministic map, but just an optimal transport plan. These methods were formulated by Benamou and Brenier (2000) and McCann (1997). Define the projections $p_1, p_2 : \mathbb{R}^n \times \mathbb{R}^n \to \mathbb{R}^n$ by $p_1(x, y) = x$, $p_2(x, y) = y$. Then let $\pi \in \Pi(\mu_0, \mu_1)$ be an optimal transport plan between μ_0 and μ_1. Define an interpolation between μ_0 and μ_1 by

$$\mu_t = ((1 - t)p_1 + tp_2)_{\#} \pi. \tag{3.156}$$

If the Wasserstein distance is defined using the quadratic cost, it is shown in Ambrosio *et al.* (2008) that

$$W_2(\mu_s, \mu_t) = |t - s| W_2(\mu_0, \mu_1)\ \forall (s,t) \in [0,1]. \tag{3.157}$$

Then a form of convexity called λ-*convexity* is required, defined by

Definition 3.11. Let $H : \mathcal{P}_2(\mathbb{R}^n) \to (-\infty, \infty)$ be proper (i.e. has some finite values) and let $\lambda \in \mathbb{R}$. Then H is λ-convex if for every $(\mu_0, \mu_1) \in \mathcal{P}_2(\mathbb{R}^n)$ and every optimal transport plan $\pi \in \Pi(\mu_0, \mu_1)$:

$$H(\mu_t) \le (1-t)H(\mu_0) + tH(\mu_1) - \frac{\lambda}{2} t(1-t) W_2^2(\mu_0, \mu_1)\ \forall t \in [0,1]. \tag{3.158}$$

Then it is proved by Ambrosio and Gangbo (2008) that

Theorem 3.23. *Equation (3.155) can be solved given*

(i) A growth condition. There are constants $C_0 \in (0, +\infty)$, $R_0 \in (0, +\infty]$ such that for all $\mu \in \mathcal{P}_2^a(\mathbb{R}^n)$ with $W_2^2(\mu, \bar{\mu}) < R_0$, $\mu \in D(H)$, $\partial H(\mu) \neq 0$ and $|\nabla H(\mu)(z)| \le C_0(1 + |z|)$ for μ-almost every $z \in \mathbb{R}^n$.

(ii) A continuity property of the gradient. If $\mu = \rho \mathcal{L}^n$, $\mu_n = \rho_n \mathcal{L}^n \in \mathcal{P}_2^a(\mathbb{R}^n)$, $\sup_n W_2^2(\mu_n, \mu) < R_0$ and $\mu_n \to \mu$ narrowly, then there exists a subsequence $n(k)$ and functions $\mathbf{w}_k, \mathbf{w} : \mathbb{R}^n \to \mathbb{R}^n$ such that $\mathbf{w}_k = \nabla H(\mu_{n(k)}), \mu_{n(k)}$ a.e., $\mathbf{w} = \nabla H(\mu), \mu$ a.e. and $\mathbf{w}_k \to \mathbf{w}\mathcal{L}^n$ a.e. in $\mathbb{R}^n$ as $k \to +\infty$.

Ambrosio and Gangbo (2008) proved another theorem under weaker assumptions on the differentiability of H. Given an initial measure $\bar{\mu} \in \mathcal{P}_2(\mathbb{R}^n)$, find a path $t \to \mu_t \in \mathcal{P}_2(\mathbb{R}^n)$ and vector fields $\mathbf{v}_t \in L^2(\mu_t : \mathbb{R}^n)$ such that

$$\begin{aligned} \frac{d\mu_t}{dt} + \nabla \cdot (J\mathbf{v}_t \mu_t) = 0,\ t \in (0,T) \\ \mu_0 = \bar{\mu}, \\ \mathbf{v}_t \in T_{\mu_t}\mathcal{P}_2(\mathbb{R}^n) \cap \partial H(\mu_t) \text{ for a.e. } t, \end{aligned} \tag{3.159}$$

where $T_{\mu_t}\mathcal{P}_2(\mathbb{R}^n)$ is the tangent space to $\mathcal{P}_2(\mathbb{R}^n)$ at μ_t. See Ambrosio *et al.* (2008) for a full characterisation.

Then the following theorem holds:

Theorem 3.24. *Equation (3.159) can be solved given*

(i) A growth condition. There are constants $C_0 \in (0, +\infty)$, $R_0 \in (0, +\infty]$ such that for all $\mu \in \mathcal{P}_2(\mathbb{R}^n)$ with $W_2^2(\mu, \bar{\mu}) < R_0$, $\mu \in D(H)$, $\partial H(\mu) \neq 0$ and $\|\nabla H(\mu)(z)\|_{L^2(\mu)} \le C_0$.

(ii) *A continuity property of the gradient. If* $\sup_n W_2^2(\mu_n, \mu) < R_0$ *and* $\mu_n \to \mu$ *narrowly, then the limit point of convex combinations of* $\{\nabla H(\mu_n)\mu_n\}_{n=1}^{\infty}$ *for the weak* topology are representable as* $\mathbf{w}_\mu$ *for some* $\mathbf{w} \in \cdot\partial H(\mu) \cap T_\mu \mathcal{P}_2(\mathbb{R}^n)$.

Full explanations of the technical language are given in Ambrosio and Gangbo (2008) and Ambrosio *et al.* (2008).

It has also been proved by Cullen *et al.* (2007c) that solutions using the 'geometric' method with purely discrete μ, as used in sections 3.4 and 5.1.2, converge to solutions of (3.155) with absolutely continuous μ. This justifies using the method to solve the semi-geostrophic equations. The theorem applies for very general cases. The presentation below is thus given for a simple two-dimensional version of the semi-geostrophic model (3.45) and the notation of Cullen *et al.* (2007c) is changed for consistency with this model.

Let Φ be the Lagrangian flow in $\Sigma \subset \mathbb{R}^2$ with coordinates (X, Y). Then (3.35), (3.36) and (3.40) give

$$\dot{\Phi} = \mathbf{U} = (U, V) = f(y - Y, X - x) = fJ(\mathbf{X} - \nabla R). \tag{3.160}$$

Suppose the initial data is a probability measure σ_0 in Σ. Then the measure σ_t will evolve in time according to (3.45) giving

$$\frac{\partial}{\partial t}\sigma_t + \nabla \cdot (\sigma_t \mathbf{U}) = 0. \tag{3.161}$$

It is assumed that $\mathbf{U}$ is uniformly bounded for all t.

Assume that σ_0 is a probability measure of bounded support on $\mathbb{R}^2$ and that $T > 0$. Represent the initial measure σ_0 by Dirac masses at random positions ξ_i in Σ. This is achieved by considering a space of probability measures $(\Omega, \Lambda, \mathbb{P})$. Ω is a space whose members are sets of random particle positions. Thus any $\omega \in \Omega$ represents a set of n independent identically distributed random variables $(\xi_i(\omega)) : \Omega \to \mathbb{R}^2$ such that $(\xi_i)_\# \mathbb{P} = \sigma_0$. This condition states that any choice of such variables must represent the initial measure σ_0. Then an empirical measure at time t can be written as

$$\sigma_t^{n,\omega} = \frac{1}{n}\sum_{i=1}^{n} \delta_{\Phi_i^n(t,\xi^n(\omega))}. \tag{3.162}$$

Similarly, define a discrete measure ν in the region Γ in (x, y) space in which (3.45) is solved. This is derived by solving at each time t a two-dimensional version of the optimal transport problem solved in section 3.2 with σ_t given by (3.162). As shown in section 3.4, this gives a convex R

with ∇R piecewise constant on subdomains $\Gamma_{i,t}$ of Γ. Define $c_{i,t}^n \in \Gamma$ as the centroid of $\Gamma_{i,t}$. This is called the *barycentric projection*, and will be used in several results later in this chapter. It is fully defined in Ambrosio *et al.* (2008). Then define a measure ν_t^n by

$$\nu_t^n = \sum_{i=1}^{n} \delta_{c_{i,t}^n}.$$

The evolution equations (3.160) and (3.161) are now represented by the discrete equations

$$\frac{\partial}{\partial t}\Phi_{i,t}^n = J(\Phi_i^n(t,\xi^n(\omega)) - c_{i,t}^n), \tag{3.163}$$

$$\frac{\mathrm{d}}{\mathrm{d}t}\sigma_t = \frac{1}{n}\sum_{i=1}^{n}\frac{\mathrm{d}}{\mathrm{d}t}\Phi_i^n(t,\xi^n(\omega))\delta_{\Phi_i^n(t,\xi^n(\omega))}.$$

The theorem of Cullen *et al.* (2007c) applied to the semi-geostrophic system is then as given below.

Theorem 3.25. *Assume that $\sigma_0 = \rho_0\mathcal{L}^2$ and that ρ_0 is a bounded function. Assume that $\{\nu^n\}_{n=1}^{\infty}$ converges to ν in the Wasserstein distance. Then there exists a $\mathbb{P}$-measurable set $\Omega' \subset \Omega$ such that $\mathbb{P}|\Omega'| = 1$ and for every $\omega \in \Omega'$ the following hold:*

(i) $t \to \Phi^n(t,\xi^n(\omega))$ is well defined and absolutely continuous on $[0,T]$.
(ii) There exists a sequence $\{m_k(\omega)\}_{k=1}^{\infty}$ (depending on ω) and for each $t \in [0,T]$ there exists a probability density $\mu_t^\omega << \mathcal{L}^2$ such that the empirical measures $\{\mu_t^{m_k(\omega),\omega}\}_{k=1}^{\infty}$ converge to μ_t^ω in the metric W_2.
(iii) There exists a constant $a(\omega) < +\infty$ such that $W_2(\mu_t^\omega,\mu_s^\omega) \le a(\omega)|t-s|$ for all $s,t \in [0,T]$.
(iv) $\mu_0^\omega = \mu_0$.
(v) There exists convex, uniformly Lipschitz functions $R_t^\omega : \mathbb{R}^2 \to \mathbb{R}$ such that $(\nabla R_t^\omega)_{\#}\sigma_t = \nu$ and (3.160) and (3.161) hold in the sense of distributions.
(vi) $H[\sigma_t^\omega] = H[\sigma_0]$ for all $t \in [0,T]$, where $H[\sigma] = \int_\Sigma (X - \nabla R)^2 \mathrm{d}\sigma$.

Many specific results for the semi-geostrophic equations were derived prior to these results. Later work has exploited Theorems 3.23 and 3.24. Weak existence results for equations (3.45) were first obtained by Benamou and Brenier (1998) by smoothing $\mathbf{U}$ to give sufficient regularity. An improved result below is given in Cullen and Feldman (2006) for $\sigma \in L^p, p > 1$. Assume that σ is a probability measure on $\mathbb{R}^3$ whose integral is the volume of Γ. Assume that the physical domain Γ is open, bounded and connected. Work in the space $C\left([0,T);L^p_w(\mathbb{R}^3)\right)$ of all σ on $\mathbb{R}^3 \times [0,T)$ such that $\sigma(t,\cdot) \in L^p(\mathbb{R}^3)$ for any $t \in [0,T)$, and for any $\{t_k\}_{k=1}^{\infty}, t^* \in [0,T)$ satisfying $\lim_{k\to\infty} t_k = t^*$, $\sigma(t,\cdot)$ converges weakly to $\sigma(t^*,\cdot)$ in $L^p(\mathbb{R}^3)$. Let $B(0,r) \subset \mathbb{R}^3$ be the open ball $|\mathbf{X}| \leq r$.

Theorem 3.26. *Let $p > 1$. For any $T > 0$ and $\sigma(0,\cdot) \in L^p(\mathbb{R}^3)$ with compact support $\Sigma(0)$ there exist functions*

$$\begin{aligned} \sigma &\in L^\infty\left([0,T);L^p(\mathbb{R}^3)\right) \cap C[0,T);L_{W^p(\mathbb{R}^3)}, \\ P &\in L^\infty\left([0,T);W^{1,\infty}(\Gamma)\right) \cap C[0,T);W^{1,q}(\Gamma), \\ R &\in L^\infty\left([0,T);W^{1,\infty}(\mathbb{R}^3)\right) \cap C[0,T);W^{1,q}(B(0,r)), \end{aligned} \tag{3.164}$$

where q is any number in $[1,\infty)$, both P and R are convex functions of their spatial variables, satisfying (3.45) where the evolution equation for σ and the initial data $\sigma(0,\cdot)$ are understood in the weak sense: that is for any $\varrho \in C^1_c([0,T) \times \mathbb{R}^3)$

$$\begin{aligned} &\int_{\mathbb{R}^3\times[0,T)} \left(\frac{\partial \varrho}{\partial t} + \mathbf{U}\cdot\nabla\varrho\right)\sigma \mathrm{d}X\mathrm{d}Y\mathrm{d}Z\mathrm{d}t + \\ &\qquad \int_{\mathbb{R}^3} \sigma(0,\cdot)\varrho(0,\cdot)\mathrm{d}X\mathrm{d}Y\mathrm{d}Z = 0. \end{aligned} \tag{3.165}$$

The space C^1_c contains all functions with a continuous first derivative which have compact support in $\mathbb{R}^3$. Moreover there exists $r > 0$ such that

$$\Sigma(t) \subset B(0,r) \text{ for all } t \in [0,T), \tag{3.166}$$

and $\nabla R(X,Y,Z) \in \Gamma$ for each $t \in [0,T)$ and almost every $(X,Y,Z) \in \mathbb{R}^3$.

This theorem shows that there are functions $P(\mathbf{x},t)$ and $R(\mathbf{X},t)$ which have the regularity associated with convex functions, but no more, and are continuous in time and solve equation (3.45) in an integrated sense. Therefore $P(\mathbf{x},t)$ and $R(\mathbf{X},t)$ can have discontinuous derivatives, but the derivatives cannot oscillate. Conditions (3.166) means that the support Σ of σ remains bounded. Thus ∇P is bounded, which ensures that $\mathbf{U} = (f(y-Y), f(X-x), 0)$ as defined in (3.45) remains bounded.

Weak existence of solutions to the shallow water semi-geostrophic equations using the dual variable formulation (3.100) was proved by Cullen and Gangbo (2001). Assume that σ is a probability measure defined on $\mathbb{R}^2$ and h is a probability measure defined on the physical domain Γ, which is assumed to be open, bounded and connected. The characterisation of σ and

h as probability measures enforces conservation of mass in both (X, Y) and (x, y) coordinates.

Theorem 3.27. *Let $1 < p < \infty$. Let $B(0, r)$ be a ball of radius r centred at the origin. Assume $\Gamma \subset B(0, r)$. Given $\sigma(0, \cdot) \in L^p(\mathbb{R}^3)$ with compact support $\Sigma(0) \subset B(r_0)$ for some r_0. Then, for any $T > 0$, there exist functions*

$$\begin{aligned} \sigma &\in L^\infty\left([0,T]; L^p(\mathbb{R}^3)\right) \cap C[0,T); L^p_w(\mathbb{R}^3), \\ h &\in L^\infty\left([0,T]; W^{1,\infty}(\Gamma)\right) \cap C[0,T); W^{1,q}(\Gamma), \\ R &\in L^\infty\left([0,T]; W^{1,\infty}(\mathbb{R}^3)\right) \cap C[0,T); W^{1,q}(B(0,r)), \end{aligned} \tag{3.167}$$

where q is any number in $[1, \infty)$, both $P = f^{-2}gh + \frac{1}{2}(x^2 + y^2)$ and R are convex in their spatial variables, satisfying (3.100) where the evolution equation for σ and the initial data $\sigma(0, \cdot)$ are understood in the weak sense. The solution satisfies the estimate

$$\int_{\mathbb{R}^2\times\mathbb{R}^2} |\nabla R(t_1, \cdot) - \nabla R(t_2, \cdot)|\sigma \mathrm{d}X\mathrm{d}Y \leq fr(2 + fT)\|\sigma(0, \cdot)\|_{L^1(\mathbb{R}^2)}|t_1 - t_2|, \tag{3.168}$$

for all $t_1, t_2 \in [0, T]$. It also satisfies the estimate

$$\|\nabla R\|_{BV(\Sigma(t))} \leq cr(r_0 + frT)(1 + r_0 + frT), \tag{3.169}$$

where c is a constant independent of T.

The proof proceeds by first showing that, given $\sigma(t, .)$ satisfying the assumptions above, there is a unique probability measure h. The estimate (3.168) shows that, in an integrated sense, the position in Γ corresponding to any (X, Y) varies in a continuous way in time. This gives some information about the physical velocity $\mathbf{u}$. The estimate (3.169) of ∇R gives an estimate of $\mathbf{U}$ by using (3.45). It reflects the fact that $\Sigma(t)$ can only grow at a bounded rate. To illustrate this, plot the points (X, Y) in (x, y) space. Then $\mathbf{U}$ is at right angles to the line connecting (X, Y) to its image point with coordinates $\nabla R \in \Gamma$. It has been assumed that Γ contains the origin. Suppose A with coordinates (X, Y) is the point of $\Sigma(t)$ furthest from Γ so that $\max_{\mathbf{x}\in\Gamma}|\mathbf{X} - \mathbf{x}|$ is largest. Let this maximum difference be between A and B. Then this maximum difference can increase with the component of $\mathbf{U}$ that is parallel to BA. This is illustrated in Fig. 3.15. It can be seen that the magnitude of this component is less than fr_0, so that, if $\Sigma(0)$ is initially within $B(r_0)$, after time t it is within $B(r_t)$ where $r_t \leq r_0 + frt$. This growth bound is required to satisfy the conditions of Theorem 3.23.

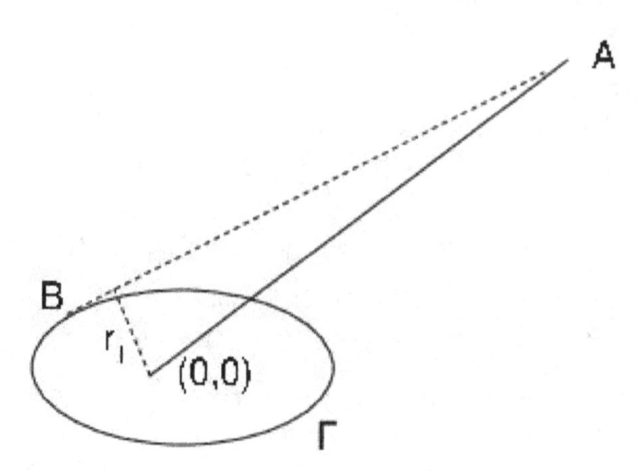

Fig. 3.15 Estimating the rate of growth of $\Sigma(t)$. The distance r_1 from the origin to the line BA is less than r_0, the radius of the ball containing Γ.

This theorem shows the existence of a time-continuous weak solution, characterised by a depth function $h(x, y)$ with $f^{-2}gh + \frac{1}{2}(x^2 + y^2)$ convex. Thus h itself does not have to be convex, but its curvature is restricted. The depth h may vanish over part of the domain Γ as shown in Fig. 3.5. The theorem covers the resulting free boundary problem.

Both Theorems 3.26 and 3.27 are restricted to σ in L^p, $p > 1$. The result of Lopes Filho and Nussenzveig Lopes (2002) extends these results to $p = 1$.

The extension of the free upper boundary case to three dimensions is discussed in sections 3.3.5 and 3.3.6. Proofs of existence for the incompressible case in dual variables, using equations (3.45) and (3.108, are given by Cullen *et al.* (2015b) and Cullen *et al.* (2019). These were extended to physical space in Cheng (2017) as discussed in section 3.5.4. The result of Cullen *et al.* (2019) is stated below. This was proved for the ocean case, so that the physical domain is $\Gamma \times [0, h)$ with $h \geq 0$.

Use the definitions from section 3.3.5 adapted to the ocean case, so Γ_h is the physical domain with free surface h, $\mathcal{L}_h$ the Lebesgue measure on Γ_h, $G(\nu_h, \sigma)$ the set of transport plans γ between ν_h and σ and $C(\gamma)$ the cost to be minimised. The result below is proved using Theorem 3.23, and some of the definitions set out before the statement of that theorem are used in stating the result.

Theorem 3.28. *Let B be an open set in $\mathbb{R}^2$ with smooth boundary. Let*

$1 \le p \le \infty$, *and let* $\sigma_0 \in \mathcal{P}_2^a(\mathbb{R}^3)$ *be a compactly supported density in* $L^p(\mathbb{R}^3)$. *Let* $h_0 \in W^{1,\infty}(B)$ *be given, and satisfy*

$$h_0 = \arg \min_{h \in A_*} \left(\min_{\gamma \in G(\mathcal{L}_h, \sigma_0)} C(\gamma) \mathrm{d}\mathbf{x} \right),$$

where

$$A_* := \{\Gamma_\eta \subset \mathbb{R}^3 : \eta \in C^0(\bar{B}), \eta \ge 0 \text{ and } \int_B \eta = 1\}.$$

Then there is a global in time weak solution (h, σ) *of (3.45) and (3.108).*

The result for the compressible case, using equations (3.123) and the dual variables (3.126) is given by Cullen *et al.* (2015b). As in section 3.3.6, define $\mathcal{L}_p$ as Lebesgue measure restricted to the domain Γ_p bounded by $p = p_s$, and σ the mass measure in dual variables. Remaining notation is as in section 3.3.6.

Theorem 3.29. *Let* $1 < r < \infty$ *and let* $\sigma_0 \in L^r(\Sigma_0)$, *where* Σ_0 *is a a bounded open set in* $\mathbb{R}^3$. *Let* T *be the optimal map from* $\mathcal{L}_p$ *to* σ *with cost* E *given by (3.127). Then equations (3.123), rewritten in the dual variables (3.126), have a stable solution* (p_s, T) *where* $p_s(t, \cdot) \in W^{1,\infty}(B(0, S))$ *for some* $S > 0$.

For $\sigma = T_\# \mathcal{L}_h$ *and* $\mathbf{U}$ *as in (3.45), the solution satisfies*

(i) $\sigma(\cdot, \cdot) \in L^r((0, \tau) \times \Sigma), \|\sigma(t, \cdot)\|_{L^r(\Sigma)} \le \|\sigma_0(\cdot)\|_{L^r(\Sigma_0)} \forall t \in [0, \tau]$, *where* Σ *is a bounded open domain contaning the support of* $\sigma(t, \cdot)$.

(ii) $\varphi(t, \cdot) \in W^{1,\infty}(\Gamma \times [0, p_s]), \|\varphi(t, \cdot)\|_{W^{1,\infty}(\Gamma \times [0,p_s])} \le C = C(\Gamma \times [0, p_s], \Sigma, E), \forall t \in [0, \tau]$,

(iii) $\|\mathbf{U}\|_{L^\infty(\Sigma)} \le C = C(\Gamma \times [0, p_s], \Sigma), \forall t \in [0, \tau]$.

The admissibility principle for semi-geostrophic solutions introduced in section 3.1.3 only requires convexity of P. This means that σ can be any probability measure, no regularity is required. Ideally, it should be possible to solve the semi-geostrophic equations with no further restrictions on σ. In addition, for physical applicability it is necessary to treat cases where there is a region of physical space with uniform X, Y or Z, for instance regions of uniform potential temperature. This is expected to be more difficult because it cannot be expected that it will be possible to define a deterministic trajectory in physical space generated by a measure-preserving mapping. It is, however, possible to make progress in dual variables by extending the results to allow σ to be measure-valued. Such an extension is given by

Loeper (2006). First define a solution of (3.45) in a *weak measure* sense, which replaces (3.165) for $\varrho \in C_c^\infty([0,T) \times \mathbb{R}^3)$ by

$$\int_{[0,T)\times\mathbb{R}^3} \frac{\partial \varrho}{\partial t} \sigma \mathrm{d}X\mathrm{d}Y\mathrm{d}Z\mathrm{d}t + \int_{[0,T)\times\Gamma} f(-Y,X,0)\cdot\nabla\varrho(t,X,Y,Z)\mathrm{d}t\mathrm{d}x\mathrm{d}y\mathrm{d}z + \int_{\mathbb{R}^3} \sigma(0,\cdot)\varrho(0,\cdot)\mathrm{d}X\mathrm{d}Y\mathrm{d}Z = 0. \tag{3.170}$$

Replacing the second term in (3.165) by an integral over Γ avoids the problem that $\mathbf{U}$ is multi-valued at points (X,Y,Z) where σ is unbounded. The natural physical interpretation is that at such singular points, the image of (X,Y,Z) is a set $D \subset \Gamma$. $\mathbf{U}$ is calculated by replacing ∇R at such points by the coordinates of the centre of mass of D. This is the barycentric projection introduced before Theorem 3.25. Loeper then proves

Theorem 3.30. *Let $\sigma(0,\cdot)$ be a probability measure with compact support Σ. Then there exists a weak measure solution of equations (3.45) with the properties that $\sigma \in C[0,T)$, there exists $r(t)$ such that for all $t \in [0,T], \Sigma(t) \subset B(0,r(t))$, and (3.170) holds for all $\varrho \in C^\infty(\mathbb{R}^3 \times [0,T))$. In addition, for any $T > 0$, if $\{\sigma_n\}$ is a sequence of weak measure solutions with initial data $\sigma_n(0,\cdot)$ supported on $\Sigma_n(0) \subset B(0,r)$, for r independent of n, then the sequence is precompact and every converging subsequence of such solutions converges to a weak measure solution of (3.45).*

The second part of the theorem shows that it is possible to find solutions by taking the limit of sequences of approximations, even in this more difficult case where $\mathbf{U}$ can be multi-valued. This property is necessary for physical usefulness of the equations.

All the theorems stated so far in this subsection are valid for arbitrarily large times, showing that the semi-geostrophic system defines a slow manifold. Physical usefulness of this, however, requires the extension to spherical geometry discussed in sections 4.2 and 4.3. The results would also be much more useful in defining limit solutions for the Navier-Stokes equations if uniqueness could be proved. Otherwise there is the possibility of convergence to different limit solutions in a stochastic manner, which would imply very low predictability.

Thus it is of interest to consider the case where the solutions are smooth. As shown in Theorem 3.20, it is possible to prove that (X,Y) are differentiable functions of (x,y) in the periodic case $\mathbb{T}^2$. Consider this problem in the three-dimensional periodic case used in Theorem 3.18. Equations (3.45) are to be solved with periodic boundary conditions on $R - \frac{1}{2}(X^2 + Y^2 + Z^2), \sigma, x, y$ and z. If smooth solutions can be maintained for large times in this geometry, it can be expected that the interior flow will remain smooth in the three-dimensional case with a finite domain Γ, since then non-smooth behaviour can only emanate from the boundaries. A theorem of Loeper (2006), using the condition (3.151) for R to be twice differentiable, is

Theorem 3.31. *Let $\sigma(\cdot, 0)$ be a probability measure on $\mathbb{T}^3$, such that σ is strictly positive and satisfies equation (3.151) everywhere. Then there exists $T > 0$ and c_1, c_2 depending on $\sigma(0, \cdot)$ such that on $[0, T]$ there exists a solution $\sigma(t, \cdot)$ of (3.45) that satisfies for all $t \in [0, T]$*

$$\int_0^1 \frac{\omega(t, r)}{r} dr < \infty, \quad \|R(t, \cdot)\|_{C^2(\mathbb{T}^3)} \leq c_2, \tag{3.171}$$

where $\omega(r)$ is as defined before equation (3.151).

The estimates in Theorem 3.22 are used by O'Neill (2020) to show that the time of existence of classical solutions can be extended indefinitely by imposing conditions on σ. This yields

Theorem 3.32. *Let $\sigma_0 \in \mathcal{C}^{1/2}(\Sigma)$ be such that $\|\sigma_0 - 1\| \leq \epsilon$. Suppose also that*

$$\epsilon \|\mathbf{U}(0, \mathbf{X}\|_{L^\infty(\Sigma)} = \|\nabla_{\mathbf{X}} R(0, \mathbf{X}) - \mathbf{X}\|_{L^\infty(\Sigma)} \leq \epsilon$$

and $R(0, 0) = 0$. then there exists a time $T = C \ln(\epsilon^{-1/2}) > 0$ such that for all $t \in [0, T]$ there exists a solution $\sigma \in C([0, T], \mathcal{C}^{1/2}(\Sigma))$ to (3.45). Moreover, for all $t \in [0, T]$,

$$\|\sigma\|^{1/2}_{\mathcal{C}^{1/2}(\Sigma)} \leq \|\sigma_0\|^{1/2}_{\mathcal{C}^{1/2}(\Sigma)} + C$$

for a constant $C \simeq O(1)$ that is independent of ϵ.

Note that the condition $\sigma \in C([0, T], \mathcal{C}^{1/2}(\Sigma))$ implies continuity of σ, as explained in O'Neill (2020), which together with the assumed bounds on σ implies that the solutions are classical. However it has not yet been shown whether the regularity associated with $\mathcal{C}^{1/2}$ is enough for uniqueness.

None of the results quoted so far give uniqueness or continuous dependence on the data, even though the rearrangement problem solved in Theorem 3.13 has a unique solution which depends continuously on the data. This is because they rely on the construction of an approximating sequence of convex functions $R_n(t)$ which can be proved to have a convergent subsequence with a limit which solves the equations. However, the choice of approximating sequence and of any convergent subsequence is not unique. To prove uniqueness, it is necessary to consider the case where two solutions R_1, R_2 of (3.45), with associated velocity fields $\mathbf{U}_1, \mathbf{U}_2$ and potential densities $\sigma_1(t,\cdot), \sigma_2(t,\cdot)$, evolve from the same initial data $\sigma(0,\cdot)$. It is necessary to show that, if $|\sigma_1 - \sigma_2|$ is $\mathrm{O}(\delta t)$ then $|R_1 - R_2| = \mathrm{O}(\delta t)$. Alternatively, if the equations are used in the real space form (3.30), it would be necessary to show that if $\mathbf{X}_1 - \mathbf{X}_2 = \mathrm{O}(\delta t)$ then $P_1 - P_2 = \mathrm{O}(\delta t)$. Though Theorem 3.13 gives continuous dependence on the data, it does not give as strong an estimate of the differences as this requires. If R and P are twice differentiable, however, the estimate can be obtained. In Loeper (2006), Theorem 3.31 is exploited to give a short time uniqueness result.

Theorem 3.33. *Let $\sigma(\cdot, 0)$ be a probability measure on $\mathbb{T}^3$, such that $0 < \lambda \leq \sigma \leq \Lambda$ and $\sigma \in C^\alpha(\mathbb{T}^3)$ for some $\alpha > 0$. From Theorem 3.31, for some $T > 0$ there exists a solution $\bar{\sigma}$ to (3.45) in $L^\infty([0, T'], C^\beta(\mathbb{T}^3))$. Then every solution of (3.45) in $L^\infty([0, T'], C^\beta(\mathbb{T}^3))$ for $T' > 0$, $\beta > 0$ with the same initial data coincides with $\bar{\sigma}$ on $[0, \inf\{T, T'\}]$.*

Note that the theorem does not prevent the existence of other solutions for which σ is not in C^β for any β.

3.5.4 *Solutions in physical variables*

The aim of the study of the semi-geostrophic equations is to approximate the Navier-Stokes equations in some asymptotic limit. It is therefore necessary to solve the equations written in a similar form to the Navier-Stokes equations, so that the differences can be estimated. The theorems in the previous subsection all apply to the equations in dual variables, (3.45) or (3.100). This has no natural counterpart in the Navier-Stokes equations. In this section the results are extended to the Lagrangian equations in physical space, using the forms (3.30) or (3.90), including the case of the free surface equation (3.108). Since the semi-geostrophic approximation is purely Lagrangian, as described in section 2.5.2, it is appropriate that the governing equations take this form. Recent results show that the Eulerian form of the semi-geostrophic equations makes sense except in the important case of boundaries, where discontinuities can form. In section 3.2.3 it was shown that the formation of discontinuities can lead to fluid initially at the boundary moving away from it, which is hard to handle in an Eulerian formulation.

In section 3.5.2 it was noted that the measure-preserving mapping defined by the polar factorisation, which defines the physical trajectory, only exists under restrictive conditions which do not cover all physically important cases. These restrictions can be removed by defining a relaxed form of Lagrangian solution as described below.

In the absence of boundaries, Eulerian weak solutions can be defined. These are still not in general regular enough for uniqueness to be proved. In smooth cases it should be possible to find classical solutions and prove uniqueness and continuous dependence on the data.

First conside the case of Theorem 3.26 for the incompressible Boussinesq Lagrangian semi-geostrophic equations (3.30) in a bounded domain $\Gamma \subset \mathbb{R}^3$. The necessary definitions and results were obtained by Cullen and Feldman (2006). The first step is to define the Lagrangian flow.

Definition 3.12. Let $\Gamma \subset \mathbb{R}^3$ be an open set. Let $P(0,\cdot)$ be a bounded convex function in Γ. Then for any $T > 0$, a pair $(P(t,\cdot), F(t,\cdot))$, where

$$P \in L^\infty\left([0,T); W^{1,\infty}(\Gamma)\right) \cap C[0,T); W^{1,q}(\Gamma), \tag{3.172}$$

$P(t,\cdot)$ is convex in Γ for each $t \in [0,T]$,

$$F : [0,T) \times \Gamma \to \Gamma,\ F \in C[0,T); L^q(\Gamma), \tag{3.173}$$

where q is any number in $[1,\infty)$, is called a weak Lagrangian solution of equations (3.30) if the following hold.

(i) $\mathbf{F}(0,x,y,z) = (x,y,z)$.

(ii) For any $t \in [0,T]$ the mapping $(x,y,z) \to \mathbf{F}(t,x,y,z)$ is measure-preserving in the sense that $\mathbf{F}(t,\cdot)_\# \mathcal{L} = \mathcal{L}$, where $\mathcal{L}$ is the Lebesgue measure on Γ.

(iii) For every $t \in (0,T)$ there exists a mapping $\mathbf{F}^*(t,\cdot)$ such that $\mathbf{F}^*(t,\cdot)_\# \mathcal{L} = \mathcal{L}$ and satisfies $\mathbf{F}^*(t,\cdot) \circ \mathbf{F}(t,x,y,z) = (x,y,z)$, $\mathbf{F}(t,\cdot) \circ \mathbf{F}^*(t,x,y,z) = (x,y,z)$ for almost every (x,y,z).

(iv) The function $\mathbf{X}(t,\cdot) = \nabla P(t,\cdot)$ satisfies the following. Define

$$\mathbf{Z}(t,\cdot) = \mathbf{X}(t,\mathbf{F}(t,\cdot)) \equiv \nabla P(t,\mathbf{F}(t,\cdot)). \tag{3.174}$$

Then $\mathbf{Z} \in L^\infty([0,T] \times \Gamma)$. Writing $\mathbf{Z} = (Z_1, Z_2, Z_3)$ and $\mathbf{F} = (F_1, F_2, F_3)$, $\mathbf{Z}$ is a weak solution of

$$\begin{aligned} \frac{\partial Z_1}{\partial t} &= f(F_2 - Z_2), \\ \frac{\partial Z_2}{\partial t} &= f(Z_1 - F_1), \\ \frac{\partial Z_3}{\partial t} &= 0, \end{aligned} \tag{3.175}$$

in $[0,T)\times\Gamma$ with $\mathbf{Z}(0,\cdot)=\nabla P(0,\cdot)$. The weak solution is in the sense that for any $\varrho\in C_c^1([0,T)\times\Gamma)$

$$\int_{[0,T)\times\Gamma}\left(Z_1\frac{\partial\varrho}{\partial t}+f\varrho(F_2-Z_2)\right)\mathrm{d}x\mathrm{d}y\mathrm{d}z\mathrm{d}t+ \qquad (3.176)$$

$$\int_\Gamma\frac{\partial P(0,\cdot)}{\partial x}\varrho(0,\cdot)\mathrm{d}x\mathrm{d}y\mathrm{d}z,$$

with similar equations for the other two components of (3.175).

It can be shown that this type of solution is equivalent to a conventional solution if all the fields are smooth enough. Then Cullen and Feldman (2006) prove

Theorem 3.34. *Let Γ be open and bounded, and the closure of Γ be contained in $B=B(0,r)$. Let $P(\cdot,0)$ be a convex function in $W^{1,\infty}(B)$, and that its Legendre transform R satisfies*

$$\det\text{Hess}\,R\in L^p\left(\nabla P(0,\cdot)(\Gamma)\right) \qquad (3.177)$$

for some $p>1$. Then for any $T>0$ there exists a weak Lagrangian solution $(P,\mathbf{F})$ of (3.30) in $[0,T)\times\Gamma$, where (3.172) and (3.173) are satisfied for any $q\in[1,\infty)$. Moreover, the function $\mathbf{Z}$ defined by (3.174) satisfies, possibly after modification on a negligible set, $\mathbf{Z}\in W^{1,\infty}([0,T);\mathbb{R}^3)$ for almost all $\mathbf{x}\in\Gamma$, and (3.175) is satisfied, in addition to the weak sense (3.176), for almost all (t,x,y,z) in $[0,T)\times\Gamma$.

The proof uses the Lagrangian trajectory Φ generated by the solution of the dual variable formulation solved in Theorem 3.26. The existence of this trajectory follows from results of Ambrosio (2004). This trajectory is then projected to physical space giving the Lagrangian flow $\mathbf{F}$ of Definition 3.12. Theorem 3.34 was extended to the case $p=1$ by Faria *et al.* (2009). It shows that the three-dimensional incompressible Boussinesq Lagrangian semi-geostrophic equations in physical space can be solved for arbitrarily large times, subject to suitable initial data satisfying (3.177) which is a convexity condition on $P(0,\cdot)$. They therefore form a slow manifold. It is possible to prove an equivalent result for the shallow water semi-geostrophic equations (3.90). Since the definitions have to be modified to allow for the 'compressibility' of shallow water flow, they are given below. Note that, in this context, a measure-preserving mapping becomes a mass-preserving mapping to enforce the mass conservation property on the Lagrangian flow.

Definition 3.13. Let $\Gamma\subset\mathbb{R}^2$ be an open set. Let $P(0,\cdot)$ be a convex function in $W^{1,\infty}(\Gamma)$. Then for any $T>0$, a pair $(P(t,\cdot),\mathbf{F}(t,\cdot))$, where P

is convex in Γ and

$$P \in L^\infty([0,T); W^{1,\infty}(\Gamma \times [0, h(\cdot, x, y)])) \cap$$
$$C[0,T); W^{1,q}(\Gamma \times [0, h(\cdot, x, y)]), \qquad (3.178)$$
$$\mathbf{F} : [0,T) \times \Gamma \to \Gamma,\ \mathbf{F} \in C[0,T); L^q(\Gamma \times [0, h(\cdot, x, y)]),$$

where q is any number in $[1, \infty)$, is called a weak Lagrangian solution of equations (3.90) if the following hold.

(i) $\mathbf{F}(0, x, y) = (x, y)$.

(ii) For any $t > 0$ the mapping $(x, y) \to \mathbf{F}(t, x, y)$ satisfies $\mathbf{F}(t, \cdot)_{\#}h(0, \cdot) = h(t, \cdot)$.

(iii) For every $t \in (0,T)$ there exists a mapping $\mathbf{F}^*(t, \cdot)$ satisfying $\mathbf{F}^*(t, \cdot)_{\#}h(t, \cdot) = h(0, \cdot)$ and $\mathbf{F}^*(t, \cdot) \circ \mathbf{F}(t, x, y) = (x, y)$, $\mathbf{F}(t, \cdot) \circ \mathbf{F}^*(t, x, y) = (x, y)$ for almost every $\mathbf{x}$.

(iv) The function $\mathbf{X}(t, \cdot) = \nabla P(t, \cdot)$ satisfies the following. Define

$$\mathbf{Z}(t, \cdot) = \mathbf{X}(t, \mathbf{F}(t, \cdot)) \equiv \nabla P(t, \mathbf{F}(t, \cdot)) \qquad (3.179)$$

Then $\mathbf{Z} \in L^\infty([0, t] \times \Gamma)$. Writing $\mathbf{Z} = (Z_1, Z_2)$ and $\mathbf{F} = (F_1, F_2)$, $\mathbf{Z}$ is a weak solution of

$$\frac{\partial Z_1}{\partial t} = f(F_2 - Z_2), \qquad (3.180)$$
$$\frac{\partial Z_2}{\partial t} = f(Z_1 - F_1)$$

in $[0,T) \times \Gamma$ with $\mathbf{Z}(0, \cdot) = \nabla P(0, \cdot)$. The weak solution is in the sense that for any $\varrho \in C_c^1([0,T) \times \Gamma)$

$$\int_{[0,T)\times\Gamma} \left(Z_1 \frac{\partial \varrho}{\partial t} + f\varrho(F_2 - Z_2) \right) h(0, \cdot) dx dy dt + \qquad (3.181)$$
$$\int_\Gamma \frac{\partial P(0, \cdot)}{\partial x} \varrho(0, \cdot) h(0, \cdot) dx dy,$$

with a similar equation for the other component of (3.180).

Theorem 3.35. *Let Γ be open and bounded, and the closure of Γ be contained in $B = B(0, r)$. Let $h(0, \cdot) \geq 0$ be such that $P(0, \cdot) = f^{-2}gh(0, \cdot) + \frac{1}{2}(x^2 + y^2)$ is a convex bounded function on B, and assume that*

$$\nabla P(0, \cdot)_{\#}h(0, \cdot) \in L^p(\nabla P(0, \cdot) \qquad (3.182)$$

for some $p > 1$. Then for any $T > 0$ there exists a weak Lagrangian solution $(\mathbf{P}, \mathbf{F})$ of (3.90) in $[0,T) \times \Gamma$, where (3.178) is satisfied for any $q \in [1, \infty)$. Moreover, the function $\mathbf{Z}$ defined by (3.174) satisfies, possibly after modification on a negligible set, $\mathbf{Z} \in W^{1,\infty}([0,T); \mathbb{R}^2)$ for almost all $\mathbf{x} \in \Gamma$, and (3.180) is satisfied, in addition to the weak sense (3.181), for almost all (t, x, y) in $[0,T) \times \Gamma$.

This theorem was also extended to the case $p = 1$ by Faria *et al.* (2009).

Next extend Theorem 3.34 to include a free surface governed by equation (3.108). This is done by Cheng (2017) by using the result quoted in Theorem 3.28 and transferring it to physical space in the same way as in Cullen and Feldman (2006). As with Theorem 3.28, the result was proved for the ocean case with $\Gamma_h = \Gamma \times [0, h]$ with $h \geq 0$. The Lagrangian flow $\mathbf{F}$ and the potential P must satisfy the assumptions in Definition 3.13 and the function $\mathbf{Z}$ has to be defined by (3.174). The definition of a weak Lagrangian solution requires properties (i)-(iii) of Definition 3.13. Property (iv) is replaced by the property in the theorem below. The parameter η defines *a priori* bounds on $\partial P/\partial z$ given by eq. (3.115).

Theorem 3.36. *Let $T > 0, 1 < p < \infty$ and admissible initial data (h_0, P_0) in the sense of Definition 3.6 in section 3.3.5 be given. Suppose that*

$$\sigma_0 := \nabla P_{0\#}\nu_{h_0} \in L^p(\mathbb{R}^3),$$

with ν_{h_0} defined as after Definition 3.6. Suppose also that

$$H > \frac{2}{L^2(\Gamma)} + \frac{2diam\Gamma}{\eta}\left[\|\nabla P_0\|_{L^\infty}T + \max_{\Gamma}|\mathbf{x}|(T+2) + 2\right], \qquad (3.183)$$

then there exists a weak Lagrangian solution (h, P, F) on $[0, T] \times \Gamma_h$. Moreover, the function $\mathbf{Z}(\cdot, \mathbf{x}) \in W^{1,\infty}(\mathbb{R}^3)$ for a.e. $\mathbf{x} \in \Gamma_{h_0}$ and the equations are satisfied in the following sense:

$$\partial_t \mathbf{Z}(t, \mathbf{x}) = J(\mathbf{Z}(t, \mathbf{x}) - \mathbf{F}(t, \mathbf{x}))\ L^4 - a.e.\ in\ (t, \mathbf{x}) \in [0, T] \times \Gamma_{h_0} \qquad (3.184)$$
$$\mathbf{Z}(0, \mathbf{x}) = \nabla P_0(\mathbf{x})\ L^3 - a.e.\ in\ \mathbf{x} \in \Gamma_{h_0}.$$

Now consider extending Theorem 3.30 on the existence of weak measure solutions in dual variables to physical space. It can be expected that such an extension is possible by replacing the space C_c^1 by C_c^∞ in the definitions (3.176) and (3.181) of weak Lagrangian solutions. Assuming that this can be done, it would imply the existence of solutions where the physical velocity (u, v, w) can be measure-valued. The mapping of parcels with given (X, Y, Z) into physical space, and thus the trajectory, is still well-defined.

In Feldman and Tudorascu (2013), a solution of (3.45) was obtained when σ_0 is a combination of an absolutely continuous probability measure with respect to Lebesgue measure and a finite set of Dirac masses. This means that a point in $\mathbf{X}$ space will correspond to a set in physical space, as in the discrete problem solved in section 3.4. The calculation of $\mathbf{U}$ required to solve (3.45) and define a dual space trajectory Φ is carried out at the centroid of the sets, as before Theorem 3.25 and also after (3.170). Definition 3.12 is replaced by:

Definition 3.14. Let $P : [0,\infty) \times \Gamma \rightarrow \mathbb{R}$ be such that $P \in C([0,\infty); H^1(\Gamma))$ and $P(t,\cdot)$ is convex in Γ for each $t \in [0,\infty)$. Let $\mathbf{F} : [0,\infty) \times \Gamma \rightarrow \Gamma$ be a weakly P–continuous Borel map. Let $\gamma_t := (\nabla P_t \times Id)_{\#}\mathcal{L}(\Gamma)$ be the joint probability measure associated with the optimal plan between Γ and Σ. Let $\bar{\gamma}$ be its barycentric projection as defined before Theorem 3.25, and set

$$\sigma_t := \nabla P_{t\#}\mathcal{L}(\Gamma),\ \mu_t = \bar{\gamma}_{t\#}\sigma_t.$$

Then the pair (P, F) is called a weak Lagrangian solution of (3.45) in $[0,T)\times\Gamma$ if

(i) $\mathbf{F}(0,x,y,z) = (x,y,z)$ and $P(0,x,y,z) = P_0(x,y,z)$ for all $(x,y,z) \in \Gamma$.
(ii) For any $t > 0$ the mapping $\mathbf{F}_t := \mathbf{F}(t,\cdot) : \Gamma \rightarrow \Gamma$ satisfies $\mathbf{F}(t,\cdot)_{\#}\mathcal{L}(\Gamma) = \mu_t$ and $\mathbf{F}(t,\cdot)_{\#}\mu_0 = \mu_t$.
(iii) The function $Z : (0,T) \times \Gamma \rightarrow \mathbb{R}^3$ defined by

$$Z_t = \nabla P_t \circ \mathbf{F}_t$$

lies, along with $\mathbf{F}$, in $L^\infty(0,T : L^2(\Gamma,\mathbb{R}^3))$ and is a distributional solution of (3.180) in the sense of (3.181)

This result still does not allow σ to be an arbitrary probability measure, which is all that is required by the admissibility principle of section 3.1.3. Theorem 3.30 does this in dual variables, and correspondingly a *relaxed Lagrangian solution* in physical space was defined in Feldman and Tudorascu (2015). This allows the initial potential P to be any convex function, with no additional regularity. Definition 3.13 is therefore replaced by a relaxed version in terms of probability measures.

Define a measure ν on $(0,T) \times \Gamma \times \Gamma$ by

$$\int_0^T \int_\Gamma \int_\Gamma \xi(t,\mathbf{x},\mathbf{y})\nu(dt,d\mathbf{x},d\mathbf{y}) = \int_0^T \int_\Gamma \xi(t,\mathbf{x},\mathbf{F}(t,\mathbf{x}))\mathrm{d}\mathbf{x}\mathrm{d}t, \tag{3.185}$$

for all $\xi \in C_b((0,T)\times\Gamma\times\Gamma)$ and where $\mathbf{F}$ is a measure-preserving mapping as defined in Definition 3.12. The function ξ gives the solution a renormalisation property which allows a proof that the relaxed solutions conserve energy. The measure-preserving property means that (3.185) becomes

$$\int_0^T \int_\Gamma \int_\Gamma \xi(t,\mathbf{x},\mathbf{y})\nu(dt,d\mathbf{x},d\mathbf{y}) = \int_0^T \left(\int_\Gamma \xi(t,\mathbf{x},\mathbf{y})\nu_t(d\mathbf{x},d\mathbf{y}\right)\mathrm{d}t, \tag{3.186}$$

for all $\xi \in C_b((0,T) \times \Gamma \times \Gamma)$, where $[0,T] \ni t \to \nu_t = (Id \times \mathbf{F}_t)_{\#}\mathcal{L}(\Gamma)$ is a Borel family of measures on $\Gamma \times \Gamma$ such that

$$p_2\nu_t = p_3\nu_t = \mathcal{L}(\Gamma), \forall t \in [0,T], \tag{3.187}$$

where p_2, p_3 are analogous to the projections defined before eq. (3.156). Then, if $\mathbf{Z}$ is defined by (3.174),

$$\begin{aligned}\int_0^T \int_\Gamma \int_\Gamma \Big(\xi(\nabla P_t(\mathbf{y}))\partial_t\zeta(t,\mathbf{x}) - f\nabla\xi(\nabla P_t(\mathbf{y}))\cdot \\ J(\mathbf{y} - \nabla P_t(\mathbf{y}))\zeta(t,\mathbf{x})\Big)\nu(\mathrm{d}t, d\mathbf{x}, d\mathbf{y}) + \int_\Gamma \xi(\nabla P_0(\mathbf{x}))\zeta(0,\mathbf{x})\mathrm{d}\mathbf{x} = 0,\end{aligned} \tag{3.188}$$

for all $\xi \in C_b^1(\mathbb{R}^3)$ and all $\zeta \in C_c^1([0,T) \times \Gamma)$. Then a relaxed Lagrangian solution of (3.30) is defined by

Definition 3.15. Let $P_0 \in H^1(\Gamma)$ be convex. Consider a family of convex functions $P \in L^\infty(0,T : H^1(\Gamma))$ and a Borel family of measures $(0,T) \ni t \to \nu_t \in \mathcal{M}(\Gamma \times \Gamma)$, where $\mathcal{M}(\cdot,\cdot)$ is a class of joint probability measures. Let ν be given by $d\nu = d\nu_t dt$ as in (3.185). Then (P,ν) is a relaxed Lagrangian solution of (3.30) with initial data P_0 if (3.187) and (3.188) are satisfied.

Feldman and Tudorascu (2015) then prove

Theorem 3.37. *Let $P_0 \in H^1(\Gamma), \{P_0^n\}_n \subset H^1(\Gamma)$ be convex and such that $\nabla P_0^n \to \nabla P_0$ in $L^2(\mathbb{R}^3)$. Assume that (P^n, ν^n) are relaxed solutions to (3.30) corresponding to the initial data P_0^n. Then, possibly up to a subsequence, (P^n, ν^n) converges to a relaxed solution (P, ν) corresponding to the initial data P_0. The convergence is in the following sense:*

(i) $\nabla P_t^n \to \nabla P_t$ in $L^p(\Gamma, \mathbb{R}^3)$ for all $t \in [0,T)$ and all $1 \le p < 2$.
(ii) ν^n converges weakly as measures to ν.

Furthermore, the corresponding dual space solutions satisfy $W_p(\sigma_t^n, \sigma_t) \to 0$ for all $t \in [0,T)$ and all $1 \le p < 2$. Thus there exists a relaxed Lagrangian solution (P, ν) to (3.30) with initial data P_0.

Item (ii) is a form of time continuity of the measure ν. Physically, this means that the position of an individual fluid particle is only known probabilistically at a particular time. However, the Eulerian state of the fluid, as defined by P_t, is known deterministically and satisfies the time continuity given by item (i). This is an example of the degeneracy of the

Lagrangian description of the flow.This result is extended to the shallow water case in Feldman and Tudorascu (2015a).

A further physically relevant case is a situation where mass is transported by a finite distance in zero time. Clearly this cannot happen as an exact solution of the Navier-Stokes equations. The effect of the semi-geostrophic approximation is to collapse time-scales faster then f^{-1} to zero; so that if the real system transports mass on faster time-scales, this will be represented as instantaneous transport in the semi-geostrophic system. An example is the mass transport in convective clouds. In reality, this takes a time up to 1-2 hours. In a semi-geostrophic model it is instantaneous, as will be seen in section 6.2.4. Another example occurs where the trajectory is constrained by mountain ranges, as shown in section 5.5. The only rigorous treatment of such a case is the one-dimensional moist rearrangement problem solved in section 6.2.6. However, as discussed in section 6.2.8, it is not yet possible to build this rigorously into a multi-dimensional time evolution problem.

Cheng (2017) shows that relaxed Lagrangian solutions can also be found in the free surface case, eq. (3.30) with (3.108). The definition of a relaxed Lagrangian solution follows Definition 3.15 with the additions below, and the parameter η defines the *a priori* bounds on $\partial P/\partial z$ given by eq. (3.115) as before. The definition of admissible generalised data is

Definition 3.16. Let $P_0 : \Gamma_\infty \to \mathbb{R}$ be a convex function, and $h_0 \in L^2(\Gamma)$. Then (P_0, h_0) is admissible generalised data if:

(i) $\nabla P_0 \in L^2(\Gamma_{h_0})$ and $-\frac{1}{\eta} \leq \partial P_0/\partial z \leq \eta$,
(ii) $h_0 > 0$ and $\int_\Gamma h_0 = 1$,
(iii) $P_0(x, y, h_0(x, y)) = \frac{1}{2}(x^2 + y^2)\ \forall (x, y) \in \Gamma$.

Then the result is:

Theorem 3.38. *Let (P_0, h_0) be admissible generalised data for (3.30) and (3.108). Then there exists a relaxed Lagrangian solution (P, h, ν) having (P_0, h_0) as initial data. In addition, there is the following continuity in time:*

(i) $h \in C([0, T] : L^2(\Gamma))$,
(ii) $\xi(P, \nabla P)\nu_h \in C([0, T] : L^1(\Gamma_\infty)), \forall \xi \in C_b(\mathbb{R} \times \mathbb{R}^3)$.

Next consider the existence of Eulerian solutions. This requires more regularity. The result of De Philippis and Figalli (2012) achieves this in cases without boundaries, as shown in Theorems 3.20 and 3.21. These results are then exploited to prove existence of Eulerian solutions. In the two-dimensional periodic case, Ambrosio *et al.* (2012), the semi-geostrophic equations (3.2) become

$$\begin{aligned} \frac{\mathrm{D}u_g}{\mathrm{D}t} + \frac{\partial \varphi}{\partial x} - fv = 0, \\ \frac{\mathrm{D}v_g}{\mathrm{D}t} + \frac{\partial \varphi}{\partial y} + fu = 0, \\ \nabla \cdot \mathbf{u} = 0. \end{aligned} \tag{3.189}$$

Note that these equations are not physically relevant, as they imply an infinite Rossby radius of deformation. As discussed in sections 2.3.4-2.3.6, these equations are only an accurate approximation to the Euler equations on scales larger than the deformation radius. However, they are often used to illustrate mathematical techniques appropriate to physically relevant cases. As shown in section 3.6.2, they still approximate the two-dimensional incompressible Euler equations to $\mathrm{O}(Ro)$.

Equation (3.189) can be written in dual variables in the manner of (3.160) and (3.161) as

$$\begin{aligned} \frac{\partial \sigma}{\partial t} + \mathbf{U} \cdot \nabla \sigma = 0, \\ \det(\text{Hess } R) = \sigma, \\ (x, y) = \nabla R, \\ (U, V) = (f(y - Y), f(X - x)). \end{aligned} \tag{3.190}$$

Theorem 3.20 shows that (X, Y) are differentiable functions of (x, y) because of the periodic boundary conditions, and the first equation of (3.190) shows that any initial bounds on σ required by Theorem 3.20 will be preserved in time. Suppose $\lambda \le \sigma_0(\cdot) \le \Lambda < \infty$. Then Ambrosio *et al.* (2012), Theorem 3.1, use this to prove long time existence of solutions of (3.190) which satisfy $\lambda \le \sigma(t, \cdot) \le \Lambda$.

Weak Eulerian solutions of (3.189) are defined by

Definition 3.17. Let $\varphi : (0, \infty) \times \mathbb{T}^2 \to \mathbb{R}$ and $\mathbf{u} : (0, \infty) \times \mathbb{T}^2 \to \mathbb{R}^2$. Then $(\varphi, \mathbf{u})$ is a *weak Eulerian solution* of (3.189) if

(i) $|\mathbf{u}| \in L^\infty((0, \infty), L^1(\mathbb{T}^2)), \varphi \in L^\infty((0, \infty), W^{1,\infty}(\mathbb{T}^2))$, and $\varphi(t, \cdot) + |\mathbf{x}|^2/2$ is convex for any $t \ge 0$.

(ii) For any $\phi \in C_c^\infty([0, \infty) \times \mathbb{T}^2)$:

$$\begin{aligned} \int_0^\infty \int_{\mathbb{T}^2} J\nabla\varphi(t, \mathbf{x})\{\partial_t \phi(t, \mathbf{x}) + \mathbf{u}(t, \mathbf{x}) \cdot \nabla \phi(t, \mathbf{x})\} - \\ f\{\nabla \varphi(t, \mathbf{x}) + J\mathbf{u}(t, \mathbf{x})\}\phi(t, \mathbf{x})\mathrm{d}\mathbf{x}\mathrm{d}t + \int_{\mathbb{T}^2} J\nabla\varphi(0, \mathbf{x})\phi(0, \mathbf{x})\mathrm{d}\mathbf{x} = 0. \end{aligned} \tag{3.191}$$

(iii) For a.e. $t \in (0, \infty)$: $\int_{\mathbb{T}^2} \nabla\psi(\mathbf{x} \cdot \mathbf{u}(t, \mathbf{x}))d\mathbf{x} = 0$, $\forall\psi \in C^\infty(\mathbb{T}^2)$.

Then Ambrosio *et al.* (2012) prove

Theorem 3.39. *Let $\varphi(0, \cdot)$ be a $\mathcal{Z}^2$ periodic function such that $\varphi(0, \mathbf{x}) + |\mathbf{x}|^2/2$ is convex, and assume that the measure $(Id + \nabla\varphi(0, \cdot))_{\#}\mathcal{L}^2 = \sigma_0\mathcal{L}^2$, where σ_0 and $1/\sigma_0$ are both in $L^\infty(\mathbb{R}^2)$. Let $\sigma(t, \cdot)$ be the solution of (3.190) discussed above. Let $P(t, \mathbf{x})$ be the unique convex function such that $P(t, 0) = 0$ and $\nabla P(t, \cdot)_{\#}\mathcal{L}^2 = \sigma(t, \cdot)\mathcal{L}^2$. Then $(\varphi(t, \cdot), \mathbf{u}(t, \cdot))$ defined by $\mathbf{X}(t, \cdot) = \nabla P(t, \cdot), \mathbf{x}(t, \cdot) = \nabla R(t, \cdot)$ and*

$$\varphi(t, \mathbf{x}) = P(t, \mathbf{x} - |\mathbf{x}|^2/2, \mathbf{u}(t, \mathbf{x} = \partial_t\mathbf{x}(\mathbf{X}) + \nabla^2 R(t, \mathbf{X})J(\mathbf{X} - \mathbf{x}),$$

is a weak Eulerian solution of (3.189) in the sense of Definition 3.17.

Now consider the three-dimensional problem in a bounded physical domain as in Theorem 3.21 with σ defined on the whole of $\mathbb{R}^3$. As discussed there, this is unphysical, but provides useful information about the behaviour of interior solutions of a problem where σ has compact support. Ambrosio *et al.* (2014) use Theorem 3.21 to prove in their Theorem 3.1 the existence of solutions $\sigma(t, \cdot)$ of (3.45) which satisfy the assumed initial bounds $\lambda \leq \sigma_0 \leq \Lambda$ on σ.

Ambrosio *et al.* (2014) then prove the existence of weak Eulerian solutions of (3.30). These are defined as follows:

Definition 3.18. Let $P : [0, \infty) \times \Gamma \to \mathbb{R}$ and $\mathbf{u} : [0, \infty) \times \Gamma \to \mathbb{R}^3$. Then $(P, \mathbf{u})$ is a *weak Eulerian solution* of (3.30) if

(i) $|\mathbf{u}| \in L^\infty_{loc}((0, \infty), L^1(\Gamma)), P \in L^\infty_{loc}((0, \infty), W^{1,\infty}(\Gamma))$, and $P(t, \cdot)$ is convex for any $t \geq 0$.

(ii) For any $\phi \in C_c^\infty([0, \infty) \times \Gamma)$:

$$\int_0^\infty \int_\Gamma \nabla P(t, \mathbf{x})\{\partial_t\phi(t, \mathbf{x}) + \mathbf{u}(t, \mathbf{x}) \cdot \nabla\phi(t, \mathbf{x})\} + \qquad (3.192)$$
$$f\ J\{\nabla P(t, \mathbf{x}) - \mathbf{u}(t, \mathbf{x})\}\phi(t, \mathbf{x})d\mathbf{x}dt + \int_\Gamma J\nabla P(0, \mathbf{x})\phi(0, \mathbf{x})d\mathbf{x} = 0.$$

(iii) For a.e. $t \in (0, \infty)$: $\int_\Gamma \nabla\psi(\mathbf{x} \cdot \mathbf{u}(t, \mathbf{x}))d\mathbf{x} = 0$, $\forall\psi \in C_c^\infty(\Gamma)$.

The main existence result is then

Theorem 3.40. *Let $\Gamma \subseteq \mathbb{R}^3$ be a convex bounded set, and let $\mathcal{L}$ be normalised Lebesgue measure, so that $\mathcal{L}(\Gamma) = 1$. Let σ_0 be a probability density on $\mathbb{R}^3$ such that $\sigma_0 \in L^\infty(\mathbb{R}^3)$ and $1/\sigma_0 \in L^\infty(\mathbb{R}^3)$ and*

$$\limsup_{|\mathbf{x}| \to \infty}(\sigma_0(\mathbf{x})|\mathbf{x}|^K) < \infty$$

for some $K > 4$. *Let* $\sigma(t, \cdot)$ *be a solution of (3.45) as discussed above, and* $R(T, \cdot) : \mathbb{R}^3 \to \mathbb{R}$ *the unique convex function so that*

$$R(t, 0) = 0 \text{ and } R(t, \cdot)_\# \sigma(t, \cdot) = \mathcal{L}(\Gamma).$$

For each t, *let* $P(t, \cdot)$ *be the Legendre transform of* $R(t, \cdot)$. *Then the vector field* $\mathbf{u}(t, \cdot)$ *defined in equation (3.192) applied to* $\mathbb{R}^3$ *is well defined and* $(P(t, \cdot), \mathbf{u}(t, \cdot))$ *is a weak Eulerian solution of (3.30) in the sense of Definition 3.18.*

Theorem 3.32 of O'Neill (2020) proves the existence of classical solutions to the dual space equations (3.45). Since these solutions are classical, they automatically apply to the physical space equations (3.30) as well. They are achieved by restricting the range of values of σ_0.

Results on the uniqueness of solutions are limited to cases where there are classical solutions. Hence the short time uniqueness result of Loeper (2006) in Theorem 3.33 and of Cheng *et al.* (2018) in section 4.3.2. However, even in this cases it is not clear whether non-smooth solutions can also develop from the same data.

Another set of results on uniqueness has been proved by Feldman and Tudorascu (2016). Two of the results are quoted here. The first applies when the initial data for a relaxed Lagrangian solution in the sense of Theorem 3.37) is $P_0 \in W^{1,\infty}(\Gamma)$. The second applies in the two-dimensional periodic case.

Theorem 3.41. *Let* (P, ν) *be a relaxed Lagrangian solution in the sense of Theorem 3.37 to (3.30) in* $[0, T) \times \Gamma$, *with* Γ *bounded, satisfying the following properties*

(i) For each $t \in [0, T)$ *there exists an extension of the dual potential* R *from* $\nabla P(t, \Gamma)$ *to the whole of* $\mathbb{R}^3$ *such that* $R \in C^1([0, T]; C^2(\mathbb{R}^3))$.

(ii) There exists $\lambda \in (0, \infty)$ *such that*

$$\lambda I_3 \leq \nabla^2 R(t, \mathbf{X}) \text{ for all } t \in [0, T) \text{ and all } \mathbf{X} \in \mathbb{R}^3.$$

(iii) $\nabla^2 R \in L^\infty(0, T; W^{1,\infty}(\mathbb{R}^3)) \cap W^{1,\infty}(0, T; L^\infty(\mathbb{R}^3))$.

Then there exists a real constant C *depending only on* P *such that for any relaxed Lagrangian solution* $(\tilde{P}, \tilde{\nu})$ *of (3.30), the inequality*

$$\|\nabla P(t, \cdot) - \nabla \tilde{P}(t, \cdot)\|_{L^2(\Gamma; \mathbb{R}^3)} \leq C \|\nabla P_0 - \nabla \tilde{P}_0\|_{L^2(\Gamma; \mathbb{R}^3)} \tag{3.193}$$

holds for all $t \in [0, T)$.

Thus relaxed Lagrangian solutions satisfying assumptions (i), (ii) and (iii) in Theorem 3.41 above are unique among relaxed Lagrangian solutions originating from the same initial data.

The result for the two-dimensional periodic case is as follows:

Theorem 3.42. *Let $(P, \mathbf{F})$ be a weak Lagrangian solution in the sense of Theorem 3.34 to (3.189) in $[0, T) \times \mathbb{T}^2$. Assume that*

(i) There exists $\lambda \in (0, \infty)$ such that

$$\lambda I_2 \leq \nabla^2 R(t, \mathbf{X}) \text{ for all } t \in [0, T) \text{ and all } \mathbf{X} \in \mathbb{T}^2.$$

(ii) $R \in W^{1,\infty}(0, T; W^{2,\infty}(\mathbb{T}^2)) \cap L^\infty(0, T; W^{3,\infty}(\mathbb{T}^2))$.

Then there exists a real constant C depending only on P such that for any relaxed Lagrangian solution $(\tilde{P}, \tilde{\mathbf{F}})$ of (3.189), the inequality

$$\|\nabla P(t, \cdot) - \nabla \tilde{P}(t, \cdot)\|_{L^2(\mathbb{T}^2;\mathbb{T}^2)} \leq C \|\nabla P_0 - \nabla \tilde{P}_0\|_{L^2(\mathbb{T}^2;\mathbb{T}^2)} \tag{3.194}$$

holds for all $t \in [0, T)$.

Thus weak Lagrangian solutions satisfying assumptions (i) and (ii) from Theorem 3.42 above are unique among weak Lagrangian solutions with the same initial data.

3.6 Approximation of Euler solutions by semi-geostrophic solutions

3.6.1 *General principles*

In order to prove that the semi-geostrophic equations define a slow manifold, it is necessary first to show that the equations can be solved either globally in time, or for at least as long a time as they are physically relevant. It is then necessary to show that they observe the assumed scaling, which is expressed by a parameter ε as in Chapter 2. Finally it is necessary to show that the solutions of the semi-geostrophic equations converge to those of the equivalent Euler equations at the expected rate as $\varepsilon \to 0$. In that case, the semi-geostrophic solution gives a strong control on the behaviour of the Euler solution. This is illustrated by the results quoted for the quasi-geostrophic equations in section 2.4.5. The last step is very challenging, because the Euler solutions in three dimensions have a wide variety of solutions on different space and time scales. Thus only a few rigorous results have been obtained to date. However, the results illustrated in section 2.7.4 show that a semi-geostrophic model derived from the form of the

Euler equations used by the Met Office UM can reproduce its large-scale behaviour, so the approximation appears to be valid.

The solutions of systems of partial differential equations will in general depend on dimensionless parameters such as ε. Thus in proving convergence, it is necessary to consider a sequence of semi-geostrophic solutions and a sequence of Euler solutions with various choices of ε. In the semi-geostrophic scaling, $\varepsilon = Ro = Fr^2$, the limit can be realised by keeping the pressure constant, and letting the rotation rate tend to infinity. Then the geostrophic wind will decrease with Ro. It is likely that the solution of the Euler equations with reducing velocities will be better behaved for a longer time, so there is some hope of successful results.

The only results proved to date are for situations where the semi-geostrophic solutions are smooth, and associated with strictly convex potentials P. Convergence of Euler solutions to weak solutions of the semi-geostrophic system, which are physically important, would require considering equivalent weak solutions of the Euler system.

The first example in this section is the solution of the two-dimensional Euler equations. This is an untypical case, because the two-dimensional Euler equations have very well-controlled behaviour, as discussed in section 2.3.4. It is also not physically significant, because the semi-geostrophic model is shown only to approximate the solution to $O(Ro)$, and in any case there is no point in approximating it as the Euler equations have simple solutions which can be fully analysed. In this case, Loeper (2006) proves that the semi-geostrophic solutions converge to the Euler solutions.

The second example is a two-dimensional model in a vertical cross-section. In this case the Euler solutions contain gravity waves, and are thus much more complicated. Therefore it is best to prove that the Euler solutions converge to semi-geostrophic solutions. This problem is unusual in that the semi-geostrophic solution is independent of Ro. The results are due to Brenier and Cullen (2009) and Brenier (2013).

The final example is a result for the three-dimensional Euler incompressible Boussinesq system by Cullen *et al.* (2020).

The second and third results use the three-dimensional non-hydrostatic Boussinesq form of the Euler equations, (3.20). It was shown in section 2.2.4 that the hydrostatic version of these equations, (3.1), is only known to be well-posed if viscosity is included. This means that they should not be used in this type of analytic convergence study unless viscosity is also included in the semi-geostrophic model. However, all analyses of the semi-geostrophic model are inviscid, which is to be expected from a model which can only describe large scales.

3.6.2 *Convergence of the two-dimensional semi-geostrophic equations to the Euler equations*

The two-dimensional incompressible equations (2.66) were derived in section 2.3.4 as the limit equations for the shallow water equations with $Fr \ll 1$ and $Fr < Ro$. Periodic boundary conditions are assumed. Existence of solutions to these equations is known under various conditions reviewed in Chemin (2000). The semi-geostrophic approximation to (2.66) is (3.189). In dual variables it is (3.190). Both are considered with periodic boundary conditons. Existence of Eulerian solutions to these equations is given by Theorem 3.39 under suitable conditions. The difference is that the vorticity ζ is replaced by the potential density σ, and the physical space velocity $\mathbf{u}$ by the dual space velocity $\mathbf{U}$. Both velocities are non-divergent in their respective coordinates, and can thus be written in terms of stream-functions ψ and Ψ respectively. The difference is that in equation (2.66)

$$\nabla^2 \psi = \zeta, \tag{3.195}$$

with periodic boundary conditions, while for (3.190)

$$f^2 R = \frac{1}{2} f^2 (X^2 + Y^2) - \Psi, \tag{3.196}$$
$$\det Hess R = \sigma,$$

Thus we are comparing a Monge-Ampère equation with a Poisson equation which is the linearisation of the Monge-Ampère equation, see Roulstone and Norbury (1994). The expected difference is thus the relative size of the nonlinear term, $O(Ro)$. The results are due to Loeper (2006). The function spaces and norms are as defined in section 2.4.5. The first result is an energy estimate.

Theorem 3.43. *Let (σ, Ψ) be a weak solution of (3.190) in $[0,T] \times \mathbb{T}^2$, and (ζ, ψ) be a smooth $C^3([0,T] \times \mathbb{T}^2)$ solution of (2.66). Choose initial data given by a stream-function $\psi(0,\cdot)$ for (2.66) and by a geopotential $\varphi(0,\cdot)$, such that $(fv_g, -fu_g) = \nabla\varphi$, for the semi-geostrophic equations. Then (3.55) implies that*

$$-\varphi(t,x,y) + \Psi(t,X,Y) = \frac{1}{2} f^2 \left((x-X)^2 + (y-Y)^2\right). \tag{3.197}$$

Use this to calculate $\Psi(0,\cdot)$ for (3.190). Define H_ϵ by

$$H_\epsilon(t) = \frac{1}{2} \int_{\mathbb{T}^2} |\nabla\psi - f^{-1}\nabla\varphi|^2 dxdy. \tag{3.198}$$

Then

$$H_\epsilon(t) \leq \left(H_\epsilon(0) + C\epsilon^{2/3}(1+t)\right)\exp(Ct), \tag{3.199}$$

where C depends on $|\psi|_{3,t}$ and $|\partial\psi/\partial t|_{2,t}|$, and $\epsilon = Ro$.

The second result is derived by assuming the solution of (3.190) is a perturbation to the solution of (2.66), suitably rescaled. Set $\epsilon = Ro$. Then assume σ takes the form $\sigma_\epsilon = 1 + \epsilon\hat{\sigma}_\epsilon$ and $\Psi = \epsilon\hat{\Psi}$.

Theorem 3.44. *Let (ζ, ψ) be a solution of (2.66) such that $\zeta \in C^2(\mathbb{R}^+ \times \mathbb{T}^2)$. Let $\sigma_\epsilon(0,\cdot)$ be a sequence of initial data for (3.190) such that $(\hat{\sigma}_\epsilon(0,\cdot) - f^{-1}\zeta(0,\cdot))/\epsilon$ is bounded in $W^{1,\infty}(\mathbb{T}^2)$. Then there exists a sequence $(\sigma_\epsilon, \Psi_\epsilon)$ of solutions to (3.190) such that, for all $T > 0$, there exists $\epsilon_T > 0$, such that the sequence*

$$\left(\frac{\hat{\sigma}_\epsilon - f^{-1}\zeta}{\epsilon}, \frac{\nabla\hat{\Psi} - \nabla\psi}{\epsilon}\right), \tag{3.200}$$

for $0 < \epsilon < \epsilon_T$, is uniformly bounded in $L^\infty([0,T], W^{1,\infty}(\mathbb{T}^2))$.

Proof This result exploits the facts that, for small geostrophic winds, (3.27) shows that $|\mathbf{x} - \mathbf{X}| = \mathrm{O}(\epsilon)$, that the potential vorticity (3.70) reduces to

$$\mathcal{Q}_{SG} = \frac{\partial(X,Y)}{\partial(x,y)} = 1 + f^{-1}\zeta_g + f^{-2}\mathrm{O}(|\zeta_g|^2), \tag{3.201}$$

and that $\sigma \simeq 2 - \mathcal{Q}_{SG}$ if $\mathcal{Q}_{SG} \simeq 1$. □

3.6.3 *Convergence for the Eady problem*

Details of the derivation of the governing equations from (3.20) are given by Visram *et al.* (2014) as discussed in section 5.3. All variables are functions of (x, z, t) except for a background state which is a function of (y, z). The domain Γ is a vertical slice $[-L, L] \times [0, H]$. The velocity is (u, v, w). Boundary conditions are periodic in the x direction and $w = 0$ at $z = 0, H$.

The potential temperature and geopotential in (3.20) are written as $\Theta(x, y, z, t)$ and $\Phi(x, y, z, t)$. The potential temperature is decomposed as

$$\Theta(x, y, z, t) = \theta_0 + \bar{\theta}(y) + \theta(x, z, t). \tag{3.202}$$

The background state consists of the first two terms and is assumed to be in hydrostatic balance. Thus the geopotential is decomposed as

$$\Phi(x, y, z, t) = \varphi_0(z) + \bar{\varphi}(y, z) + \varphi(x, z, t). \tag{3.203}$$

The background zonal wind is assumed to be in geostrophic balance, so that

$$U(x,y,z,t) = \bar{u}(z) + u(x,z,t), \qquad (3.204)$$
$$\bar{u}(z) = -\frac{g}{f\theta_0}\frac{\partial \bar{\theta}}{\partial y}\left(z - \frac{H}{2}\right).$$

The Lagrangian time derivative is defined as

$$\frac{\mathrm{D}}{\mathrm{D}t} \equiv \frac{\partial}{\partial t} + u\frac{\partial}{\partial x} + w\frac{\partial}{\partial z}.$$

Viscous terms are included, as in Brenier (2013). They are multiplied by f so that they scale correctly as $Ro \to 0$, and by a small constant κ. The governing Navier-Stokes equations derived from (3.20) are then

$$\begin{aligned}
\frac{\mathrm{D}u}{\mathrm{D}t} - fv + \frac{\partial \varphi}{\partial x} &= \kappa f \nabla^2 u,\\
\frac{\mathrm{D}v}{\mathrm{D}t} + fu + \frac{g}{\theta_0}\frac{\partial \bar{\theta}}{\partial y}\left(z - \frac{H}{2}\right) &= 0,\\
\frac{\mathrm{D}w}{\mathrm{D}t} - \frac{g\theta}{\theta_0} + \frac{\partial \varphi}{\partial z} &= \kappa f \nabla^2 w, \qquad (3.205)\\
\frac{\mathrm{D}\theta}{\mathrm{D}t} + v\frac{\partial \bar{\theta}}{\partial y} &= 0,\\
\frac{\partial u}{\partial x} + \frac{\partial w}{\partial z} &= 0.
\end{aligned}$$

The semi-geostrophic, hydrostatic and inviscid approximation to (3.205) is

$$\begin{aligned}
-fv + \frac{\partial \varphi}{\partial x} &= 0,\\
\frac{\mathrm{D}v}{\mathrm{D}t} + fu + \frac{g}{\theta_0}\frac{\partial \bar{\theta}}{\partial y}\left(z - \frac{H}{2}\right) &= 0,\\
-\frac{g\theta}{\theta_0} + \frac{\partial \varphi}{\partial z} &= 0, \qquad (3.206)\\
\frac{\mathrm{D}\theta}{\mathrm{D}t} + v\frac{\partial \bar{\theta}}{\partial y} &= 0,\\
\frac{\partial u}{\partial x} + \frac{\partial w}{\partial z} &= 0.
\end{aligned}$$

A feature of eqs. (3.206) is that the solutions are invariant under the rescaling $x \to \beta x_0, u \to \beta u_0, f \to f_0/\beta$. If the Rossby number Ro for eqn. (3.206) is defined as U/fL, where U is a typical velocity in the $x-$ direction and L is a length-scale in the $x-$ direction, then the rescaling replaces

Ro by βRo. Thus taking the limit as $Ro \to 0$ can be replaced by taking the limit as $\beta \to 0$. Applying this rescaling to the terms omitted in the semi-geostrophic approximation, the Lagrangian Rossby number, which is the ratio of $\mathrm{D}u/\mathrm{D}t$ to fv, is decreased by β^2. Thus the semi-geostrophic system (3.206) approximates an inviscid hydrostatic version of (3.205) to $\mathrm{O}(Ro^2)$. It is possible to extend this argument to the non-hydrostatic case.

Energy conservation for (3.206) is deduced by defining the energy density $\tilde{E}$ as

$$\tilde{E} = \frac{1}{2}v^2 - g\Theta(z - H/2)/\theta_0 \tag{3.207}$$

and calculating from (3.206)

$$\begin{aligned}\frac{\mathrm{D}\tilde{E}}{\mathrm{D}t} &= \frac{g}{\theta_0}\frac{\partial\bar{\theta}}{\partial y}(z - H/2)v - fuv - \frac{g}{\theta_0}\left(v\frac{\partial\bar{\theta}}{\partial y}(z - H/2) + \Theta w\right), \\ &= -u\frac{\partial\Phi}{\partial x} - w\frac{\partial\Phi}{\partial z}.\end{aligned} \tag{3.208}$$

Note that the energy has to be defined in terms of the total Θ. The last expression integrates to zero because of the condition $\frac{\partial u}{\partial x} + \frac{\partial w}{\partial z} = 0$ and the boundary conditions.

Equations (3.206) can be rewritten using (3.27), thus setting $X = f^{-1}v + x, Z = \frac{g\Theta}{f^2\theta_0}$, as

$$\begin{aligned}-fv + \frac{\partial\varphi}{\partial x} &= 0, \\ \frac{\mathrm{D}X}{\mathrm{D}t} + \frac{g}{f\theta_0}\frac{\partial\bar{\theta}}{\partial y}\left(z - \frac{H}{2}\right) &= 0, \\ \frac{\mathrm{D}Z}{\mathrm{D}t} + \frac{g}{f^2\theta_0}\frac{\partial\bar{\theta}}{\partial y} &= 0, \\ \frac{\partial u}{\partial x} + \frac{\partial w}{\partial z} &= 0. \\ P &= \frac{1}{2}x^2 + f^{-2}\Phi, \\ \nabla P &= (X, Z).\end{aligned} \tag{3.209}$$

Then set $\mathbf{U} \equiv (U, W) = -\frac{g}{f\theta_0}\frac{\partial\bar{\theta}}{\partial y}\left(z - \frac{H}{2}, X - x\right)$. Using (3.40) it can be shown that $\nabla \cdot \mathbf{U} = 0$, so (3.43) shows that the potential density $\sigma = \partial(x, z)/\partial(X, Z)$ is conserved by (3.206) in a Lagrangian sense, and then (3.70) shows that the potential vorticity $\partial(X, Z)/\partial(x, z)$ is also conserved in a Lagrangian sense. The conserved potential vorticity for the inviscid version of (3.205) includes an extra term and is

$$q = \left(\frac{\partial \bar{\theta}}{\partial y}\left(\frac{\partial u}{\partial z} - \frac{\partial w}{\partial x}\right) + f\left(\frac{\partial X}{\partial x}\frac{\partial \Theta}{\partial z} - \frac{\partial X}{\partial z}\frac{\partial \Theta}{\partial x}\right)\right). \tag{3.210}$$

Existence of solutions to these equations can be inferred from Theorem 3.34. In general they contain physically important discontinuities, as will be illustrated in section 5.3. The potential density σ will be compactly supported in $\mathbb{T} \times \mathbb{R}$, so Theorem 3.40 cannot be used to prove the existence of Eulerian solutions.

The convergence result of Brenier and Cullen (2009) contained an error in the proof, and was restated with different (periodic) boundary conditions in Brenier (2013). The results require strict convexity of the potential P in 3.209. The use of periodic boundary conditions, though unphysical in the vertical, means that Eulerian solutions should exist according to Theorem 3.39 and the strict convexity will be maintained.

First rewrite (3.205) in the same form as (3.209) and rearrange both to match the result of Brenier (2013), (Theorem 3.1). Set $\epsilon = f^{-1}$ so that the limit $Ro \to 0$ becomes $\epsilon \to 0$. (3.205) becomes

$$\begin{aligned} \mathbf{X} &= \nabla P + \epsilon^2 \frac{\mathrm{D}\mathbf{u}}{\mathrm{D}t} - \epsilon \nabla^2 \mathbf{u}, \\ \nabla \cdot \mathbf{u} &= 0, \\ \frac{\mathrm{D}\mathbf{X}}{\mathrm{D}t} = -\frac{g}{f\theta_0}\frac{\partial \bar{\theta}}{\partial y}&\left(z - \frac{H}{2}, X - x\right) \equiv G(t, \mathbf{x}). \end{aligned} \tag{3.211}$$

(3.209) becomes

$$\begin{aligned} \mathbf{X} &= \nabla P, \\ \nabla \cdot \mathbf{u} &= 0, \\ \frac{\mathrm{D}\mathbf{X}}{\mathrm{D}t} = -\frac{g}{f\theta_0}\frac{\partial \bar{\theta}}{\partial y}&\left(z - \frac{H}{2}, X - x\right) \equiv G(t, \mathbf{x}). \end{aligned} \tag{3.212}$$

Then Brenier (2013) proves

Theorem 3.45. *Let $\Gamma = \mathbb{R}^2/\mathbb{Z}^2.n$. Assume G is smooth with bounded derivatives up to second order. Let $(\mathbf{X}^\epsilon, \mathbf{u}^\epsilon, P^\epsilon)$ be a Leray-type solution to (3.211). Let $(\mathbf{X} = \nabla P, \mathbf{u})$ be a smooth solution to (3.212) on a given finite time interval $[0, T]$. Assume that $P(t, \cdot)$ is strictly convex for all $t \in [0, T]$ in the sense that*

$$\lambda Id < Id + \det \text{Hess}\left(P - \frac{1}{2}x^2\right) < \lambda^{-1} Id$$

for some $\lambda > 0$.

Then the L^2 *distance between* $\mathbf{X}^\epsilon$ *and* $\mathbf{X}$ *stays uniformly of order* $\sqrt{\epsilon}$ *as* ϵ *tends to zero, uniformly in* $t \in [0,T]$, *provided it does so at* $t = 0$ *and the initial velocity* $\mathbf{u}^\epsilon(0,\mathbf{x})$ *stays uniformly bounded in* L^2.

The proof uses a 'relative entropy' to compare solutions of (3.212) with (3.211). This is defined by

$$\int_\Gamma (K(t,\mathbf{X}^\epsilon(t,\mathbf{x}),\mathbf{X}(t,\mathbf{x})) + \epsilon^2|\mathbf{u}^\epsilon - \mathbf{u}|^2)\mathrm{d}\mathbf{x}, \tag{3.213}$$

where

$$K(t,\mathbf{X}^\epsilon,\mathbf{X}) = R(t,\mathbf{X}^\epsilon) - R(t,\mathbf{X}) - \nabla R(t,\mathbf{X})\cdot(\mathbf{X}^\epsilon - \mathbf{X}) \simeq |\mathbf{X}^\epsilon - \mathbf{X}|^2. \tag{3.214}$$

Note that this estimate is well short of the expected ϵ^2 convergence, which will be demonstrated numerically in section 5.3.

3.6.4 *Convergence to the solution of the three-dimensional Euler equations*

The aim is to prove convergence of solutions of the Euler equations (3.20) to solutions of the semi-geostrophic equations (3.2) as $Ro = Fr^2 \to 0$. As found in section 2.4.6, the error is a second Lagrangian time derivative, and convergence can only be achieved if there is a sequence of solutions of the Euler equations with reducing Ro where the second time derivatives can be controlled. This is unknown, and is thus an assumption made in the analysis.

The first step is to rewrite (3.20) using some of the dual variables defined in (3.27). In addition, use a tilde to denote variables generated by the Euler solution to distinguish them from those generated by the semi-geostrophic solution. Setting $\tilde{P} = \frac{1}{2}(x^2 + y^2) + f^{-2}\tilde{\varphi}$ as in 3.28 gives

$$\begin{aligned}
\frac{\mathrm{D}\tilde{\mathbf{u}}}{\mathrm{D}t} + fJ\tilde{\mathbf{u}} + f^2\tilde{\mathbf{X}} = (f^2x, f^2y, \frac{g\tilde{\theta}}{\theta_0}),& \\
\tilde{\mathbf{X}} = \nabla\tilde{P},& \\
\nabla\cdot\tilde{\mathbf{u}} = 0,& \\
\frac{\mathrm{D}\tilde{\theta}}{\mathrm{D}t} = 0,& \\
J = \begin{pmatrix} 0 & -1 & 0 \\ 1 & 0 & 0 \\ 0 & 0 & 0 \end{pmatrix}.&
\end{aligned} \tag{3.215}$$

These equations are to be solved in a bounded region $\Gamma \subset \mathbb{R}^3$ with $\tilde{\mathbf{u}} \cdot \mathbf{n} = 0$ on the boundary. Following the derivation of (2.156), take the Lagrangian time derivative of the first equation. This gives, using the last equation:

$$\frac{\mathrm{D}^2\tilde{\mathbf{u}}}{\mathrm{D}t^2} + fJ\frac{\mathrm{D}\tilde{\mathbf{u}}}{\mathrm{D}t} + f^2\frac{\mathrm{D}\tilde{\mathbf{X}}}{\mathrm{D}t} = f^2(\tilde{u}, \tilde{v}, 0). \tag{3.216}$$

Use the first equation of (3.215) to replace the second term in (3.216). Note that J^2 is a 2×2 identity matrix with a zero third row and column, so multiplying by J^2 eliminates the third component of the vector. Cancelling further terms gives the system

$$\begin{aligned} f^{-2}\frac{\mathrm{D}^2\tilde{\mathbf{u}}}{\mathrm{D}t^2} + fJ(\mathbf{x} - \tilde{\mathbf{X}}) + \frac{\mathrm{D}\tilde{\mathbf{X}}}{\mathrm{D}t} &= 0, \\ \tilde{\mathbf{X}} &= \nabla\tilde{P}, \\ \nabla \cdot \tilde{\mathbf{u}} &= 0. \end{aligned} \tag{3.217}$$

In order to compare with (3.30), remove the tildes. Then, with the exception of the first term, this is exactly (3.30) with $\mathbf{u}_g$ defined in terms of $\mathbf{x}$ and $\mathbf{X}$ by (3.27).

Now suppose (P, F) is a weak Lagrangian solution of (3.30) as in Theorem 3.34 so that $P_0 \in H^1(\Gamma)$. Use the notation as in section 3.5.4. Then Cullen *et al.* (2020) prove

Theorem 3.46. *Let $(P, \mathbf{F})$ be a weak Lagrangian solution of (3.30) in $[0, T) \times \Gamma$ in the sense of Definition 3.12 such that*

(i) There exists $\mathcal{R} \in (0, \infty)$ and $0 < \delta < \mathcal{R}$ such that $\overline{\nabla P(t, \Gamma)} \subset B(0, \mathcal{R} - \delta)$.

(ii) For each $t \in [0, T)$ there exists an extension of $R(t, \cdot)$ from $\nabla P(t, \Gamma)$ to $B(0, \mathcal{R})$ and there exists $\lambda \in (0, \infty)$ such that

$$\lambda I_3 \leq \nabla^2 R(t, \mathbf{X}) \text{ for all } t \in [0, T) \text{ and all } \mathbf{X} \in B(0, \mathcal{R}).$$

(iii) $R \in W^{1,\infty}((0, T); W^{2,\infty}(B(0, \mathcal{R}))) \cap L^\infty((0, T); W^{3,\infty}(B(0, \mathcal{R})))$.

Then there exists a real constant C depending only on P such that for any classical solution $(\tilde{P}, \tilde{\mathbf{u}})$ of (3.217) which satisfies $\overline{\nabla\tilde{P}(t, \Gamma)} \subset B(0, \mathcal{R} - \delta)$ for all $t \in [0, T)$, the inequality

$$\begin{aligned} &\|\nabla P(t, \cdot) - \nabla\tilde{P}(t, \cdot)\|_{L^2(\Gamma;\mathbb{R}^3)} \leq \\ &C\|\nabla P_0 - \nabla\tilde{P}_0\|_{L^2(\Gamma;\mathbb{R}^3)} + \|\mathrm{D}_t^2\tilde{\mathbf{u}}\|_{L^2(\Gamma;\mathbb{R}^3)}, \end{aligned} \tag{3.218}$$

holds for all $t \in [0, T)$.

Note that Theorem 3.41 also holds with the assumptions above. The proof uses the relative entropy of Brenier (2013) given by (3.214) with appropriate changes of notation.

Cullen *et al.* (2020) also prove a similar result for the two-dimensional periodic case. This requires less regularity than the result of Loeper (2006) described in section 3.6.2, but no quantitative convergence estimate can be obtained.

Chapter 4

Solution of the Semi-geostrophic and Related Equations in More General Cases

4.1 Solution of the semi-geostrophic equations for compressible flow

4.1.1 *The compressible equations in Cartesian geometry*

This section considers the fully compressible semi-geostrophic equations (2.166)-(2.168) as discussed in section 3.3.6, In that section, the hydrostatic approximation was exploited to allow the equations to be rewritten and solved with pressure as a vertical coordinate. This has the advantage that the continuity equation takes the same form as for the incompressible equations. It is also possible to use zero pressure as the upper boundary condition with an evolution equation for the surface pressure. The system is then very similar to a three-dimensional incompressible system with a free surface, as described in section 3.3.5.

In the present section, the compressible semi-geostrophic equations are solved directly. This has the advantage that the solutions can be compared directly with those of the compressible Euler equations. A rigid upper boundary condition is used in the rigorous proof in section 4.1.5. The formal arguments showing that the solution can be regarded as a sequence of minimum energy states apply using a physically realistic decay condition for the upper boundary. Subsequent research has suggested that this could be used in the rigorous argument also. The analysis follows Shutts and Cullen (1987). While the rotation rate is still assumed to be uniform, it is no longer required to be about an axis parallel to the gravitational acceleration. Choose the z axis to be in the direction of the axis of rotation. The compressible Navier-Stokes equations (2.1) are then modified to replace the gravitational term by the gradient of a geopotential which does not have to be in the z direction. The moisture equation and the viscous, diffusive, and source terms are omitted, giving the compressible Euler equations. Then (2.1), using (2.3) and (2.4), become

$$\begin{aligned}
\frac{\mathrm{D}\mathbf{u}}{\mathrm{D}t} + (-fv, fu, 0) + C_p\theta\nabla\Pi + \nabla\Phi = 0,\\
\frac{\partial\rho}{\partial t} + \nabla\cdot(\rho\mathbf{u}) = 0,\\
\frac{\mathrm{D}\theta}{\mathrm{D}t} = 0,\\
\theta = T(p/p_{ref})^{-R/C_p} \equiv T/\Pi,\\
p = \rho RT,
\end{aligned} \tag{4.1}$$

where Φ is the geopotential and the Coriolis parameter f is equal to 2Ω. There are eight equations for the unknowns $(u, v, w, p, \Pi, \rho, \theta, T)$. At various stages in the argument it is easier to work with the pair (T, p) than (θ, Π), so both options are retained. The thermodynamic equation can be written in the alternative form

$$C_v\frac{\mathrm{D}T}{\mathrm{D}t} - \frac{p}{\rho^2}\frac{\mathrm{D}\rho}{\mathrm{D}t} = 0. \tag{4.2}$$

The domain Γ is assumed to take the form $\Gamma_z \times [0, \infty)$ and the boundary conditions are that $\mathbf{u}\cdot\mathbf{n} = 0$ on all finite parts of the boundary and that p, $\rho \to 0$ as $z \to \infty$.

Define the geostrophic wind by

$$\begin{aligned}
-fv_g + C_p\theta\frac{\partial\Pi}{\partial x} + \frac{\partial\Phi}{\partial x} = 0,\\
fu_g + C_p\theta\frac{\partial\Pi}{\partial y} + \frac{\partial\Phi}{\partial y} = 0.
\end{aligned} \tag{4.3}$$

The compressible semi-geostrophic equations are then derived by making the hydrostatic and geostrophic momentum approximations in (4.1). The first two components of the first equation of (4.1) are replaced by

$$\begin{aligned}
\frac{\mathrm{D}u_g}{\mathrm{D}t} - fv + C_p\theta\frac{\partial\Pi}{\partial x} + \frac{\partial\Phi}{\partial x} = 0,\\
\frac{\mathrm{D}v_g}{\mathrm{D}t} + fu + C_p\theta\frac{\partial\Pi}{\partial y} + \frac{\partial\Phi}{\partial y} = 0,
\end{aligned} \tag{4.4}$$

and the third by

$$C_p\theta\frac{\partial\Pi}{\partial z} + \frac{\partial\Phi}{\partial z} = 0. \tag{4.5}$$

It is shown in Shutts and Cullen (1987) that these equations, with the boundary conditions above, conserve the energy integral

$$E = \int_\Gamma \rho \left(\frac{1}{2}(u_g^2 + v_g^2) + C_v T + \Phi \right) dxdydz. \tag{4.6}$$

The three terms represent respectively kinetic, internal and potential energy. The equations also conserve a form of potential vorticity, discussed in section 4.1.4.

The equations can be rewritten in the form

$$\mathbf{Q}\begin{pmatrix} u \\ v \\ w \end{pmatrix} + \frac{\partial}{\partial t}\begin{pmatrix} fv_g \\ -fu_g \\ \theta \end{pmatrix} = \begin{pmatrix} f^2 u_g \\ f^2 v_g \\ 0 \end{pmatrix},$$

$$\mathbf{Q} = \begin{pmatrix} f^2 + f\frac{\partial v_g}{\partial x} & f\frac{\partial v_g}{\partial y} & f\frac{\partial v_g}{\partial z} \\ -f\frac{\partial u_g}{\partial x} & f^2 - f\frac{\partial u_g}{\partial y} & -f\frac{\partial u_g}{\partial z} \\ \frac{\partial \theta}{\partial x} & \frac{\partial \theta}{\partial y} & \frac{\partial \theta}{\partial z} \end{pmatrix}. \tag{4.7}$$

Substitute for (u_g, v_g) in the time-derivative term using (4.3) and the third equation of (4.1). The time derivative of (4.5) gives

$$\frac{\partial \theta}{\partial t}\frac{\partial \Pi}{\partial z} + \theta \frac{\partial}{\partial t}\left(\frac{\partial \Pi}{\partial z}\right) = 0. \tag{4.8}$$

Using (4.5) gives

$$\frac{1}{\theta}\frac{\partial \Phi}{\partial z}\frac{\partial \theta}{\partial t} = C_p \theta \frac{\partial}{\partial t}\left(\frac{\partial \Pi}{\partial z}\right). \tag{4.9}$$

Equation (4.7) can then be written

$$\mathbf{Q_1}\begin{pmatrix} u \\ v \\ w \end{pmatrix} + C_p \theta \frac{\partial}{\partial t}\nabla \Pi = \begin{pmatrix} f^2 u_g \\ f^2 v_g \\ 0 \end{pmatrix},$$

$$\mathbf{Q_1} = \begin{pmatrix} f^2 + f\theta\frac{\partial}{\partial x}\left(\frac{v_g}{\theta}\right) & f\theta\frac{\partial}{\partial y}\left(\frac{v_g}{\theta}\right) & f\theta\frac{\partial}{\partial z}\left(\frac{v_g}{\theta}\right) \\ -f\theta\frac{\partial}{\partial x}\left(\frac{u_g}{\theta}\right) & f^2 - f\theta\frac{\partial}{\partial y}\left(\frac{u_g}{\theta}\right) & -f\theta\frac{\partial}{\partial z}\left(\frac{u_g}{\theta}\right) \\ \frac{1}{\theta}\frac{\partial \Phi}{\partial z}\frac{\partial \theta}{\partial x} & \frac{1}{\theta}\frac{\partial \Phi}{\partial z}\frac{\partial \theta}{\partial y} & \frac{1}{\theta}\frac{\partial \Phi}{\partial z}\frac{\partial \theta}{\partial z} \end{pmatrix}. \tag{4.10}$$

Equation (4.10) can be written as

$$\mathbf{u} + \mathbf{Q_1}^{-1} C_p \theta \frac{\partial}{\partial t}\nabla \Pi = \mathbf{Q_1}^{-1}\begin{pmatrix} f^2 u_g \\ f^2 v_g \\ 0 \end{pmatrix}. \tag{4.11}$$

Now multiply by ρ and use the second equation of (4.1) to substitute for $\mathbf{u}$. To relate $\partial\rho/\partial t$ and $\partial\Pi/\partial t$, differentiate the logarithm of the equation of state with respect to time, giving

$$-\frac{C_v}{R}\frac{\partial}{\partial t}\log\Pi + \frac{\partial}{\partial t}\log\rho + \frac{\partial}{\partial t}\log\theta = 0. \tag{4.12}$$

Now use (4.9) to give

$$-\frac{C_v}{R}\frac{\partial}{\partial t}\log\Pi + \frac{\partial}{\partial t}\log\rho + \frac{C_p\theta}{\frac{\partial\Phi}{\partial z}}\frac{\partial}{\partial t}\frac{\partial\Pi}{\partial z} = 0. \tag{4.13}$$

Combining these gives a single second order equation for $\frac{\partial\Pi}{\partial t}$:

$$-\frac{C_v}{R}\frac{\rho}{\Pi}\frac{\partial}{\partial t}\Pi + \frac{\rho C_p\theta}{\frac{\partial\Phi}{\partial z}}\frac{\partial}{\partial t}\frac{\partial\Pi}{\partial z} + \nabla\cdot\rho{\mathbf{Q}_1}^{-1}C_p\theta\frac{\partial}{\partial t}\nabla\Pi = \tag{4.14}$$

$$\nabla\cdot\rho{\mathbf{Q}_1}^{-1}\begin{pmatrix} f^2u_g \\ f^2v_g \\ 0 \end{pmatrix}.$$

It can be expected that solvability of (4.14) will require that the coefficients of the highest order terms form a positive definite operator. This will be the case if $\mathbf{Q}_1$ is positive definite. To evaluate this condition, eliminate the θ derivatives from the first two rows of $\mathbf{Q}_1$ by subtracting multiples of the third row from the first two rows, and deduce

$$\det\mathbf{Q}_1 = \frac{1}{\theta}\frac{\partial\Phi}{\partial z}\det\mathbf{Q}. \tag{4.15}$$

Thus solvability will be determined by the positive definiteness of $\mathbf{Q}$. This condition takes the same form as that obtained in section 3.1.

4.1.2 *The solution as a sequence of minimum energy states*

In this section it is shown that a geostrophic and hydrostatic state can again be characterised as a stationary point of the energy with respect to a certain class of variations. More details of these arguments are given in Shutts and Cullen (1987), section 4.

Start with a state of the fluid with an associated vector field $(\tilde{u}, \tilde{v})$ and scalar fields $\tilde{\theta}, \tilde{\rho}$. Assume the equation of state is satisfied so that $\tilde{p}$ and $\tilde{T}$ can be derived from $\tilde{\theta}$ and $\tilde{\rho}$. Associate with this state an energy integral given by the formula analogous to (4.6)

$$E = \int_\Gamma \left(\frac{1}{2}(\tilde{u}^2 + \tilde{v}^2) + C_v\tilde{T} + \Phi\right)\tilde{\rho}\,\mathrm{d}x\mathrm{d}y\mathrm{d}z. \tag{4.16}$$

This is a functional of $\tilde{u}$, $\tilde{v}$, $\tilde{\theta}$ and $\tilde{\rho}$, regarded as functions of position over Γ, which has the following property.

Theorem 4.1. *The conditions for the energy E to be stationary with respect to variations $\Xi = (\xi, \eta, \chi)$ of particle positions satisfying continuity $\delta(\tilde{\rho} dx dy dz) = 0$ via*

$$\delta\tilde{\rho} = -\tilde{\rho}\nabla \cdot \Xi \tag{4.17}$$

in Γ, and

$$\delta\tilde{u} = f\eta,\ \delta\tilde{v} = -f\xi,\ \delta\tilde{\theta} = 0, \tag{4.18}$$

together with $\Xi \cdot \mathbf{n} = 0$ on the finite boundary of Γ and $\tilde{p}\Xi \to 0$ as $z \to \infty$, are that

$$(-f\tilde{v}, f\tilde{u}, 0) + \nabla\Phi + C_p\tilde{\theta}\nabla\tilde{\Pi} = 0. \tag{4.19}$$

Proof Consider δE from a Lagrangian viewpoint, and so write

$$\delta E = \int_\Gamma \delta\left(\frac{1}{2}(\tilde{u}^2 + \tilde{v}^2) + C_v\tilde{T} + \Phi\right)\tilde{\rho} dx dy dz. \tag{4.20}$$

Equation (4.2) implies that the condition $\delta\tilde{\theta} = 0$ is equivalent to $C_v\delta\tilde{T} - \frac{\tilde{p}}{\tilde{\rho}^2}\delta\tilde{\rho} = 0$. We then have

$$\begin{aligned} \delta E &= \int_\Gamma \left((\tilde{u}\delta\tilde{u} + \tilde{v}\delta\tilde{v}) - \tfrac{\tilde{p}}{\tilde{\rho}}\nabla \cdot \Xi + \Xi \cdot \nabla\Phi\right)\tilde{\rho} dx dy dz, \qquad (4.21)\\ &= \int_\Gamma \left((f\tilde{u}\eta - f\tilde{v}\xi) - \tfrac{\tilde{p}}{\tilde{\rho}}\nabla \cdot \Xi + \Xi \cdot \nabla\Phi\right)\tilde{\rho} dx dy dz. \end{aligned}$$

Integrating by parts to eliminate $\nabla \cdot \Xi$, and using the boundary conditions, gives

$$\delta E = \int_\Gamma \Xi \cdot \left((-f\tilde{v}, f\tilde{u}, 0) + \frac{1}{\tilde{\rho}}\nabla\tilde{p} + \nabla\Phi\right)\tilde{\rho} dx dy dz. \tag{4.22}$$

Use the equation of state and the definition of Π in equations (4.1) to replace $\frac{1}{\tilde{\rho}}\nabla\tilde{p}$ by $C_p\tilde{\theta}\nabla\tilde{\Pi}$. For this to vanish for any Ξ, (4.19) must be satisfied. □

Now consider the condition for the stationary point to be a minimum.

Theorem 4.2. *The condition for E to be minimised with respect to the class of variations defined in Theorem 4.1 is that the matrix $\mathbf{Q}$ defined by*

$$\mathbf{Q} = \begin{pmatrix} f^2 + f\frac{\partial\tilde{v}}{\partial x} & f\frac{\partial\tilde{v}}{\partial y} & f\frac{\partial\tilde{v}}{\partial z} \\ -f\frac{\partial\tilde{u}}{\partial x} & f^2 - f\frac{\partial\tilde{u}}{\partial y} & -f\frac{\partial\tilde{u}}{\partial z} \\ \frac{1}{\tilde{\theta}}\frac{\partial\Phi}{\partial z}\frac{\partial\tilde{\theta}}{\partial x} & \frac{1}{\tilde{\theta}}\frac{\partial\Phi}{\partial z}\frac{\partial\tilde{\theta}}{\partial y} & \frac{1}{\tilde{\theta}}\frac{\partial\Phi}{\partial z}\frac{\partial\tilde{\theta}}{\partial z} \end{pmatrix} \tag{4.23}$$

is positive definite.

Proof The proof is carried out on the assumption that f could be a function of position, as that will be needed in section 4.3. A minimum is also a stationary point, so the stationary point can be characterised by

$$(\tilde{u}, \tilde{v}, \tilde{\theta}, \tilde{\rho}, \tilde{p}, \tilde{\Pi}) = (u_g, v_g, \theta_g, \rho_g, p_g, \Pi_g), \qquad (4.24)$$
$$(-fv_g, fu_g, 0) + \nabla\Phi + C_p\theta_g\nabla\Pi_g = 0.$$

Form a second variation starting from (4.22). Using (4.24) to substitute for the terms in the integrand not perturbed by the displacement, this gives

$$\delta^2 E = \qquad (4.25)$$
$$\frac{1}{2}\int_\Gamma \Xi \cdot \delta\left\{f\left((-\tilde{v}, \tilde{u}, 0) + \frac{1}{f\tilde{\rho}}\nabla\tilde{p} + \frac{1}{f}\nabla\Phi\right)\right\}\rho_g \mathrm{d}x\mathrm{d}y\mathrm{d}z,$$

Since the term multiplying δf vanishes at the stationary point, this becomes, using (4.18),

$$\delta^2 E = \qquad (4.26)$$
$$\frac{1}{2}\int_\Gamma \Xi \cdot f\left(f(\xi, \eta, 0) + \delta\left(\frac{1}{f\tilde{\rho}}\nabla\tilde{p}\right) + \delta\left(\frac{1}{f}\nabla\Phi\right)\right)\rho_g \mathrm{d}x\mathrm{d}y\mathrm{d}z.$$

The condition $\delta\tilde{\theta} = 0$, together with the equation of state and the definition of $\tilde{\theta}$, imply that $\delta(\tilde{p}\tilde{\rho}^{(-C_p/C_v)}) = 0$, so that, using (4.17)

$$\frac{\delta\tilde{\rho}}{\tilde{\rho}} = \frac{C_v}{C_p}\frac{\delta\tilde{p}}{\tilde{p}} = -\nabla \cdot \Xi. \qquad (4.27)$$

As in the proof of Theorem 3.7, write ∂ for the change at a fixed position in space caused by a displacement. Then, also using (4.24), $\delta\tilde{p} = \partial\tilde{p} + \Xi \cdot \nabla\tilde{p} = \partial\tilde{p} + \Xi \cdot \nabla p_g$. Using (4.24) in the last part of (4.27) gives $\frac{C_v}{C_p}\delta\tilde{p} = -p_g\nabla \cdot \Xi$. Combining these gives

$$\partial\tilde{p} = -\left(\Xi \cdot \nabla p_g + p_g\frac{C_p}{C_v}\nabla \cdot \Xi\right), \qquad (4.28)$$

so that

$$\delta(\nabla\tilde{p}) = \partial(\nabla\tilde{p}) + \Xi \cdot \nabla(\nabla\tilde{p}) \qquad (4.29)$$
$$= \Xi \cdot \nabla(\nabla p_g) - \nabla\left(\Xi \cdot \nabla p_g + p_g\frac{C_p}{C_v}\nabla \cdot \Xi\right).$$

Using (4.27) and (4.29) and $\partial f = 0$, it can then be shown that

$$\rho_g\delta\left(\frac{1}{f\tilde{\rho}}\nabla\tilde{p}\right) = \rho_g\Xi \cdot \nabla\left(\frac{1}{f\rho_g}\nabla p_g\right) +$$
$$\frac{1}{f}\left(\nabla p_g\nabla \cdot \Xi - \nabla(\Xi \cdot \nabla p_g) + p_g\frac{C_p}{C_v}\nabla \cdot \Xi\right) \qquad (4.30)$$
$$= \rho_g\Xi \cdot \nabla\left(\frac{1}{f}C_p\theta_g\nabla\Pi_g\right) + \frac{1}{f}\left(\nabla p_g\nabla \cdot \Xi - \nabla(\Xi \cdot \nabla p_g) + p_g\frac{C_p}{C_v}\nabla \cdot \Xi\right).$$

Since Φ and f are time-independent, $\delta\left(\frac{1}{f}\nabla\Phi\right) = \Xi\cdot\nabla(\frac{1}{f}\nabla\Phi)$ and, using (4.24), gives

$$\delta\frac{1}{f}\nabla\Phi = \Xi\cdot\nabla\left((v_g, -u_g, 0) - \frac{C_p\theta_g}{f}\nabla\Pi_g\right). \tag{4.31}$$

The first term on the right hand side of (4.30) and the last term of (4.31) cancel when they are substituted into equation (4.25). Then

$$\delta^2 E = \frac{1}{2}\int_\Gamma \begin{pmatrix} \Xi\cdot\mathbf{Q}_2\cdot\Xi + \\ \Xi\cdot(\Xi\cdot\nabla(\nabla p_g) + (\nabla p_g)\nabla\cdot\Xi - \nabla(\Xi\cdot\nabla p_g)) \\ +\Xi\cdot\left(p_g\frac{C_p}{C_v}\nabla\cdot\Xi - C_p\theta\Xi\cdot\nabla(\nabla\Pi_g)\right) \end{pmatrix} \rho_g dxdydz, \tag{4.32}$$

where $\mathbf{Q}_2$ is defined by

$$\begin{pmatrix} f^2 + f\frac{\partial v_g}{\partial x} - C_p\frac{\partial\theta_g}{\partial x}\frac{\partial\Pi_g}{\partial x} & f\frac{\partial v_g}{\partial y} - C_p\frac{\partial\theta_g}{\partial y}\frac{\partial\Pi_g}{\partial x} & f\frac{\partial v_g}{\partial z} - C_p\frac{\partial\theta_g}{\partial z}\frac{\partial\Pi_g}{\partial x} \\ -f\frac{\partial u_g}{\partial x} - C_p\frac{\partial\theta_g}{\partial x}\frac{\partial\Pi_g}{\partial y} & f^2 - f\frac{\partial u_g}{\partial y} - C_p\frac{\partial\theta_g}{\partial y}\frac{\partial\Pi_g}{\partial y} & -f\frac{\partial u_g}{\partial z} - C_p\frac{\partial\theta_g}{\partial z}\frac{\partial\Pi_g}{\partial y} \\ -C_p\frac{\partial\theta_g}{\partial x}\frac{\partial\Pi_g}{\partial z} & -C_p\frac{\partial\theta_g}{\partial y}\frac{\partial\Pi_g}{\partial z} & -C_p\frac{\partial\theta_g}{\partial z}\frac{\partial\Pi_g}{\partial z} \end{pmatrix}. \tag{4.33}$$

It is then shown in Shutts and Cullen (1987) that, after integration by parts, (4.32) reduces to

$$\delta^2 E = \frac{1}{2}\int_\Gamma \left(\Xi\cdot\mathbf{Q}_2\cdot\Xi + p_g\frac{C_p}{C_v}\left(\nabla\cdot\Xi + \frac{C_v}{C_p p_g}\Xi\cdot\nabla p_g\right)^2\right)\rho_g dxdydz. \tag{4.34}$$

Since the second term is positive definite, the condition for E to be minimised is that $\mathbf{Q}_2$ is positive definite. It can be seen that det$\mathbf{Q}_2$ is equal to det $\mathbf{Q}$, where $\mathbf{Q}$ is as defined in (4.7), by subtracting appropriate multiples of the third row from the first two rows and using (4.24) to replace the factor $-C_p\frac{\partial\Pi_g}{\partial z}$ that multiplies the elements of the third row by the factor $\frac{1}{\theta_g}\frac{\partial\Phi}{\partial z}$ that appears in (4.7). Thus the condition for E to be minimised is the positive definiteness of $\mathbf{Q}$. □

Finally, state the *stability principle* under which the solution of the compressible semi-geostrophic equations is attempted.

Definition 4.1. An admissible solution of the compressible semi-geostrophic equations on a region Γ is one that is characterised by a function $\Pi(t)$ whose evolution satisfies (4.14) in a suitable sense and where the matrix $\mathbf{Q}$ calculated from Π using (4.7), (4.5) and (4.3) is positive definite.

4.1.3 Solution by change of variables

Next show that the compressible semi-geostrophic equations are solvable in the same manner as in section 3.3. The first step is to use the change of variables introduced by Hoskins (1975). Set

$$X = f^{-1}v_g + x,\ Y = -f^{-1}u_g + y,\ Z = \theta. \tag{4.35}$$

The compressible semi-geostrophic equations assembled from equations (4.1), (4.3), (4.4) and (4.5) can then be written as

$$\begin{aligned}
\frac{\mathrm{D}X}{\mathrm{D}t} &= f(y - Y),\\
\frac{\mathrm{D}Y}{\mathrm{D}t} &= f(X - x),\\
\frac{\mathrm{D}Z}{\mathrm{D}t} &= 0,\\
\frac{\partial\rho}{\partial t} + \nabla\cdot(\rho\mathbf{u}) &= 0,\\
f^2(x - X, y - Y, 0) + C_pZ\nabla\Pi + \nabla\Phi &= 0,\\
Z = T(p/p_{ref})^{-R/C_p} &\equiv T/\Pi,\\
p &= \rho RT.
\end{aligned} \tag{4.36}$$

This is a system of ten equations for the unknowns $(X, Y, Z, T, p, \Pi, \rho, u, v, w)$. The options of different versions of the thermodynamic variables is retained for convenience below.

Now describe the energy minimisation property in these variables. Suppose that associated with each particle is a vector field $(\tilde{X}, \tilde{Y}, \tilde{Z})$ and a scalar field $\tilde{\Pi}$. Then $\tilde{\rho}, \tilde{p}, \tilde{T}$ can be calculated from $\tilde{\Pi}$ and $\tilde{Z}$ using the last two equations of (4.36). Given these fields, define the energy integral

$$E = \int_\Gamma \left(\frac{1}{2}f^2\left((x - \tilde{X})^2 + (y - \tilde{Y})^2\right) + C_v\tilde{T} + \Phi\right)\tilde{\rho}\mathrm{d}x\mathrm{d}y\mathrm{d}z. \tag{4.37}$$

This is a functional of $(\tilde{X}, \tilde{Y}, \tilde{Z})$ and $\tilde{\Pi}$, regarded as functions of position over Γ, which has the following property.

Theorem 4.3. *The conditions for the energy E to be stationary with respect to variations Ξ of particle positions satisfying continuity $\delta(\tilde{\rho}\mathrm{d}x\mathrm{d}y\mathrm{d}z) = 0$ via*

$$\delta\tilde{\rho} = -\tilde{\rho}\nabla\cdot\Xi \tag{4.38}$$

in Γ and

$$\delta\tilde{X} = \delta\tilde{Y} = \delta\tilde{Z} = 0, \tag{4.39}$$

together with $\Xi \cdot \mathbf{n} = 0$ *on the finite boundary of* Γ *and* $\tilde{p}\Xi \to 0$ *as* $z \to \infty$, *are that*

$$f^2(x - \tilde{X}, y - \tilde{Y}, 0) + C_p \tilde{Z} \nabla \tilde{\Pi} + \nabla \Phi = 0. \tag{4.40}$$

The condition for E *to be minimised is that the matrix* $\mathbf{Q}$ *defined by*

$$\begin{pmatrix} f^2 \frac{\partial \tilde{X}}{\partial x} & f^2 \frac{\partial \tilde{X}}{\partial y} & f^2 \frac{\partial \tilde{X}}{\partial z} \\ f^2 \frac{\partial \tilde{Y}}{\partial x} & f^2 \frac{\partial \tilde{Y}}{\partial y} & f^2 \frac{\partial \tilde{Y}}{\partial z} \\ \frac{\partial \tilde{Z}}{\partial x} & \frac{\partial \tilde{Z}}{\partial y} & \frac{\partial \tilde{Z}}{\partial z} \end{pmatrix} \tag{4.41}$$

is positive definite.

The proof is a simple rewrite of that of Theorems 4.1 and 4.2. The definition of admissible solutions is a rewrite of Definition 4.1. However, it cannot be simply stated that it implies convexity of a scalar potential.

The next step, following Cullen and Maroofi (2003), is to rewrite the internal energy term in (4.37) using the last two equations of (4.36). Note that the notation of Cullen and Maroofi (2003) is changed to be consistent with the rest of this book. In particular their definitions of σ and ν have been interchanged. This gives

$$\begin{aligned} E = & \int_\Gamma \left(\frac{1}{2} f^2((x - X)^2 + (y - Y)^2) + \Phi \right) \rho \mathrm{d}x\mathrm{d}y\mathrm{d}z + \\ & \int_\Gamma C_v \left(\frac{R}{p_{ref}} \right)^{(R/C_v)} (Z\rho)^{(C_p/C_v)} \mathrm{d}x\mathrm{d}y\mathrm{d}z. \end{aligned} \tag{4.42}$$

Write $\nu = Z\rho$. Then (4.42) becomes

$$\begin{aligned} E = & \int_\Gamma \left(\frac{\frac{1}{2} f^2 ((x - X)^2 + (y - Y)^2) + \Phi}{Z} \right) \nu \mathrm{d}x\mathrm{d}y\mathrm{d}z + \\ & \int_\Gamma C_v \left(\frac{R}{p_{ref}} \right)^{(R/C_v)} \nu^{(C_p/C_v)} \mathrm{d}x\mathrm{d}y\mathrm{d}z. \end{aligned} \tag{4.43}$$

This is the equivalent of (3.92) in the shallow water case. The third and fourth equations of (4.36) imply that

$$\frac{\partial \nu}{\partial t} + \nabla \cdot (\nu \mathbf{u}) = 0. \tag{4.44}$$

(4.44) states that, since Z is conserved on particles, the conservation of mass $\rho \mathrm{d}x\mathrm{d}y\mathrm{d}z$ by the compressible equations (4.36) implies conservation of $\nu \mathrm{d}x\mathrm{d}y\mathrm{d}z$. Therefore ν can be regarded as a probability measure on Γ, which will play a similar role to h in the shallow water case.

Now suppose that the values of (X, Y, Z) as functions of (x, y, z) are given by a mapping $\mathbf{s}^{-1} : \Gamma \to \mathbb{R}^3$. Using Definition 3.3, define the potential density σ to be the push forward of the probability measure ν under the map $\mathbf{s}^{-1}$. An energy minimising state characterised by Theorem 4.3 can be sought by finding such a mapping that minimises the energy integral (4.43) under the constraints (4.38), which implies mass conservation, and (4.39), which implies invariance of σ. (4.38) implies that $\delta\nu = -\nu\nabla \cdot \Xi$. Then, given a set $B \subset \mathbb{R}^3$ and a map $\mathbf{s} : \mathbb{R}^3 \to \Gamma$, $\sigma(B) = \nu(\mathbf{s}(B))$ so that

$$\mathbf{s}_{\#}^{-1}\nu = \sigma. \tag{4.45}$$

Now rewrite Theorem 4.3.

Theorem 4.4. *Let ν be a probability measure defined on Γ. Let S be the class of maps $\mathbf{s}^{-1} : \Gamma \to \mathbb{R}^3$ satisfying (4.45). For some $\mathbf{s}^{-1} \in S$ write $\mathbf{s}^{-1}(x, y, z) = (\tilde{X}, \tilde{Y}, \tilde{Z})$. Define the associated energy integral by*

$$E = \int_\Gamma \left(\frac{\frac{1}{2} f^2((x - \tilde{X})^2 + (y - \tilde{Y})^2) + \Phi}{\tilde{Z}} \right) \nu \mathrm{d}x \mathrm{d}y \mathrm{d}z + \tag{4.46}$$
$$\int_\Gamma C_v \left(\frac{R}{p_{ref}} \right)^{(R/C_v)} \nu^{(C_p/C_v)} \mathrm{d}x \mathrm{d}y \mathrm{d}z.$$

The conditions for the energy E to be minimised over maps $\mathbf{s}^{-1} \in S$ are that (4.40) is satisfied and that the matrix $\mathbf{Q}$ defined by (4.41) is positive definite. If such a map exists it is called an optimal map and written $\mathbf{t}^{-1}$: $\Gamma \to \mathbb{R}^3$.

Proof It is necessary to identify the definition of S with the class of variations used in Theorem 4.3. The constraints (4.38) and (4.39) used in Theorem 4.3 imply that $\nu \mathrm{d}x \mathrm{d}y \mathrm{d}z$ is preserved under the displacements, so that the perturbed map $\mathbf{s}^{-1}$ stays in S. Conversely, equation (4.45) applied for a given $\tilde{Z}$ implies that (4.38) holds. □

Note that the definition of Γ is still that made in section 4.1.1 so that the upper boundary is a decay condition.

4.1.4 *The equations in dual variables*

It is not practicable to solve the problem as stated in Theorem 4.4. As in the incompressible case, the problem needs to be restated in dual variables. It is not yet known whether the decay condition at the upper boundary can be retained in the new formulation, so Γ will be redefined as a closed bounded set in $\mathbb{R}^3$.

Then exchange the definitions of dependent and independent variables as in section 3.2.2, so that (x, y, z) are regarded as functions of the independent variables (X, Y, Z). The first three equations of (4.36) are now regarded as defining a velocity in (X, Y, Z) coordinates on $\mathbb{R}^3$ given by

$$\mathbf{U} = (U, V, W) = (f(y - Y), f(X - x), 0). \tag{4.47}$$

As in section 3.2.2, $\nabla \cdot \mathbf{U} = 0$.

Now assume the converse of the conditions applied to Theorem 4.4, namely that the potential density σ is given as a function of (X, Y, Z), and the converse of equation (4.45) holds, namely

$$\mathbf{s}_{\#}\sigma = \nu, \tag{4.48}$$

or

$$\sigma \mathrm{d}X\mathrm{d}Y\mathrm{d}Z = \nu \mathrm{d}x\mathrm{d}y\mathrm{d}z. \tag{4.49}$$

Conservation of the measure ν in physical space becomes conservation of σ in (X, Y, Z) coordinates, namely

$$\frac{\partial \sigma}{\partial t} + \nabla \cdot (\sigma \mathbf{U}) = 0. \tag{4.50}$$

The system of compressible semi-geostrophic equations in dual variables can therefore be written as

$$\begin{aligned}
\frac{\partial \sigma}{\partial t} + \mathbf{U} \cdot \nabla \sigma &= 0, \\
(U, V, W) &= (f(y - Y), f(X - x), 0), \\
\nu \frac{\partial(x, y, z)}{\partial(X, Y, Z)} &= \sigma, \\
f^2(x - X, y - Y, 0) + C_p Z \nabla \Pi + \nabla \Phi &= 0, \\
Z = T(p/p_{ref})^{-R/C_p} &\equiv T/\Pi, \\
p &= \rho R T, \\
\nu &= \rho Z.
\end{aligned} \tag{4.51}$$

This is a system of twelve equations for the twelve unknowns $(\sigma, x, y, z, U, V, W, \rho, \Pi, T, p, \nu)$. In order to solve it, assume that σ at a particular time is given as a function of (X, Y, Z). Seek to find (x, y, z) as a function of (X, Y, Z) by solving the energy minimisation problem for an optimal map $\mathbf{t}$ and a measure ν. Assuming this can be done, (U, V, W) can be found from the second equation of (4.51) and the solution advanced in time.

Thus seek to construct an optimal map $\mathbf{t} : \mathbb{R}^3 \to \Gamma$. Let S be the set of such mappings. For $\mathbf{s} \in S$, write $\mathbf{s}(\mathbf{X}) = (\tilde{x}, \tilde{y}, \tilde{z})$. Then the energy integral (4.43) can be rewritten as

$$E = \int_{\mathbb{R}^3} \left(\frac{\frac{1}{2} f^2((\tilde{x} - X)^2 + (\tilde{y} - Y)^2) + \Phi(\tilde{x}, \tilde{y}, \tilde{z})}{Z} \right) \sigma \mathrm{d}X \mathrm{d}Y \mathrm{d}Z + \int_\Gamma C_v \left(\frac{R}{p_{ref}} \right)^{(R/C_v)} \nu^{(C_p/C_v)} \mathrm{d}x \mathrm{d}y \mathrm{d}z. \tag{4.52}$$

This is the equivalent of (3.102) in the shallow water case.

ν is determined from $\mathbf{s}$ by (4.48). Then. as with (3.102), the first term in the integrand of (4.52) can be written as a cost function c depending on the map $\mathbf{s}$, and the second term is a convex function of ν. The map $\mathbf{s}$ that minimises E must therefore also minimise the integral of c under the constraint (4.48).

Next rewrite the minimisation of c subject to (4.48) as a Monge-Kantorovich problem. c is defined by

$$c(\mathbf{s}(\mathbf{X}), \mathbf{X}) = \frac{\frac{1}{2} f^2 (x - X)^2 + (y - Y)^2) + \Phi}{Z}. \tag{4.53}$$

so the problem is to minimise

$$C(\mathbf{s}) = \int_{\mathbb{R}^3} c(\mathbf{s}(\mathbf{X}), \mathbf{X}) \sigma \mathrm{d}X \mathrm{d}Y \mathrm{d}Z, \tag{4.54}$$

subject to (4.48). As in section 3.2.3, this problem cannot always be solved. It is therefore necessary to solve a relaxed problem where $\mathbf{s}$ is replaced by a joint probability measure $\pi(\mathbf{x}, \mathbf{X})$, as defined in section 3.2.3, with marginals ν and σ. Then the relaxed version of (4.54) is to minimise

$$C(\pi) = \int_{\Gamma \times \mathbb{R}^3} c(\mathbf{x}, \mathbf{X}) \pi(\mathbf{x}, \mathbf{X}) \mathrm{d}\mathbf{x} \mathrm{d}\mathbf{X}, \tag{4.55}$$

over all $\pi \in \mathbb{P}$ where $\mathbb{P}$ is the set of joint probability measures π such that $\pi(A \times \mathbb{R}^3) = \nu(A)$ for all $A \subset \Gamma$ and $\pi(\Gamma \times B) = \sigma(B)$ for all $B \subset \mathbb{R}^3$.

The dual problem, Theorem 3.5, for this case is

Theorem 4.5. *Given potentials* $\Pi(\mathbf{x}), \Psi(\mathbf{X})$, *define*

$$J(\Pi, \Psi) = \int_\Gamma -C_p \Pi \nu \mathrm{d}\mathbf{x} + \int_{\mathbb{R}^3} \Psi \sigma \mathrm{d}\mathbf{X}, \; I(\pi) = \int_{\Gamma \times \mathbb{R}^3} c(\mathbf{x}, \mathbf{X}) \pi(\mathbf{x}, \mathbf{X}) \mathrm{d}\mathbf{x} \mathrm{d}\mathbf{X}. \tag{4.56}$$

Let Φ_C *be the set of* Π, Ψ *such that*

$$-C_p \Pi(\mathbf{x}) + \Psi(\mathbf{X}) \leq c(\mathbf{x}, \mathbf{X}). \tag{4.57}$$

Then

$$\inf_{\mathbb{P}(\nu, \sigma)} I(\pi) = \sup_{\Phi_C} J(\Pi, \Psi). \tag{4.58}$$

This is proved in Cullen and Maroofi (2003).

This shows the equivalence of the minimisation of $C(\mathbf{s})$ as defined in (4.54) and the maximisation of $J(\Pi, \Psi)$ under the constraint (4.57). At the maximiser, denote the maximising Π and Ψ as $\hat{\Pi}, \hat{\Psi}$. Then (4.57 gives

$$-C_p\hat{\Pi}(\mathbf{x}) + \hat{\Psi}(\mathbf{X}) = c(\mathbf{x}, \mathbf{X}). \tag{4.59}$$

This is a slightly different form of the duality relation which is again related to Legendre duality as expressed in (3.51). It still falls within the general framework discussed by Sewell (2002).

Differentiating (4.59) with respect to $\mathbf{x}$ and using the definition (4.53) of c gives

$$-C_p\nabla\hat{\Pi}(\mathbf{x}) = \frac{f^2\left((x - X), (y - Y), 0\right) + \nabla\Phi}{Z}. \tag{4.60}$$

(4.60) shows that $\hat{\Pi}$ satisfies the geostrophic and hydrostatic relations given in the fourth equation of (4.51).

Differentiating (4.59) with respect to $\mathbf{X}$ gives

$$Z\nabla\hat{\Psi}(\mathbf{X}) = f^2\left((X - x), (Y - y), -c(\mathbf{x}, \mathbf{X})\right). \tag{4.61}$$

Using the definitions of U and V from (4.51) gives

$$Z\nabla\hat{\Psi}(\mathbf{X}) = f\left(V, -U, -c(\mathbf{x}, \mathbf{X})\right), \tag{4.62}$$

so that

$$(U, V) = Z\left(-\frac{\partial\hat{\Psi}}{\partial Y}, \frac{\partial\hat{\Psi}}{\partial X}\right), \tag{4.63}$$

and $\nabla \cdot \mathbf{U} = 0$ as stated after (4.47). Therefore σ satisfies the Lagrangian conservation law stated in (4.51) as required.

As in section 3.3.4 write the minimising cost as the Wasserstein distance $W_c(\sigma, \nu)$. Then the energy (4.52) can be written as

$$E = W_c(\sigma, \nu) + \int_\Gamma C_v \left(\frac{R}{p_{ref}}\right)^{(R/C_v)} \nu^{(C_p/C_v)} dxdydz.$$

Cullen and Maroofi (2003) prove that $W_c(\sigma, \nu)$ is weakly semi-continuous as a function of ν. Therefore the energy (4.64) is strictly convex as a function of ν with σ fixed, so a unique minimiser exists. Theorem 4.4 can then be applied to characterise the minimiser, since the variations used are consistent with a fixed σ. This means that (4.40) holds. Comparing this with (4.60) shows that $\Pi = \hat{\Pi}$, so that the solution of the Monge-Kantorovich problem can be used to characterise the solution of the complete problem.

The convexity condition satisfied by the minimiser can be deduced by defining the c-transforms of Ψ and Π as in (3.66) and (3.67):

$$\Psi^c(\mathbf{x}) = \inf_{\mathbf{X}\in\mathbb{R}^3}\{c(\mathbf{X},\mathbf{x})) - \Psi(\mathbf{X})\}, \tag{4.64}$$
$$-C_p\Pi^c(\mathbf{X}) = \inf_{\mathbf{x}\in\Gamma}\{c(\mathbf{X},\mathbf{x}) + C_p\Pi(\mathbf{x})\}$$

At the solution of the dual problem, both inequalities in (4.64) will be achieved, so that (4.59) means that $\Psi^c = -C_p\Pi$ and $-C_p\Pi^c = \Psi$. Together these imply that $(\Psi^c)^c = \Psi$ and $(\Pi^c)^c = \Pi$, so that Π and Ψ are c-convex. In this problem, unlike that in section 3.2.3, it is not possible to modify Π and Ψ to create convex potentials. These conditions will be consistent with positive definiteness of the matrix $\mathbf{Q}$ in (4.41).

These results generalise the incompressible results obtained in section 3.2.2. These relations were obtained in a slightly different context by Shutts (1989), p.556, using Hamilton's principle. Note that potentials of the form P and R do not appear. Instead, potentials Π and Ψ satisfying the duality relation (4.59) are used. In the incompressible case, the potentials are φ and Ψ which satisfy (3.63). In both cases it is a pressure-related function in physical space that appears.

In the compressible case, convex potentials do not appear, so the geometrical properties of convex functions cannot be exploited. Instead, (4.59) has to be exploited directly. In Cullen and Gangbo (2001) and Cullen and Maroofi (2003) it is shown how the regularity of convex functions extends to $c-$convex functions as defined above. This is sufficient for the analysis to proceed, since in particular, it can be proved that oscillatory behaviour of Π and Ψ is restricted.

4.1.5 *Rigorous weak existence results*

It is proved by Cullen and Maroofi (2003) that the Monge-Kantorovich problem stated in Theorem 4.5 has a unique solution under suitable conditions. This generalises Theorem 3.16. Their results are quoted below. A more concise proof is given in Cullen *et al.* (2014a) exploiting the general result of Ambrosio and Gangbo (2008).

Theorem 4.6. *Given probability measures σ, ν with bounded supports $\Sigma \subset \mathbb{R}^3, \Gamma \subset \mathbb{R}^3$, where Σ takes the form $\Sigma_Z \times [\delta, \frac{1}{\delta}]$ for some $\delta > 0$ and Γ is convex. Assume that $\frac{\partial\Phi}{\partial z} \neq 0$, and that Φ is twice continuously differentiable. Then there exist optimal maps $\mathbf{t} : \Sigma \to \Gamma$ and $\mathbf{t}^{-1} : \Gamma \to \Sigma$ which are inverses, satisfy (4.48) and (4.45) respectively, and minimise the integral of (4.53). These maps are unique up to sets of measure zero. The minimum* *value of the cost is the Wasserstein distance between σ and ν associated with the cost function (4.53).*

In order to prove that weak solutions of the compressible semi-geostrophic equations exist for all finite times, it is necessary to show that $\mathbf{U}$ is bounded, and thus that the support Σ of σ remains bounded for all finite time. This is because Theorem 4.6 only applies to mappings between bounded domains. The same requirement had to be met to prove the results in section 3.5.2. In the present case, it is also necessary to prove that $c-$convex functions defined using the cost function c given by (4.53) have the same regularity properties as convex functions and to ensure that the energy integral (4.52) makes sense by making appropriate assumptions on Z. Let $B(0,r)$ be a ball of radius r centred at the origin. Assume $\Gamma \subset B(0,r)$. At $t=0$, suppose that the support $\Sigma(0)$ of σ takes the form $\Sigma_Z \times [\delta, \frac{1}{\delta}]$ with $\Sigma_Z \subset \mathbb{R}^2$ for some $\delta > 0$. Then since the component W of $\mathbf{U}$ is zero, it will remain of this form for all t. If $B_z(0,r)$ is a ball in the plane $z=$ constant, suppose that $\Sigma(t) \subset \cup_{z\in[\delta,1/\delta]} B_z(0,r(t))$. Then a similar argument to that giving (3.169) can be applied to show that $r(t)$ is bounded by $r(0)+frt$ for each t. For a given time $T \geq 0$, define $B_T = \cup_{z\in[\delta,1/\delta]} B_z(0, r(0)+frT)$.

Theorem 4.7. *Let $1 < p < \infty$ and $T > 0$. Assume that Γ and Φ satisfy the conditions of Theorem 4.6 and that the given probability measure $\sigma(0,\cdot)$ has support $\Sigma(0) \subset B_0$ where B_0 is as defined above. Let B_T be the bounded subset of $\mathbb{R}^3$ defined above. Then the system (4.51) of semi-geostrophic equations in dual variables has an admissible weak solution $(\nu, \mathbf{t}^{-1})$ such that with $\sigma(t,\cdot) = \mathbf{t}^{-1}(t,\cdot)_{\#}\nu(t,\cdot)$ and $\mathbf{U}$ as defined in (4.51):*

$$\begin{aligned} \sigma \in L^p\left((0,T)\times B_T\right), \|\sigma(t,\cdot)\|_{L^p(B_T)} &\leq \|\sigma(0,\cdot)\|_{L^p(B_T)}, \\ \nu \in W^{1,\infty}(\Gamma), \|\nu(t,\cdot)\|_{W^{1,\infty}}(\Gamma) &\leq C(\Gamma, B_T, c(\cdot,\cdot)), \\ \mathbf{U} \in L^\infty, \|\mathbf{U}\|_{L^\infty} &\leq C(\Gamma, B_T). \end{aligned} \tag{4.65}$$

where C are constants with the indicated dependencies, and $c(\cdot,\cdot)$ is as defined in (4.53).

This result confirms that the assumptions of incompressible Boussinesq flow, which are unrealistic on large scales in the atmosphere, can be relaxed and that the resulting semi-geostrophic equations still define a slow manifold. The proper boundary conditions as stated after equations (4.1) are applied, except that Γ is assumed finite with a rigid boundary all round. The generalisation to unbounded domains in z has been studied, but not yet achieved. An appropriate free boundary problem for the equations in pressure coordinates, which does not imply an unbounded domain, was solved in Theorem 3.29. The formulation of the equations and the results also relax the requirement that the axis of rotation and the gravitational acceleration are parallel. This is exploited in the next section, when spherical geometry is considered. In order to extend Theorem 4.7 to the spherical case, it would be necessary to replace the assumption on Φ with one of monotonicity in a radial coordinate. This has not yet been done.

4.2 Semi-geostrophic theory on a sphere

Now consider the appropriate form of the geostrophic momentum approximation to the deep atmosphere Navier-Stokes equations (2.1), and their shallow atmosphere counterpart (2.166), in spherical geometry. The shallow atmosphere approximation was discussed in sections 2.2.3 and 2.2.4.

Consider first the deep atmosphere equations (2.1)-(2.4) with viscous and forcing terms removed, and no moisture equation.

$$\begin{aligned}
\frac{\mathrm{D}\mathbf{u}}{\mathrm{D}t} + 2\Omega \times \mathbf{u} + C_p\theta\nabla\Pi + g\hat{\mathbf{r}} &= 0, \\
\frac{\partial \rho}{\partial t} + \nabla \cdot (\rho\mathbf{u}) &= 0, \\
\frac{\mathrm{D}\theta}{\mathrm{D}t} &= 0, \\
\theta = T(p/p_{ref})^{-R/C_p} &\equiv T/\Pi, \\
p &= \rho R T.
\end{aligned} \tag{4.66}$$

Equations (4.66) are exactly the same as equations (4.1) if the z axis is chosen in the direction of the axis of rotation, Φ is chosen to be gr, where r is the radial coordinate, and the domain Γ is a spherical semi-infinite annulus defined by $a \le r < \infty$. In practice, the equations have only been analysed using a fixed upper boundary, so that $\Gamma = \{r \in [a, a+H]\}$, see White *et al.* (2005) and Wood and Staniforth (2003).

The geostrophic momentum approximation to (4.66) can therefore be made in the same way as in (4.4). Write the equations in cylindrical polar coordinates $(\check{\lambda}, \check{r}, \check{z})$ with the $\check{z}$ axis in the direction of the axis of rotation. The associated velocity components are $(\check{u}, \check{v}, \check{w})$. The definition of the geostrophic wind is

$$\frac{1}{\check{r}}C_p\theta\frac{\partial\Pi}{\partial\check{\lambda}} + 2\Omega\check{v}_g = 0, \tag{4.67}$$

$$C_p\theta\frac{\partial\Pi}{\partial\check{r}} + \frac{g\check{r}}{\sqrt{\check{r}^2+\check{z}^2}} - 2\Omega\check{u}_g = 0.$$

The momentum equations become

$$\frac{\mathrm{D}\check{u}_g}{\mathrm{D}t} + \frac{\check{u}\check{v}_g}{\check{r}} + \frac{1}{\check{r}}C_p\theta\frac{\partial\Pi}{\partial\check{\lambda}} + 2\Omega\check{v} = 0,$$
$$\frac{\mathrm{D}\check{v}_g}{\mathrm{D}t} - \frac{\check{u}\check{u}_g}{\check{r}} + \frac{C_p\theta}{\check{r}}\frac{\partial\Pi}{\partial\check{r}} + \frac{g\check{r}}{\sqrt{\check{r}^2+\check{z}^2}} - 2\Omega\check{u} = 0, \tag{4.68}$$
$$C_p\theta\frac{\partial\Pi}{\partial\check{z}} + \frac{g\check{z}}{\sqrt{\check{r}^2+\check{z}^2}} = 0.$$

The analysis in section 4.1 is applicable to these equations, since the only change is to the coordinate system. Following the method of section 4.1.3, define new variables $(\check{\Lambda}, \check{\Upsilon}, \check{Z})$ where $(\check{\Lambda}, \check{\Upsilon}) = (\check{\lambda}, \check{r})$ if $\check{u}_g = \check{v}_g = 0$. Then (4.35) gives

$$\check{\Upsilon}\cos\check{\Lambda} = \frac{\check{u}_g}{2\Omega} + \check{r}\cos\check{\lambda},$$
$$\check{\Upsilon}\sin\check{\Lambda} = \frac{\check{v}_g}{2\Omega} + \check{r}\sin\check{\lambda}, \tag{4.69}$$
$$\check{Z} = \check{\theta}.$$

The momentum equations from (4.36) are then

$$\frac{\mathrm{D}\check{\Upsilon}}{\mathrm{D}t} + \check{u}\check{\Lambda} - \check{v}_g = 0, \tag{4.70}$$
$$\frac{\mathrm{D}\check{\Lambda}}{\mathrm{D}t} - \frac{\check{u}\check{\Upsilon}}{\check{r}^2} + \frac{\check{u}_g}{\check{r}} = 0.$$

The duality relation (4.59) becomes

$$-C_p\Pi(\check{\lambda},\check{r},\check{z}) + \Psi(\check{\Lambda},\check{\Upsilon},\check{Z}) = \tag{4.71}$$
$$\frac{2\Omega^2\left((\check{r}\cos\check{\lambda} - \check{\Upsilon}\cos\check{\Lambda})^2 + (\check{r}\sin\check{\lambda} - \check{\Upsilon}\sin\check{\Lambda})^2)\right) + g\sqrt{\check{r}^2+\check{z}^2}}{\check{Z}}.$$

The potential density σ defined by (4.49) becomes

$$\sigma\check{\Upsilon}\mathrm{d}\check{\Lambda}\mathrm{d}\check{\Upsilon}\mathrm{d}\check{Z} = \rho\theta\check{r}\mathrm{d}\check{\lambda}\mathrm{d}\check{r}\mathrm{d}\check{z}. \tag{4.72}$$

The equations are to be solved in a region Γ corresponding to the atmosphere in spherical geometry given by

$$a \le \sqrt{\check{r}^2+\check{z}^2} \le a + H. \tag{4.73}$$

The plane $\check{z} = 0$ is the equatorial plane. There will be conservation laws for energy in the form (4.6) and potential density in the form (4.50).

Note that the hydrostatic equation, which is the third equation of (4.68), is degenerate at $z = 0$, which corresponds to the equator where it reduces to $\partial\Pi/\partial\check{z} = 0$. This is the reason why the assumption $\partial\Phi/\partial z > 0$ is made in Theorems 4.6 and 4.7. It is not yet known if the results can be extended to this case. More seriously, the assumption that the kinetic energy is approximated by a geostrophic value in the equatorial plane, without regard to the local vertical, is not appropriate in shallow atmospheres, and does not correspond to observed behaviour, which shows that the wind is primarily horizontal on large scales, as discussed in section 2.2.3.

In Shutts (1989), this system was therefore further approximated by imposing the shallow atmosphere approximation as in section 2.2.3. Shutts (1989) achieved this by making the approximation within the framework of Hamilton's principle, thus ensuring that conservation laws for energy and potential density were retained. The method amounts to projecting the equations which would apply on the equatorial plane $\check{z} = 0$ onto a spherical annulus with radius a, and applying the hydrostatic approximation in the local vertical. The resulting equations are therefore written in spherical polar coordinates (λ, ϕ, r) as in section 2.1, with the transformation from cylindrical coordinates given by

$$\lambda = \check{\lambda}, \; r\cos\phi = \check{r}, \; r = \sqrt{\check{r}^2 + \check{z}^2}. \tag{4.74}$$

Under the shallow atmosphere approximation, the metric factor r is replaced by a wherever it appears.

In the projection, vectors in the $\check{r}$ direction will be multiplied by a factor $\sin\phi$, recalling that ϕ is the latitude. The derivative $\partial/\partial\check{r}$ becomes $-\frac{1}{a\sin\phi}\frac{\partial}{\partial\phi}$ and velocities $\check{v}$ in the $\check{r}$ direction becomes velocities $-v\sin\phi$ in the ϕ direction. The definitions (4.67) become

$$\begin{aligned} \frac{1}{a\cos\phi}C_p\theta\frac{\partial\Pi}{\partial\lambda} - 2\Omega v_g\sin\phi &= 0, \\ \frac{1}{a}C_p\theta\frac{\partial\Pi}{\partial\phi} + 2\Omega u_g\sin\phi &= 0. \end{aligned} \tag{4.75}$$

The momentum equations become

$$\begin{aligned} \frac{\mathrm{D}u_g}{\mathrm{D}t} - \frac{uv_g\sin\phi}{a\cos\phi} + \frac{1}{a\cos\phi}C_p\theta\frac{\partial\Pi}{\partial\lambda} - 2\Omega v\sin\phi &= 0, \\ \sin\phi\frac{\mathrm{D}v_g\sin\phi}{\mathrm{D}t} + \frac{uu_g\sin\phi}{a\cos\phi} + \frac{1}{a}C_p\theta\frac{\partial\Pi}{\partial\phi} + 2\Omega u\sin\phi &= 0, \\ C_p\theta\frac{\partial\Pi}{\partial r} + g &= 0. \end{aligned} \tag{4.76}$$

Solve these equations in the region $\Gamma : a \leq r \leq a+H$, with the boundary conditions $\mathbf{u} \cdot \mathbf{n} = 0$. The equations then conserve the energy integral

$$E = \int_{\Gamma} \rho \left(\frac{1}{2}(u_g^2 + (v_g \sin\phi)^2) + C_v T + gr \right) 4\pi a^2 \cos\phi \mathrm{d}\lambda \mathrm{d}\phi \mathrm{d}r. \quad (4.77)$$

Define new variables (Λ, Υ, Z), where (Λ, Υ) is equal to (λ, ϕ) if $u_g = v_g = 0$, by projecting the data back onto the equatorial plane. The Cartesian coordinates of the projection of the point (λ, ϕ, a) onto the equatorial plane are $(a\cos\phi\cos\lambda, a\cos\phi\sin\lambda)$. Using this, together with the other projection rules above, (4.69) becomes

$$\begin{aligned} \cos\Upsilon\cos\Lambda &= \frac{u_g}{2a\Omega} + \cos\phi\cos\lambda, \\ \cos\Upsilon\sin\Lambda &= \frac{v_g\sin\phi}{2a\Omega} + \cos\phi\sin\lambda, \\ Z &= \theta. \end{aligned} \quad (4.78)$$

These definitions mean that the kinetic energy term in (4.77) is $2\Omega^2$ times the square of the Euclidean distance between the points (λ, ϕ, a) and (Λ, Υ, a) projected onto the equatorial plane. The momentum equations (4.76) become

$$\begin{aligned} a\frac{\mathrm{D}\Upsilon}{\mathrm{D}t} + u\Lambda - v_g &= 0, \\ a\frac{\mathrm{D}\Lambda}{\mathrm{D}t} - u\Upsilon - u_g &= 0. \end{aligned} \quad (4.79)$$

The duality relation (4.59) becomes

$$-C_p\Pi(\lambda, \phi, r) + \Psi(\Lambda, \Upsilon, Z) = \quad (4.80)$$
$$\frac{2\Omega^2 a^2 \left((\cos\phi\cos\lambda - \cos\Upsilon\cos\Lambda)^2 + (\cos\phi\sin\lambda - \cos\Upsilon\sin\Lambda)^2)\right) + gr}{Z}.$$

The potential density σ defined by (4.49) becomes

$$\sigma\cos\Upsilon \mathrm{d}\Lambda \mathrm{d}\Upsilon \mathrm{d}Z = \nu\cos\phi \mathrm{d}\lambda \mathrm{d}\phi \mathrm{d}r. \quad (4.81)$$

Since the dependence of (4.80) on Λ and Υ takes the form of a squared Euclidean distance, as in equation (4.59), the arguments leading to the potential density conservation law in the first equation of (4.51) still hold.

These definitions are explored in a slightly different way by Roulstone and Sewell (1996). In particular, it is demonstrated that no change of variables of the form (4.78) is possible in spherical geometry without the device of projecting to Cartesian geometry as used here.

Next show that the same properties of energy minimisation hold for equations (4.75) and (4.76) as for (4.3) and (4.4). Given a vector field $(\tilde{u}, \tilde{v})$ and scalar fields $\tilde{\theta}$ and $\tilde{\rho}$ as for Theorem 4.1, define the energy integral

$$E = \int_\Gamma \tilde{\rho}\left(\frac{1}{2}(\tilde{u}^2 + (\tilde{v}\sin\phi)^2) + C_v\tilde{T} + gr\right) 4\pi a^2 \cos\phi \mathrm{d}\lambda \mathrm{d}\phi \mathrm{d}r. \tag{4.82}$$

Write particle positions as $(\tilde{\lambda}, \tilde{\phi}, \tilde{r})$ so that, given a displacement $\Xi = (\xi, \eta, \chi)$,

$$\xi = a\cos\phi\delta\tilde{\lambda},\ \eta = a\delta\tilde{\phi},\ \chi = \delta\tilde{r}. \tag{4.83}$$

Theorem 4.8. *The conditions for (4.82) to be minimised with respect to variations $\Xi = (\xi, \eta, \chi)$ of particle positions as specified above, which satisfy continuity in the shallow atmosphere sense, $\delta(\tilde{\rho}\cos\phi \mathrm{d}\lambda \mathrm{d}\phi \mathrm{d}r) = 0$, via*

$$\delta\tilde{\rho} = -\tilde{\rho}\nabla\cdot\Xi \tag{4.84}$$

in Γ, and

$$\begin{aligned} \delta\tilde{u} &= 2\Omega\sin\phi\eta + \frac{\tilde{v}\tan\phi}{a}\xi, \\ \sin\phi\delta(\tilde{v}\sin\tilde{\phi}) &= -2\Omega\sin\phi\xi - \frac{\tilde{u}\tan\phi}{a}\xi, \\ \delta\tilde{\theta} &= 0, \end{aligned} \tag{4.85}$$

together with $\Xi\cdot\mathbf{n} = 0$ on the boundary of Γ , are that

$$(-2\Omega\tilde{v}\sin\phi, 2\Omega\tilde{u}\sin\phi, 0) + (0, 0, g) + C_p\tilde{\theta}\nabla\tilde{\Pi} = 0, \tag{4.86}$$

and that the matrix $\mathbf{Q}$ defined by

$$\mathbf{Q} = f\begin{pmatrix} f + \frac{1}{a\cos\phi}\frac{\partial\tilde{v}\sin\tilde{\phi}}{\partial\lambda} + \frac{\tilde{u}}{a\cos\phi} & \frac{1}{a}\frac{\partial\tilde{v}\sin\tilde{\phi}}{\partial\phi} & \frac{\partial\tilde{v}\sin\tilde{\phi}}{\partial r} \\ -\frac{1}{a\cos\phi}\frac{\partial\tilde{u}}{\partial\lambda} + \frac{\tilde{v}\tan\phi}{a} & f - \frac{1}{a}\frac{\partial\tilde{u}}{\partial\phi} & -\frac{\partial\tilde{u}}{\partial r} \\ \frac{1}{\tilde{\theta}}\frac{\partial\Phi}{\partial r}\frac{1}{a\cos\phi}\frac{\partial\tilde{\theta}}{\partial\lambda} & \frac{1}{\tilde{\theta}}\frac{\partial\Phi}{\partial r}\frac{1}{a}\frac{\partial\tilde{\theta}}{\partial\phi} & \frac{1}{\tilde{\theta}}\frac{\partial\Phi}{\partial r}\frac{\partial\tilde{\theta}}{\partial r} \end{pmatrix}, \tag{4.87}$$

where $f = 2\Omega\sin\phi$, is positive definite.

Proof Most of the proof follows that of Theorems 4.1 and 4.2. Only the differences are described. Set $f = 2\Omega\sin\phi$. The quantity $(\tilde{u}\delta\tilde{u} + \tilde{v}\delta\tilde{v})$ that appears in (4.21) is replaced by $(\tilde{u}\delta\tilde{u} + \tilde{v}\sin\tilde{\phi}\delta(\tilde{v}\sin\tilde{\phi})$. Using (4.85), this becomes $\eta 2\Omega\sin\phi\tilde{u} - \xi 2\Omega\sin\phi\tilde{v}$ which is $f(\eta\tilde{u} - \xi\tilde{v})$, which is identical to the expression in the proof of Theorem 4.1. The rest of the proof is the same.

In the second variation, the expression $\delta(-\tilde{v}, \tilde{u}, 0)$ has to be evaluated to pass from (4.25) to (4.26). The second component is straightforward, because using (4.85) gives $f\eta + \frac{\tilde{v}\tan\tilde{\phi}}{a}\xi$. This leads to the extra term in the second row of (4.87). To evaluate the first component, take out a factor 2Ω rather than a factor f. Using (4.85) gives $\delta(\sin\tilde{\phi}\tilde{v}) = -2\Omega\xi - \frac{\tilde{u}}{a\cos\phi}\xi$. However, instead of $\left(\frac{1}{f}C_p\tilde{\theta}\nabla\tilde{\Pi}\right) + \frac{1}{f}\nabla\Phi$ in (4.30) and (4.31), which gives just v_g at the stationary point, this gives $\left(\frac{1}{2\Omega}C_p\tilde{\theta}\nabla\tilde{\Pi}\right) + \frac{1}{2\Omega}\nabla\Phi$ which is $v_g \sin\phi$. These changes lead to the changes in the first row of (4.87). □

Inspection of (4.87) shows that the conditions for inertial stability, given by the requirement that the diagonal elements of $\mathbf{Q}$ are positive, are respectively

$$2\Omega + \frac{1}{a\cos\phi}\frac{\partial\tilde{v}\sin\tilde{\phi}}{\partial\lambda} + \frac{\tilde{u}}{a\cos\phi} \geq 0, \qquad (4.88)$$
$$f - \frac{1}{a}\frac{\partial\tilde{u}}{\partial\phi} \geq 0.$$

After rewriting the second equation of (4.88) as $2\Omega + \frac{1}{a}\frac{\partial\tilde{u}}{\partial(\cos\phi)} \geq 0$, these can be seen to be exactly the inertial stability conditions for the flow projected onto the equatorial plane. This form of the condition is to be expected, since the equations were derived by projection onto the equatorial plane.

These equations have the disadvantage that the momentum and kinetic energy are approximated not by their geostrophic values, as in (4.4), but by the projection of their geostrophic values on the equatorial plane. If they are applied on regions of the Earth's surface away from the poles small enough that $2\Omega\sin\phi$ does not change much across the region, they will be less accurate than the f-plane version of the equations used in chapter 3. An alternative procedure is to make the approximations in the reverse order. This respects the physical situation discussed in sections 2.2.3 and 2.2.4 that the shallow atmosphere approximation is much more accurate in the Earth's atmosphere than the geostrophic momentum approximation. Thus as in section 2.5.2, make the shallow atmosphere and hydrostatic approximations first, and then make the geostrophic momentum approximation. The shallow atmosphere equations (2.166) in spherical polar coordinates, with the forcing and dissipation omitted as in equations (4.66), are

$$\frac{\mathrm{D}(u,v)}{\mathrm{D}t} + \frac{(-uv,u^2)\tan\phi}{a} + (-fv, fu) + C_p\theta\left(\frac{1}{a\cos\phi}\frac{\partial\Pi}{\partial\lambda}, \frac{1}{a}\frac{\partial\Pi}{\partial\phi}\right) = 0,$$
$$C_p\theta\frac{\partial\Pi}{\partial r} + g = 0,$$
$$f = 2\Omega\sin\phi,$$
$$\frac{\partial\rho}{\partial t} + \nabla\cdot(\rho\mathbf{u}) = 0, \tag{4.89}$$
$$\frac{\mathrm{D}\theta}{\mathrm{D}t} = 0,$$
$$\theta = T(p/p_{ref})^{-R/C_p} \equiv T/\Pi,$$
$$p = \rho RT.$$

Define the geostrophic wind by (2.167), with Π replacing Π':

$$\frac{1}{a\cos\phi}C_p\theta\frac{\partial\Pi}{\partial\lambda} - fv_g = 0, \tag{4.90}$$
$$\frac{C_p\theta}{a}\frac{\partial\Pi}{\partial\phi} + fu_g = 0.$$

Then make the geostrophic momentum approximation as in (2.168), giving

$$\frac{\mathrm{D}u_g}{\mathrm{D}t} - \frac{uv_g\tan\phi}{a} - fv + C_p\theta\frac{1}{a\cos\phi}\frac{\partial\Pi}{\partial\lambda} = 0, \tag{4.91}$$
$$\frac{\mathrm{D}v_g}{\mathrm{D}t} + \frac{uu_g\tan\phi}{a} + fu + C_p\theta\frac{1}{a}\frac{\partial\Pi}{\partial\phi} = 0.$$

Equations (4.89) with (4.91) on $\Gamma = \{r \in [a, a+H]\}$, with the boundary conditions $\mathbf{u}\cdot\mathbf{n} = 0$ on the boundary of Γ, conserve the energy integral

$$E = \int_\Gamma \rho\left(\frac{1}{2}(u_g^2 + v_g^2) + C_vT + gr\right)4\pi a^2\cos\phi\mathrm{d}\lambda\mathrm{d}\phi\mathrm{d}r. \tag{4.92}$$

The energy minimisation result of Theorems 4.1 and 4.2 also applies to this case. Given a vector field $(\tilde{u}, \tilde{v})$ and scalar fields $\tilde{\theta}$ and $\tilde{\rho}$ as for Theorem 4.1, define the energy integral

$$E = \int_\Gamma \tilde{\rho}\left(\frac{1}{2}(\tilde{u}^2 + \tilde{v}^2) + C_v\tilde{T} + gr\right)4\pi a^2\cos\phi\mathrm{d}\lambda\mathrm{d}\phi\mathrm{d}r. \tag{4.93}$$

Specify displacements as for Theorem 4.8, equation (4.83).

Theorem 4.9. *The conditions for (4.82) to be minimised with respect to variations $\Xi = (\xi, \eta, \chi)$ of particle positions, specified in spherical polar coordinates (λ, ϕ, r), which satisfy continuity in the shallow atmosphere sense, $\delta(\tilde{\rho}\cos\phi\mathrm{d}\lambda\mathrm{d}\phi\mathrm{d}r) = 0$, via*

$$\delta\tilde{\rho} = -\tilde{\rho}\nabla\cdot\Xi \tag{4.94}$$

in Γ *and*

$$\begin{aligned}\delta\tilde{u} &= f\eta + \frac{\tilde{v}\tan\phi}{a}\xi,\\ \delta\tilde{v} &= -f\xi - \frac{\tilde{u}\tan\phi}{a}\xi,\\ \delta\tilde{\theta} &= 0,\end{aligned} \tag{4.95}$$

together with $\Xi\cdot\mathbf{n}=0$ *on the boundary of* Γ *, are that*

$$(-f\tilde{v}, f\tilde{u}, 0) + (0,0,g) + C_p\tilde{\theta}\nabla\tilde{\Pi} = 0, \tag{4.96}$$

and that the matrix $\mathbf{Q}$ *defined by*

$$\mathbf{Q} = f\begin{pmatrix} f + \frac{1}{a\cos\phi}\frac{\partial\tilde{v}}{\partial\lambda} + \frac{\tilde{u}\tan\phi}{a} & \frac{1}{a}\frac{\partial\tilde{v}}{\partial\phi} & \frac{\partial\tilde{v}}{\partial r}\\ -\frac{1}{a\cos\phi}\frac{\partial\tilde{u}}{\partial\lambda} + \frac{\tilde{v}\tan\phi}{a} & f - \frac{1}{a}\frac{\partial\tilde{u}}{\partial\phi} & -\frac{\partial\tilde{u}}{\partial r}\\ \frac{g}{\tilde{\theta}}\frac{1}{a\cos\phi}\frac{\partial\tilde{\theta}}{\partial\lambda} & \frac{g}{\tilde{\theta}}\frac{1}{a}\frac{\partial\tilde{\theta}}{\partial\phi} & \frac{g}{\tilde{\theta}}\frac{\partial\tilde{\theta}}{\partial r}\end{pmatrix}, \tag{4.97}$$

is positive definite.

Proof Only the differences from the proofs of Theorems 4.1 and 4.2 are described. Using (4.95), the quantity $(\tilde{u}\delta\tilde{u} + \tilde{v}\delta\tilde{v})$ that appears in (4.21) becomes $f(\eta\tilde{u} - \xi\tilde{v})$, which is identical to the expression in the proof of Theorem 4.1. The rest of the proof is the same.

In the second variation, the expression $\delta(-\tilde{v}, \tilde{u}, 0)$ has to be evaluated to pass from (4.25) to (4.26). This becomes $\left(f\xi + \frac{\tilde{u}\xi\tan\phi}{a}, f\eta + \frac{\tilde{v}\xi\tan\phi}{a}\right)$. The extra terms that then appear in (4.26) are the extra terms that appear in $\mathbf{Q}$ in (4.97). □

Equations (4.89)-(4.91) cannot be written straightforwardly in dual variables like the other systems treated in this book. The reasons will be made clear in the analysis in section 4.3. The structure can only be retained if the equations are further approximated in some way, for instance as by Salmon (1985) and Magnusdottir and Schubert (1991). In Salmon (1985), the approximation is made within Hamilton's principle, thus ensuring that the approximated system retains conservation laws for global angular momentum, energy and potential vorticity. The effect of the approximation is that the equations cannot be written down explicitly in physical space, and so it is difficult to estimate their accuracy as a limit of the Navier-Stokes equations. However, it is unlikely that the second order accuracy in the limit $Ro = Fr^2 \to 0$ is retained. The approximation made by Magnusdottir and Schubert (1991) also retains conservation properties, though it is not made using Hamilton's principle. The physical space equations obtained are more implicit, so it is again difficult to assess their accuracy and in particular whether second order accuracy is retained in the limit above. In both cases the mapping from dual variables to physical space is defined by a potential density σ of a form similar to (4.81), but it is not clear if it is always possible to do this.

Therefore, following Cullen *et al.* (2005), seek to prove that (4.89)-(4.91) can be solved, so retaining second order accuracy despite some loss of structure. Recall that at the start of section 2.2.4 and in section 2.5.2 it was pointed out that there is a loss of rotational symmetry when the shallow atmosphere approximation is imposed, which will inevitably lead to a loss of mathematical structure. In particular, this will explain the difference between extratropical and tropical weather.

In order to compare the physical usefulness of the alternative systems (4.66) with (4.75)-(4.76) and (4.89)-(4.91), the treatment of Rossby waves by both systems is analysed. The respective analyses are given by Shutts (1989) and Mawson (1996). The equations for incompressible perturbations independent of r about a state of rest with $\theta = \theta_0$ derived from (4.75)-(4.76) are

$$\begin{aligned} \sin^2\phi\frac{\partial v_g'}{\partial t} + 2\Omega u'\sin\phi &= 2\Omega u_g'\sin\phi, \\ \frac{\partial u_g'}{\partial t} - 2\Omega v'\sin\phi &= -2\Omega v_g'\sin\phi, \\ \frac{\partial u'}{\partial\lambda} + \frac{\partial(v'\cos\phi)}{\partial\phi} &= 0, \end{aligned} \tag{4.98}$$

where

$$u_g'\sin\phi = -\frac{C_p\theta_0}{2\Omega a}\frac{\partial\Pi'}{\partial\phi},\ v_g'\sin\phi = \frac{C_p\theta_0}{2\Omega a\cos\phi}\frac{\partial\Pi'}{\partial\lambda}. \tag{4.99}$$

Shutts (1989) makes the substitution

$$\Pi' = \mathrm{Re}[G_m(\varsigma)\exp i(m\lambda - \omega t)], \tag{4.100}$$

where $G_m(\varsigma)$ is the wave amplitude and $\varsigma = \sin\phi$. Then Shutts (1989) shows that $G_m(\varsigma)$ satisfies the equation

$$(1-\varsigma)^2\frac{\mathrm{d}^2G_m}{\mathrm{d}\varsigma^2} - \frac{2}{\varsigma}\frac{\mathrm{d}G_m}{\mathrm{d}\varsigma} + \left(\alpha_m - \frac{m^2}{1-\varsigma^2}\right)G_m = 0, \tag{4.101}$$

where $\alpha_m = m^2 - 2\Omega m/\omega$. Given the boundary conditions $G_m = 0$ at the poles $\varsigma = \pm 1$, equation (4.101) is an eigenvalue problem for α_m. Shutts (1989) shows that the frequencies of the permitted solutions α_{mk} are given by

$$\omega_k = \frac{-2\Omega m}{k(k+1) - m^2}. \tag{4.102}$$

The equivalent linearisation of equations (4.89)-(4.91) is

$$\begin{aligned}
\frac{\partial v'_g}{\partial t} + 2\Omega u' \sin\phi &= 2\Omega u'_g \sin\phi,\\
\frac{\partial u'_g}{\partial t} - 2\Omega v' \sin\phi &= -2\Omega v'_g \sin\phi,\\
\frac{\partial u'}{\partial \lambda} + \frac{\partial (v'\cos\phi)}{\partial \phi} &= 0,
\end{aligned} \tag{4.103}$$

together with (4.99). A similar analysis in Mawson (1996) shows that (4.102) is replaced by

$$\omega_k = \frac{-2\Omega m}{k(k+1)}. \tag{4.104}$$

Equation (4.104) is also the dispersion relation for the unapproximated equations for incompressible flow independent of r. Therefore it can be seen that the use of equations (4.89-(4.91) is advantageous in the treatment of Rossby waves, which are the basis of the motions producing weather systems. Therefore this model is analysed further, despite its lack of mathematical structure. Theorem 4.9 shows that the characterisation of the solutions as time evolving energy minimising states still holds. This is the basis of all the versions of semi-geostrophic theory analysed in this book and gives them a sound physical basis.

4.3 The shallow water spherical semi-geostrophic equations

4.3.1 *Solutions as minimum energy states*

The analysis of equations (4.89)-(4.91) has so far only been carried out for the shallow water case. However, it is likely that the extension of any analysis to the compressible equations (4.89)-(4.91) will be straightforward. The analysis is difficult because of the lack of a simple change of variables.

The shallow water semi-geostrophic equations in spherical polar coordinates can be derived from (2.87) and (2.88). The geostrophic wind is defined by

$$\begin{aligned}
g\frac{1}{a\cos\phi}\frac{\partial h}{\partial \lambda} - fv_g &= 0, \qquad (4.105)\\
g\frac{1}{a}\frac{\partial h}{\partial \phi} + fu_g &= 0,
\end{aligned}$$

where $f = 2\Omega \sin\phi$, and the evolution equations are

$$\begin{aligned}\frac{\mathrm{D}u_g}{\mathrm{D}t} - \frac{uv_g \tan\phi}{a} + \frac{g}{a\cos\phi}\frac{\partial h}{\partial\lambda} - fv = 0,\\ \frac{\mathrm{D}v_g}{\mathrm{D}t} + \frac{uu_g \tan\phi}{a} + \frac{g}{a}\frac{\partial h}{\partial\phi} + fu = 0,\\ \frac{\partial h}{\partial t} + \frac{1}{a\cos\phi}\left(\frac{\partial}{\partial\lambda}(hu) + \frac{\partial}{\partial\phi}(hv\cos\phi)\right) = 0.\end{aligned} \tag{4.106}$$

The region of integration is the surface S^2 of the whole sphere, so no boundary conditions are required. The energy integral is given by (2.89). The following energy minimisation result holds. Given a vector field $(\tilde{u}, \tilde{v})$ and a scalar field $\tilde{h}$, define the energy integral

$$\int_{S^2} \left(\frac{1}{2}\tilde{h}\left(\tilde{u}^2 + \tilde{v}^2\right) + \frac{1}{2}g\tilde{h}^2\right) 4\pi a^2 \cos\phi \mathrm{d}\lambda \mathrm{d}\phi. \tag{4.107}$$

Write particle positions as $(\tilde{\lambda}, \tilde{\phi})$ so that, given a displacement $\Xi = (\xi, \eta)$,

$$\xi = a\cos\tilde{\phi}\delta\tilde{\lambda}, \ \eta = a\delta\tilde{\phi}. \tag{4.108}$$

Theorem 4.10. *The conditions for (4.107) to be minimised with respect to the variations in particle positions specified in (4.108), which satisfy continuity* $\delta(\tilde{h}\cos\tilde{\phi}\mathrm{d}\lambda\mathrm{d}\phi) = 0$, *via*

$$\delta\tilde{h} = -\tilde{h}\nabla\cdot\Xi \tag{4.109}$$

in Γ, *and*

$$\begin{aligned}\delta\tilde{u} = f\eta + \frac{\tilde{v}\tan\phi}{a}\xi,\\ \delta\tilde{v} = -f\xi - \frac{\tilde{u}\tan\phi}{a}\xi,\end{aligned} \tag{4.110}$$

are that

$$(-f\tilde{v}, f\tilde{u}) + g\nabla\tilde{h} = 0, \tag{4.111}$$

and that the matrix $\mathbf{Q}$ *defined by*

$$\mathbf{Q} = f\begin{pmatrix} f + \frac{1}{a\cos\tilde{\phi}}\frac{\partial\tilde{v}}{\partial\lambda} + \frac{\tilde{u}\tan\tilde{\phi}}{a} & \frac{1}{a}\frac{\partial\tilde{v}}{\partial\phi} \\ -\frac{1}{a\cos\tilde{\phi}}\frac{\partial\tilde{u}}{\partial\lambda} + \frac{\tilde{v}\tan\tilde{\phi}}{a} & f - \frac{1}{a}\frac{\partial\tilde{u}}{\partial\phi} \end{pmatrix} \tag{4.112}$$

is positive definite.

Proof This is a restatement of Theorem 3.7 in spherical polar coordinates, noting that Theorem 3.7 was proved for the case where f is a function of position. The changes to deal with spherical coordinates were described in the proof of Theorem 4.9. □

This argument works when f is a smooth function of position bounded away from zero because the terms $-fv, fu$ in the momentum equations in (4.105) can be represented to $O(\delta t)$ in the variational argument above by $-f\eta, f\xi$. The variations of f only enter at $O(\delta t^2)$. In Cullen *et al.* (2005) and Cullen (2008a), the time evolution of the semi-geostrophic system was modelled by timestepping using the same linearisation. In addition, the matrix $\mathbf{Q}$ was not varied. However, if $\mathbf{Q}$ is allowed to vary, it may not stay positive definite. In Cheng *et al.* (2018), it was shown using elliptic regularity theory that $\mathbf{Q}$ does stay positive definite for a (short) finite time under suitable conditions. The result is described in the next subsection.

All the rigorous long time existence results for the semi-geostrophic equations described in Chapter 3 and section 4.1 rely on rewriting the equations in dual variables and exploiting Monge-Kantorovich theory to characterise energy minimisers as the solution of an optimal transport problem. These solutions are usually weak. The difficulty in finding dual variables for (4.105)-(4.106) was analysed by Roulstone and Sewell (1996) and Roulstone and Sewell (1997), eq. (7.20). It arises because the variations (4.110) are non-integrable. Integrability is essential if (4.110) is to be written in the form $\delta\tilde{X} = \tilde{\delta Y} = 0$, as in Theorem 3.8. For integrability, a finite displacement satisfying (4.110) has to give a well-defined change to $\tilde{\mathbf{u}}$. However, given $\tilde{u} = \tilde{v} = 0$ at $(0, 0)$, displace a particle at that position to the point $(\pi/4, \pi/4)$, and calculate the change in $\tilde{\mathbf{u}}$ using (4.110). If the displacement proceeds via the point $(0, \pi/4)$, the result is $(\Omega a, -\Omega a(1 + \pi/4))$. If it is via $(\pi/4, 0)$, the result is $(\Omega a\sqrt{2}, 0)$. A method of formulating the problem as a Monge-Kantorovich problem on individual timesteps is described in section 4.3.3. In principle this could give long time existence of weak solutions.

4.3.2 *Classical solutions for a short time interval*

This analysis was carried out by Cheng *et al.* (2018). They also solved the two-dimensional incompressible case. Equations (4.105)-(4.106) were solved in plane geometry using periodic boundary conditions.. The domain of integration is referred to as the flat torus $\mathbb{T}^2$. The variable Coriolis parameter f was not specified analytically. The first step is to write the equations in Lagrangian form.

$$J = \begin{pmatrix} 0 & -1 \\ 1 & 0 \end{pmatrix}$$

$$\begin{aligned} \mathbf{u}_g &= f^{-1} J g \nabla h, \\ \mathrm{D}_t \mathbf{u}_g - f J \mathbf{u}_g + f J \mathbf{u} &= 0, \\ \partial_t h + \nabla \cdot (h\mathbf{u}) &= 0. \end{aligned} \tag{4.113}$$

Write $\mathbf{F}$ for the Lagrangian flow generated by $\mathbf{u}$. Then

$$\partial_t \mathbf{F} = \mathbf{u}, \ \mathbf{F}(0, \mathbf{x}) = \mathbf{x}. \tag{4.114}$$

Next define Lagrangian variables defined on fluid particles. Write $\mathbf{x}_0$ for particle positions at $t = 0$. Thus write

$$\mathbf{v}_g(t, \mathbf{x}_0) = \mathbf{u}_g(t, \mathbf{F}(t, \mathbf{x}_0)). \tag{4.115}$$

The fluid depth generates a probability measure h, since its integral over $\mathbb{T}^2$ is conserved and normalised to 1. Set this to h_0 at $t = 0$. Then the third and fourth equations of (4.113) become

$$\begin{aligned} \partial_t \mathbf{v}_g(t, \mathbf{x}_0) - f(\mathbf{F}(t, \mathbf{x}_0)) J \mathbf{v}_g(t, \mathbf{x}_0) + f(\mathbf{F}(t, \mathbf{x}_0)) J \partial_t \mathbf{F}(t, \mathbf{x}_0) &= 0, \\ \mathbf{F}(t, \cdot)_{\#} h_0 &= h(t, \mathbf{x}). \end{aligned} \tag{4.116}$$

The second equation defines $h(t, \mathbf{x})$. Write D for the Hessian matrix. Then the second equation of (4.116) becomes

$$\det D\mathbf{F}^{-1} = \frac{h(t, \cdot)}{h(0, \mathbf{F}^{-1}(t, \mathbf{x}))} \tag{4.117}$$

in the smooth case.

The next step is to construct a formal discretisation of the first equation of (4.116) in time. Subsequently it is proved that the time-discrete solutions exist, and that they converge to a solution of the original equation. Thus write

$$\begin{aligned} \mathbf{v}_g(t + \delta t, \mathbf{x}_0) - \mathbf{v}_g(t, \mathbf{x}_0) - f(\mathbf{F}(t, \mathbf{x}_0)) J \mathbf{v}_g(t, \mathbf{x}_0) \delta t + \\ f(\mathbf{F}(t, \mathbf{x}_0)) J (\mathbf{F}(t + \delta t, \mathbf{x}_0) - \mathbf{F}(t, \mathbf{x}_0)) = 0. \end{aligned} \tag{4.118}$$

Note that the discretisation of the term $f(\mathbf{F}(t, \mathbf{x}_0)) J \partial_t \mathbf{F}(t, \mathbf{x}_0)$ is explicit in f, which makes the solution much easier. Now note that

$$\mathcal{R}_{f(\mathbf{F}(t,\mathbf{x}_0))\delta t} \mathbf{v}_g(t, \mathbf{x}_0) = \mathbf{v}_g(t, \mathbf{x}_0) + f(\mathbf{F}(t, \mathbf{x}_0)) J \mathbf{v}_g(t, \mathbf{x}_0) \delta t + \mathrm{O}(\delta t)^2, \tag{4.119}$$

where $\mathcal{R}_\alpha$ is a matrix defining a rotation by angle α. Then (4.118) becomes

$$\mathbf{v}_g(t+\delta t,\mathbf{x}_0)-\mathcal{R}_{f(\mathbf{F}(t,\mathbf{x}_0))\delta t}\mathbf{v}_g(t,\mathbf{x}_0)+f(\mathbf{F}(t,\mathbf{x}_0))J(\mathbf{F}(t+\delta t,\mathbf{x}_0)-\mathbf{F}(t,\mathbf{x}_0))=0. \tag{4.120}$$

Solving this equation allows the solution to be advanced from time t to $t+\delta t$. Now compress the notation to apply to this timestep. Set $\mathbf{x}=\mathbf{F}((n+1)\delta t,\mathbf{x}_0)$. Write $\mathbf{F}_{n+1}$ for the flow map connecting time $n\delta t$ to $(n+1)\delta t$, h_n for $h(n\delta t,\cdot)$ and h_{n+1} for $h((n+1)\delta t,\cdot)$. Also write f_{n+1} for $f(\mathbf{x})$ and f_n for $f(\mathbf{F}_{n+1}^{-1}(\mathbf{x}))$. Then, noting from (4.113) that $\mathbf{v}_g(n\delta t,\mathbf{x}_0)=f_n gJ\nabla h(n\delta t,\mathbf{F}(t,\mathbf{x}_0))=f_n gJ\nabla h_n$, the equation to be solved is

$$\mathbf{x}+gf_{n+1}^{-1}f_n^{-1}\nabla h_{n+1}=\mathbf{F}_{n+1}^{-1}(\mathbf{x})+gf_n^{-2}\mathcal{R}_{f_n\delta t}\nabla h_n(\mathbf{F}_{n+1}^{-1}(\mathbf{x})). \tag{4.121}$$

This equation has to be solved, subject to the condition $\mathbf{F}_{n+1\#}h_n=h_{n+1}$, for h_{n+1} and $\mathbf{F}_{n+1}$. In section 4.3.3 it is solved by formulating a Monge-Kantorovich problem. In the present case, sufficient smoothness is assumed to allow (4.117) to be used to give

$$\frac{\det(I+gf_{n+1}^{-1}\nabla h_{n+1}\otimes\nabla f_n^{-1}+gf_n^{-2}D^2h_{n+1}+A(\mathbf{x}))}{\det(I+gf_n^{-1}\nabla h_n\otimes\nabla f_n^{-1}+gf_n^{-2}D^2h_n+B(\mathbf{x}))}=\frac{h_{n+1}(\mathbf{x})}{h_n(\mathbf{F}^{-1}(\mathbf{x}))}, \tag{4.122}$$

where $A(\mathbf{x})$ and $B(\mathbf{x})$ are known functions derived from (4.121).

Cheng *et al.* (2018) use the implicit function theorem to solve (4.121) and (4.122) for $(h_{n+1},\mathbf{F}_{n+1})$ for each n. Then they prove convergence of the timestepping to establish existence of a unique solution to the evolution equations (4.116) for a short time. The result is

Theorem 4.11. *Let $k\geq 2$ be an integer. Let the Coriolis parameter f be a member of $C^{k,\alpha}(\mathbb{T}^2)$ with $f>0$ on $\mathbb{T}^2$. Let $h_0\in C^{k+2,\alpha}(\mathbb{T}^2)$ with $\int_{\mathbb{T}^2}h_0(\mathbf{x})\mathrm{d}\mathbf{x}=1$. Suppose that h_0 satisfies the following convexity and positivity conditions:*

$$I+gf^{-1}D(f^{-1}Dh_0)>c_0I,\ \ h_0>c_1 \text{ on } \mathbb{T}^2 \text{ for some } c_0,c_1>0. \tag{4.123}$$

The convexity condition is a strict version of (4.112). Then there exists $T_0>0$, depending on $\|h_0\|_{k+2,\alpha},c_0,c_1,f$ and k, such that there exists a solution $(h,\mathbf{u}_g,\mathbf{u})$ to (4.113) with initial data h_0 on $[0,T_0]\times\mathbb{T}^2$ which satisfies

$$I+gf^{-1}D(f^{-1}Dh)\geq 0,\ \ h\geq 0 \text{ on } [0,T_0]\times\mathbb{T}^2, \tag{4.124}$$

$$\int_{\mathbb{T}^2}h(t,\mathbf{x})\mathrm{d}\mathbf{x}=1\ \forall\, t\in[0,T_0],$$

and the following regularity

$$\partial_t^m h \in L^\infty(0,T_0;C^{k+2-m,\alpha}(\mathbb{T}^2)) \text{ for } 0 \le m \le k+1, \qquad (4.125)$$
$$\mathbf{u} \in L^\infty(0,T_0;C^{k,\alpha}(\mathbb{T}^2)).$$

Moreover, any solution $(h,\mathbf{u}_g,\mathbf{u})$ *to (4.113) with initial data* h_0 *defined on* $[0,T]\times\mathbb{T}^2$ *for some* $T>0$ *which satisfies (4.124) and has regularity* $h\in L^\infty(0,T;C^3(\mathbb{T}^2))$, $\partial_t h \in L^\infty(0,T;C^2(\mathbb{T}^2))$ *is unique.*

4.3.3 *Formulation of the problem as a Monge-Kantorovich problem*

The lack of integrability means that the existence of weak Lagrangian solutions or of relaxed Lagrangian solutions as described in section 3.5.4 is impossible, because the integration by parts needed to define the solution cannot be carried out. The procedure is therefore to establish the long-time existence of a time dependent h using optimal transport methods, together with a joint probability measure as in Definition 3.15, which can be characterised at each time as a local-in-time solution of the equations. The use of optimal transport guarantees the maintenance of the convexity condition (4.123). The procedure is validated by showing that it reproduces the classical solutions for smooth data derived in Theorem 4.11 and, if f is a constant, willl reproduce the weak measure solutions defined in Definition 3.15. It also shows that the energy is non-increasing. While formal arguments in section 4.2 show that the energy is conserved, it is not clear whether this can be proved rigorously in the present case given the likelihood that only very weak solutions exist.

Start from equation (4.121) which applies to solving a discrete form of the equations for a single timestep.

$$\mathbf{x} + g f_{n+1}^{-1} f_n^{-1}\nabla h_{n+1} = \mathbf{F}_{n+1}^{-1}(\mathbf{x}) + g f_n^{-2}\mathcal{R}_{f_n\delta t}\nabla h_n(\mathbf{F}_{n+1}^{-1}(\mathbf{x})). \qquad (4.126)$$

Rearrange (4.126) to give ∇h_{n+1}:

$$g f_{n+1}^{-1}\nabla h_{n+1} = f_n\mathbf{F}_{n+1}^{-1}(\mathbf{x}) - f_n\mathbf{x} + g f_n^{-1}\mathcal{R}_{f_n\delta t}\nabla h_n(\mathbf{F}_{n+1}^{-1}(\mathbf{x})). \qquad (4.127)$$

h_{n+1} and $\mathbf{F}_{n+1}^{-1}$ are unknowns which have to be found simultaneously as in Section 4.3.2. To distinguish the known and unknown variables, set $\mathbf{y} = \mathbf{F}_{n+1}^{-1}(\mathbf{x})$, giving

$$g f^{-1}(\mathbf{x})\nabla h_{n+1} = f(\mathbf{y})(\mathbf{y}-\mathbf{x}) + g f^{-1}(\mathbf{y})\mathcal{R}_{f(\mathbf{y})\delta t}\nabla h_n(\mathbf{y}). \qquad (4.128)$$

This assumes, as in (4.118), that the term $fJ\partial_t\mathbf{F}$ in (4.116) is discretised as $f(\mathbf{y})J(\mathbf{x}-\mathbf{y})$. This is accurate to $O(\delta t^2)$ in the smooth case discussed in section 4.3.2, which is sufficient for the short time argument in Cheng *et al.* (2018). However, if optimal transport methods are used to determine h_{n+1} and $\mathbf{F}_{n+1}^{-1}$, this will require minimising a cost function determined by the right hand side of (4.128) which does not involve $f(\mathbf{x})$. This will then give a solution where $gf^{-1}(\mathbf{y})\nabla h_{n+1} = -J\mathbf{u}_{g(n+1)}$. This is not consistent with (4.128) by an amount $\mathrm{O}|f(\mathbf{y})-f(\mathbf{x})|$ which will be of a similar magnitude to $(\mathbf{x}-\mathbf{y})$ assuming, as in section 4.3.1, that f is a smooth function of position bounded away from zero. This inconsistency will not necessarily be small or even well-defined. Thus adding successive timesteps together will lead to a wrong solution over a finite time interval. The implicit function theorem approach used by Cheng *et al.* (2018) avoids this problem since it solves for $f^{-1}(\mathbf{x})\nabla h_{n+1}$ directly.

Thus replace $f(\mathbf{y})(\mathbf{y}-\mathbf{x})$ in (4.128) by $-\mathbf{d}(\mathbf{x},\mathbf{y})$ where $\mathbf{d}(\mathbf{x},\mathbf{y}) = \int f\mathrm{d}\mathbf{F}$ along a trajectory $\mathbf{F}$ linking $\mathbf{y}$ to $\mathbf{x}$ where $\mathbf{F}$ is chosen to minimise $|\mathbf{d}|$. This resolves the ambiguity caused by the non-integrability of this term. Then $f(\mathbf{y})\mathbf{x}$ in (4.128) is replaced by $f(\mathbf{y})\mathbf{y}+\mathbf{d}(\mathbf{x},\mathbf{y})$. The derivative of this term with respect to $\mathbf{x}$ is $f(\mathbf{x})$, and so it will be shown that, at a solution of the optimal transport problem, $-f(\mathbf{x})J\mathbf{u}_{g(n+1)} = g\nabla h_{n+1}$ as required.

Note that the minimising $\mathbf{F}$ is a geodesic on $\mathbb{T}^2$ conformally rescaled by f, an interpretation explored in Cullen *et al.* (2005). McCann (2001) showed that the optimal transport problem could be solved using $|\mathbf{d}|^2$ as the cost function. However, this does not cover the present case.

Eq. (4.128) then becomes

$$gf^{-1}(\mathbf{x})\nabla h_{n+1} = -\mathbf{d}(\mathbf{x},\mathbf{y}) + gf^{-1}(\mathbf{y})\mathcal{R}_{f(\mathbf{y})\delta t}\nabla h_n(\mathbf{y}). \tag{4.129}$$

Now define $\mathbf{Y} \in \mathbb{T}^2$ by

$$\mathbf{Y} = f(\mathbf{y})\mathbf{y} + gf^{-1}(\mathbf{y})\mathcal{R}_{f(\mathbf{y})\delta t}\nabla h_n(\mathbf{y}). \tag{4.130}$$

Then

$$-J\mathbf{u}_{g(n+1)} = gf^{-1}(\mathbf{x})\nabla h_{n+1} = -\mathbf{d}(\mathbf{x},\mathbf{y}) - f(\mathbf{y})\mathbf{y} + \mathbf{Y}. \tag{4.131}$$

Note that $\mathbf{d}(\mathbf{x},\mathbf{y}) + f(\mathbf{y})\mathbf{y}$ is an approximation to $f(\mathbf{x})\mathbf{x}$. If f is constant, the right hand side of (4.131) is just $\mathbf{Y} - f(\mathbf{x})\mathbf{x}$, which is the same as in the standard solution procedure of section 3.2 with different notation. It would reduce to $\mathbf{Y} - f(\mathbf{y})\mathbf{x}$ if $\mathbf{d}(\mathbf{x},\mathbf{y})$ was replaced by $f(\mathbf{y})(\mathbf{x}-\mathbf{y})$ as in (4.128).

Now set up the optimal transport argument, following the procedure of section 3.3.4. Define a measure σ on $\mathbb{T}^2$ as

$$\sigma = \mathbf{Y}_{\#}h_n. \tag{4.132}$$

For a general h, the energy (4.107) in the plane geometry used in this section becomes

$$\int_{\mathbb{T}^2} \frac{1}{2}\left(h\mathbf{u}_g{}^2 + gh^2\right) \mathrm{d}\mathbf{x} = \int_{\mathbb{T}^2} \frac{1}{2}\left(hf^{-2}g^2(\nabla h)^2 + gh^2\right) \mathrm{d}\mathbf{x}. \tag{4.133}$$

Theorem 4.10 characterises the required solution h_{n+1} as an energy minimiser, so that (4.133) has to be minimised for maps $\mathbf{s} : \mathbb{T}^2 \to \mathbb{T}^2$, where $\mathbf{s}_{\#}\sigma = h$. Using (4.132) shows that $(\mathbf{s} \circ \mathbf{Y})_{\#}h_n = h$. Since, at a solution, h will define h_{n+1}, $\mathbf{s} \circ \mathbf{Y}$ will represent the Lagrangian flow $\mathbf{F}_{n+1}$ used in (4.127) as required. As in section 3.3.3, the minimisation problem is not an optimal transport problem because h is unknown.

Use (4.131) to rewrite (4.133) as

$$\int_{\mathbb{T}^2} \frac{1}{2}\left(h\left(-\mathbf{d}(\mathbf{x},\mathbf{y}) - f(\mathbf{y})\mathbf{y} + \mathbf{Y}\right)^2 + gh^2\right) \mathrm{d}\mathbf{x}. \tag{4.134}$$

Suppose h is a minimiser. Then, as with the constant rotation case (3.104), the second term of the integrand does not depend explicitly on the map $\mathbf{s}$, so the minimising map must also minimise

$$\int_{\mathbb{T}^2} \frac{1}{2}\left(-\mathbf{d}(\mathbf{x},\mathbf{y}) - f(\mathbf{y})\mathbf{y} + \mathbf{Y}\right)^2 h\mathrm{d}\mathbf{x} = \int_{\mathbb{T}^2} \frac{1}{2}\left(-\mathbf{d}(\mathbf{x},\mathbf{y}) - f(\mathbf{y})\mathbf{y} + \mathbf{Y}\right)^2 \sigma\mathrm{d}\mathbf{Y} \tag{4.135}$$

with respect to maps that satisfy $\mathbf{s}_{\#}\sigma = h$. This is an optimal transport problem.

There is no reason why σ as defined by (4.132) should be absolutely continuous with respect to Lebesgue measure. Thus the solution of the optimal transport problem will be a joint probability measure rather than a deterministic map.

The first step is to solve the optimal transport problem for a given h. Write the cost function to be minimised as

$$c(\mathbf{x},\mathbf{Y}) = \frac{1}{2}(\mathbf{Y} - f(\mathbf{y})\mathbf{y} - \mathbf{d}(\mathbf{x},\mathbf{y}))^2, \qquad C = \int_{\mathbb{T}^2} c(\mathbf{x},\mathbf{Y})\pi\mathrm{d}\mathbf{x}\mathrm{d}\mathbf{Y}, \tag{4.136}$$

where π is a joint probability measure on $\mathbb{T}^2 \times \mathbb{T}^2$ with marginals h and σ. Let $\mathbb{P}$ be the set of all such probability measures.

Follow the standard procedure for the Monge-Kantorovich problem as in Theorem 4.5. Seek potentials φ, Ψ which maximise the functional

$$J(\varphi, \Psi) = \int_{\mathbb{T}^2} -\varphi h \mathrm{d}\mathbf{x} + \int_{\mathbb{T}^2} \Psi \sigma \mathrm{d}\mathbf{Y}, \tag{4.137}$$

subject to the constraint

$$-\varphi(\mathbf{x}) + \Psi(\mathbf{Y}) \leq c(\mathbf{x}, \mathbf{Y}). \tag{4.138}$$

Let Φ_C be the set of φ, Ψ satisfying (4.138). Define

$$I(\pi) = \int_{\mathbb{T}^2 \times \mathbb{T}^2} c(\mathbf{x}, \mathbf{Y}) \pi \mathrm{d}\mathbf{x} \mathrm{d}\mathbf{Y}. \tag{4.139}$$

Then, if the Kantorovich duality result (3.62) can be applied to this case,

$$\inf_{\mathbb{P}(h,\sigma)} I(\pi) = \sup_{\Phi_C} J(\varphi, \Psi). \tag{4.140}$$

At the solution write $\varphi = \hat{\varphi}, \Psi = \hat{\Psi}$, giving the duality relation

$$-\hat{\varphi}(\mathbf{x}) + \hat{\Psi}(\mathbf{Y}) = c(\mathbf{x}, \mathbf{Y}). \tag{4.141}$$

Then, differentiating with respect to $\mathbf{x}$ gives, using (4.131)

$$\nabla_{\mathbf{x}} \hat{\varphi} = -f(\mathbf{x}) J \hat{\mathbf{u}}_g, \tag{4.142}$$

Differentiating with respect to $\mathbf{Y}$ does not give a useful expression, reflecting the lack of a dual variable for this problem. The convexity property follows from the c–convexity discussed after (4.64). Thus $(\hat{\varphi}^c)^c = \hat{\varphi}$, where

$$\hat{\varphi}^c(\mathbf{Y}) = \inf_{\mathbf{x} \in \mathbb{T}^2} \{c(\mathbf{x}, \mathbf{Y}) - \hat{\varphi}(\mathbf{x})\} \tag{4.143}$$

This requires $\hat{\varphi}$ to be locally more convex than $-\frac{1}{2} f^2 \mathbf{x}^2$.

The next step is to consider the minimisation of the whole energy integral (4.133). Define the minimising value of $c(\mathbf{x}, \mathbf{Y})$ as a Wasserstein distance $W_c(\sigma, h)$. Then (4.133) becomes

$$W_c(\sigma, h) + \int_{\mathbb{T}^2} \frac{1}{2} g h^2 \mathrm{d}\mathbf{x}. \tag{4.144}$$

If $W_c(\sigma, h)$ can be proved to be lower-semi-continuous with respect to h then, since the additional term $\frac{1}{2}h^2$ is strictly convex in h, there will be a unique minimiser h_{n+1}. This will satisfy the property (4.111) that $fJ\mathbf{u}_g + g\nabla h = 0$ derived in Theorem 4.10. Thus $\hat{\varphi}$ in (4.142) can be identified with gh_{n+1}. Then h_{n+1} will satisfy the convexity property (4.143) which is equivalent to the convexity condition (4.112).

Now consider the evolution over multiple timesteps. The procedure above yields a solution of (4.128) allowing $f_{n+1}^{-1} \nabla h_{n+1}$ to be calculated from

$f_n^{-1}\nabla h_n$ together with a joint probability measure $\pi(\mathbf{x},\mathbf{y})$ which gives the probability that a particle at $\mathbf{y}$ at time $n\delta t$ will be at position $\mathbf{x}$ at time $(n+1)\delta t$. Since the deterministic trajectory $\mathbf{F}$ used in section 4.3.2 has been replaced by the measure π, deterministic Lagrangian variables as defined in (4.115) are no longer useful, so the geostrophic wind has to be represented as the Eulerian variable $\mathbf{u}_g$ as defined in (4.113). The convexity condition (4.112) implies that $\frac{\partial v_g}{\partial x} > -f$ and that $\frac{\partial u_g}{\partial y} < f$. The periodic boundary conditions mean that $\int f v_g(x,y)\mathrm{d}x$ is zero for each y and $\int f u_g(x,y)\mathrm{d}y$ is zero for each x. Together, these imply that $|\mathbf{u}_g|$ is bounded by some constant U.

The solutions for each timestep can now be added together, giving a sequence $h(n\delta t,\cdot)$ with the implied $|\mathbf{u}_g|$ uniformly bounded and each h satisfying (4.112). There will also be a sequence of measures $\pi_n(\mathbf{x},\mathbf{x}_0)$ which give the probability that a particle at $\mathbf{x}_0$ at the initial time will be at position $\mathbf{x}$ at time $n\delta t$. Convergence of the sequence $h(n\delta t,\cdot)$ with $n = T/\delta t$ to $h(t,\cdot)$ for all $t \in [0,T]$ as $\delta t \to 0$ requires a stability result for the solutions obtained for each timestep. In Feldman and Tudorascu (2015a), this result is proved for relaxed Lagrangian solutions of the sort derived here by bounding the Wasserstein distance between σ_n and σ_{n+1}, where σ_n is the dual space measure at timestep n. In the present case there is no dual space equation, so it is necessary to work from the definition of $\mathbf{Y}$. Since h_n is an energy minimiser, set δt to zero in the definition (4.130) of $\mathbf{Y}$ giving $\mathbf{Y}_0 = f(\mathbf{y})\mathbf{y} + gf^{-1}(\mathbf{y})\nabla h_n(\mathbf{y})$. Then (4.131) is satisfied with $h_{n+1} = h_n$ and $\mathbf{x} = \mathbf{y}$. Let $\sigma^* = \mathbf{Y}_{0\#}h_n$. σ is defined by (4.132). $\mathbf{Y}$ is obtained from $\mathbf{Y}_0$ by rotating $\mathbf{u}_g$ by an angle $f\delta t$. Thus the Wasserstein distance $W_2(\sigma^*,\sigma)$ defined after (3.104) is bounded by $U\delta t$, and since h_n is derived from σ^* and h_{n+1} is derived from σ it should be possible to bound the difference between h_{n+1} and h_n. In Cullen and Gangbo (2001), the strict convexity of the energy integral for the shallow water case was exploited to improve the stability result.

It is then plausible that a function $h(t,\cdot)$ can be found as a limit of a sequence of approximations as $\delta t \to 0$. In the shallow water case, Cullen and Gangbo (2001) prove that h is continuous in time. Uniqueness cannot be expected. A convergent sequence of discrete approximations as $\delta t \to 0$ will generate a convergent sequence of measures as in Feldman and Tudorascu (2015a), so that the limiting $\pi_t(\mathbf{x},\mathbf{x}_0)$ gives the probability that a particle at $\mathbf{x}_0$ at the initial time will be at position $\mathbf{x}$ at time t.

This h will correspond to a solution obtained by the method of Cullen and Gangbo (2001) if f is constant, because the solution procedures are then the same. It will also correspond to that obtained by Cheng *et al.* (2018) if smooth initial data is specified. This is because both solutions are obtained from (4.121). The energy will be non-increasing, because the replacement of $\mathbf{v}_g$ by $\mathcal{R}(\mathbf{v_g})$ in the timestepping (4.120) does not change the energy, and the energy is then minimised so cannot increase. Note that if the solution is smooth enough so that $h((n+1)\delta t, \cdot) - h(n\delta t, \cdot) = \mathrm{O}(\delta t)$, then the loss in energy will be $\mathrm{O}(\delta t^2)$, and integrating over $\mathrm{O}(\delta t^{-1})$ timesteps and letting $\delta t \to 0$ will give energy conservation. However, this degree of regularity is not obtained by Cullen and Gangbo (2001) in the constant f case, so cannot be expected. Conservation would still be obtained if $h((n+1)\delta t, \cdot) - h(n\delta t, \cdot) = \mathrm{O}(\delta t)^{\frac{1}{2}+\alpha}$ for $\alpha > 0$. Though it is unlikely that this can be proved to hold everywhere, it may only be violated on a small set. Then energy conservation may hold.

The final step is to try and characterise this solution as a weak solution of the original equations (4.113) in an appropriate sense. However, as noted above, the lack of integrability means that the solution cannot be a relaxed Lagrangian solution in the sense of Theorem 3.37. The best that can be done is to show that the solution represents an envelope of relaxed Lagrangian solutions at each t, in the sense defined below. This is equivalent to the idea that a smooth trajectory can be regarded as an envelope of tangents.

The tangents are created by assuming that f is a function of the initial position of the trajectories, not the actual position. This makes the equations integrable along trajectories, so that relaxed Lagrangian solutions can be defined. Pick arbitrary times $t_1, t_2 \in [0, T)$ with $t_2 > t_1$. Set $\mathbf{x}_0$ to be the $\mathbf{x}$ coordinate at time t_1. In the converged solution of (4.113),

$$-f(\mathbf{x})J\mathbf{u}_g(t_1, \mathbf{x}) = g\nabla h(t_1, \mathbf{x})).$$

To justify the formulation, consider first the smooth case where $\mathbf{F}(t, \cdot)$ represents the Lagrangian flow. As in (3.174), define a Lagrangian variable $\mathbf{Z}(t, \mathbf{x}_0) = (x + f^{-1}(\mathbf{x}_0)v_g, y - f^{-1}(\mathbf{x}_0)u_g)$. Let $\mathbf{Z}$ satisfy

$$\frac{\partial \mathbf{Z}}{\partial t} + f(\mathbf{x_0})J(\mathbf{x} - \mathbf{Z}) = 0, \tag{4.145}$$

as in (3.175). This can be derived from (4.113) under the assumption that $f(\mathbf{x}) = f(\mathbf{F}^{-1}(\mathbf{x}))$. For the general case, initialise a relaxed Lagrangian solution at time t_1 and let $\zeta(t, \mathbf{x}_0)$ be a smooth Lagrangian variable. Then

(3.188) becomes

$$\int_{t_1}^{t_2}\int_{\mathbb{T}^2}\int_{\mathbb{T}^2}\Big(Z(t,\mathbf{x})\partial_t\zeta(t,\mathbf{x}_0) - f(\mathbf{x}_0)J(\mathbf{x}-Z(t,\mathbf{x}))\zeta(t,\mathbf{x}_0)\Big)\mu(\mathrm{d}t,d\mathbf{x},d\mathbf{x}_0) + \int_{\mathbb{T}^2} Z_0(\mathbf{x})\zeta(t_1,\mathbf{x}_0)\mathrm{d}\mathbf{x}_0 = 0, \tag{4.146}$$

where μ is a joint probability measure on $[t_1,t_2)\times\mathbb{T}^2\times\mathbb{T}^2$. The map Z to $\mathbf{x}$ at each time has to be found by minimising

$$\int_{\mathbb{T}^2} f(\mathbf{x}_0)^2(Z(t,\mathbf{x}_0)-\mathbf{F}(t,\mathbf{x}_0))^2 h_0\mathrm{d}\mathbf{x}_0.$$

Then the method of Feldman and Tudorascu (2015a) should show that a solution of 4.146 exists for $t\in[t_1,t_2]$.

Now show that this defines a 'tangent' to the solution of the original problem. Given that both solutions are expressed by probability measures, this can only be done by comparing the first timestep (4.129) of the time discretisation of (4.113) with (4.128) which is the first timestep of a time discretisation of (4.145). It can be seen that, for a smooth case, the two solutions converge at an $\mathrm{O}(\delta t^2)$ rate as $\delta t\to 0$. In the probabilistic case, the measures $\mu(t_1+\delta t,\mathbf{x},\mathbf{x}_0)$ from the solution of (4.145) will converge to $\pi_{t_1+\delta t}(\mathbf{x},\mathbf{x}_0)$ from the solution of (4.113).

4.4 The theory of axisymmetric flows

4.4.1 *Forced axisymmetric flows*

Semi-geostrophic theory takes advantage of the fact that straight flow in geostrophic balance is an exact steady-state solution of the Euler equations. The condition that the Lagrangian Rossby number is small requires that variations in the flow direction are small. A straight flow is stable to parcel displacements if it satisfies the stability condition derived in section 3.1.3. The other case where a simple flow is an exact solution to the inviscid equations is the axisymmetric vortex. The stability of such vortices was studied by Fjortoft (1946) and Eliassen and Kleinschmidt (1957) by considering variations conserving angular momentum. A theory parallel to the semi-geostrophic theory of Cullen and Purser (1984) was developed by Shutts *et al.* (1988) which characterised a stable vortex as an energy minimiser. It was made mathematically rigorous by Cullen and Sedjro (2014), using methods described by Cullen *et al.* (2015a). The description of the problem here largely follows the latter paper.

The vortex is a steady state solution of the dynamical equations. To study the stability of the vortex, or to consider its evolution under forcing, extra terms have to be included in the equations. Start with the hydrostatic Boussinesq Euler equations (3.1). Use cylindrical coordinates (λ, r, z) with associated velocity components (u, v, w). The angular velocity of the system rotation is Ω. All variables are assumed to be independent of λ. The forcing is described by terms F, S which are also independent of λ. These could represent the averaged effects of λ-dependent disturbances to the vortex, or other external forcing. Then the equations become

$$\begin{aligned}\frac{\mathrm{D}u}{\mathrm{D}t} + \frac{uv}{r} + 2\Omega v &= \frac{1}{r}F(t,r,z),\\ \frac{\mathrm{D}\theta}{\mathrm{D}t} &= S(t,r,z),\\ \frac{u^2}{r} + 2\Omega u &= \frac{\partial\varphi}{\partial r},\\ \frac{1}{r}\frac{\partial}{\partial r}(rv) + \frac{\partial w}{\partial z} &= 0,\\ \frac{\partial\varphi}{\partial z} - g\frac{\theta}{\theta_0} &= 0.\end{aligned} \tag{4.147}$$

The third of these equations is only valid if $u \to 0$ as $r \to 0$. Following Shutts *et al.* (1988), to avoid the coordinate singularity at the vortex centre, a small region $r < r_0$ with $u = 0$ is excluded from the problem. Thus assume that equations (4.147) define the flow in an isolated vortex occupying a region Γ_ς defined by

$$\Gamma_\varsigma = [0, 2\pi] \times [r_0, \varsigma(t,z)] \times [0, H]. \tag{4.148}$$

The vortex is assumed to be sitting in an ambient fluid at rest with constant potential temperature θ_0. Thus $\frac{\partial\varphi}{\partial z} = g$ in the ambient fluid. Redefine θ as $\theta - \theta_0$ and φ as $\varphi = \varphi - gz$. Assume for this analysis that $\theta > 0$. It may be possible to relax this assumption. The boundary conditions are then

$$\begin{aligned} w(t,r,0) = w(t,r,H) &= 0,\\ v(t,r_0,z) &= 0,\\ \varphi(t,\varsigma(t,z),z) &= 0 \text{ for } \varsigma > r_0.\end{aligned} \tag{4.149}$$

The last of these conditions is required to make φ well-defined.

The boundary condition $v(t, r_0, z) = 0$ is physically appropriate if $u(t, r_0, z) \geq 0$, because then the assumption $u = 0$ for $r < r_0$ is consistent with stability of the vortex. Since imposing this extra condition makes the problem overdetermined, it implies a restriction on the choice of initial data which will be discussed later.

The boundary $r = \varsigma(t,z)$ is a material surface so that

$$\frac{\partial \varsigma}{\partial t} + w\frac{\partial \varsigma}{\partial z} = v. \tag{4.150}$$

Physically relevant solutions will require this vortex to be stable to internal axisymmetric perturbations. Similarly, the vortex is required to retain its identity, so that there is no mixing between the vortex and the surrounding fluid. This will require the boundary $r = \varsigma$ to be stable against axisymmetric perturbations also involving the surrounding fluid. It is then possible to justify only considering the fluid in the vortex and ignoring the motion of the ambient fluid.

4.4.2 *The vortex as a minimum energy state*

Consider first the vortex with a fixed boundary at $r = r_1$. The steady state is given by the third and fifth equations of (4.147). The angular momentum associated with this state is $M = ur + \Omega r^2$. Then the third equation of (4.147) can be written as

$$\frac{M^2}{r^3} = \frac{\partial}{\partial r}\left(\varphi + \frac{1}{2}\Omega^2 r^2\right). \tag{4.151}$$

This suggests a change of coordinate to replace r by

$$s = \frac{1}{2}\left(\frac{1}{r_0^2} - \frac{1}{r^2}\right). \tag{4.152}$$

Then $r^{-3}\mathrm{d}r = \mathrm{d}s$. (4.151) also suggests defining a new potential by

$$P = \varphi + \frac{1}{2}\Omega^2 r^2. \tag{4.153}$$

Then (4.151) and the fifth equation of (4.147) become

$$M^2 = \frac{\partial P}{\partial s}; \quad g\frac{\theta}{\theta_0} = \frac{\partial P}{\partial z}. \tag{4.154}$$

For this analysis, the domain Γ_ς defined in (4.148) is replaced by the two-dimensional domain

$$\Gamma = [r_0, r_1] \times [0, H]. \tag{4.155}$$

The energy integral associated with the vortex is

$$E = \int_\Gamma \left(\frac{1}{2}u^2 - g\theta z/\theta_0\right) r\mathrm{d}r\mathrm{d}z, \tag{4.156}$$

$$= \int_\Gamma \left(\frac{1}{2}\left(\frac{M}{r} - \Omega r\right)^2 - g\theta z/\theta_0\right) r\mathrm{d}r\mathrm{d}z.$$

Then the results of Fjortoft (1946) and Eliassen and Kleinschmidt (1957) can be stated as follows. Given a state $(\tilde{M}, \tilde{\theta})$. Calculate the energy using the second equation of (4.156). Define variations $\Xi = (\eta, \chi)$ of particle positions $(\tilde{r}, \tilde{z})$ satisfying

$$\eta = \delta\tilde{r},\ \chi = \delta\tilde{z}. \tag{4.157}$$

Theorem 4.12. *The conditions for the energy E to be minimised with respect to the displacements Ξ defined in (4.157), satisfying continuity $\delta(\tilde{r}\mathrm{d}r\mathrm{d}z) = 0$ via*

$$\frac{1}{\tilde{r}}\frac{\partial(\tilde{r}\eta)}{\partial r} + \frac{\partial\chi}{\partial z} = 0 \tag{4.158}$$

in Γ, and conserving angular momentum and potential temperature via

$$\delta\tilde{M} = 0,\ \delta\tilde{\theta} = 0, \tag{4.159}$$

together with $\Xi \cdot \mathbf{n} = 0$ on the boundary of Γ, are that

$$\tilde{M}^2 = \frac{\partial\tilde{P}}{\partial s}, \tag{4.160}$$

$$g\frac{\tilde{\theta}}{\theta_0} = \frac{\partial\tilde{P}}{\partial z}.$$

for some $\tilde{P}$, and

$$\mathbf{Q} = \begin{pmatrix} \frac{\partial\tilde{M}^2}{\partial s} & \frac{\partial\tilde{M}^2}{\partial z} \\ \frac{g}{\theta_0}\frac{\partial\tilde{\theta}}{\partial s} & \frac{g}{\theta_0}\frac{\partial\tilde{\theta}}{\partial z} \end{pmatrix} \tag{4.161}$$

is positive definite.

Proof Applying the perturbations gives

$$\delta E = \int_\Gamma \left(\left(-\frac{\tilde{M}\eta}{\tilde{r}^2} - \Omega\eta \right) \left(\frac{\tilde{M}}{\tilde{r}} - \Omega\tilde{r} \right) - g\tilde{\theta}\chi/\theta_0 \right) 2\pi\tilde{r}\mathrm{d}r\mathrm{d}z, \tag{4.162}$$

$$= \int_\Gamma \left(-\eta \left(\frac{\tilde{M}^2}{\tilde{r}^3} - \Omega^2\tilde{r} \right) - \chi g\tilde{\theta}/\theta_0 \right) \tilde{r}\mathrm{d}r\mathrm{d}z.$$

If this to be zero for any Ξ satisfying (4.158), then

$$\left(\left(\frac{\tilde{M}^2}{\tilde{r}^3} - \Omega^2\tilde{r} \right), g\tilde{\theta}/\theta_0 \right) = \nabla\tilde{\varphi} \tag{4.163}$$

for some $\tilde{\varphi}$. This is exactly the condition given by (4.151) and the fifth equation of (4.147). Rewriting using the coordinate s and defining $\tilde{P}$ from $\tilde{\varphi}$ using (4.153) gives (4.160).

As in the proof of Theorem 3.2, characterise the values at the stationary point as $(\tilde{M},\tilde{\theta}) = (M_g,\theta_g)$ satisfying (4.160) with $\tilde{P} = P_g$. Then

$$\delta E = \int_\Gamma -\left(\frac{\eta}{r^3},\chi\right)\cdot\left(\tilde{M}^2 - \frac{\partial P_g}{\partial s}, g\tilde{\theta}/\theta_0 - \frac{\partial P_g}{\partial z}\right)\tilde{r}\mathrm{d}r\mathrm{d}z. \tag{4.164}$$

Now take a second variation, noting that $\eta/\tilde{r}^3 = \delta\tilde{s}$. This gives, using the vanishing of the second term in the integrand at the stationary point,

$$\delta^2 E = \int_\Gamma -(\delta\tilde{s},\chi)\cdot\delta\left(\tilde{M}^2 - \frac{\partial P_g}{\partial s}, g\tilde{\theta}/\theta_0 - \frac{\partial P_g}{\partial z}\right)\tilde{r}\mathrm{d}r\mathrm{d}z. \tag{4.165}$$

Since $\delta\tilde{M} = \delta\tilde{\theta} = 0$, this reduces to

$$\delta^2 E = \int_\Gamma ((\delta\tilde{s},\chi)\cdot\mathbf{Q}\cdot(\delta\tilde{s},\chi))\,\tilde{r}\mathrm{d}r\mathrm{d}z,$$

where $\mathbf{Q}$ is calculated from (4.161) with $\tilde{M} = M_g, \tilde{\theta} = \theta_g$. Thus the condition for a minimiser is that $\mathbf{Q}$ is positive definite. □

Using (4.160), the condition that $\mathbf{Q}$ is positive definite is equivalent to convexity of P as a function of (s,z). Using (4.154) and the convexity property shows that necessary conditions for a minimum energy state are that M^2 is monotonically increasing in r and θ is monotonically increasing in z. These are the conditions given by Fjortoft (1946) and Eliassen and Kleinschmidt (1957). In addition, Shutts *et al.* (1988) show that the convexity of P is equivalent to the statement

$$\nabla P_t \qquad \text{is invertible.} \tag{4.166}$$

4.4.3 *Solution by change of variables*

Now show how the energy minimisation principle described above can be formalised as an optimal transport problem, as in section 3.2 for the semi-geostrophic case. Axisymmetry is assumed, so that only the radial cross-section is analysed. The domain occupied by the vortex, Γ_ς, defined in (4.148) written in (s,z) coordinates is then

$$D_\rho = [0,\rho(t,z)]\times[0,H]. \tag{4.167}$$

The boundary condition satisfied by φ in (4.149) becomes

$$P(\rho_t(z),z) = \frac{\Omega^2 r_0^2}{2(1-2r_0^2\rho(z))} \quad \text{on} \quad \{\rho>0\}, \tag{4.168}$$

where

$$2\rho = r_0^{-2} - \varsigma^{-2}. \tag{4.169}$$

Assuming that the angular momentum $M > 0$, dual variables can be defined as introduced by Schubert and Hack (1983) and Shutts *et al.* (1988):

$$\begin{aligned}\Omega R^2 &= M = ur + \Omega r^2, \\ \Upsilon &= \Omega^2 R^4, \\ Z &= g\frac{\theta}{\theta_0}.\end{aligned} \tag{4.170}$$

Then (4.154) gives

$$\nabla P = (\Upsilon, Z). \tag{4.171}$$

The monotonicity condition (4.166) means that Υ is monotonically increasing in r and that Z is monotonically increasing in z.

Writing the evolution equations (4.147) in these variables and (s, z) coordinates gives

$$\begin{aligned}\frac{1}{2\sqrt{\Upsilon}}\frac{\mathrm{D}\sqrt{\Upsilon}}{\mathrm{D}t} &= F(t, r, z), \\ \frac{\mathrm{D}Z}{\mathrm{D}t} &= gS(t, r, z)/\theta_0, \\ \nabla P &= (\Upsilon, Z), \\ \frac{1}{e(s)}\frac{\partial}{\partial s}(e(s)\varpi) + \frac{\partial w}{\partial z} &= 0,\end{aligned} \tag{4.172}$$

where ϖ is the radial velocity in s coordinates, and the volume measure $r\mathrm{d}r\mathrm{d}z$ in (r, z) coordinates becomes $e(s)\mathrm{d}s\mathrm{d}z$ in (s, z) coordinates where

$$e(s) = r_0^4/(1 - 2sr_0^2)^2 \quad \text{for} \quad 0 \le 2r_0^2 s < 1. \tag{4.173}$$

The boundary conditions become

$$\begin{cases} w(t, s, 0) = w(t, s, H) = 0, \\ \varpi(t, 0, z) = 0, \\ \varphi(t, \rho(t, z), z) = 0. \end{cases} \tag{4.174}$$

where

$$\frac{\partial \rho}{\partial t} + w\frac{\partial \rho}{\partial z} = \varpi. \tag{4.175}$$

Next introduce angular momentum and isentropic coordinates. The monotonicity condition (4.166) means that these will be well-defined. Define

$$\Psi(\Upsilon, Z) = s\Upsilon + zZ - P. \tag{4.176}$$

This means that $\Psi(\Upsilon, Z)$ is the Legendre transform of $P(s, z)$ for each t as in (3.37) in the semi-geostrophic case. Then it can be shown that Ψ is also convex as a function of (Υ, Z) and $\nabla\Psi = (s, z)$. Thus the map from physical coordinates (s, z) to angular momentum and potential temperature coordinates (Υ, Z) is generated by ∇P and its inverse is $\nabla\Psi$.

Now recognise that these two coordinates are defined in different units, which is important when proving that the transformation can be constructed and in generating examples. To illustrate this, apply this transformation to a state of rest in hydrostatic balance, as considered in Shutts *et al.* (1988). Eq. (4.170) shows that this corresponds to choosing $\Upsilon = \Omega^2 r^4$, while Z can be any positive function of z. The definition of s in (4.152) shows that the domain $r_0 \le r < \infty$ transforms to the finite domain $0 \le s \le \frac{1}{2r_0^2}$, but the associated steady state value of Υ will be in the domain $\Omega^2 r_0^4 \le \Upsilon < \infty$. Since the aim is to show that a unique vortex with prescribed angular momentum and potential temperature can be constructed, it is necessary to restrict the values of Υ to a finite range. Noting the requirement $u \ge 0$ at $r = r_0$ discussed in section 4.4.1, write this range as $\Upsilon \in [\Omega^2 r_0^4, \Upsilon_m]$ for some Υ_m. Then map this into a finite physical region as in the definition (4.167) where the boundary ρ has to be determined as part of the solution. The definition of $e(s)$ in (4.173) and the definition of s in (4.152) show that $e(s) = r^4$. Thus a state of rest corresponds to $\Upsilon_m = \Omega^2 e(\rho)$.

4.4.4 *Generation of a mass conservation equation in the new variables*

The next step is to show that equations (4.172) in dual variables can imply a conservation of mass in dual coordinates. This is exactly parallel to the method used in section 3.2.2 for the semi-geostrophic equations.

The mass in physical space is given by the integral of $r\chi_{\Gamma_\varsigma(r,z)}\mathrm{d}r\mathrm{d}z$, where χ_{Γ_ς} is the characteristic function of Γ_ς as defined in (4.148). In dual variables, the mass is the integral of

$$e(s)\chi_{D_\rho}\mathrm{d}s\mathrm{d}z, \tag{4.177}$$

where χ_{D_ρ} is the characteristic function of D_ρ, the domain occupied by the fluid in (s, z) coordinates.

The mass in (Υ, Z) coordinates is given by the potential density σ, where

$$\sigma\mathrm{d}\Upsilon\mathrm{d}Z = e(s)\mathrm{d}s\mathrm{d}z, \tag{4.178}$$

Since $(\Upsilon, Z) = \nabla P(s,z)$, equation (4.178) gives

$$\sigma \frac{\partial(\Upsilon, Z)}{\partial(s,z)} = \sigma \det(\partial^2 P) = e(s), \tag{4.179}$$

where $\partial^2 P$ is the Hessian matrix of P.

Using the notation of Definition 3.3, (4.179) can be written as

$$\sigma = \nabla P \# e(s)\chi_{D_\rho} \tag{4.180}$$

As ∇P is invertible with inverse $\nabla\Psi$, σ is equivalently defined by

$$e(\partial_\Upsilon \Psi)\det(\partial^2\Psi) = \sigma, \qquad \nabla\Psi(\Delta) = D_h, \tag{4.181}$$

where $s = \partial_\Upsilon \Psi = \partial\Psi/\partial\Upsilon$ and Δ is the region of $\mathbb{R}^2_+$ where σ is non-zero. This is a form of Monge-Ampère equation.. (4.181) can be written as

$$\nabla\Psi \# \sigma = e(s)\chi_{D_\rho}. \tag{4.182}$$

Conservation of mass then requires that

$$\frac{\partial \sigma}{\partial t} + \nabla\cdot(\sigma V) = 0 \qquad (0,T)\times\mathbb{R}^2 \tag{4.183}$$

$$\sigma|_{t=0} = \sigma_0,$$

where V is the velocity in (Υ, Z) coordinates, so that

$$V = \left(\frac{\mathrm{D}\Upsilon}{\mathrm{Dt}}, \frac{\mathrm{DZ}}{\mathrm{Dt}}\right). \tag{4.184}$$

These equations take the same form as (3.43) and (3.35) in the semi-geostrophic case. Equations (4.147) give, using the definitions (4.170),

$$\frac{1}{2\sqrt{\Upsilon}}\frac{\mathrm{D}\sqrt{\Upsilon}}{\mathrm{Dt}} = F(t,r,z), \tag{4.185}$$

$$\frac{\mathrm{DZ}}{\mathrm{Dt}} = \frac{gS(t,r,z)}{\theta_0}.$$

Writing F and S as functions of (Υ, Z), using the relation $\nabla\Psi = (s,z)$ and the definition of s from (4.152) gives

$$V = \left(2\sqrt{\Upsilon}F\left(t, \frac{r_0}{\sqrt{1-2r_0^2\frac{\partial\Psi}{\partial\Upsilon}}}, \frac{\partial\Psi}{\partial Z}\right),\right. \tag{4.186}$$

$$\left.\frac{g}{\theta_0}S\left(t, \frac{r_0}{\sqrt{1-2r_0^2\frac{\partial\Psi}{\partial\Upsilon}}}, \frac{\partial\Psi}{\partial Z}\right)\right).$$

Solving (4.185) requires calculating Ψ given σ. If this can be done, then (4.182) determines ρ and (4.176) is used to calculate P. Equation (4.154) is then used to calculate u and θ, giving a complete solution for the vortex. The radial and vertical velocities (ϖ, w) can be determined by transforming V as defined in (4.184) into (s, z) coordinates using the map $\nabla\Psi$, allowing for the time dependence of Ψ.

The result will correspond to a solution of (4.172) if Ψ is convex in space, since then the change of coordinates is valid. The first two equations of (4.172) correspond to the definition of V, (4.186), the third equation is just equation (4.171), and the fourth equation, together with (4.175), corresponds to equation (4.183). The boundary equation (4.175) corresponds to (4.182) which determines ρ.

4.4.5 *Solution using optimal transport*

In this section it is shown rigorously that the energy can be minimised under rearrangements of the fluid conserving angular momentum, represented by Υ, and potential temperature Z. This will confirm the classical calculus of variations argument in section 4.4.2 and extend it to the free boundary case. The method is to specify the mass of the fluid in angular momentum and isentropic coordinates and then minimise the energy with respect from maps from (Υ, Z) to physical coordinates (s, z) which conserve mass. In order to confirm the conditions found by Fjortoft (1946) and Eliassen and Kleinschmidt (1957), it is necessary to show that this minimising state is characterised by a map $\nabla\Psi$ with Ψ convex. This will then generate a solution of (4.172) for the reasons above.

The problem above can be formulated as an optimal transport problem. In section 3.2.3, the solution of the semi-geostrophic equations was obtained by solving an optimal transport problem where the cost function is the energy integral. The cost is minimised with respect to mass conservation. In the rigorous analysis described in section 3.5, the mass is treated as a probability measure. The same approach can be applied to the axisymmetric problem. The extra difficulty is the presence of the free boundary defined by (4.150). This is similar to the issues caused by the free upper boundary in sections 3.3.5 and 3.3.6.

Suppose that the mass in (Υ, Z) is σ, which is a probability measure with support $\Delta \subset \mathbb{R}^2_+$. Assume additionally that Δ is compact (i.e. closed and bounded). Then finding the map from Δ to D_ρ which conserves mass while minimising the energy corresponds to finding an invertible map $\nabla\Psi$, together with a function $\rho(z)$ which defines the region D_ρ in physical space occupied by the vortex. A proof that the energy minimiser is unique will show that a unique stable axisymmetric vortex with a free boundary with suitably prescribed angular momentum and potential temperature on fluid parcels can be embedded in an ambient isentropic fluid at rest.

The total energy to be minimised is given by (4.156). In the s coordinate this becomes

$$E = \int_{D_\rho} \left(\frac{1}{2}u^2 - \frac{g\theta z}{\theta_0} \right) e(s)\mathrm{d}s\mathrm{d}z. \tag{4.187}$$

The energy density $\frac{1}{2}u^2 - \frac{g\theta z}{\theta_0}$ can be rewritten using (4.170) as

$$\frac{r_0^2\Omega^2}{2(1-2r_0^2 s)} - s\Upsilon - zZ + \frac{\Upsilon}{2r_0^2} - \Omega\sqrt{\Upsilon}. \tag{4.188}$$

Now minimise (4.187) over maps from Δ to D_ρ for a given σ. This gives an optimal transport problem. In order to solve it, it is necessary to solve a Monge-Kantorovich problem as in section 3.2.3. This seeks a joint probability measure γ with marginals σ and $e(s)\chi_{D_\rho}$, where χ_{D_ρ} is the characteristic function of D_ρ, such that, for any functions $\phi(s,z), \psi(\Upsilon, Z)$,

$$\int_{\Delta\times D_\rho} (\psi(\Upsilon, Z) + \phi(s,z))\mathrm{d}\gamma = \tag{4.189}$$

$$\int_{\Delta} \psi\sigma \mathrm{d}\Upsilon\mathrm{d}Z + \int_{D_\rho} \phi e(s)\mathrm{d}s\mathrm{d}z.$$

Write such a joint probability measure as

$$\gamma \in G(\sigma_t, e(s)\chi_{D_\rho}), \tag{4.190}$$

where G is the set of joint probability measures on $\mathbb{R}_+^2 \times \mathbb{R}_+^2$. In the special case where this corresponds to a map from Δ to D_ρ, (4.190) reduces to

$$\gamma = \sigma\delta[(s,z) = \nabla\Psi], \tag{4.191}$$

where δ denotes the Dirac delta function.

Now minimise (4.187) over choices of γ as in (4.190) and choices of ρ which define the vortex domain D_ρ. Write (4.187) using (4.188) in the form

$$I(\rho,\gamma) := \int_{D_\rho\times\Delta} \left(-s\Upsilon - zZ + \frac{\Upsilon}{2r_0^2} - \Omega\sqrt{\Upsilon} + \frac{r_0^2\Omega^2}{2(1-2r_0^2 s)} \right) \mathrm{d}\gamma. \tag{4.192}$$

First assume a given choice $\tilde{\rho}$ of ρ, and minimise (4.192) over choices of γ. The result can be written as

$$\bar{I}[\sigma](\tilde{\rho}) := \inf_{[\gamma\in G(\sigma_t, e(s)\chi_{D_{\tilde{\rho}}})]} \left\{ \int_{\Delta\times D_{\tilde{\rho}}} (-s\Upsilon - zZ)\,\mathrm{d}\gamma \right\} + \tag{4.193}$$
$$\int_{D_{\tilde{\rho}}} \left(\frac{\Omega^2 r_0^2}{2(1-2sr_0^2)} \right) e(s)\mathrm{d}s\mathrm{d}z + \int_{\mathbb{R}^2} \left(\frac{\Upsilon}{2r_0^2} - \Omega\sqrt{\Upsilon} \right) \sigma \mathrm{d}\Upsilon \mathrm{d}Z.$$

The second term in (4.193) depends only on the choice of $D_{\tilde{\rho}}$ and the third term depends only on σ.

In order to prove that the first term in (4.193) can be uniquely minimised, write it as a dual problem, as in section 3.2.3. It can be proved that

$$\int_{\mathbb{R}^2} -\tilde{\Psi}\sigma\mathrm{d}\Upsilon\mathrm{d}Z - \int_0^H \int_0^{\tilde{\rho}(z)} \tilde{P}(s,z)e(s)\mathrm{d}s\mathrm{d}z, \tag{4.194}$$

can be uniquely maximised over the set of continuous functions $\tilde{\Psi}, \tilde{P}$ such that

$$\tilde{P}(s,z) + \tilde{\Psi}(\Upsilon, Z) \geq s\Upsilon + zZ \text{ for all } ((s,z),(\Upsilon,Z)) \in D_{\tilde{\rho}} \times \mathbb{R}^2_+. \tag{4.195}$$

At the solution the inequality in (4.195) becomes an equality and the value is the value of $s\Upsilon + zZ$ to be used in the first term of (4.193), so that $\bar{I}[\sigma](\tilde{\rho})$ is defined for a given $\tilde{\rho}$.

The complete minimization problem can then be written as the calculation of

$$H_*(\sigma) = \inf_{\tilde{h}} \bar{I}[\sigma](\tilde{h}). \tag{4.196}$$

Using (4.193) and (4.194) gives

$$H_*(\sigma) = \sup \left(\int_{\mathbb{R}^2} \left(\frac{\Upsilon}{2r_0^2} - \Omega\sqrt{\Upsilon} - \tilde{\Psi} \right) \sigma\mathrm{d}\Upsilon\mathrm{d}Z + \right. \tag{4.197}$$
$$\left. \inf_{\tilde{\rho}\in\mathcal{H}_\prime} \int_0^H \int_0^{\tilde{\rho}(z)} \left(\frac{\Omega^2 r_0^2}{2(1-2sr_0^2)} - \tilde{P}(s,z) \right) e(s)\mathrm{d}s\mathrm{d}z \right).$$

Here, $\mathcal{H}_\prime$ consists of all measurable functions $\tilde{\rho} : [0, H] \longmapsto [0, 1/(2r_0^2)]$. The supremum in (4.197) is taken over the same set of continuous functions $\tilde{\Psi}, \tilde{P}$ as is defined in (4.195).

It is proved in Cullen and Sedjro (2014) that there is a unique minimiser in (4.196). The proof is carried out by solving the maximisation problem (4.197). Write ρ for the minimiser in (4.196) for a given σ and Ψ, P for the functions that maximise (4.197). It is proved that the maximiser corresponds to equality in (4.195), so that Ψ, P are Legendre transforms of each other as required by (4.176) in section 4.4.3. It is proved that Ψ satisfies (4.181) together with (4.168). It is also proved that Ψ, P are convex in space and so $\nabla_{s,z}P$ is invertible almost everywhere. Provided that σ is bounded, so that the non-degeneracy condition that a finite amount of mass cannot be associated with a single value of (Υ, Z) is satisfied, the solutions satisfy (4.180) and the inverse relation (4.182). This means that the mass of fluid specified in (Υ, Z) coordinates completely fills the physical domain D_ρ. However, ρ may be zero for some values of $z \in [0, H]$. It is proved that ρ is monotonically increasing in z if the region Δ in which the mass is specified is contained in $Z > 0$. This corresponds to the original assumption $\theta > 0$ in (4.147). In the original coordinates (r, z) this means that the boundary $r = \varsigma(z)$ is monotonically increasing in z.

In Craig (1991), eq. (64), the structure of the vortex was calculated by solving an elliptic equation in physical space. He used the boundary condition that φ tends to a horizontally uniform reference profile as $r \to \infty$. He then proposed a solution of the 'almost' axisymmetric equations obtained by using initial data which depends on λ in (4.147) and allowing all the variables to depend on λ. The method required finding a structure in (r, z) separately for each λ. This amounts to letting σ depend on λ. In order to solve the non-axisymmetric equations, it would be necessary to show that the Ψ obtained by solving the Monge-Ampère equation (4.181) separately for each λ had at least as much regularity in λ as σ. However, this is unlikely, noting the results quoted in section 3.5, and so the resulting evolution equations are likely to be ill-posed

4.4.6 *Properties of the vortex*

The analysis above shows that, given a mass of fluid with specified bounded values of angular momentum and potential temperature, the fluid can be uniquely arranged to give an axisymmetric vortex in an unbounded ambient fluid at rest. However, the vortex may not fill the depth of the domain.

It is simplest to study the implications of the results using the original physical coordinate r. Then the radial coordinate of the free boundary becomes ς. First consider the implications of the result at the boundary $r = \varsigma$ of the vortex Γ_ς. The discussion at the end of the previous section shows that ρ, and thus ς, is monotonically increasing in z. Outside the vortex, $\varphi = 0$. Then (4.153) implies, using (4.170), that $\nabla P = (\Omega^2 r^4, 0)$. In general, (4.154) implies that $\nabla P = (\Upsilon, Z) = (M^2, g\theta/\theta_0)$. The assumption that the support of σ, $\Delta \subset \mathbb{R}^2_+$ is compact means that Υ is less than some

Υ_m for all points in Δ. Thus the vortex in physical space is bounded by $r = r_m$ where $\Omega^2 r_m^4 = \Upsilon_m$. Since θ has been chosen to be greater than zero, the invertibility of ∇P at $r = \varsigma$ implies that ς is monotonically nondecreasing in z. It also implies that $\varphi > 0$ within Γ_ς. Thus $\partial\varphi/\partial r$ will be negative as r approaches ς, which implies that $u < 0$. Thus the flow must become anticyclonic at the vortex boundary. This is an artificial restriction resulting from embedding the vortex in an ambient fluid at rest. More realistically, u would be less cyclonic at the boundary than the ambient flow.

First illustrate the nature of a stable vortex with a free boundary. At $z = 0$, construct a vortex with radius $\varsigma(0) = r_1$. Using (4.170), choose $\Upsilon = \Upsilon_1$ throughout the vortex, where $\Upsilon_1 = \Omega^2 r_1^4$. Then

$$u = \frac{1}{r}\Omega(r_1^2 - r^2). \tag{4.198}$$

Eq. (4.147) gives

$$\frac{\partial\varphi}{\partial r} = \frac{\Omega^2}{r^3}(r_1^4 - r^4). \tag{4.199}$$

Integrating this, and using the boundary condition (4.149), gives

$$\varphi(r,0) = \Omega^2\left(r_1^2 - \frac{1}{2}\left(\frac{r_1^4}{r^2} + r^2\right)\right). \tag{4.200}$$

Now extend this to a three-dimensional vortex by setting $Z = g\theta/\theta_0 = g\frac{\hat{\theta}z}{\theta_0 H}$, where $\hat{\theta}$ is a constant, so that the static stability $\partial\theta/\partial z$ is constant. Then

$$\varphi(r,z) = \varphi(r,0) + \frac{1}{2}\frac{g\hat{\theta}}{\theta_0 H}z^2 : \text{ for } r \leq r_1. \tag{4.201}$$

Using (4.200), (4.201) and the boundary condition $\varphi = 0$ at $r = \varsigma(z)$ gives

$$\Omega^2\left(r_1^2 - \frac{1}{2}\left(\frac{r_1^4}{\varsigma(z)^2} + \varsigma(z)^2\right)\right) + \frac{g\hat{\theta}}{2\theta_0 H}z^2 = 0, \tag{4.202}$$

which determines $\varsigma(z)$.

This solution is illustrated in Fig. 4.1 which shows plots of u against r and ς against z. r_1 is chosen to be 100km which is typical of the radius of the damaging axisymmetric inner core of a tropical cyclone, Smith and Montgomery (2016). The solution at $z = 0$ shows azimuthal winds of about 50ms^{-1}, which is realistic. (4.198) shows that u will become increasingly negative in the outer part of the vortex $r > r_1$ as z increases. This is an artefact of the simple data used to construct the illustration.

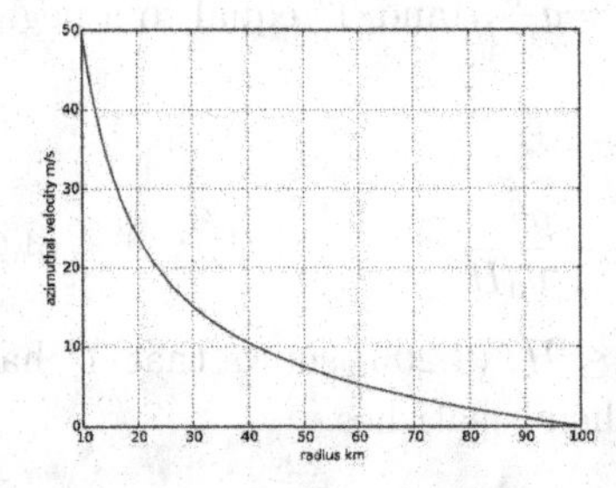

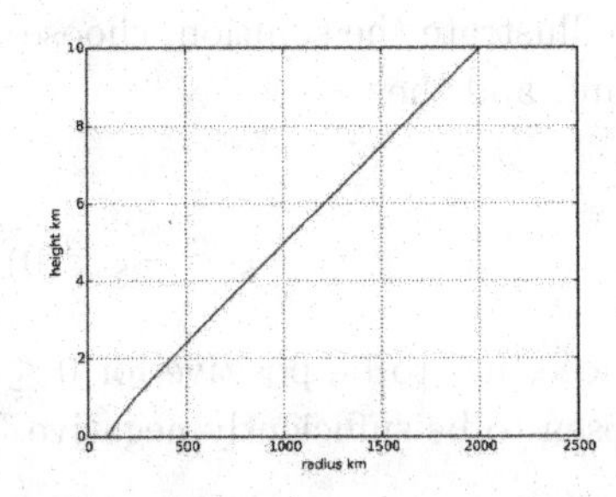

Fig. 4.1 Left: Azimuthal velocity profile (ms^{-1}) at $z = 0$ given by eq. (4.198) with $\Omega = 0.5 \times 10^{-4}\mathrm{s}^{-1}$, $r_0 = 10$km, $r_1 = 100$km. Right: Vertical profile of ς (km) given by eq. (4.202) with additionally g=10 ms^{-2}, $\hat{\theta} = 30°$K, $\theta_0 = 300°$K, H=10 km. From Cullen *et al.* (2015a). ©Crown Copyright, Met Office.

Next illustrate the solution procedure based on the construction of a vortex where the mass σ is given as a function of Υ and Z. It is then necessary to find $\varsigma(z)$. The total mass of the fluid has to be the same in both (Υ, Z) and physical (r, z) coordinates. Since the vortex fills D_ς, this means that

$$\int_\Delta \sigma \mathrm{d}\Upsilon \mathrm{d}Z = \int_{D_\varsigma} r\mathrm{d}r\mathrm{d}z. \tag{4.203}$$

However, ς may be zero for some values of z.

Next consider the nature of ς. The condition $\varphi = 0$ applied at $r = \varsigma(z)$ implies

$$\frac{\partial \varphi}{\partial r}\frac{\partial \varsigma}{\partial z} + \frac{\partial \varphi}{\partial z} = 0. \tag{4.204}$$

Using (4.147) then gives

$$\left(\frac{u^2}{\varsigma} + 2\Omega u\right)\frac{\partial \varsigma}{\partial z} + g\frac{\theta}{\theta_0} = 0.$$

This can be rewritten as

$$\left(\frac{1}{\varsigma^3}\Upsilon - \Omega^2\varsigma\right)\frac{\partial \varsigma}{\partial z} + Z = 0.$$

At a state of rest, $\Upsilon = \Omega^2 r^4$ so at $r = \varsigma$ write $\Upsilon' = \Upsilon - \Omega^2\varsigma^4$. Then

$$\frac{\partial}{\partial z}\varsigma^{-2} = 2\frac{Z}{\Upsilon'}. \tag{4.205}$$

If ς increases with z then Υ' has to be negative, as expected from the condition $u < 0$ at $r = \varsigma$ as found above.

To illustrate the solution, choose $Z = g\frac{\hat{\theta}z}{\theta_0 H}$ and Υ' equal to a negative constant, and then

$$\varsigma^{-2} = \varsigma^{-2}(0) + \frac{g\hat{\theta}}{\Upsilon'\theta_0 H}z^2. \tag{4.206}$$

Since ς has to be positive for $0 \le z \le H$, (4.206) shows that Υ' has to be chosen to be sufficiently negative. Then σ satisfies

$$\int_\Delta \sigma \mathrm{d}\Upsilon \mathrm{d}Z = \frac{1}{2}\int_0^H \left(\frac{\Upsilon'\theta_0 H}{\Upsilon'\theta_0 H\varsigma^{-2}(0) + gz^2}\right)\mathrm{d}z - \frac{1}{2}r_0^2 H. \tag{4.207}$$

Noting that $\varsigma(0) \ge r_0$ and $\Upsilon' < 0$, $\varsigma(0)$ can be found to ensure positivity of (4.207) if $-gH > \Upsilon'\theta_0 r_0^{-2}$. The right hand side of (4.207) can then be made arbitrarily large by letting $\varsigma(0)$ approach $\sqrt{-gH/(\Upsilon'\theta_0)}$. Thus $\varsigma(0)$ can be chosen to fit any desired value of the total mass on the left hand side provided that

$$\int_\Delta \sigma \mathrm{d}\Upsilon \mathrm{d}Z \ge \frac{1}{2}\int_0^H \left(\frac{\Upsilon'\theta_0 H}{\Upsilon'\theta_0 H r_0^{-2} + gz^2}\right)\mathrm{d}z. \tag{4.208}$$

If the total mass is smaller than the expression on the right hand side of (4.208), then set

$$\begin{aligned} \varsigma &= r_0 : \quad z < z_1, \\ \varsigma &= \sqrt{\frac{\Upsilon'\theta_0 H}{\Upsilon'\theta_0 H r_0^{-2} + g(z^2 - z_1^2)}} : \quad z \ge z_1, \end{aligned} \tag{4.209}$$

which is consistent with (4.205).

The effect of this is illustrated in Fig. 4.2. Choosing $\hat{\theta} = 30°$K, which implies a potential temperature excess of 30°K at the top of the troposphere, and $\Upsilon' = -2.5 \times 10^{12}\mathrm{m}^4\mathrm{s}^{-2}$, which implies a 5% angular momentum deficit compared to the rest state value at a radius of about 200km, means that the rate of increase of vortex diameter with height is very sensitive to the radius. The situation where the vortex does not reach the ground is illustrated. For a more realistic choice of 100km for the bottom level radius a much larger value of the angular momentum deficit would be required to allow (4.208) to be solved. If a value corresponding to a 5% deficit at a 500km radius is used, the radius at the top of the troposphere becomes 330km. The vertical profile of ς shown in Fig. 4.2 is quite different from that in the example of

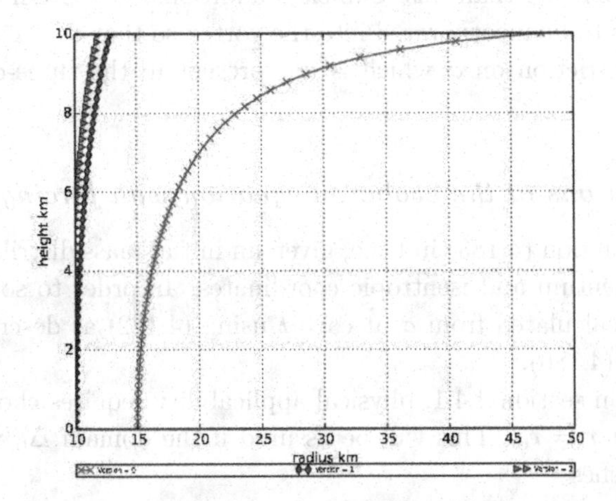

Fig. 4.2 Vertical profiles of vortex radius $\varsigma(z)$ (km) given by eq. (4.208) for various choices of total mass, setting $\Omega = 0.5 \times 10^{-4}\mathrm{s}^{-1}$, $\hat{\theta} = 30°\mathrm{K}$ and $\Upsilon' = -2.5 \times 10^{12}\mathrm{m}^4\mathrm{s}^{-2}$. From Cullen *et al.* (2015a).

Fig. 4.1 because the assumption of uniform Υ is replaced by the assumption that the angular momentum deficit Υ' is uniform with height.

These examples are limited by the need to choose a uniform angular momentum or angular momentum deficit in order to allow analytic solution. This is not very realistic. In order to simulate a realistic hurricane-like vortex which extends through the vertical domain, it is necessary to choose a sufficiently large total σ, as illustrated above, and to choose σ small for small Υ so that large values of the angular momentum Υ are mapped onto small values of r.

In order to understand the implications further, consider the two-dimensional case where there is no variation in z. The variational problem solved by Cullen and Sedjro (2014) becomes highly degenerate in this case. The angular momentum is given by $\sqrt{\Upsilon}$. A stable vortex is given by choosing s to be a monotonically increasing function of Υ such that $e(s)\partial s/\partial\Upsilon = \sigma$. Suppose the maximum angular momentum of the fluid specified to be in the vortex is $\sqrt{\Upsilon_m}$. The boundary of the vortex is given by $r = \varsigma$, and so the area of the vortex will be given by $\frac{1}{2}\varsigma^2$. This must be equal to the integral of σ over $\Delta_t = [0, \Upsilon_m]$. The angular momentum of the ambient fluid is Ωr^2 for $r \geq \varsigma$. Applying the condition that the angular momentum increases with r at the vortex boundary $r = \varsigma$ implies that $\Upsilon_m \leq \Omega^2\varsigma^4$, so that $\varsigma \geq \Omega^{-\frac{1}{2}}\Upsilon_m^{\frac{1}{4}}$. If the specified total mass is too small, then ς will be smaller than this. Sufficient additional mass from the ambient fluid has then to be incorporated into the vortex so that $\varsigma = \Omega^{-\frac{1}{2}}\Upsilon_m^{\frac{1}{4}}$. This imposes a restriction on σ which is not present in the three-dimensional problem.

4.4.7 *Solutions of the evolution equation with forcing*

Now solve equation (4.183) in time, given an initial mass distribution σ_0 in angular momentum and isentropic coordinates. In order to solve (4.183), V has to be calculated from σ at each t using (4.182) as described in the derivation of (4.186).

As noted in section 4.4.1, physical applicability requires choosing σ_0 so that $u \geq 0$ at $r = r_0$. This will be assured if the domain Δ_0 where σ_0 is non-zero satisfies

$$\Delta_0 = [(\Upsilon, Z) : \Omega^2 r_0^4 \leq \Upsilon \leq \Upsilon_m < \infty, 0 \leq Z \leq Z_m < \infty], \qquad (4.210)$$

for some Υ_m, Z_m.

In Cullen and Sedjro (2014), equation (4.183) is solved for particular classes of angular momentum forcing F and thermal forcing S. The main difficulty is that the forcing terms are specified as functions of physical space but, as shown in (4.186), have to be applied as functions of angular momentum and potential temperature, Υ and Z. Since the mapping from (Υ, Z) to (r, z) is unknown until the problem has been solved, the results are highly non-trivial. However, some types of physical forcing, for instance those resulting from air-sea interaction, are naturally specified in physical coordinates. The only results which can be proved come from the requirements that Υ is an inherently positive quantity and that Z has been assumed to be positive to distinguish it from the ambient fluid. It is likely that the restriction on Z could be relaxed, though the proof of the results in section 4.4.5 would have to be modified, and the nature of $\varsigma(z)$ would change.

Some types of forcing, such as latent heat release, could be more naturally imposed as a function of Υ and Z. In that case it is quite easy to solve (4.183) because V as defined in (4.184) will be known explicitly from (4.186).

Equation (4.183) is solved under two different sets of assumptions. In the first case, $\sigma(0, \cdot)$ is assumed to be bounded, so that the results of section 4.4.5 mean that (4.182) holds, as well as (4.180). Essentially this means that the transformation between physical coordinates (s, z) and angular momentum and potential temperature coordinates (Υ, Z) is invertible so that $\sigma(0,\cdot)$ must not contain Dirac masses, which would mean that there are values of (Υ, Z) where a single value of angular momentum and potential temperature is mapped to a finite region of physical space. Cullen and Sedjro (2014) then prove in their Theorem 6.2 that

Theorem 4.13. *Given $\sigma(0,\cdot)$ which is absolutely continuous with respect to Lebesgue measure and has bounded support, and forcing terms F, S in (4.172), then if*

(i) $0 \leq F, \frac{g}{\theta_0} S \leq M$ *for some positive constant* M,
(ii) $\frac{\partial F}{\partial z} = \frac{\partial S}{\partial r} = 0$,
(iii) $\frac{\partial F}{\partial r}, \quad \frac{\partial S}{\partial z} > 0$,

then σ remains absolutely continuous with respect to Lebesgue measure with bounded support as it evolves in time according to (4.183) with the forcing terms F and S.

These restrictions are very artificial. For instance, they require that diabatic heating increases with height which is the reverse of the usual case. If these restrictions are not observed, it is possible that Dirac masses will be created in σ, corresponding to well-mixed layers of uniform potential temperature and absolute angular momentum. This is definitely possible in the real system and is illustrated in the computations of Shutts *et al.* (1988). If this happens, then Cullen and Sedjro prove in their Theorem 6.7 that

Theorem 4.14. *Given any probability measure $\sigma(0,\cdot)$ with bounded support, (4.183) can still be solved, and the solutions will respect (4.180), if F and S are continuous, bounded and non-negative. The solution will take the form of a potential $P(t,\cdot)$ which is convex in space and a joint probability measure $\gamma(t,(\Upsilon, Z),(s,z))$. The restrictions that $F, S \geq 0$ are required because Υ is an inherently positive quantity and Z is assumed to be positive as discussed above.*

Shutts *et al.* (1988) solved eq. (4.147) using the geometric algorithm described in section 5.3.2. Fig. (4.3) shows the initial element configuration, which is regular in (r, z) coordinates. The annular domain extends from an inner radius of 50km to an outer radius of 1000km. The depth was 10km and the Brunt-Väisälä frequency was $1.3\times10^{-2}\mathrm{s}^{-1}$. The rotation rate Ω was $3\times10^{-5}\mathrm{s}^{-1}$. The data defines a state of rest in hydrostatic balance in a fixed domain $D_{\varsigma_0} = [r_0, \varsigma_0] \times [0, H]$. This corresponds to choosing $\sigma_0 = \frac{1}{4\Omega\sqrt{\Upsilon}}$

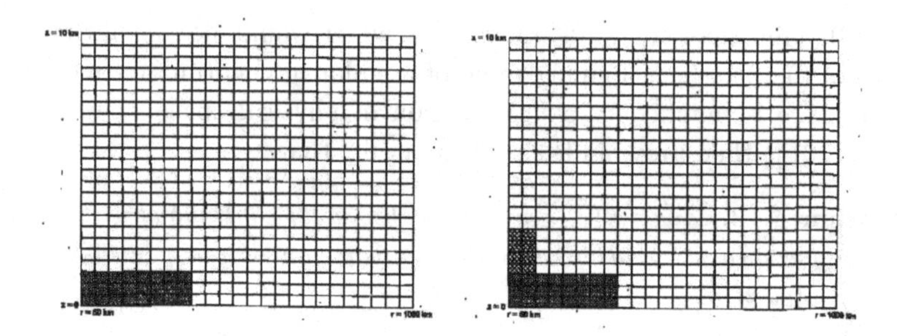

Fig. 4.3 Initial element geometries for the convective mass transfer experiments. The shaded elements are moist. In the right hand panel, the stippled region is to be cooled. From Shutts *et al.* (1988). ©Crown Copyright, Met Office.

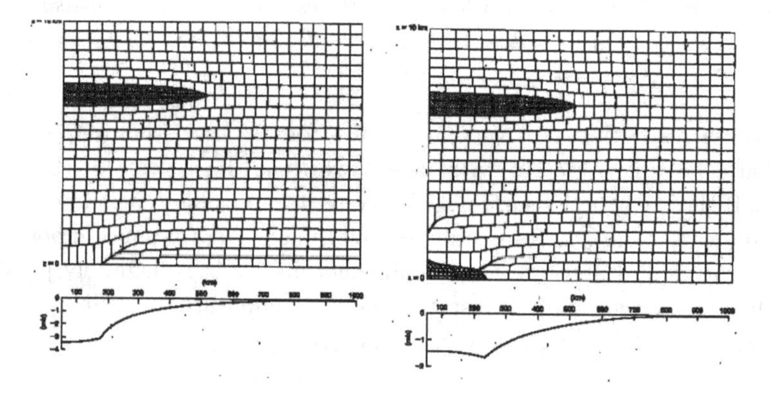

Fig. 4.4 The final states after the corresponding shaded elements in Fig. 4.3 have convected. The pressure perturbation is shown below. Bold lines correspond to frontal surfaces. From Shutts *et al.* (1988). ©Crown Copyright, Met Office.

over a domain $\Delta_0 = [\Omega^2 r_0^4, \Upsilon_m] \times [0, H]$. This was represented discretely by assigning masses σ_i to points $(\Upsilon_i(0), Z_i(0))$ in Δ_0. The solution was then advanced to time t by adding a single time increment S to Z_i at points i with small values of r and z, so that $Z_i(t) = Z_i(0) + S$. This gives a discrete solution of (4.183) with $F = 0$ and with $S > 0$ for small values of r and z. This choice is consistent with the assumptions in Theorem 4.14. Mapping the solution at t back to physical space produced the solution shown in Fig. 4.4. The heating S was such that the elements convected up to about 7km, forming a 'lens'. An eye-wall discontinuity is formed at a radius of 175km. Within the eye, the surface pressure perturbation is about -3.5hpa, and the temperature perturbation about +6°K. The warm core arises from the subsiding of warmer air from above to replace the air that has been heated and convected away. In the second experiment, shown in the right hand panels of Fig. 4.3 and Fig. 4.4, some elements indicated by stippling are cooled so that $S < 0$. The values of S were chosen to ensure that Z_i remained positive which is required for the results of Cullen and Sedjro (2014) to be valid. This represents the effect of the evaporation of rainfall, which will be large in the region of the eye-wall. This creates a surface cold dome, which displaces the eye-wall outwards. These effects are qualitatively like those that occur in real tropical cyclones, though the latter are much more intense. The experiments are discussed in more detail in Shutts *et al.* (1988) where a number of experiments were carried out with choices of region with different aspect ratios where S was non-zero.

In Smith *et al.* (2018) the structure of an axisymmetric vortex was also found by solving an elliptic equation, though the derivation is rather different. A simple boundary layer was included. The outer boundary of the vortex satisfied a Neumann condition for the transverse stream-function, implying zero vertical velocity. When integrating with physically realistic idealised forcing, it was necessary to regularise the elliptic equation, and even then the integration could not be continued for large times. This forcing was not consistent with the conditions of Theorem 4.13 so that the optimal transport solution only exists as a probability measure as in Theorem 4.14. Results for the semi-geostrophic system discussed in section 5.1.4 suggest that it should be possible to continue the solution indefinitely with an appropriate regularisation. However, the resulting solution is likely to be very diffusive.

Observed tropical cyclones typically exhibit non-axisymmetric structure in their outer parts, but it is also believed that intense extra-tropical storms and polar lows often develop a central core with a near-axisymmetric flow regime as a result of convection, Reed and Albright (1986), Rasmussen (1985). Theories of almost axisymmetric vortices have been developed, though no rigorous results are available. As noted above, the equations proposed by by Craig (1991) appear to be ill-posed. The almost axisymmetric model of Shapiro and Montgomery (1993) may be well-posed because the elliptic equation to be solved is three-dimensional, and the coefficients of the highest order terms come from azimuthally meaned values. Thus there will be much better regularity of the solutions in the λ direction.

4.4.8 *Stability of the vortex*

The requirement that absolute angular momentum increases with radius ensures that a vortex whose evolution is described by (4.183) is stable to axisymmetric perturbations. In order for the evolution to be physically relevant, it is also necessary for the solutions to be stable to non-axisymmetric disturbances. Therefore consider what distributions of σ as a function of Υ and Z correspond to vortices which are stable to non-axisymmetric disturbances. The study of Schubert *et al.* (1999) shows that the evolution of a barotropic vortex in a non-divergent rotating fluid from initial data consisting of an annulus of vorticity with a small azimuthal perturbation goes towards a vortex concentrated at the origin. However, conservation of energy and angular momentum require that some of the initial vorticity is left behind as filamentary structures outside the core vortex. This evolution was discussed in terms of either a minimum enstrophy hypothesis or a maximum entropy hypothesis.

Burton and Nycander (1999) analysed the nonlinear stability of a three-dimensional quasi-geostrophic vortex to perturbations which rearranged and/or mixed the potential vorticity. The inclusion of mixing is necessary to make the extremisation problem well posed. In the case where there is no background shear flow, the stable states are axisymmetric and represent maximisers of the energy. If an axisymmetric vortex is stable to axisymmetric perturbations, but unstable to non-axisymmetric perturbations, it is expected that non-axisymmetric perturbations will grow, equilibriate and the solution will evolve to a new axisymmetric state with the vorticity mixed, as discused by Smith and Montgomery (1995) and illustrated in the computations of Schubert *et al.* (1999). Thus, in particular, a stable state can be characterised as an extremum of the energy with respect to axisymmetric perturbations which rearrange and/or mix vorticity and conserve total angular momentum. Similar results were obtained by Cullen and Douglas (2003) for straight geostrophic flows as described in section 4.6.

In the barotropic case, which was analysed by Burton and Nycander (1999), a maximum energy state corresponding to a cyclonic vortex is obtained by rearranging the vorticity to be axisymmetric and monotonically decreasing, which allows the largest values of u to be obtained over the maximum region, given $u = 0$ at $r = 0$. It is like the end state of the simulations in Schubert *et al.* (1999), though this state could not be reached because the energy would have been larger than the initial energy. The initial data used by Schubert *et al.* (1999) and the vortex analysed by Smyth and McWilliams (1998) do not satisfy this monotonicity condition. This is consistent with the instability to non-axisymmetric disturbances found by Smyth and McWilliams (1998). The stability condition derived by Montgomery and Shapiro (1995) for three-dimensional vortices using linear theory reduces to the requirement that $\frac{\partial\sqrt{\Upsilon}}{r\partial r}$ is either monotonically increasing or decreasing in r, which is consistent with Burton and Nycander's result.

Schubert *et al.* (1999)'s results do not suggest why there is an 'eye' with small relative vorticity at the centre of the vortex, which would correspond to a vortex which was unstable to non-axisymmetric perturbations. An eye was found in the three-dimensional vortex of Shutts *et al.* (1988) illustrated in section 4.4.7. The analysis of Montgomery and Shapiro (1995) suggests that stability of such a vortex requires the potential vorticity to vary monotonically with r on each isentropic surface. It would be desirable to extend the nonlinear stability analysis discussed above to this case using potential vorticity instead of vorticity. However, there is now the additional difficulty that the analysis would have to be carried out with the full Euler equations, whose solutions cannot be completely described by the potential vorticity. In particular, the stability of an energy maximiser with respect to rearrangements of potential vorticity could be compromised by the radiation of energy in inertia-gravity waves. This possibility is discussed by Schecter and Montgomery (2006), who derive conditions under which such radiation is inhibited. This requires regimes where there is a time-scale separation between the evolution of perturbations to the vortex and the frequency of inertio-gravity and acoustic waves. Burton and Nycander (1999) state that their quasi-geostrophic analysis is only valid in regimes where the quasi-geostrophic equations remain accurate for an extended time, so that gravity wave radiation is small.

4.5 Zonal flows on the sphere

4.5.1 *Governing equations*

The same methods as used in section 4.4 to describe an axisymmetric vortex can be used to describe purely zonal flows on a sphere. This is illustrated in particular by Schubert *et al.* (1991). Start with the hydrostatic Boussinesq Euler equations (3.1) as in section 4.4, and use spherical polar coordinates (λ, ϕ, z) with associated velocity components (u, v, w). For later convenience, write $\mu = \sin\phi, U = (u/a)\cos\phi$ and $V = (v/a)\cos\phi$ where a is the radius of the Earth. All variables are then assumed to be functions of (μ, z) and independent of λ. Then (4.147) is replaced by

$$\begin{aligned}\frac{\mathrm{D}U}{\mathrm{D}t} - 2\Omega V\mu &= \frac{1}{a^2}F(t,\mu,z),\\ \frac{\mathrm{D}\theta}{\mathrm{D}t} &= S(t,\mu,z),\\ U^2\mu + 2\Omega U\mu(1-\mu^2) &= -(1-\mu^2)^2\frac{1}{a^2}\frac{\partial\varphi}{\partial\mu},\\ \frac{\partial V}{\partial\mu} + \frac{\partial w}{\partial z} &= 0,\\ \frac{\partial\varphi}{\partial z} - g\frac{\theta}{\theta_0} &= 0.\end{aligned} \tag{4.211}$$

Unlike section 4.4, no uniform state is subtracted from θ and φ. Equations (4.211) are the same as the tropical long wave equations (2.162) written in dimensional form in spherical polar coordinates and assuming that the variables are independent of λ. The terms F, S, which are also independent of λ, could either represent the averaged effects of $\lambda-$dependent disturbances to the zonal flow, or other external forcing.

Equations (4.211) define the flow on the whole sphere excluding small polar caps where $|\mu| > \mu_1$ with μ_1 close to 1. This avoids the coordinate singularity at the poles. While the exclusion of the region $r < r_0$ is not significant in modelling tropical cyclones, as the central 'eye' region is typically inactive, it undermines the credibility of the axisymmetric assumption when modelling flows in spherical geometry, as there is clearly nothing special about atmospheric behaviour near the poles. However, using a zonal mean to characterise atmospheric behaviour is much less useful near the poles. Since only the meridional cross-section is considered, the region of integration is

$$\Gamma = [-\mu_1, \mu_1] \times [0, H]. \tag{4.212}$$

The boundary conditions are

$$\begin{aligned} w(t,\mu,0) = w(t,\mu,H) &= 0,\\ v(t,-\mu_1,z) = v(t,\mu_1,z) &= 0.\end{aligned} \tag{4.213}$$

4.5.2 *Zonal flow as a minimum energy state*

First consider zonal flow on the Northern hemisphere Γ_N. The steady state is given by the third and fifth equations of (4.211). The absolute angular momentum associated with this state is

$$M = a^2(U + \Omega(1 - \mu^2)). \tag{4.214}$$

Then the third equation of (4.211) can be written as

$$\begin{aligned} \mu\frac{M^2}{a^4} = & -(1-\mu^2)^2\tfrac{1}{a^2}\tfrac{\partial\varphi}{\partial\mu} + \mu\Omega^2(1-\mu^2)^2 \\ = & -(1-\mu^2)^2\tfrac{1}{a^2}\tfrac{\partial}{\partial\mu}(\varphi + \tfrac{1}{2}a^2\Omega^2(1-\mu^2)). \end{aligned} \tag{4.215}$$

The first equation of (4.215) can also be written

$$\frac{\partial\varphi}{\partial\mu} = -\mu\left(\frac{M^2}{a^2(1-\mu^2)^2} - a^2\Omega^2\right). \tag{4.216}$$

The second equation of (4.215) suggests defining a new latitude coordinate s and a potential P by

$$s = -\frac{1}{2a^2}\frac{\mu^2}{1-\mu^2}, \tag{4.217}$$

so

$$\mathrm{d}s = -\frac{1}{a^2}\mu\mathrm{d}\mu/(1-\mu^2)^2, \tag{4.218}$$

and

$$P(s,z) = \varphi + \frac{1}{2}\Omega^2a^2(1-\mu^2). \tag{4.219}$$

Define $s_1 = -\frac{1}{2a^2}\frac{\mu_1^2}{1-\mu_1^2}$. Then the domain Γ_N maps into the region $[-s_1, 0]\times[0, H]$.

Using these definitions, the third and fifth equations of (4.211) give

$$\nabla P = (M^2, g\theta/\theta_0). \tag{4.220}$$

The energy integral associated with the flow is

$$\begin{aligned} E = & \int_{\Gamma_D}\left(\tfrac{1}{2}u^2 - g\theta z/\theta_0\right)a^2\cos\phi\mathrm{d}\phi\mathrm{d}z, \\ = & \int_{\Gamma_D}\left(\tfrac{1}{2}\left(\tfrac{M}{a\sqrt{(1-\mu^2)}} - \Omega a\sqrt{(1-\mu^2)}\right)^2 - g\theta z/\theta_0\right)a^2\mathrm{d}\mu\mathrm{d}z. \end{aligned} \tag{4.221}$$

Then, as in section 4.4.2, given a state $(\tilde{M}, \tilde{\theta})$, calculate the energy using the second equation of (4.221). Define variations $\Xi = (\eta, \chi)$ of particle positions $(\tilde{\mu}, \tilde{z})$ satisfying

$$\eta = \delta\tilde{\mu},\ \chi = \delta\tilde{z}. \tag{4.222}$$

Then the following result holds:

Theorem 4.15. *The conditions for the energy E to be minimised with respect to the displacements Ξ defined in (4.222), satisfying continuity $\delta(\mathrm{d}\mu \mathrm{d}z) = 0$ via*

$$\frac{\partial \eta}{\partial \mu} + \frac{\partial \chi}{\partial z} = 0 \tag{4.223}$$

in Γ_D, and conserving angular momentum and potential temperature via

$$\delta \tilde{M} = 0, \; \delta \tilde{\theta} = 0, \tag{4.224}$$

together with $\Xi \cdot \mathbf{n} = 0$ on the boundary of Γ_D, are that

$$\tilde{M}^2 = \frac{\partial \tilde{P}}{\partial s}, \tag{4.225}$$
$$g\frac{\tilde{\theta}}{\theta_0} = \frac{\partial \tilde{P}}{\partial z}.$$

for some $\tilde{P}$ with s defined by (4.217), and

$$\mathbf{Q} = \begin{pmatrix} \frac{\partial \tilde{M}^2}{\partial s} & \frac{\partial \tilde{M}^2}{\partial z} \\ \frac{g}{\theta_0}\frac{\partial \tilde{\theta}}{\partial s} & \frac{g}{\theta_0}\frac{\partial \tilde{\theta}}{\partial z} \end{pmatrix} \tag{4.226}$$

is positive definite.

Proof Applying the perturbations gives

$$\delta E = \int_{\Gamma_D} \Bigg(\left(\frac{\tilde{M}\eta\tilde{\mu}}{a\sqrt{(1-\tilde{\mu}^2)}^3} + \frac{\Omega \eta a \tilde{\mu}}{\sqrt{(1-\tilde{\mu}^2)}} \right)$$
$$\left(\frac{\tilde{M}}{a\sqrt{(1-\tilde{\mu}^2)}} - \Omega a \sqrt{(1-\tilde{\mu}^2)} \right) - g\tilde{\theta}\chi/\theta_0 \Bigg) \mathrm{d}\mu \mathrm{d}z, \tag{4.227}$$
$$= \int_{\Gamma_D} \left(\eta\tilde{\mu} \left(\frac{\tilde{M}^2}{a^2(1-\tilde{\mu}^2)^2} - a^2\Omega^2 \right) - \chi g\tilde{\theta}/\theta_0 \right) \mathrm{d}\mu \mathrm{d}z.$$

If this to be zero for any Ξ satisfying (4.223), then

$$\left(-\tilde{\mu} \left(\frac{\tilde{M}^2}{a^2(1-\tilde{\mu}^2)^2} - a^2\Omega^2 \right), g\tilde{\theta}/\theta_0 \right) = \nabla \tilde{\varphi} \tag{4.228}$$

for some $\tilde{\varphi}$. This is exactly the condition given by (4.216) and the fifth equation of (4.211). These then imply (4.225) by the changes of variable (4.217) and (4.219).

As in the proof of Theorem 3.2, characterise the values at the stationary point as $(\tilde{M}, \tilde{\theta}) = (M_g, \theta_g)$ satisfying (4.225) with $\tilde{P} = P_g$. Then (4.228) becomes

$$\delta E = \int_{\Gamma_D} - \left(-\frac{\tilde{\mu}\eta}{a^2(1-\tilde{\mu}^2)^2}, \chi \right) \cdot \left(\tilde{M}^2 - \frac{\partial P_g}{\partial s}, g\tilde{\theta}/\theta_0 - \frac{\partial P_g}{\partial z} \right) \mathrm{d}\mu \mathrm{d}z. \tag{4.229}$$

Now take a second variation, noting from (4.218) that $-\eta\tilde{\mu}/(a^2(1-\tilde{\mu}^2)^2) = \delta\tilde{s}$. This gives, using the vanishing of the second term in the integrand at the stationary point,

$$\delta^2 E = \int_\Gamma -(\delta\tilde{s},\chi)\cdot\delta\left(\tilde{M}^2 - \frac{\partial P_g}{\partial s}, g\tilde{\theta}/\theta_0 - \frac{\partial P_g}{\partial z}\right)\mathrm{d}\mu\mathrm{d}z. \qquad (4.230)$$

Since $\delta\tilde{M} = \delta\tilde{\theta} = 0$, this reduces to

$$\delta^2 E = \int_\Gamma ((\delta\tilde{s},\chi)\cdot\mathbf{Q}\cdot(\delta\tilde{s},\chi))\,\mathrm{d}\mu\mathrm{d}z,$$

where $\mathbf{Q}$ is calculated from (4.226) with $\tilde{M} = M_g, \tilde{\theta} = \theta_g$. Thus the condition for a minimiser is that $\mathbf{Q}$ is positive definite. □

Using (4.225), the condition that $\mathbf{Q}$ is positive definite is equivalent to convexity of P as a function of (s,z). Using (4.220 and the convexity property shows that necessary conditions for a minimum energy state are that M^2 is monotonically increasing in s and θ is monotonically increasing in z. In addition,

$$\nabla P_t \qquad \text{is invertible.} \qquad (4.231)$$

4.5.3 *Global solutions*

The condition that M^2 is monotonically increasing in s is natural on a hemisphere, where s increases from the pole to the equator. However, there is an immediate problem extending this to the whole sphere, since M^2 is naturally maximised at the equator and becomes smaller at both poles. This it cannot be monotonic in any coordinate. Solutions are found by minimising the energy subject to rearrangements of angular momentum and potential temperature as shown above. However, if such a minimisation is carried out over the whole sphere, it would inevitably give a solution where the Southern hemisphere circulation was a mirror image of the Northern hemisphere circulation. This is not realistic, and shows that a global minimisation principle is not physically appropriate.

The theory of Schubert and Hack (1983) was exploited by Schubert *et al.* (1991) to model the axisymmetric response to tropical forcing in spherical geometry. They defined a potential latitude coordinate as a dual variable. This is the value of μ at which v is zero for a given angular momentum M, and so is equal to $\cos^{-1}\left(\sqrt{\frac{M}{\Omega a}}\right)$. They then noted that a solution containing cross-equatorial flow could be obtained by allowing the equator in the potential latitude coordinate to be displaced from the physical equator.

In order to achieve this capability in the present formulation, the two hemispheric problems have to be regarded as separate, each satisfying its own energy principle, with some form of continuity between them. If this is to be the case, M^2 must vary monotonically with latitude in each hemisphere. While the maximum value of M^2 is typically reached close to the equator, it does not have to lie on the equator. The method of Schubert *et al.* (1991) allows fluid to cross the equator, while retaining its identity as Northern or Southern. A similar idea is adopted here.

Thus consider the spherical problem as a two-fluid system. Both fluids obey (4.211). The interface between them is defined by $\mu = \varrho(z)$, where $\int_0^H \varrho(z)\mathrm{d}z = 0$. This ensures mass conservation. ϱ has to be determined as part of the solution, as in section 4.4 for the vortex problem. At the interface, P and its derivative along the interface must be continuous, but there can be a jump in the normal component of ∇P. The interface must also be stable in the same sense as the separate zonal flows in each hemisphere are stable. Continuity of P requires

$$\frac{\partial \varrho}{\partial z} = \eta \frac{ga^2(1-\mu^2)^2}{\mu\theta_0}\frac{[\theta]}{[M^2]} \tag{4.232}$$

with $\eta = 1$ in the Northern hemisphere and $\eta = -1$ in the Southern hemisphere. The notation $[\cdot]$ indicates the jump across the interface, which is in the positive μ direction for $[M]$ and the positive z direction for $[\theta]$. This ensures that the change in M^2 and θ across the interface is consistent with (4.218) and (4.220). The stability condition consistent with the monotonicity of M in s and θ with z in each hemisphere requires that

$$\begin{aligned} [M]\frac{\partial \varrho}{\partial z} &\geq 0 \text{ in the Northern hemisphere} \\ &\leq 0 \text{ in the Southern hemisphere.} \end{aligned} \tag{4.233}$$

Note that these conditions are written in terms of μ because the definition of s will be different in each hemisphere. This is consistent with (4.232).

Now generalise (4.217) as follows:

$$\begin{aligned} s_N &= -\tfrac{1}{2a^2}\tfrac{\mu^2}{1-\mu^2}, \quad \text{for} \quad \mu \geq 0,\ \mu \geq \varrho(z), \\ &= \tfrac{1}{2a^2}\tfrac{\mu^2}{1-\mu^2}, \quad \text{for} \quad \mu < 0,\ \mu \geq \varrho(z), \\ s_S &= -\tfrac{1}{2a^2}\tfrac{\mu^2}{1-\mu^2}, \quad \text{for} \quad \mu \leq 0,\ \mu \leq \varrho(z), \\ &= \tfrac{1}{2a^2}\tfrac{\mu^2}{1-\mu^2}, \quad \text{for} \quad \mu > 0,\ \mu \leq \varrho(z). \end{aligned} \tag{4.234}$$

This ensures that s_N increases monotonically from $\mu = \mu_1$ to $\mu = \varrho$, and s_S increases monotonically from $\mu = -\mu_1$ to $\mu = \varrho$. This property is essential to make the argument work.

Then define dual variables

$$\Upsilon = M^2, \quad Z = g\theta/\theta_0. \tag{4.235}$$

Then (4.220) becomes

$$\nabla P = (\Upsilon, Z). \tag{4.236}$$

The monotonicity requirements established in section 4.5.2 require that Υ is monotonically increasing in s and Z is monotonically increasing in z.

Writing the evolution equations (4.211) in dual variables and (s, z) coordinates gives

$$\begin{aligned} \frac{\mathrm{D}\sqrt{\Upsilon}}{\mathrm{D}t} &= F(t,\mu,z), \\ \frac{\mathrm{D}Z}{\mathrm{D}t} &= S(t,\mu,z), \\ \nabla P &= (\Upsilon, Z), \\ \frac{1}{e(s)}\frac{\partial}{\partial s}(e(s)\varpi) + \frac{\partial w}{\partial z} &= 0, \end{aligned} \tag{4.237}$$

where ϖ is the meridional velocity in s coordinates and the volume measure $a^2\mathrm{d}\mu\mathrm{d}z$ in (μ, z) coordinates becomes $e(s)\mathrm{d}s\mathrm{d}z$ in (s, z) coordinates where

$$e(s) = \frac{a^2}{(2a^2 s - 1)^2}\sqrt{1 - \frac{1}{2a^2 s}} \quad \textit{for} \quad s_1 \leq s \leq 0. \tag{4.238}$$

Note that $e(s)$ tends to infinity as $s \to 0$ which corresponds to the equator. Similarly $\frac{\varpi}{v} \to 0$ as $s \to 0$. Thus a non-zero v at the equator will not be visible in these coordinates. Where $s > 0$, $e(s) = e(-s)$. The boundary condition at $s = s_1$ is $\varpi = 0$. The rest of the boundary conditions are the same as in (4.213).

Define

$$\begin{aligned} \rho &= s(\varrho) : \quad \varrho \geq 0, \\ &= -s(\varrho) : \quad \varrho < 0. \end{aligned} \tag{4.239}$$

This is consistent with the monotonicity established by (4.234) and applies to both $s = s_N$ and $s = s_S$, giving ρ_N and ρ_S. ρ_N and ρ_S have opposite signs and the same magnitude. Write the two fluid domains as $\Gamma_N = [-s_1, \rho_N] \times [0, H]$ and $\Gamma_S = [-s_1, \rho_S] \times [0, H]$.

Solutions of (4.237) with (4.236) and (4.234) are then characterised as follows:

Definition 4.2. An admissible solution of (4.237), (4.236) and (4.234) is given by convex functions P_N on Γ_N and P_S on Γ_S and an interface $\varrho(\mu)$. The continuity conditions $P_N(\rho_N, \cdot) = P_S(\rho_S, \cdot)$ and the stability conditions (4.232) and (4.233) must be satisfied.

4.5.4 *Generation of a mass conservation equation in angular momentum and potential temperature coordinates*

Next introduce angular momentum and isentropic coordinates. The monotonicity condition stated after (4.220) means that these will be well-defined. Define

$$\Psi(\Upsilon, Z) = s\Upsilon + zZ - P. \tag{4.240}$$

This means that $\Psi(\Upsilon, Z)$ is the Legendre transform of $P(s, z)$ as in (3.37) in the semi-geostrophic case. Then it can be shown that Ψ is also convex and $\nabla\Psi = (s, z)$. The map from physical coordinates (s, z) to angular momentum and isentropic coordinates (Υ, Z) is generated by ∇P and its inverse is $\nabla\Psi$.

Now recognise that, as in section 4.4.3, these two coordinates are defined in different units, which is important when proving that the transformation can be constructed and in generating examples. To illustrate this, apply this transformation to a state of rest in hydrostatic balance. Eqs. (4.214) and (4.235) show that this corresponds to choosing $\Upsilon = a^4\Omega^2(1 - \mu^2)^2$, while Z can be any positive function of z. It was shown after (4.239) that $s \in [s_1, \rho]$. However, if $\rho > 0$, this state violates the condition that Υ is monotone as a function of s. This will be the case in one or other hemisphere if $\rho \neq 0$. Therefore there can only be a state of rest if $\rho = 0$ so that the associated value of Υ will be in the range $0 \leq \Upsilon \leq \Omega^2 a^4$. Larger values are possible if $U \neq 0$ at the equator. The value of U at $s = \rho$ must then be consistent with monotonicity of Υ as a function of s in both hemispheres.

The region Γ_N in physical space can be written in μ coordinates as $\Gamma_N = [\varrho(z), \mu_1] \times [0, H]$. The definition of $e(s)$ in (4.238) and the definition of s in (4.217) show that

$$e(s) = \frac{a^2}{\mu}(1 - \mu^2)^2.$$

The next step is to show that equations (4.237) can be written as a conservation of mass in the new variables as in section 4.4.4 for the axisymmetric vortex. Schubert *et al.* (1991) did this in their equation (3.9) using their variables. It is necessary to do this separately for the two hemispheres. Consider the Northern hemisphere.

The mass in physical space is given by the integral of $a^2\chi_{\Gamma_N}\mathrm{d}\mu\mathrm{d}z$, where χ_{Γ_N} is the characteristic function of Γ_N. In (s,z) coordinates, the mass is the integral of

$$e(s)\chi_{\Gamma_N}\mathrm{d}s\mathrm{d}z. \tag{4.241}$$

The mass in (Υ, Z) coordinates is given by the potential density σ as in section 4.4, eq.(4.178), so that

$$\sigma\mathrm{d}\Upsilon\mathrm{d}Z = e(s)\mathrm{d}s\mathrm{d}z, \tag{4.242}$$

Since $(\Upsilon, Z) = \nabla P(s,z)$, equation (4.242) gives

$$\sigma\frac{\partial(\Upsilon, Z)}{\partial(s,z)} = \sigma\det(\partial^2 P) = e(s), \tag{4.243}$$

where $\partial^2 P$ is the Hessian matrix of P.

Using the notation of Definition 3.3, (4.243) can be written as

$$\sigma = \nabla P \# e(s)\chi_{\Gamma_N} \tag{4.244}$$

As ∇P is invertible with inverse $\nabla\Psi$, σ is equivalently defined by

$$e(\partial_\Upsilon\Psi)\det(\partial^2\Psi) = \sigma,\ \nabla\Psi(\Delta) = \Gamma_N, \tag{4.245}$$

where $s = \partial_\Upsilon\Psi = \partial\Psi/\partial\Upsilon$ and Δ is the region of $\mathbb{R}^2_+$ where σ is non-zero. (4.245) can be written as

$$\nabla\Psi\#\sigma = e(s)\chi_{\Gamma_N}. \tag{4.246}$$

Now consider the time dependence. Conservation of mass requires that

$$\frac{\partial\sigma}{\partial t} + \nabla\cdot(\sigma V) = 0 \qquad (0,T)\times\mathbb{R}^2 \tag{4.247}$$

$$\sigma|_{t=0} = \sigma_0,$$

where V is the velocity in (Υ, Z) coordinates, so that

$$V = \left(\frac{\mathrm{D}\Upsilon}{\mathrm{D}t}, \frac{\mathrm{D}Z}{\mathrm{D}t}\right). \tag{4.248}$$

Equations (4.237) give,

$$\frac{1}{2\sqrt{\Upsilon}}\frac{\mathrm{D}\sqrt{\Upsilon}}{\mathrm{D}t} = F(\mu, z), \tag{4.249}$$

$$\frac{\mathrm{D}Z}{\mathrm{D}t} = S(\mu, z).$$

Writing F and S as functions of (Υ, Z), using the relation $\nabla\Psi = (s, z)$ and the definition of s from (4.217) gives

$$V = \left(2\sqrt{\Upsilon} F\left(\tan^{-1}\left(\sqrt{-2a^2\frac{\partial\Psi}{\partial\Upsilon}}\right), \frac{\partial\Psi}{\partial Z}\right), \right. \tag{4.250}$$
$$\left. \frac{g}{\theta_0} S\left(\tan^{-1}\left(\sqrt{-2a^2\frac{\partial\Psi}{\partial\Upsilon}}\right), \frac{\partial\Psi}{\partial Z}\right)\right).$$

Solving (4.249) requires calculating Ψ given σ. If this can be done, then (4.246) determines ρ and (4.240) is used to calculate P. Equation (4.225) is then used to calculate u and θ, giving a complete solution for the zonal flow. The radial and vertical velocities (ϖ, w) can be determined by transforming V as defined in (4.248) into (s, z) coordinates using the map $\nabla\Psi$, allowing for the time dependence of Ψ.

The result will correspond to a solution of (4.237) if Ψ is convex, since then the change of coordinates is valid. The first two equations of (4.237) correspond to the definition of V (4.249), the third equation is equation (4.236), and the fourth equation corresponds to equation (4.247).

4.5.5 *Solution procedure*

As in the axisymmetric case with a free boundary, the first step will be to show that the solution in each hemisphere can be constructed separately with the boundary condition $v = 0$ at $\mu = 0$. This problem is essentially the same as that solved in section 4.4 with no free boundary. The method will be to write an optimal transport problem with the cost equal to the energy. The energy in (4.221) written in terms of Υ and Z is

$$E = \int_{\Gamma_D} \left(\frac{1}{2}\left(\frac{\Upsilon}{a^2(1-\mu^2)} - 2\sqrt{\Upsilon}\Omega + \Omega^2(1-\mu^2)\right) - Zz\right) a^2 \mathrm{d}\mu \mathrm{d}z. \tag{4.251}$$

It can be seen that varying μ and z for fixed Υ and Z only affects the terms which couple μ and Υ or z and Z. Thus the first term in the integrand will be minimised by maximising the negative correlation between μ and Υ, and the last term will be minimised by maximising the correlation between z and Z. This leads to the optimal transport problem solved in section 3.5.2.

The proof of existence of a global solution satisfying the conditions of Definition 4.2, in particular determining ϱ, is much harder and unsolved at the time of writing. However, to show that this is feasible, consider the simple case where Z is uniform, but different, in each hemisphere and

M is different in the extratropics and the tropics. Assume typical values M_E and M_T in the extratropics and tropics. If there is no difference in M, (4.232) gives $\frac{\partial \varrho}{\partial z} = \infty$. which would mean that the warm air in the Northern hemisphere would spread out over the whole sphere at upper levels. Since the latitudinal variation of M is dominated by the term $a^2\Omega \cos^2 \phi$, which peaks at the equator, a southward spread of warm air from the Northern hemisphere at upper levels, and a compensating northward spread of air at low levels, will lead to a difference of M in the positive μ direction across the interface of $M_E - M_T$ in the Northern hemisphere, which is negative, and of $M_T - M_E$ in the Southern hemisphere, which is positive. ϱ will be given by (4.232) as

$$\frac{\partial \varrho}{\partial z} = \eta \frac{a^2(1-\mu^2)^2}{\mu} \frac{Z_N - Z_S}{\Upsilon_E - \Upsilon_T}, \tag{4.252}$$

where $\eta = 1$ in the Northern hemisphere and -1 in the Southern hemisphere and $\int_0^H \varrho \mathrm{d}z = 0$. This solution is illustrated in Fig. 4.5. The chosen $[Z]$ implies a temperature difference of about 30°K and the chosen angular momentum difference implies a zonal wind difference of about 20ms^{-1}.

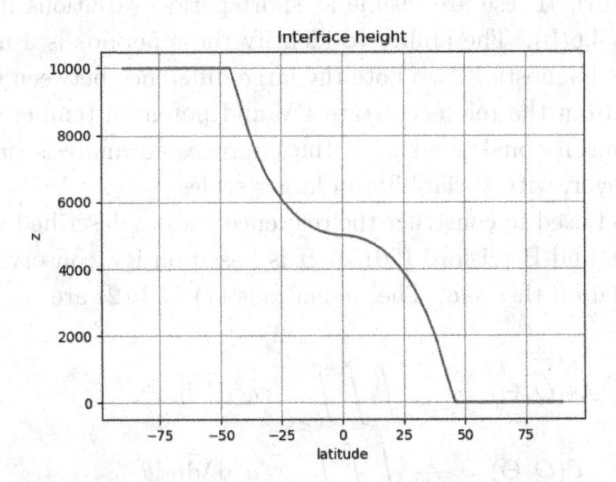

Fig. 4.5 Vertical profile of the interface between air from the two hemispheres given by eq. (4.252), setting $\Omega = 0.7 \times 10^{-4}\ \text{s}^{-1}$, $a = 6.37 \times 10^6\text{m}$, $a^{-4}\Omega^{-2}(\Upsilon_E - \Upsilon_T) = 0.05$ and $Z_N - Z_S = 1$. Source: Met Office. ©Crown Copyright, Met Office.

The air is assumed to be warmer in the Northern hemisphere and the angular momentum is assumed to be greater in the tropics. The penetration of the warmer air into the Southern hemisphere at upper levels, with the converse at lower levels, is clearly seen. The interface is flat at the equator because the Coriolis parameter which appears in (4.252) is zero there.

4.5.6 *Application*

Methven and Berrisford (2015) developed a diagnostic of the long time evolution of the atmosphere which worked by constructing a zonally symmetric reference state by rearranging the potential vorticity (PV) and potential temperature from the three-dimensional evolution. The reference state varies much less than the three-dimensional state. The claim is that significant changes in the circulation are picked out by changes in the reference state. Since potential temperature and PV are both conserved by the adiabatic dynamics, the changes in the reference state must result from diabatic processes.

An example is shown in Fig. 4.6. The data used are from the ERA interim reanalysis. The method of constructing the reference state is summarised below. It uses the hydrostatic equations in pressure coordinates, which give accurate results for real data. The labels A to F define significant changes at the tropopause level, which are discussed in Methven and Berrisford (2015). These are visible as short period variations in the PV shown in Fig. 4.6(b). The ability to identify these periods is a major advantage of the diagnostic. Also note the large difference between the zonal wind inferred from the reference state PV and potential temperature and the Eulerian mean zonal wind from the unprocessed analysis data. The former is stronger, with variability on larger scales.

The method used to construct the reference state is described in section 2.1 of Methven and Berrisford (2015). It is based on PV conservation and Kelvin's circulation theorem. Their equations (1) and (2) are

$$\mathcal{M}(Q,\Theta) = \frac{1}{\Delta\Theta}\int\int\int_{q\geq Q} ra^2 \mathrm{d}\lambda\mathrm{d}\mu\mathrm{d}\theta, \tag{4.253}$$

$$\begin{aligned}\mathcal{C}(Q,\Theta) &= \frac{1}{\Delta\Theta}\int\int\int_{q\geq Q} \zeta a^2 \mathrm{d}\lambda\mathrm{d}\mu\mathrm{d}\theta + \\ &\quad \frac{1}{\Delta\Theta}\int\int_{\partial\Theta} (\mathbf{u} + \Omega\mathbf{k}\times\mathbf{r})\cdot \mathbf{l}\mathrm{d}l\mathrm{d}\theta.\end{aligned} \tag{4.254}$$

Here $r = -\frac{1}{g}\frac{\partial p}{\partial \theta}$, $q = \zeta/r$ and

$$\zeta = \Omega\mu - \frac{\partial U}{\partial \mu} + \frac{1}{1-\mu^2}\frac{\partial V}{\partial \lambda}, \tag{4.255}$$

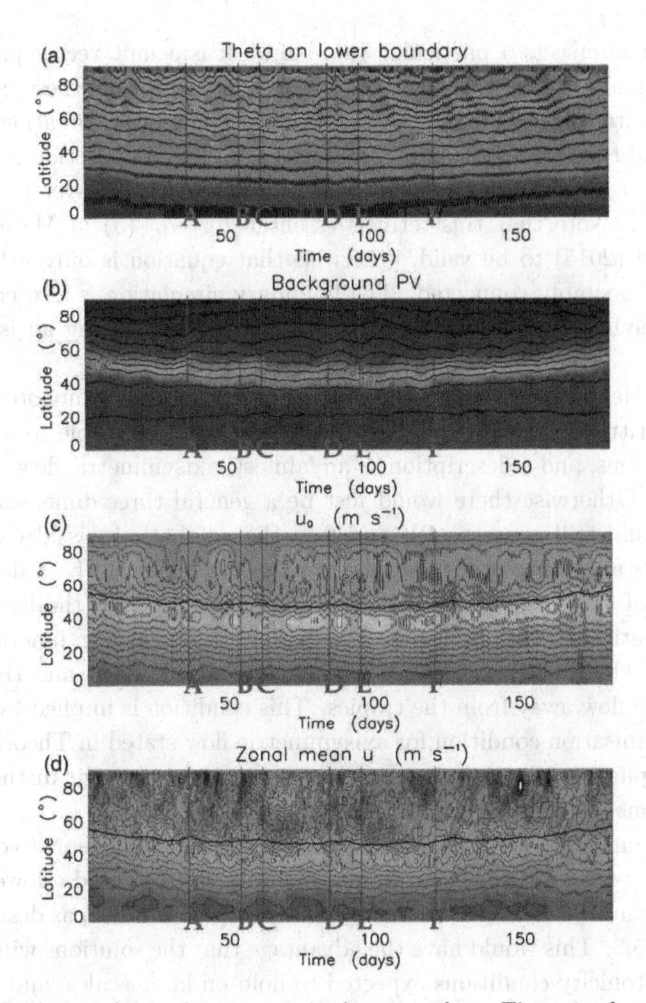

Fig. 4.6 Evolution of the reference state in the troposphere. Time runs from 00UTC 1 November 2009 to 00UTC 1 May 2010. (a) Potential temperature on the lower boundary (°K, contour interval 5K) with gradient $-\partial\theta/\partial\phi$ shaded. (b) Potential vorticity (PV) on the 311°K surface, which crosses the tropopause in mid-latitudes, with meridional gradient shaded. (c) Zonal wind, (ms^{-1}), contour interval 2.5 ms^{-1} with the PV= 2 contour in bold. (d) Eulerian zonal mean wind (same scale as (c)). From Methven and Berrisford (2015).

where Θ is a value of θ representing a layer in isentropic coordinates, and Q is a particular value q of the PV. U, V and μ are defined as in section 4.5.1. In the boundary term in (4.254), $\partial\Theta$ represents the part of the boundary intersecting the chosen θ layer, and $\mathbf{l}$ represents the contour where the boundary intersects a particular value of θ. $\mathbf{k}$ is a unit vector parallel to the rotation axis and $\mathbf{r}$ is the position vector. Note that these equations are not quite correct unless the shallow atmosphere approximation is withdrawn and the full 3-d Coriolis term is used, as noted in section 2.2.4. The boundary term may have several contributions if the θ layer is multiply connected. Note that this term is essential for eq. (5) in Methven and Berrisford (2015) to be valid, otherwise that equation is only valid if the θ surface is simply connected. The boundary circulation is also conserved under Kelvin's circulation theorem as it is calculated along an isentropic surface.

These definitions are motivated by the case where q is a monotonic function of latitude. If this is the case, a zonal flow will be stable to non-zonal perturbations, and a description as an 'almost' axisymmetric flow would be justified. Otherwise there would just be a general three-dimensional flow which would still conserve PV and θ in the adiabatic inviscid case. The definitions above can still be applied in this case. $\mathcal{M}$ will be a decreasing function of Q provided that $r < 0$, which is required for the hydrostatic approximation to be appropriate. $\mathcal{C}$ will be a decreasing function of Q if $\zeta > 0$, which will be the case for inertially stable flows and thus most large-scale flow away from the tropics. This condition is implied by the energy minimisation condition for axisymmetric flow stated in Theorem 4.15, which implies stability of the zonal flow to zonally symmetric disturbances. It is assumed that this is the case.

Methven and Berrisford (2015) do not use optimal transport to calculate the reference state. Only a hemispheric solution is attempted. However, the problem can be formulated as an optimal transport problem as described in section 4.5.5. This would have the advantage that the solutions will respect the monotonicity conditions expected to hold on large scales, and allows a straightforward formulation of the boundary conditions.

The formulation of section 4.5.5 requires the zonally symmetric background state to be defined in terms of the mass as a function of angular momentum Υ and θ. However, angular momentum as a function of λ is not a conserved quantity in three-dimensional flow, so the definition of $\mathcal{C}$ for the reference state in terms of the three-dimensional state in (4.254) is used. Then, comparing the definitions of Υ in (4.236) and of circulation $\mathcal{C}$ for a zonally symmetric flow in eq. (5) of Methven and Berrisford (2015) gives $\mathcal{C} = 2\pi M$. In order to generate an equation for Υ, first note that the monotonicity of $\mathcal{C}$ as a function of Q means that (4.254) can be replaced by

$$\mathcal{N}(\mathcal{C},\Theta)=\frac{1}{\Delta\Theta}\int\int\int^{C\leq\mathcal{C}(Q,\Theta)} r\zeta^{-1}a^2\mathrm{d}\lambda\mathrm{d}\mu\mathrm{d}\theta. \tag{4.256}$$

Note that $r\zeta^{-1}=q^{-1}$ so that the inverse PV represents the mass in angular momentum and isentropic coordinates. This is the same argument that gives eq. (27) in Methven and Berrisford (2015). The potential density σ defined in (4.242) is also the inverse of the semi-geostrophic PV as noted in eq. (3.70) in the three-dimensional case.

Now express σ as a function of Υ and θ. Since $\Upsilon=\frac{1}{4\pi^2}\mathcal{C}^2$, $\mathrm{d}\Upsilon=\frac{1}{2\pi^2}\mathcal{C}\mathrm{d}\mathcal{C}$. Then discretise (4.256) in $\mathcal{C}$ to give

$$\sigma(\Upsilon,\Theta)=\frac{2\pi^2}{\mathcal{C}\Delta\Theta\Delta\mathcal{C}}\int\int\int_{C\leq\mathcal{C}(Q,\Theta)-\Delta\mathcal{C}}^{C\leq\mathcal{C}(Q,\Theta)} r\zeta^{-1}a^2\mathrm{d}\lambda\mathrm{d}\mu\mathrm{d}\theta. \tag{4.257}$$

The optimal transport method in section 4.5.5 can now be used to calculate the reference state from σ.

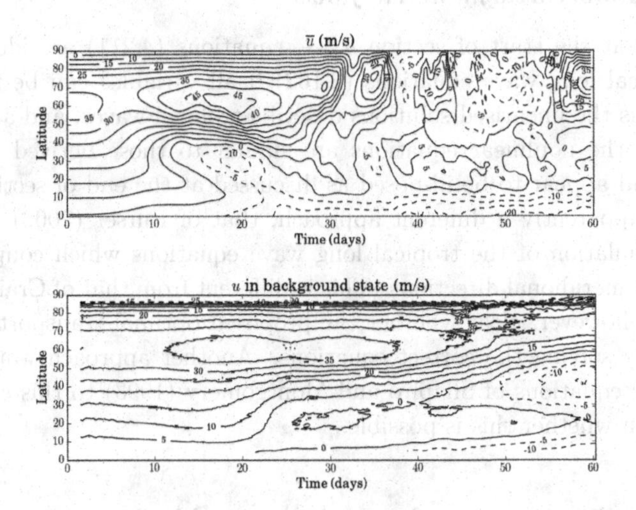

Fig. 4.7 Evolution of the reference state for the stratospheric shallow water simulation over 60 days. (a) Eulerian zonal mean (ms^{-1}). (b) Zonal wind from reference state, (ms^{-1}). From Mansfield (2017). ©Laura Mansfield

So far, this formulation has only been used in the shallow water case. Mansfield (2017) describes the results for an idealised stratospheric vortex case shown in Fig. 4.7. In the shallow water case, the reference state is calculated from the mass $\mathcal{M}(Q)$ and circulation $\mathcal{C}(Q)$ within each PV contour. Since these are both conserved by smooth solutions of the shallow water equations, the reference state should be independent of time. However, some artifical smoothing has to be included in the model to control numerical errors, and this results in some long-period variation in the reference state. The plots are time-latitude diagrams using data from the model at 6 hour intervals. The two types of zonal mean are similar in the first few days. The reference state can only evolve because of the artificial smoothing, and this becomes noticeable in the second half of the experiment. The effect is that the tropics relax to a state of solid body rotation. In the Eulerian mean, large values of the zonal wind disappear earlier in the integration, again starting with the tropics, but no large values remain at any latitude later in the experiment. Thus using the reference state illustrates that the model can 'remember' conserved quantities for a long time away from the tropics, while the Eulerian mean hides this fact.

4.5.7 *Almost axisymmetric flows*

As noted at the start of section 4.5.1, equations (4.211) are identical to the tropical long wave equations (2.162) if all variables can be functions of λ. Thus the linearised solutions describe tropical waves, and are useful. However, the nonlinear equations are similar to those derived by Craig (1991) and appear to be ill-posed as discussed at the end of section 4.4.5. Though apparently a different approach, that of Purser (2002) derives a dual formulation of the tropical long wave equations which couples both zonal and meridional directions. This is different from that of Craig (1991). It is not, however, clear whether the proposed optimal transport solution is actually solving the correct equations. Another approach would be to adapt the equations of Shapiro and Montgomery (1993) to this case. It is not known whether this is possible.

4.6 Stability theorems for semi-geostrophic flow

4.6.1 *Extremising the energy by rearrangement of the potential density*

This section discusses the qualitative behaviour of large-scale atmospheric circulations using semi-geostrophic theory.

This is appropriate if we assume that the effect of physical forcing is to create air masses with different properties, but that the rate of change of air mass properties is slow compared with the advective time-scale. The internal dynamics is then regarded as semi-geostrophic, which is energy conserving, and conserves potential temperature and potential density following fluid particles. The minimum energy state that is consistent with potential temperature conservation is a 'rest' state, where the air masses are rearranged so that the potential temperature surfaces are horizontal. It is well known that the energy available to the internal dynamics is the difference between the actual energy and the energy of the rest state. In this section, the additional restriction that the available energy is only the excess over that of a minimum energy state obtained by rearranging both potential density and potential temperature is explored. Such a minimum energy state will be nonlinearly stable using Kelvin's principles, Thomson (1910): firstly that steady states are stationary points of the energy under rearrangements of the potential density and potential temperature, and secondly that stable steady states are extrema of the energy under these rearrangements. Application of these ideas to the atmosphere originated with Fjortoft (1950). The reference state on a hemisphere described in section 4.5.6 is an example of the rearrangement procedure. The use of optimal transport in constructing it ensures that it represents a minimum energy state. However, if the potential density σ, regarded as a function of Υ, is then rearranged on θ surfaces as discussed below, a lower energy may be obtained.

The problem, however, of minimising the energy with respect to simultaneous rearrangements of potential density and potential temperature is not generally well-posed. It is possible for there to be a sequence of rearrangements with successively smaller energies which do not converge to a limit. This corresponds to increasingly fine-scale filamentation of the potential density. The limiting state is actually a mixed state that is not a rearrangement. The widely used energy-Casimir method of analysing nonlinear stability, e.g. Kushner and Shepherd (1995a), Kushner and Shepherd (1995b), avoids this difficulty by replacing the rearrangement constraint by conservation of a particular function, the 'Casimir', of the potential density. An invariant is constructed using this function and the energy. The function is chosen so that if the basic state is perturbed, the change to the invariant is sign definite. This proves stability of the basic state to variations which conserve the invariant. The class of variations allowed is now larger than the rearrangements, so the stability results may be pessimistic. In the present paper, the alternative approach of considering the smallest (weakly) compact class of perturbations that includes the rearrangements is adopted. Compactness ensures that the minimisation problem will be well-posed. In particular, the class includes all states reached as the limit of a sequence of rearrangements. The energy-Casimir method cannot take account of these limit states, as they do not conserve the Casimirs.

This method has been used by Burton and Nycander (1999) to analyse the stability of a localised potential vorticity anomaly in a uniformly sheared environment using three-dimensional quasi-geostrophic theory. They have to allow for the possibility of mixing when posing the variational problem. However, the maximum energy states which they study are achieved by rearrangements and do not involve mixing. Similarly, it can be shown that the solution to the problem of minimising the energy with respect to rearrangements of potential temperature is a rearrangement with potential temperature monotonically increasing with height. Mixed solutions have a higher energy, see Douglas (2002).

The advantage of using rearrangement methods is that stability can be established with respect to all displacements, whether smooth or not. It is also possible to derive results by geometrical arguments which may be difficult to establish by algebraic methods. For instance, Ren (2000) shows that it is very difficult to apply Arnold's stability methods to semi-geostrophic theory because of the nonlinear form of the potential density. However, the natural geometric interpretation of potential density can be readily used with rearrangement methods.

In this section, results obtained by Cullen and Douglas (2003) using rearrangement theory are summarised. A formal definition of a rearrangement is given in section 3.5.2 and more details of results from rearrangement theory are given in Douglas (2002). The Boussinesq incompressible semi-geostrophic equations (3.2) are solved in a domain Γ which is a channel of width $2D$ and height H, with periodicity $2L$ in the x-direction. Thus the boundary conditions are $v = 0$ on $y = \pm D$, $w = 0$ on $z = 0, H$, with all variables periodic in x.

Using the change of variables (3.27), the evolution equations are given by (3.30), repeated below:

$$\begin{aligned}
\frac{\mathrm{D}X}{\mathrm{D}t} &= u_g, \\
\frac{\mathrm{D}Y}{\mathrm{D}t} &= v_g, \\
\frac{\mathrm{D}Z}{\mathrm{D}t} &= 0, \\
\nabla \cdot \mathbf{u} &= 0.
\end{aligned} \tag{4.258}$$

The periodicity condition means that

$$\frac{\partial}{\partial t}\int_{\Gamma} Y\mathrm{d}x\mathrm{d}y\mathrm{d}z = \int_{\Gamma} v_g \mathrm{d}x\mathrm{d}y\mathrm{d}z, \tag{4.259}$$
$$= \int_{\Gamma} f^{-1}\frac{\partial \varphi}{\partial x}\mathrm{d}x\mathrm{d}y\mathrm{d}z = 0.$$

This expresses the conservation of the momentum integral. The energy integral is given by (3.47)

$$E = \int_{\mathbb{R}^3} f^2\left(\frac{1}{2}\left((x-X)^2 + (y-Y)^2\right) - zZ\right)\sigma \mathrm{d}X\mathrm{d}Y\mathrm{d}Z. \tag{4.260}$$

The evolution can be written in dual variables in (X, Y, Z) coordinates on $\mathbb{T}\times\mathbb{R}^2$. Following (3.44), the evolution of potential density is given by

$$\frac{\partial \sigma}{\partial t} + \mathbf{U}\cdot\nabla\sigma = 0, \tag{4.261}$$
$$\mathbf{U} = (f(y-Y), f(X-x), 0).$$

According to Theorem 3.26, extended to the periodic case, (4.261) can be solved for initial data $\sigma(0,\cdot)$ satisfying the compatibility condition

$$\sigma(X+L, Y) = \sigma(X-L, Y), \tag{4.262}$$
$$\int_{-L}^{L}\int_{-\infty}^{\infty}\int_{-\infty}^{\infty} \sigma \mathrm{d}X\mathrm{d}Y\mathrm{d}Z = \quad 4LDH.$$

Equation (4.261) shows that $\mathbf{U}$ is non-divergent and has no component in the Z direction. Thus the evolution can be considered as a rearrangement of σ on Z-surfaces in $\mathbb{T}\times\mathbb{R}^2$. The theorems discussed in section 3.5.3 show that the solutions are sufficiently regular for this statement to make sense.

Kelvin's principle applied to equation (4.261) says that steady states are stationary points of the energy with respect to rearrangements of σ along Z surfaces. Stable steady states correspond to maxima or minima of the energy under such rearrangements. It is likely that there will only be a small number of globally stable steady states corresponding to the maximum and the minimum energy that are obtainable over the whole class of rearrangements, including mixing, but there can also be a large class of locally stable states which are extrema of the energy subject to physically reasonable displacements.

4.6.2 *Properties of rearrangements*

Work with the definition of a rearrangement given in Definition 3.7. Write the set of rearrangements of a given potential density as $\mathcal{R}(\sigma)$. Consider the problem of finding the maximum or minimum of the energy which can be obtained by rearranging a given potential density distribution on isentropic surfaces. A classical approach would be to choose a maximising (or minimising) sequence, and extract a subsequence converging to a maximiser (or a minimiser). However, this limit might not be a rearrangement, there is the possibility of 'mixing'. It is possible to construct an increasingly fine-grained sequence of rearrangements, which converge in a weak sense to a smoothed potential density distribution which is not a rearrangement. A simple one-dimensional example is given by the function ϱ_0 on $[0,1]$:

$$\varrho_0(x) = \begin{cases} 0 \text{ if } & x \in [0, \frac{1}{2}], \\ 1 \text{ if } & x \in [\frac{1}{2}, 1]. \end{cases} \tag{4.263}$$

Define, for n a positive integer,

$$\varrho_n(x) = \begin{cases} 0 \text{ if } & x = 0, \\ 0 \text{ if } & x \in (m/n, (2m+1)/2n], \\ 1 \text{ if } & x \in ((2m+1)/2n, (m+1)/n], \end{cases} \tag{4.264}$$

where $m = 0, 1, ..., n-1$. The functions ϱ_3 and ϱ_8 are illustrated in Figure 4.8. For each integer n, ϱ_n is equal to zero on a set of length $\frac{1}{2}$, and equal to 1 on a set of length $\frac{1}{2}$. Therefore ϱ_n is a rearrangement of ϱ_0. However, given any $\vartheta \in L^2(0,1)$, it may be shown that $\int_0^1 \varrho_n \vartheta \mathrm{d}x \to \frac{1}{2}\int_0^1 \vartheta \mathrm{d}x$ as $n \to \infty$, that is ϱ_n converges weakly to the constant function with value $\frac{1}{2}$, which is not a rearrangement of ϱ_0.

This occurs in situations like the filamentation of potential vorticity at the stratospheric vortex edge. Physically, including these limit functions can be thought of as allowing for a small but finite viscosity or conductivity. The energy will be almost the same whether the fine-scale potential vorticity filaments are present or are averaged out.

Now attempt to solve the energy extremisation problem by extracting weakly convergent subsequences from an energy extremising sequence. In Douglas (2002), section 2.4, and Burton and Nycander (1999) it is shown that these limit solutions have to be included to ensure convergence. (In the particular problem solved by Burton and Nycander (1999), additional work showed that there is a solution which is a rearrangement.) A set is *weakly sequentially compact*, if for any sequence composed of elements of the set, it is possible to find a subsequence which converges weakly to an element of the set. For a given σ, seek the smallest such set which contains $\mathcal{R}(\sigma)$. It may be characterised, see for example Ryff (1970), as the closed

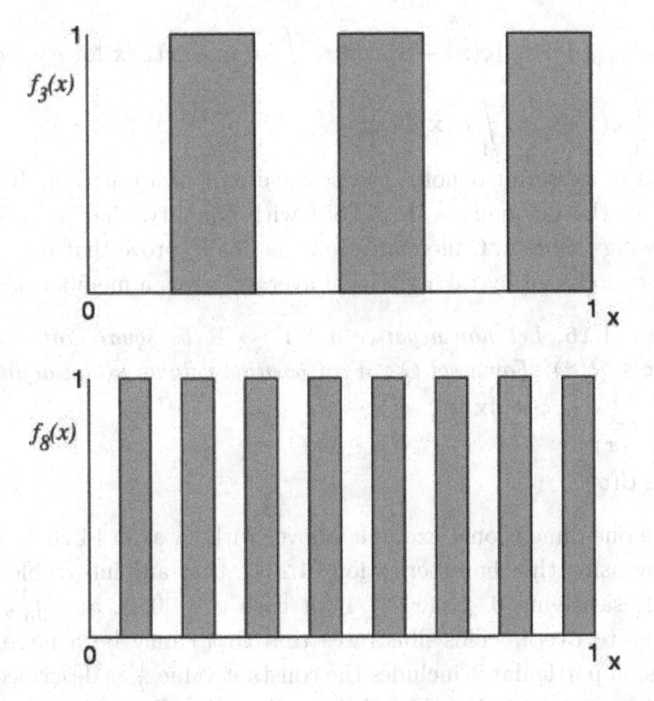

Fig. 4.8 Graphs of $\varrho_3(x)$ and $\varrho_8(x)$ as defined in (4.264). From Cullen and Douglas (2003). ©Royal Meteorological Society, Reading, U.K., 2003.

convex hull of the set $\mathcal{R}(\sigma)$, which is the intersection of all the (strongly) closed convex sets that contain $\mathcal{R}(\sigma)$; denote this set $\mathcal{C}(\sigma)$.

Choosing this set ensures that the most optimistic possible assessment of stability is made, since the energy extremisation is over the smallest possible set of perturbations which allow a solution. Now characterise this set. Douglas (1994) showed that $\varrho \in \mathcal{C}(\sigma)$ implies $\left(\int_\Gamma \varrho(\mathbf{x})^p d\mathbf{x}\right)^{1/p} \leq \left(\int_\Gamma \sigma(\mathbf{x})^p d\mathbf{x}\right)^{1/p}$ for every $p > 1$. This shows that the higher order moments of σ will, in general, be decreased when the limit solutions are included. This expresses the lack of robustness of the higher order moments of the potential density as constants of the motion, and that a state which extremises the energy may be obtained by selective decay of the higher order moments. These issues are discussed, for instance, by Robert and Sommeria (1991) and Larichev and McWilliams (1991).

Next, make use of the following characterisation of $\mathcal{C}(\sigma)$ by Douglas (1994):

$$\mathcal{C}(\sigma) = \Big\{\varsigma \geq 0 : \int_\Gamma (\varsigma(\mathbf{x}) - \alpha)_+ d\mathbf{x} \leq \int_\Gamma (\sigma(\mathbf{x}) - \alpha)_+ d\mathbf{x} \text{ for each } \alpha > 0,$$
$$\int_\Gamma \varsigma(\mathbf{x}) d\mathbf{x} = \int_\Gamma \sigma(\mathbf{x}) d\mathbf{x}.\Big\}, \tag{4.265}$$

where the + subscript denotes the positive part of a function. If $\varsigma \in \mathcal{C}(\sigma)$ satisfies all the inequalities in (4.265) with equality, then $\varsigma \in \mathcal{R}(\sigma)$. In the following theorem, Cullen and Douglas (2003) prove that rearranging a function σ, followed by taking a local average, gives a member of $\mathcal{C}(\sigma)$.

Theorem 4.16. *Let non-negative $\sigma : \Gamma \to \mathbb{R}$ be square integrable, and suppose $\varsigma \in \mathcal{R}(\sigma)$. For a set $G \subset \Gamma$ of positive volume $\mu(G)$, define*

$$\varrho(x) = \begin{cases} \frac{1}{\mu(G)} \int_G \varsigma(\mathbf{x}) \mathrm{d}\mathbf{x} \equiv \overline{\varsigma} & \text{if } \mathbf{x} \in G, \\ \varsigma(x) & \text{if } x \in \Gamma \backslash G. \end{cases}$$

Then $\varrho \in \mathcal{C}(\sigma)$.

In the one-dimensional example above, with ϱ_0 as in (4.263), it can be shown, by using the characterisation (4.264), that any integrable function ς on $[0,1]$ satisfying $0 \leq \varsigma(x) \leq 1$ for each $x \in [0,1]$, and $\int_0^1 \varsigma(x)dx = \frac{1}{2}$, belongs to $\mathcal{C}(\varrho_0)$. This illustrates that $\mathcal{C}(\varrho_0)$ may be a large class of functions, in particular it includes the constant value $\frac{1}{2}$ as described earlier.

Another useful result, which follows from the characterisation of $\mathcal{C}(\sigma)$ as the closed convex hull of $\mathcal{R}(\sigma)$, is that $\mathcal{R}(\sigma)$ is weakly dense in $\mathcal{C}(\sigma)$ so that, for every $\varsigma \in \mathcal{C}(\sigma)$, it is possible to find a sequence $(\varsigma_n) \subset \mathcal{R}(\sigma)$ which converges weakly to ς, see Douglas (1994).

In multi-dimensional problems, it is typical that the extremising states can be proved to be independent of one or more spatial coordinates. In the shear-flow problems described in this section, it can be proved that the extremising states are independent of X. In Burton and Nycander (1999) it is shown that the extremising states for an almost axisymmetric quasi-geostrophic vortex are independent of λ. It is then sufficient to analyse the properties of one-dimensional rearrangements. Therefore it is worth deriving some one-dimensional results. Take σ to be a non-negative function on $[0,1]$ as illustrated in Fig. 3.14. It is stated in Douglas (2002) that there is an (essentially) unique rearrangement of σ which is an increasing function. Write it as $\tilde{\sigma}$. Similarly, σ has an (essentially) unique rearrangement which is decreasing; denote it $\hat{\sigma}$. Now state the following result which follows from standard rearrangement inequalities:

Theorem 4.17. *Let non-negative $\varrho, \sigma : [0,1] \to \mathbb{R}$ be square integrable, and suppose ϱ is increasing. Then $\int_0^1 \varrho(x)\varsigma(x)\mathrm{d}x$ is maximised for $\varsigma \in \mathcal{C}(\sigma)$*

by the increasing rearrangement of σ, and is minimised by the decreasing rearrangement of σ. The reverse statements apply if $\varrho(x)$ is non-negative and decreasing.

Another useful result concerns the stability of straight semi-geostrophic flows in the X-direction. For this case, y is a function of Y only, and the potential density σ is $\partial y/\partial Y$. The velocity U is $f(y-Y)$ and (4.260) shows that the kinetic energy density is $\frac{1}{2}f^2(y-Y)^2\sigma$. Take as an example, (Fig. 4.9), the case where the physical domain is $0 \le y \le \frac{1}{2}$ and $\sigma(Y)$ is a function of the same form as ϱ_0 defined in (4.264), so that $\sigma(Y) = 0 : 0 \le Y \le \frac{1}{2}; = 1 : \frac{1}{2} < Y \le 1$. The mean value of σ is $\frac{1}{2}$, so that the compatibility condition (4.262) is satisfied. The velocity U is derived by first calculating $y(Y) = \int_0^Y \sigma(Y')\mathrm{d}Y'$, and then setting $U = f(y-Y)$. It is intuitively clear that the largest energy for $\varsigma \in \mathcal{R}(\sigma)$ will be obtained by choosing ς to be either the monotonically increasing or decreasing rearrangement of σ. Moreover, no larger value will be achieved by any other $\varsigma \in \mathcal{C}(\sigma)$. The energy will be minimised over $\varsigma \in \mathcal{R}(\sigma)$ by choosing ς to oscillate about $\frac{1}{2}$, the mean value of σ. Define $\varsigma_n(Y) = \begin{cases} \varrho_n(Y + 1/(2n)) & \text{if } 0 \le Y \le 1, \\ 0 & \text{if } Y > 1, \end{cases}$ where ϱ_n is as defined in (4.264), then the energy is $\mathrm{O}(\frac{1}{n^2})$. Note that a minimum energy state is not attained within $\mathcal{R}(\sigma)$, but that zero energy is achieved within $\mathcal{C}(\sigma)$.

This example shows that the minimiser will typically involve mixing while the maximiser will be a strict rearrangement. The physically important case for the global stability problem is usually an energy minimiser, because the global maximiser will often correspond to an unreachable or unphysical state. However, the local stability problem may well be solved by an energy maximiser, as in the study of Burton and Nycander (1999).

Now consider general $\sigma(Y)$ with the same physical domain $0 \le y \le \frac{1}{2}$. The compatibility condition (4.262) requires that $\int_{-\infty}^{\infty} \sigma(Y)dY = \frac{1}{2}$. Let $y(Y) = \int_0^Y \sigma(Y')\mathrm{d}Y'$. Theorem 4.16 shows that $\mathcal{C}(\sigma)$ contains all local averages. Thus in particular, given any Y_1, Y_2, if $y(Y)$ is replaced by its linear interpolant between Y_1 and Y_2, then the associated $\varsigma = \partial y/\partial Y$ will be in $\mathcal{C}(\sigma)$ because $\sigma(Y)$ has been replaced by its average value over the range (Y_1, Y_2). In the example shown in Fig. 4.9, the linear interpolant between 0 and 1 gives the function $y(Y) = Y/2$. Thus $\frac{1}{2} \in \mathcal{C}(\sigma)$ and the sequence $\varsigma_n(x)$ in $\mathcal{R}(\sigma)$ has the weak limit $\varsigma(x) = \frac{1}{2}$, giving zero energy.

In order to obtain useful nonlinear stability results, it is necessary to require the potential density to have compact support. The same condition

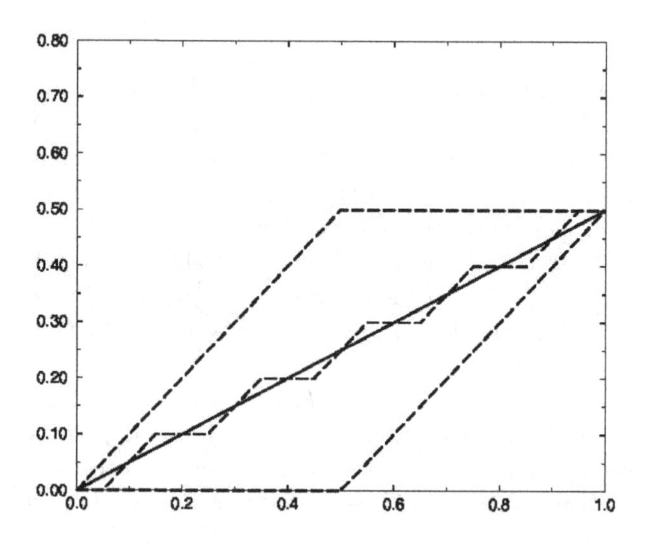

Fig. 4.9 Graphs of $y(Y) = Y/2$ (solid line) and various choices of $y(Y)$ obtained by setting $\varsigma(Y)$ to be a rearrangement of $\sigma(Y)$ of the form (4.263). These choices are the monotonically increasing and decreasing rearrangements, and the choice $\varsigma(x) = \varrho_5(Y + 0.1)$ as defined in (4.264). From Cullen and Douglas (2003). ©Royal Meteorological Society, Reading, U.K., 2003.

was required to prove most of the results in section 3.5. In the following theorem, it is assumed without loss of generality that the finite region is $[0, 1]$.

Theorem 4.18. *Let $\sigma : [0,1] \to \mathbb{R}$ be square integrable. Given $\varsigma \in \mathcal{C}(\sigma)$, set $y_\varsigma(Y) = \int_0^Y \varsigma(Y')\mathrm{d}Y'$. Then, for each $Y \in [0,1]$, $y_{\hat{\sigma}}(Y) \geq y_\varsigma(Y)$ for every $\varsigma \in \mathcal{C}(\sigma)$, where $\hat{\sigma}$ denotes the decreasing rearrangement of σ. Consequently $\int_0^1 y_{\hat{\sigma}}(Y)dY \geq \int_0^1 y_\varsigma(Y)dY$.*

4.6.3 *Analysis of semi-geostrophic shear flows*

Now use these results to analyse the stability of semi-geostrophic shear flows. In order to apply Kelvin's principle , seek a class of perturbations which is dynamically consistent with (4.261). It is shown by Cullen and Douglas (2003) that if the support of σ at $t = 0$ is within an interval $[-S, S]$, the support at time T will be contained within $[-(S+fLT), S+fLT]$. Therefore assume that the perturbations are within the class $\mathcal{C}_h(\sigma)$, defined by

$$\varsigma \in \mathcal{C}_h(\sigma) \text{ if } \begin{cases} \varsigma(.,Z) \in \mathcal{C}(\sigma(.,Z)) \text{ for almost all } Z \\ \int Y\varsigma \mathrm{d}X\mathrm{d}Y\mathrm{d}Z = \int Y\sigma \mathrm{d}X\mathrm{d}Y\mathrm{d}Z \\ \text{supp } \varsigma \subset \Sigma = [-(S+fLT), S+fLT] \end{cases} \tag{4.266}$$

The first condition restricts the rearrangements of σ to the (X,Y) variables only, and includes the weak limits. This is similar to the space of 'stratified' rearrangements used by Burton and Nycander (1999). The additional condition is that the mean Y over the particles cannot be changed. This corresponds to the conservation of the momentum integral (4.259). Write $\mathcal{R}_h(\sigma)$ for functions ρ satisfying (4.266) with $\mathcal{R}$ replacing $\mathcal{C}$.

The problem of extremising the energy with respect to perturbations in $\mathcal{C}_h(\sigma)$ only makes physical sense if weak convergence of a sequence of potential density distributions to a limit implies convergence of the energy associated with the distributions. It is shown in Cullen and Douglas (2003) that, given a sequence $\varsigma_n \in \mathcal{R}_h(\sigma)$ with energies E_n converging weakly to a member ς of $\mathcal{C}_h(\sigma)$ with energy E, E_n converges to E. Then given a maximising or minimising sequence for the energy, it is possible to extract a weakly convergent subsequence converging to some $\varsigma \in \mathcal{C}_h(\sigma)$, by weak sequential compactness of $\mathcal{C}_h(\sigma)$.

Next characterise the steady states of (4.261) as stationary points of the energy with respect to appropriate variations. Given compactly supported $\sigma = \sigma(X,Y,Z)$ satisfying (4.262), seek necessary conditions for $\delta E = 0$ given perturbations satisfying $\sigma + \delta\sigma \in \mathcal{R}_h(\sigma)$. Generate such variations by keeping σ fixed on particles in $\mathbf{X}$ space and perturbing X and Y with an incompressible displacement field $\Xi = (\xi, \eta)$. As the incompressible displacement is infinitesimal, it must also be non-divergent, so that $(\xi,\eta) = (-\partial\vartheta/\partial Y, \partial\vartheta/\partial X, 0)$ for some arbitrary function $\vartheta(X,Y)$, and must satisfy the periodicity condition $\vartheta(X-L,Y) = \vartheta(X+L,Y)$ The restriction that the mean Y cannot be changed is then automatically enforced.

In both original and perturbed solutions (x,y,z) is defined as a function of (X,Y,Z) by an optimal map, respectively $\mathbf{t}(\mathbf{X})$ and $\mathbf{t}'(\mathbf{X})$. Write the changes to $\mathbf{x}$ following a particle as $(\delta x, \delta y, \delta z)$. Since the value of σ is not changed following particles, their volume in physical space is preserved, and so $\nabla \cdot (\delta x, \delta y, \delta z) = 0$. Then integrating over particles gives

$$\delta E = f^2 \int_{-L}^{L} \int_{-\infty}^{\infty} \int_{-\infty}^{\infty} ((X - x)\xi + (Y - y)\eta - X\delta x - Y\delta y - Z\delta z)\sigma \mathrm{d}X\mathrm{d}Y\mathrm{d}Z. \quad (4.267)$$

The condition that the original $\mathbf{x}$ is given as a function of $\mathbf{X}$ by an optimal map means that

$$f^2 \int_{-L}^{L} \int_{-\infty}^{\infty} \int_{-\infty}^{\infty} (-X\delta x - Y\delta y - Z\delta z)\sigma \mathrm{d}X\mathrm{d}Y\mathrm{d}Z = 0. \quad (4.268)$$

By substituting (4.268) into (4.267), using the definition (4.261) of $\mathbf{U}$, and after some further manipulations, Cullen and Douglas (2003) show that the condition for a stationary point is

$$U\frac{\partial \sigma}{\partial X} + V\frac{\partial \sigma}{\partial Y} = 0, \quad (4.269)$$

which is the condition for the flow to be steady.

Next consider the stability of barotropic shear flows where σ is a function of (X, Y) only, and the energy is

$$E = \frac{1}{2} f^2 \int_{-L}^{L} \int_{-\infty}^{\infty} (x - X)^2 + (y - Y)^2 \sigma \mathrm{d}X\mathrm{d}Y. \quad (4.270)$$

The maximum energy attainable for $\varsigma \in \mathcal{C}_h(\sigma)$ is a function of the assumed maximum value of $|Y|$ in the support Σ of σ. Using (4.266), this can be calculated as $2f^4L^3TD$. It is therefore only physically meaningful to seek minimum energy states. In the stability problem for two-dimensional incompressible flow, however, the maximum energy state is physically meaningful. This is because the evolution equation is written in physical space and so the displacements have to be within the physical domain. This problem was treated by Burton and McLeod (1991).

Given σ, seek to minimise E for $\sigma + \delta\sigma \in \mathcal{C}_h(\sigma)$. (4.260) shows that the minimum energy is attained by making the map $\mathbf{t}$ from (X, Y) to (x, y) as close as possible to the identity. Therefore it can be expected that the minimum energy will be achieved by mixing the values of σ to give a potential density σ_0 defined by

$$\sigma_0 = 1 : |Y - Y_0| \leq D, X \leq |L| \text{ and } \sigma_0 = 0 \text{ elsewhere}, \quad (4.271)$$

with Y_0 chosen to satisfy the second equation of (4.266). Then (4.261) shows that $\mathbf{U} = (-fY_0, 0)$ for all (X, Y). This will give the lowest energy consistent with the requirement that the mean velocity and hence the momentum integral is specified. This distribution is not always achievable by mixing the given σ, as Cullen and Douglas (2003) show in the following theorem:

Theorem 4.19. *If Σ has area greater than $4DL$, then σ_0 as defined by (4.271) is not in $\mathcal{C}_h(\sigma)$.*

It follows from Theorem 4.19 that, if the mean value $\overline{\sigma}$ of σ is less than 1 over Σ, then $\sigma_0 \notin \mathcal{C}_h(\sigma)$. Fig. 4.10 shows three possibilities for $\overline{\sigma}$, assuming $Y_0 = 0$. Recalling that σ is the inverse of the potential vorticity, the anticyclonic case corresponds to $\overline{\sigma} > 1$. The distribution shown can be mixed to give σ_0. The cyclonic case has $\overline{\sigma} < 1$, Theorem 4.19 applies, and σ cannot be mixed to give σ_0. In that case, the minimum energy will be obtained by getting as close to σ_0 as possible. The following stability result for shear flows independent of Z is then proved by Cullen and Douglas (2003):

Theorem 4.20. *Given $\sigma(X, Y)$ satisfying (4.261) with compact support $\Sigma \subset \mathbb{R}^2$ with area μ. If $\mu > 4DL$, the minimiser of E over $\mathcal{C}_h(\sigma)$ takes the form*

$$\varsigma(X,Y) = \begin{cases} 1 & \text{if } |Y - Y_0| \leq Y_1, \\ \tilde{\sigma} & \text{if } Y_1 < |Y - Y_0| \leq \mu/4L, \\ 0 & \text{otherwise,} \end{cases} \tag{4.272}$$

where $\tilde{\sigma}$ is a monotonically decreasing function of $|Y - Y_0|$ and $0 \leq Y_1 < \mu/4L$. Y_0 is chosen to satisfy the momentum integral constraint. If $\mu \leq 4DL$ the minimiser is defined by (4.271).

Theorem 4.20 shows that the only minimum energy states are distributions independent of X with $\sigma \leq 1$ everywhere. Thus there can be no stable states with $\sigma > 1$ at any point. Since σ is an inverse potential vorticity, this condition excludes values of potential vorticity less than 1, which correspond to anticyclonic relative vorticity. This agrees with the results of Kushner and Shepherd (1995a), Kushner and Shepherd (1995b), that there were no stable shear flows with anticyclonic shear. This argument also shows that no steady states with anticyclonic relative vorticity can be stable in any limited domain with rigid boundary conditions under

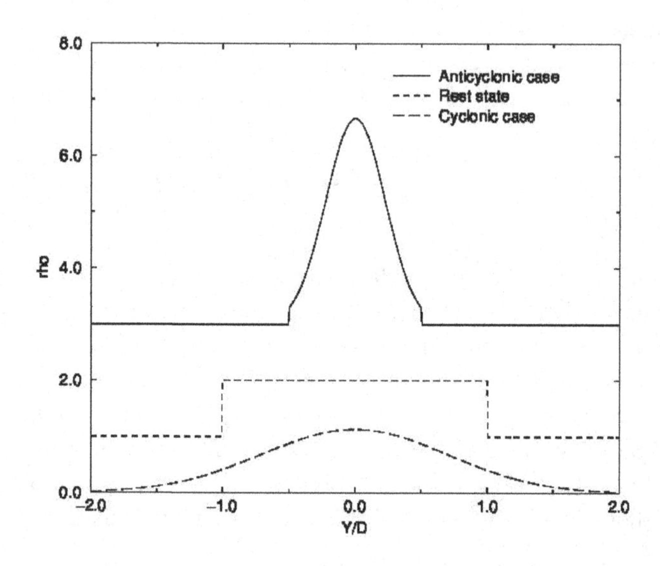

Fig. 4.10 Graphs of $\sigma(Y)$ against Y for the cyclonic and anticyclonic cases defined in the text and the rest state $\sigma_0(Y) = 1, |Y| \leq D$. The base values are shifted for clarity. From Cullen and Douglas (2003). ©Royal Meteorological Society, Reading, U.K., 2003.

semi-geostrophic dynamics. They could be stable in doubly periodic flows, because there is then no region with $\sigma = 0$ in the (X, Y) plane to mix with the non-zero values. Ren (2000) discusses the physical relevance of this stability condition, suggesting that it may be most relevant in the baroclinic case discussed next.

Now consider the baroclinic case where σ depends on Z. Write, for a given Z, $\mu(Z) = \frac{1}{2}\max(S(Z), 2D)$, where $2LS(Z)$ is the area of $\Sigma \cap \{Z\}$. Set $M = \max_Z(\mu(Z))$, for all Z. Since σ can only be rearranged on Z surfaces, the energy minimiser ς is expected to be obtained by first assuming a zero momentum integral on each Z surface, and then minimising the energy on each Z surface separately using Theorem 4.20. It may then be advantageous to remove the Z dependence by mixing σ uniformly over the whole region $|Y| \leq M$. The momentum constraint, which is vertically integrated, is then used to displace the entire solution by some distance Y_0 in Y. For each Z, the area of the support of ς cannot be less than the area of the support of σ. However, zero values can be mixed in to increase the size of the set to $4DL$ for each Z. If $M \leq D$, σ can be mixed to give ς as a function of Z only whose energy will be the minimum rest state potential energy. However, if $M > D$, this state is not in $\mathcal{C}_h(\sigma)$. There then has to be kinetic energy in the minimum energy state. These situations are illustrated schematically in Fig. 4.11.

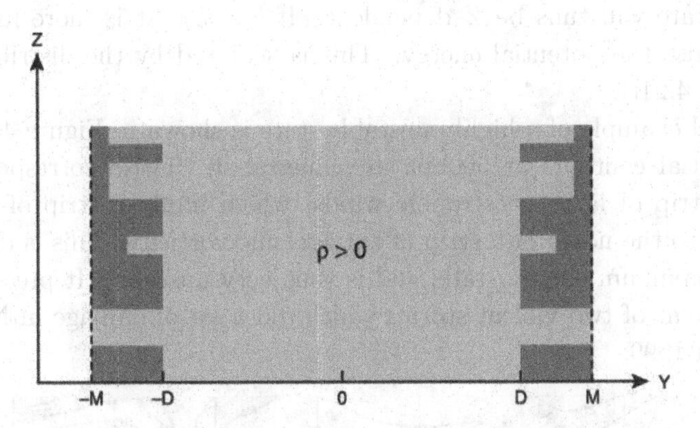

Fig. 4.11 Schematic illustration of the region of the (Y, Z) plane for which ς is non-zero in the energy minimising state for the two extreme cases; total shaded area: $fl/NH \gg 1$, light shaded area: $fl/NH \ll 1$. From Cullen and Douglas (2003). ©Royal Meteorological Society, Reading, U.K., 2003.

These procedures are formalised in:

Theorem 4.21.

(i) If $M \leq D$, $\varsigma = \{\text{constant}, |Y| \leq D, = 0, |Y| > D\}$ is in $\mathcal{C}_h(\sigma)$ and minimises E over $\mathcal{C}_h(\sigma)$. The geostrophic wind then takes a uniform value on Γ. This value is equal to the minimum rest state potential energy if the specified momentum integral is also zero.

ii) If $M > D$, and the specified momentum integral is zero, the minimum possible kinetic energy achievable for $\varsigma \in \mathcal{C}_h(\sigma)$ is $4L \int_D^{\mu(Z)} \int_{-\infty}^{\infty} (D - \hat{Y})^2 \hat{\sigma}(Y, Z) dY dZ$, where $\hat{\sigma}(Y, Z)$ takes the form (4.272) for each Z and $\hat{Y}(y)$ is defined by $dY^*/dy = (\hat{\sigma})^{-1}$.

(iii) The potential energy is minimised for $\varsigma \in \mathcal{C}_h(\sigma)$ by choosing ς to take a uniform value for $|Y - Y_0| \leq M$ for each Z.

If $M > D$, then the solution will contain kinetic energy even if the specified momentum integral is zero. The solution will be scale-dependent, according to whether potential or kinetic energy perturbations contribute more to the total energy. As discussed in section 2.4, this depends on whether the horizontal length scale l is smaller than the Rossby radius of deformation L_R. If $l < L_R$, it is most important to minimise the kinetic energy. The distribution of Theorem 4.21(ii) requires there to be less kinetic energy than that of Theorem 4.21(iii), since there only has to be kinetic energy associated with values of Z for which $\mu(Z) > D$. The minimum energy state will thus be Z dependent. If $l > L_R$, it is more important to minimise the potential energy. This is achieved by the distribution of Theorem 4.21(iii).

A real example of a highly unstable state is shown in Figure 4.12. The geopotential contours at 500hpa are illustrated. These correspond to a narrow strip of large geostrophic winds, which imply a strip of cyclonic vorticity to the north of a strip of anticyclonic vorticity. This is close to a locally maximum energy state, and is thus very unstable. It preceded the development of two violent storms which did a lot of damage in Northern France in 1999.

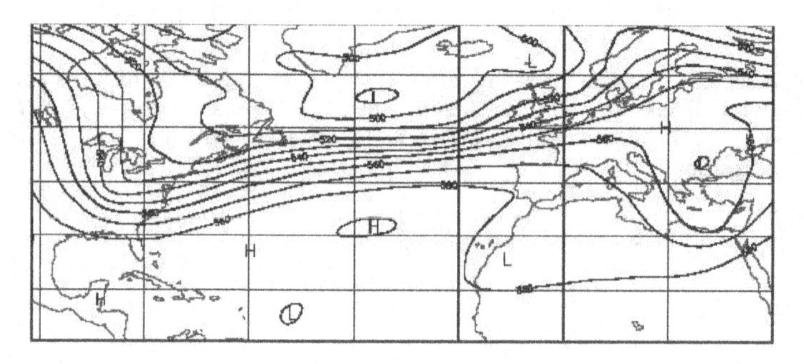

Fig. 4.12 500hpa height over part of the Northern hemisphere at 1200UTC on 24 December 1999. Source: ECMWF.

In general, it has been shown that there are minimum energy states with more energy than the rest state potential energy. This gives more control over the possible dynamic evolution of the system, since only the excess energy above this minimum value is available for transient motion. It will be interesting to see if these results can be extended to spherical geometry or almost axisymmetric flows.

Chapter 5

Properties and Validation of Asymptotic Limit Solutions

The main idea set out in Chapter 1 is that the behaviour of a general set of equations in a particular regime can be studied by first identifying a simpler set of equations which are specific to the regime, and a good approximation to the general equations in that regime, and then studying the properties of the simpler equations. In this chapter this idea is exploited and validated using the semi-geostrophic approximations to the shallow water equations and the three-dimensional Boussinesq equations in a vertical slice. Published comparisons using three-dimensional models are reviewed.

5.1 Numerical methods for solving the semi-geostrophic equations

5.1.1 *Solutions using the geostrophic coordinate transformation*

Most numerical solutions of the three-dimensional semi-geostrophic equations have been produced by the geostrophic coordinate transformation introduced by Hoskins (1975). An early example of such a solution was the simulation of a growing baroclinic wave by Hoskins and West (1979). The description of the method here follows Snyder *et al.* (1991).

Consider the shallow atmosphere hydrostatic incompressible Boussinesq Euler equations with constant rotation, eq. (3.1), together with the semi-geostrophic approximation to them, (3.2). For convenience, both are restated below. The Euler equations are

$$
\begin{aligned}
\frac{\mathrm{D}u}{\mathrm{D}t} + \frac{\partial \varphi}{\partial x} - fv &= 0, \\
\frac{\mathrm{D}v}{\mathrm{D}t} + \frac{\partial \varphi}{\partial y} + fu &= 0, \\
\frac{\mathrm{D}\theta}{\mathrm{D}t} &= 0, \\
\frac{\partial \varphi}{\partial z} &= \frac{g\theta}{\theta_0}, \\
\nabla \cdot \mathbf{u} &= 0.
\end{aligned} \tag{5.1}
$$

and the semi-geostrophic approximation is given by

$$
\begin{aligned}
\frac{\mathrm{D}u_g}{\mathrm{D}t} + \frac{\partial \varphi}{\partial x} - fv &= 0, \\
\frac{\mathrm{D}v_g}{\mathrm{D}t} + \frac{\partial \varphi}{\partial y} + fu &= 0, \\
\frac{\mathrm{D}\theta}{\mathrm{D}t} &= 0, \\
\nabla \varphi &= \left(fv_g, -fu_g, \frac{g\theta}{\theta_0} \right), \\
\nabla \cdot \mathbf{u} &= 0.
\end{aligned} \tag{5.2}
$$

Both systems are solved in a region $\Gamma = \mathbb{T}^2 \times [0, H]$, so that there are horizontal boundaries at $z = 0, H$ where $w = 0$. The non-hydrostatic version of (5.1) is given in (3.20).

The potential vorticity equation is given by (3.70). Transform the equations to the geostrophic coordinates (X, Y, Z_g), where X and Y are defined by (3.27) and $Z_g = z$. Then (3.70) becomes

$$
\begin{aligned}
\frac{f^2}{Q_{SG}} \frac{\partial^2 \Psi}{\partial Z_g^2} + \frac{\partial^2 \Psi}{\partial X^2} + \frac{\partial^2 \Psi}{\partial Y^2} - \frac{1}{f^2} J_{XY}\left(\frac{\partial \Psi}{\partial X}, \frac{\partial \Psi}{\partial Y} \right) &= f^2, \\
\text{where } J_{cd}(a, b) \equiv \frac{\partial a}{\partial c}\frac{\partial b}{\partial d} - \frac{\partial a}{\partial d}\frac{\partial b}{\partial c}, \\
\Psi = \frac{f^2}{2}\left((x - X)^2 + (y - Y)^2 \right).
\end{aligned} \tag{5.3}
$$

The definition of Ψ is similar to that given by assuming equality in (3.55). The difference arises because the vertical coordinate has not been transformed. The hydrostatic relation transforms to

$$
\frac{\partial \Psi}{\partial Z_g} = g\frac{\theta}{\theta_0}. \tag{5.4}
$$

In the cases studied by Hoskins and West (1979) and Snyder *et al.* (1991), $\mathcal{Q}_{SG}$ was assumed uniform initially, and so it is assumed it will remain uniform. Therefore equation (3.70) does not have to be solved. It is shown in section 3.2.3 that this property will in general be lost if there is frontogenesis. It is not clear how much this would alter the conclusions of these papers. Given this assumption, a solution can be obtained by predicting θ on the horizontal boundaries. Since $w = 0$ at $z = Z_g = (0, H)$ and using (5.4), this requires solving

$$\frac{\partial}{\partial t}\frac{\partial \Psi}{\partial Z_g} = \frac{1}{f}\left(\frac{\partial \Psi}{\partial Y}\frac{\partial}{\partial X} - \frac{\partial \Psi}{\partial X}\frac{\partial}{\partial Y}\right)\frac{\partial \Psi}{\partial Z_g}, \tag{5.5}$$

at $Z_g = 0, H$.

The solutions in Hoskins and West (1979) were obtained by neglecting the nonlinear term in the first equation of (5.3). This equation is then a linear constant coefficient elliptic equation for Ψ, which was solved by using a second order spatial finite difference representation, fast Fourier transforms in the horizontal, and a tri-diagonal inversion in the vertical using the solutions of (5.5) as boundary conditions. Equation (5.5) was integrated forward in time using a positive-definite advection scheme. The equations were solved in (X, Y, Z_g) coordinates and only transformed back to physical coordinates for plotting purposes. If the nonlinear terms in (5.3) are included, the solution has to be iterated, using the previous iterate to calculate the nonlinear terms. It was found by Snyder *et al.* (1991) that neglecting the nonlinear terms increased the difference between semi-geostrophic solutions and solutions of (3.20).

While this method appears much simpler than the solution procedure described in section 3.2.2 using the Monge-Ampère equation, it relies on the transformation back to physical space being single-valued. It is shown in Hoskins and West (1979) that this is not the case in general. However, the consequences were assumed small in these studies. The resulting error is not known.

The geostrophic coordinate transformation has also been used when using semi-geostrophic theory to derive a 'balanced' part of the solutions of the Euler equations (5.1). It is thus equivalent to the procedure described in section 2.3.2, equation (2.54) for the shallow water equations. It was shown by Hoskins and Draghici (1977) that, under this transformation, the semi-geostrophic omega equation takes the same form as the quasi-geostrophic omega equation in physical coordinates. In Pedder and Thorpe (1999), this was exploited to diagnose the ageostrophic circulation predicted by semi-geostrophic theory from fields predicted by a solution of (5.1). The method is summarised below.

The omega equation for the three-dimensional quasi-geostrophic equations (2.137) and (2.138) rewritten in dimensional form is

$$N^2\nabla_h^2 w + f^2\frac{\partial^2 w}{\partial z^2} = 2\nabla_h\cdot\mathbb{Q}, \quad (5.6)$$

$$\mathbb{Q} = \left\{-\frac{g}{\theta_0}\left(\frac{\partial u_g}{\partial x}\frac{\partial\theta}{\partial x} + \frac{\partial v_g}{\partial x}\frac{\partial\theta}{\partial y}\right), -\frac{g}{\theta_0}\left(\frac{\partial u_g}{\partial y}\frac{\partial\theta}{\partial x} + \frac{\partial v_g}{\partial y}\frac{\partial\theta}{\partial y}\right)\right\}.$$

∇_h is the gradient with respect to the horizontal coordinates. If the nonlinear term in equation (5.3) is neglected, then the semi-geostrophic omega equation can be written in (X, Y, Z_g) coordinates as

$$\left(\frac{g}{f\theta_0}\nabla_h^2 Q_{SG} + f^2\frac{\partial^2}{\partial Z_g^2}\right)\left(w\frac{\partial(x,y)}{\partial(X,Y)}\right) = 2\nabla_h\cdot\left(\mathbb{Q}\frac{\partial(x,y)}{\partial(X,Y)}\right), \quad (5.7)$$

where Q_{SG} is the semi-geostrophic potential vortcity defined by (3.70), and $\mathbb{Q}$ is defined in (5.6).

In Pedder and Thorpe (1999), equation (5.7) is solved with $\mathbf{u}_g, \mathbb{Q}$ and $\frac{\partial(x,y)}{\partial(X,Y)}$ calculated from data from a solution of (5.1) on a regular grid in (x, y, z) coordinates. Values X_j and Y_j are also calculated on this grid. A regular grid (X_i, Y_i, Z_{gi}) is then chosen in (X, Y, Z_g) coordinates, and the corresponding values of (x, y, z) determined by iteration. Assuming the values of z and Z_g correspond, only interpolation in (x, y) needs to be considered. The first guess is obtained by interpolating $\mathbf{u}_g$ from (X_j, Y_j) to (X_i, Y_i), and then calculating

$$x_i = X_i - f^{-1}v_{gi}, y_i = Y_i + f^{-1}u_{gi}. \quad (5.8)$$

$\mathbf{u}_g$ can then be interpolated from the original grid to the points (x_i, y_i) and used in (5.8) to update the values of (x_i, y_i). Then interpolate the values of $\mathbb{Q}$ and $\frac{\partial(x,y)}{\partial(X,Y)}$ to the points (x_i, y_i) and solve (5.7).

The method of Pedder and Thorpe (1999) works if $fQ_{SG} > 0$ and $\frac{\partial(x,y)}{\partial(X,Y)} > 0$. These are implied by the convexity condition in Definition 3.1. When (5.7) is used diagnostically with the right hand side calculated from data from a solution of (5.1), it is unlikely that this condition will be satisfied. In Pedder and Thorpe (1999), a case of this is demonstrated, and resolved by local modifications to φ. These are calculated by solving the inverse problem $\frac{\partial(X,Y)}{\partial(x,y)} = \max(\mathcal{Q}, \mathcal{Q}_0)$ for φ at each z, where $\mathcal{Q}$ is the original value of $\frac{\partial(X,Y)}{\partial(x,y)}$ and $\mathcal{Q}_0$ a small positive value. Any changes made to φ are applied equally to all values of z. The procedure is iterated over the levels till no further changes result. An application of this method due to Thorpe and Pedder (1999) is shown in section 5.3.1. Recall that a similar regularisation was used in applying a semi-geostrophic diagnostic to the Met Office UM in section 2.7.4.

5.1.2 *The geometric method*

The geometric algorithm used in sections 3.4.2 and 5.3.5 is based on the explicit construction procedure of Theorem 3.12. It represents the construction of an optimal map from a set of Dirac masses σ_i in dual variables to physical space. The image of the ith Dirac mass is a hyperface with volume σ_i. Convergence of this method to a 'continuous' solution of (5.2) is proved in Theorem 3.25 in section 3.5.3. It has so far only been implemented in two space dimensions. The original implementation was by Chynoweth (1987). A novel recent implementation is described by Egan *et al.* (2021).

The notation of section 3.4.1 is used. In principle, the solution can be obtained by iteration on the intersection p_i of the face A_i defined by the equation $p = xX_i + yY_i + p_i$ with $x = y = 0$. Convergence is guaranteed by Theorem 3.12. The cost of the original algorithm was $O(N^3)$ operations, where N is the number of faces.

An efficient implementation was developed by R. J. Purser (private communication). This has a cost of $O(N \log N)$ operations . Given data $(X_i, Y_i, \sigma_i) : i = 1, N$, the efficiency of the algorithm depends on the sorting procedure which uses the convex hull algorithm developed by Preparata and Hong (1977).

The first step is to choose values p_i such that all faces of the polyhedron

$$p(x,y) = \sup_i (xX_i + yY_i + p_i), \tag{5.9}$$

have a non-zero area. This is achieved by the Voronoi solution, a simple form was shown in (3.133). More generally, suppose $\Gamma = [a,b] \times [c,d]$ and that $(X_i, Y_i) \in [A,B] \times [C,D]$. Then set

$$\begin{aligned} p_i = -\frac{1}{2}\left(\frac{b-a}{B-A}X_i^2 + \frac{d-c}{D-C}Y_i^2\right) &+ \left(a - A\frac{b-a}{B-A}\right)X_i \\ &+ \left(c - C\frac{d-c}{D-C}\right)Y_i. \end{aligned} \tag{5.10}$$

Then use the *divide and conquer* algorithm of Preparata and Hong (1977) to calculate the geometry of the intersections between the planes $p(x,y) = xX_i + yY_i + p_i$. Sort (X_i, Y_i) values into groups of up to four elements by successive binary subdivision using alternately the X_i and Y_i values. Construct the geometry of a four-element patch. If the faces of the polyhedron associated with two pairs (X_i, Y_i) and (X_j, Y_j) have a common edge, draw a link between the pairs as shown in Figure 5.1. Figure 5.1 shows the two possibilities within a four element patch.

Next knit two disjoint patches together using the algorithm of Preparata and Hong (1977). If elements 3 and 2^* have a common edge, join the two patches by an arc from 3 to 2^*. Now test that the new link is consistent with convexity of the combined patch. The links have thus to reflect the monotonic change of X and Y as x and y vary across the polyhedron. In the example shown, a link from 3 to 2^* will clearly be consistent with this, while joining 2 and 3^* will not. Then complete the connections between the patches to form a joint convex hull, and repeat up through the binary tree, building up the whole solution.

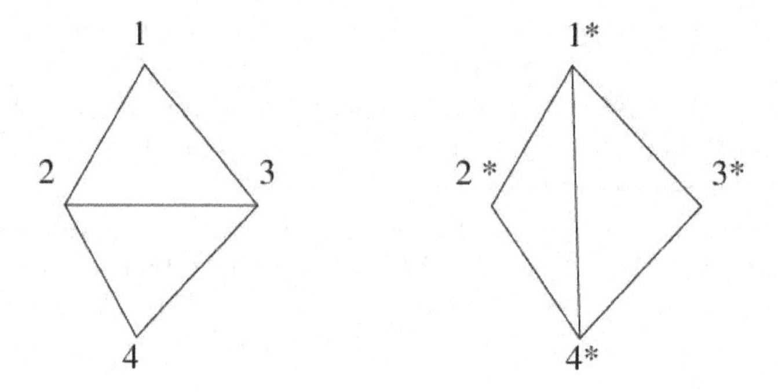

Fig. 5.1 Two patches of four elements, showing the possible linkages between vertices, and defining the notation for the discussion in the text.

At some points, this successive build-up may lead to incorrect linkages, so that the surface defined by assuming corresponding intersections between planes is not convex. The *panel-beater* algorithm is used to resolve these. Typically it will involve exchanging links between the two formats shown in Fig. 5.1. The metrical properties of the polyhedron are now calculated. In particular the area of each face in (x, y) space and its rate of change with respect to its value of p_i are needed. Increasing a value of p_i will increase the area of the corresponding face at the expense of its neighbours. The iteration to the correct area σ_i thus has the same structure as the discrete solution of an elliptic partial differential equation, and so can be handled by conjugate gradient or multigrid methods.

5.1.3 *Finite difference methods*

First describe the method of Mawson (1996) for the shallow water semi-geostrophic equations (3.72), (3.73). This will be demonstrated in section 5.2.2. The same method, with some refinements, was used by Cloke and Cullen (1994) as described in section 3.4.3, for the Eady problem in section 5.3, and the semi-geostrophic diagnostic calculation in section 2.7.4. Finite difference methods for the equations expressed as a mass transport problem have been developed by Benamou and Brenier (1997) and Angenent *et al.* (2003).

For convenience, eqs. (3.72)-(3.75) are restated below:

$$\frac{\partial h}{\partial x} - f v_g = 0, \qquad (5.11)$$

$$\frac{\partial h}{\partial y} + f u_g = 0.$$

$$\begin{aligned} \frac{\partial u_g}{\partial t} + u\frac{\partial u_g}{\partial x} + v\frac{\partial u_g}{\partial y} + g\frac{\partial h}{\partial x} - fv = 0, \\ \frac{\partial v_g}{\partial t} + u\frac{\partial v_g}{\partial x} + v\frac{\partial v_g}{\partial y} + g\frac{\partial h}{\partial y} + fu = 0, \\ \frac{\partial h}{\partial t} + \frac{\partial}{\partial x}(hu) + \frac{\partial}{\partial y}(hv) = 0. \end{aligned} \qquad (5.12)$$

$$\begin{aligned} \mathbf{Q}\begin{pmatrix} u \\ v \end{pmatrix} + g\frac{\partial}{\partial t}\nabla h = fg\begin{pmatrix} -\frac{\partial h}{\partial y} \\ \frac{\partial h}{\partial x} \end{pmatrix}, \\ \mathbf{Q} = \begin{pmatrix} f^2 + f\frac{\partial v_g}{\partial x} & f\frac{\partial v_g}{\partial y} \\ -f\frac{\partial u_g}{\partial x} & f^2 - f\frac{\partial u_g}{\partial y} \end{pmatrix}. \end{aligned} \qquad (5.13)$$

$$\frac{\partial h}{\partial t} - \nabla\cdot\left(h\mathbf{Q}^{-1}g\frac{\partial}{\partial t}\nabla h\right) = -\nabla\cdot\left(h\mathbf{Q}^{-1}fg\begin{pmatrix} -\frac{\partial h}{\partial y} \\ \frac{\partial h}{\partial x} \end{pmatrix}\right). \qquad (5.14)$$

The data (h, u, v, u_g, v_g) are represented on a grid as shown in Fig. 5.2. Thus u_g is held at a different point from u and v_g at a different point from v. This choice is natural for the definition of the geostrophic wind, (5.11), and for the continuity equation in (5.12). In the spherical case described by Mawson (1996), depth values are stored on the equator. This is because semi-geostrophic solutions, as characterised by Theorem 4.10, do not allow h to vary along the equator. The constraint is easier to enforce with this choice of grid.

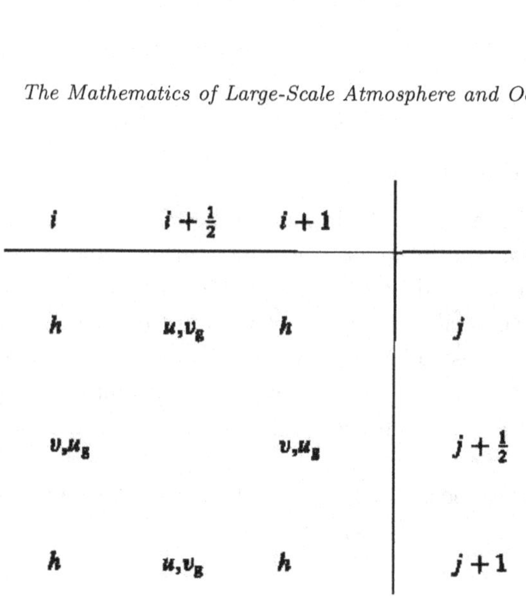

i	$i+\frac{1}{2}$	$i+1$	
h	u, v_g	h	j
v, u_g		v, u_g	$j+\frac{1}{2}$
h	u, v_g	h	$j+1$

Fig. 5.2 Grid staggering for the solution of the semi-geostrophic shallow water equations. From Mawson (1996). ©Crown Copyright, Met Office.

The solution can be obtained using a semi-implicit time differencing scheme. Given initial data for (h, u, v, u_g, v_g), h, u_g and v_g can be advanced in time for a step δt by a standard numerical scheme using a discrete form of the evolution equations (5.12). This gives $h = h^*, u_g = u_g^*, v_g = v_g^*$. The result will not in general satisfy (5.11). This guess is therefore corrected by using a discrete form of (5.13) and the third equation of (5.12).

$$
\begin{aligned}
\mathbf{Q}\delta t \begin{pmatrix} \delta u \\ \delta v \end{pmatrix} + g\nabla\left(h^{t+\delta t} - h^*\right) &= 0, \\
h^{t+\delta t} - h^* + \delta t \nabla \cdot (h^* \delta \mathbf{u}) &= 0, \\
\mathbf{Q} = \begin{pmatrix} f^2 + f\frac{\partial v_g}{\partial x} & f\frac{\partial v_g}{\partial y} \\ -f\frac{\partial u_g}{\partial x} & f^2 - f\frac{\partial u_g}{\partial y} \end{pmatrix}&.
\end{aligned} \tag{5.15}
$$

As in the derivation of (5.14) in section 3.3.1, these equations can be converted into a single equation for $h^{t+\delta t}$ by eliminating δu and δv:

$$
(h^{t+\delta t} - h^*) - \nabla \cdot h^* \mathbf{Q}^{-1} g \nabla (h^{t+\delta t} - h^*) = 0. \tag{5.16}
$$

If $\mathbf{Q}$ is positive definite, (5.16) is elliptic and can be solved for $h^{t+\delta t}$. Then the first equation of (5.15) can be solved for $(\delta u, \delta v)$.

This procedure has to be iterated, because calculation of $\mathbf{Q}$ from the values obtained for $u_g(t + \delta t), v_g(t + \delta t)$ will not necessarily give a positive definite matrix, though (5.11) will be satisfied with values at $t + \delta t$. While this should work in a smooth case, it may not work in a non-smooth case. This is discussed in section 5.1.4.

If the matrix $\mathbf{Q}$ is derived in discrete form from the discretisation of (5.12), with the variables arranged on the grid as shown in Fig. 5.12, then the various terms in $\mathbf{Q}$ are naturally calculated at different points, and $\mathbf{Q}$ cannot be inverted. In Mawson (1996), an iterative method was therefore used where the kth iterate $h^{(k)}$ to $h^{t+\delta t}$ is defined by solving

$$\begin{aligned}(h^{(k)} - h^*) - \nabla \cdot h^* \mathbf{Q}_1^{-1} g \nabla (h^{(k)} - h^*) = \\ \nabla \cdot h^* (\mathbf{Q}^{-1} - \mathbf{Q}_1^{-1}) g \nabla (h^{(k-1)} - h^*).\end{aligned} \tag{5.17}$$

The matrix $\mathbf{Q}_1$ is the diagonal part of $\mathbf{Q}$, so that the positive definiteness of $\mathbf{Q}$ implies the positive definiteness of $\mathbf{Q}_1$. The term $f^2 + f\frac{\partial v_g}{\partial x}$ is computed at v_g points and the term $f^2 - f\frac{\partial u_g}{\partial y}$ is computed at u_g points. The corresponding entries in the inverse matrix $\mathbf{Q}_1^{-1}$ are therefore also held at v_g and u_g points respectively. The use of an iterative method allows the flux-limiting required in the outcropping problem of Cloke and Cullen (1994) to be incorporated, ensuring that h cannot become negative.

Though Theorem 3.35 shows that the positive definiteness of $\mathbf{Q}$ is maintained in the analytic solution, it may not be maintained by the discrete solution because of numerical errors, particularly close to the equator. This was resolved in Mawson (1996) by adding a selective smoothing term to the right hand side of (5.12), chosen to be large wherever det$\mathbf{Q}$ is small. This regularisation will result in a loss of energy and a less accurate solution. The issue is discussed further in section 5.1.4.

The discretisation as given will not work at the equator where $f = 0$ and $\mathbf{Q}$ cannot be inverted. h is required to be a constant h_e along the equator. Therefore first calculate the zonal mean solution $\overline{h}^{t+\delta t}$ from the integral of (5.16) round lines of latitude. In particular, this gives $h_e = h^{t+\delta t}$ at the equator. Then solve (5.16) for $h^{t+\delta t}$ separately in each hemisphere, with the boundary condition $h^{t+\delta t} = h_e$ at the equator. The first equation of (5.15) is then solved for δu and δv.

Now consider a discrete solution of the semi-geostrophic equations in a vertical slice. In Cullen (1989b), solutions were obtained to the frontogenesis problem described in section 3.4.2 and the flow over a mountain ridge described in section 5.5. In both cases, the solution contains discontinuities, and it is not clear whether an Eulerian finite difference method should be able to find the correct solution. This is discussed in section 5.1.4. The equations derived from (5.2) are

$$\begin{aligned}
-fv_g + \frac{\partial \varphi}{\partial x} &= 0,\\
\frac{\mathrm{D}v_g}{\mathrm{D}t} + fu + \frac{\partial \varphi_0}{\partial y} &= 0,\\
\frac{\mathrm{D}\theta}{\mathrm{D}t} + v_g\frac{\partial \vartheta}{\partial y} &= 0,\\
\frac{\partial \varphi}{\partial z} - g\frac{\theta}{\theta_0} &= 0,\\
\frac{\partial \varphi_0}{\partial z} - g\frac{\vartheta}{\theta_0} &= 0,\\
\nabla\cdot\mathbf{u} &= 0.
\end{aligned} \tag{5.18}$$

Here $\varphi_0(y,z)$ and $\vartheta(y)$ are given functions. $\mathbf{u} = (u,w)$ and $\mathbf{u}, v_g, \theta$ and φ are functions of (x,z) only. (5.18) is to be solved in a closed region Γ with $\mathbf{u}\cdot\mathbf{n} = 0$ on the boundary.

Equations (5.18) can be discretised using the arrangement of variables on the grid shown in Fig. 5.3. Semi-implicit time integration can be applied as in the shallow water case. Given u, w, v_g, θ and φ at time t, explicit estimates v_g^* and θ^* of the values $v_g^{t+\delta t}$ and $\theta^{t+\delta t}$ can be made using discrete approximations to the second and third equations of (5.18). These values will not satisfy the geostrophic and hydrostatic relations. Then calculate $\delta u, \delta w$ to enforce these relations by solving

$$f\frac{\partial}{\partial z}\left(\delta\mathbf{u}\cdot\nabla v_g + f\delta u\right) - \frac{g}{\theta_0}\frac{\partial}{\partial x}\left(\delta\mathbf{u}\cdot\nabla\theta\right) = f\frac{\partial v_g^*}{\partial z} - \frac{g}{\theta_0}\frac{\partial\theta^*}{\partial x}. \tag{5.19}$$

The left hand side of (5.19) can be written as $\mathbf{Q}\begin{pmatrix}\delta u\\ \delta w\end{pmatrix}$, where $\mathbf{Q}$ is a two-dimensional version of the matrix defined in (3.3). To solve (5.19), impose the constraint $\frac{\partial}{\partial x}(\delta u) + \frac{\partial}{\delta z}(\delta w) = 0$ by setting $\delta u = -\frac{\partial}{\partial z}(\delta\psi), \delta w = \frac{\partial}{\partial x}(\delta\psi)$. The boundary condition for (5.18) implies that $\delta\psi$ is constant on the boundary of Γ. Equation (5.19) then becomes a second order equation for $\delta\psi$, which is elliptic if $\mathbf{Q}$ is positive definite and can be solved.

In Cullen (1989b) this procedure was found satisfactory for continuous solutions, but when discontinuities occurred, the positive definiteness of $\mathbf{Q}$ was violated in the discrete solution. The remedy was to solve (5.19) iteratively be a similar method to (5.17), only retaining the diagonal terms of $\mathbf{Q}$ on the left hand side. A small amount of smoothing had to be added to the right hand side of (5.19) to control numerical errors, as in the shallow water case.

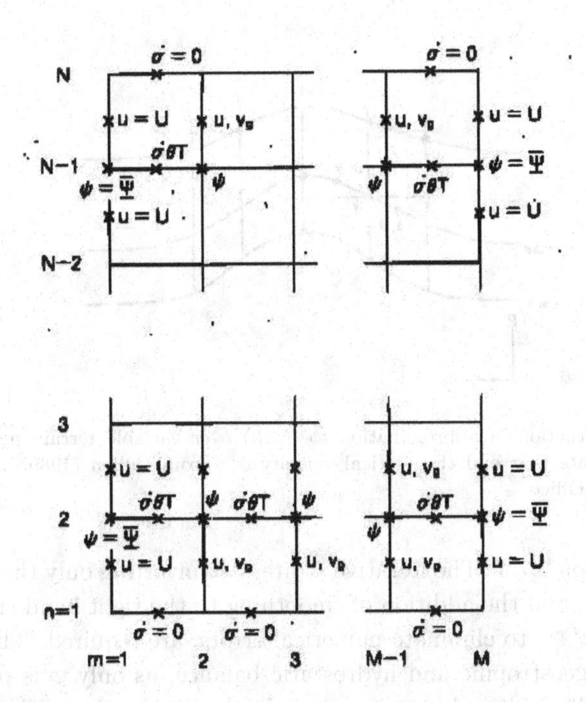

Fig. 5.3 Arrangement of variables on grid in (x, z) plane for the solution of (5.18). The variable $\dot{\sigma}$ in the diagram is equivalent to the variable w in (5.18). From Cullen (1989b). ©Crown Copyright, Met Office.

When the method is used to solve for the flow over a ridge as in section 5.5, terrain-following coordinates are used as shown in Fig. 5.14. Such coordinates are standard in weather forecasting models. The semi-implicit procedure above is followed, with the refinement that the derivative $\partial\delta\psi/\partial x$, that appears when (5.19) is written as an equation for $\delta\psi$, is calculated at constant z, rather than on the sloping coordinate surfaces. The equation for $\delta\psi$ therefore involves 7 points, as shown in Fig. 5.4, rather than the 5 that would be required when the coordinate surfaces were orthogonal.

There have been relatively few three-dimensional solutions using finite difference techniques in physical coordinates. Cullen and Mawson (1992) solved the equations by applying the procedure for the vertical slice in an alternating direction formulation, where the problem was successively solved in the (x, z) and (y, z) planes. An alternative is to solve (5.2) for $\varphi^{t+\delta t}$ directly using a formulation based on (3.4). This was used by Wakefield and Cullen (2005) and also in the diagnostic calculations of Cullen (2018) illus-

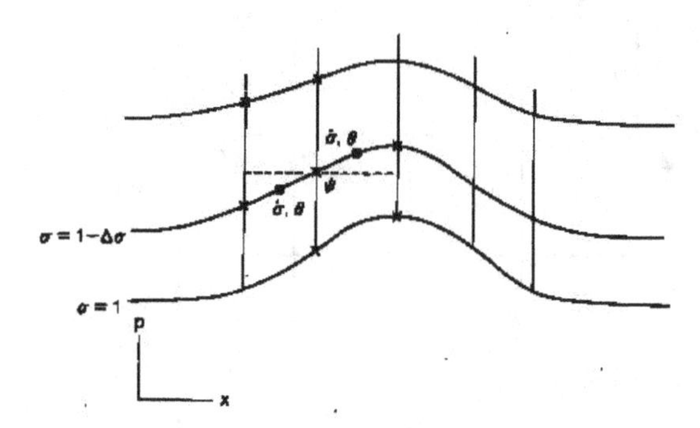

Fig. 5.4 Construction of approximation to (5.19) over variable terrain height. The vertical coordinate is σ and the vertical velocity $\dot{\sigma}$. From Cullen (1989b). ©Crown Copyright, Met Office.

trated in section 2.7.4. The iterative strategy of inverting only the diagonal elements of $\mathbf{Q}$, and the addition of smoothing to the right hand side of the equation for $\varphi^{t+\delta t}$ to eliminate numerical errors, are required. This procedure ensures geostrophic and hydrostatic balance, as only φ is predicted. However it will result in lower accuracy. In the diagnostic application this limits the information that can be obtained.

5.1.4 *Approximation of weak solutions given by Lagrangian conservation laws*

A recurring theme in sections 5.1.1 and 5.1.3 is the need to regularise numerical semi-geostrophic solutions because $\mathbf{Q}$ does not remain positive definite. However, the theory set out in section 3.5.4 ensures the long time existence of solutions to these problems which conserve energy. The issue is that finite difference schemes are justified by Taylor series methods, and will thus work for smooth solutions. There is no guarantee that they will work for weak solutions. In particular, they have no knowledge of the convexity condition. Only optimal transport methods such as those in section 5.1.2 ensure that this is satisfied.

The iteration (5.15)-(5.16) should converge because the proof of Theorem 4.11 shows that, given data with $\mathbf{Q}$ strictly positive definite, there will be a short finite time while it stays positive definite. However, once the solutions become non-smooth, this is not guaranteed. Though the conservation of potential vorticity, (3.70), ensures that det$\mathbf{Q}$ will remain positive, the convexity condition requires all the eigenvalues of $\mathbf{Q}$ to be non-negative. This may fail unless the algorithm is designed to ensure convexity, as in the optimal transport method of section 5.1.2. Thus finite difference methods may not be able to capture weak solutions of the semi-geostrophic equations. If finite difference methods are regularised in such a way that they lose energy, then they are unlikely to fail, because a non-convex solution implies an increase in energy.

A similar situation arises in the forced axisymmetric vortex problem described in section 4.4. Theorem 4.14 ensures the long time existence of a weak solution. However, in Smith *et al.* (2018), the failure of a numerical solution was quoted as evidence that no solution existed, so that a balanced vortex could not survive the type of axisymmetric forcing that was applied. This forcing did not satisfy Theorem 4.13, so a measure-valued solution is to be expected. The regularisation strategies used in that paper were different from those described in section 5.1.3. Had the latter been applied, it is likely that the solution could have been continued, but would have been dissipative.

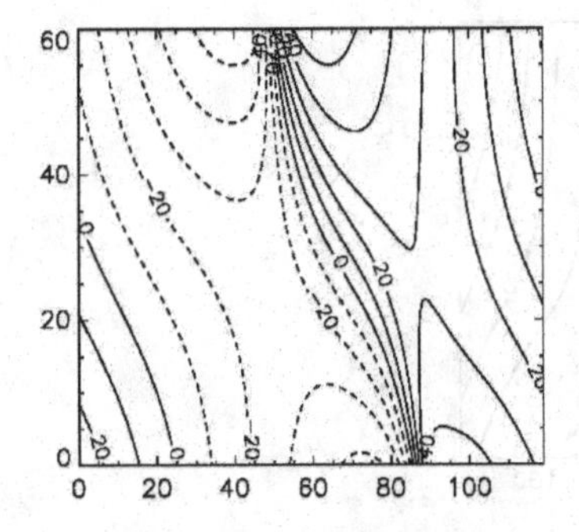

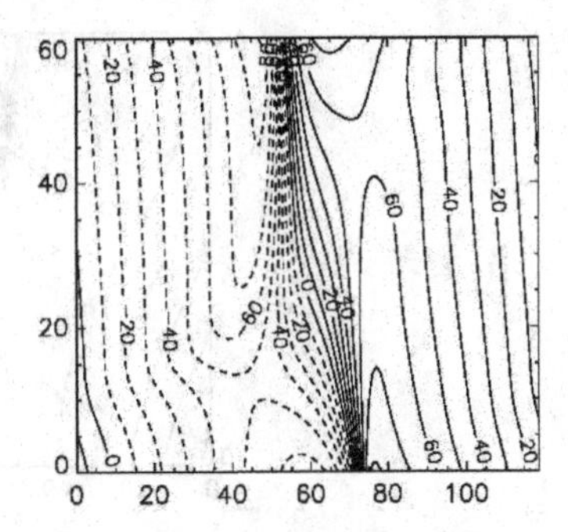

Fig. 5.5 Plots of the v-component of wind from the semi-geostrophic solution. The vertical and horizontal axes are in units of grid-point number; the total dimension of the plot is 2000 km$\times$10 km. Contour interval is 10 ms^{-1}; (a) day 7, (b) day 9. From Cullen (2008b). ©Crown Copyright, Met Office.

In Cullen (2008b), this issue was studied in a compressible version of the Eady problem which will be discussed further in section 5.3. The full compressible Navier-Stokes equations (4.1) were solved in a vertical cross-section, using the integration scheme from the Met Office UM at the time, Davies *et al.* (2005). These are compared with a solution of a fully compressible semi-geostrophic model, (4.4), obtained by making the geostrophic and hydrostatic approximations within the UM formulation. The resulting equations were solved using the formulation (4.7) with regularisation of the **Q** matrix. Both models were solved for data similar to that used in section 5.3. The semi-geostrophic solution after 7 and 9 days is illustrated in Fig. 5.5. Areas of strong gradients develop at both boundaries.

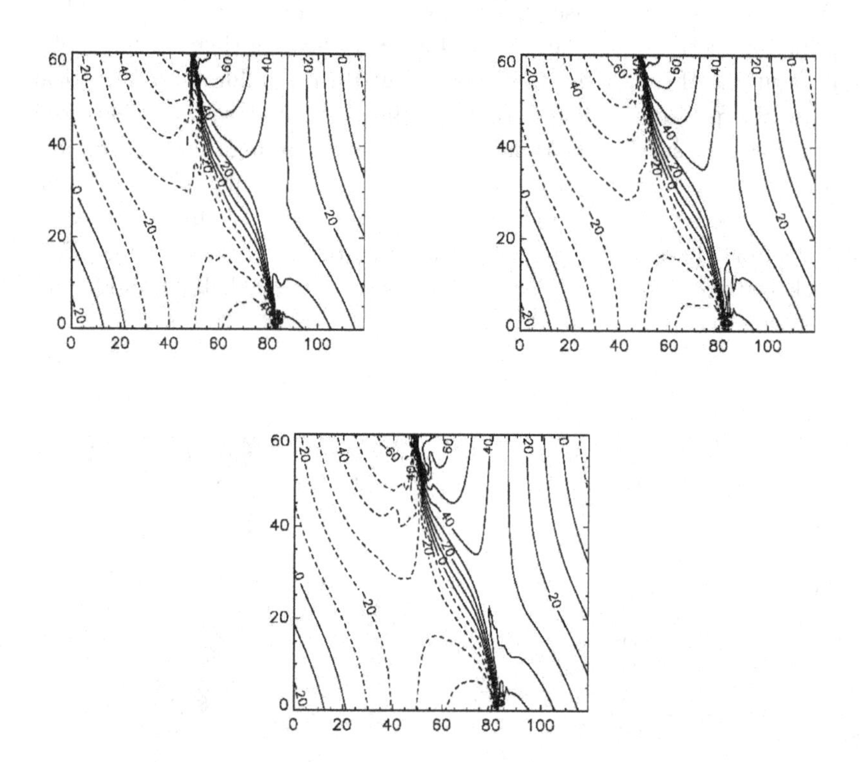

Fig. 5.6 Plots of the UM solution at day 7 with quasi-monotone interpolation in the semi-Lagrangian advection. Notation as in Figure 5.5. (a) Ro = 0.125, (b) Ro = 0.062, (c) Ro = 0.031. From Cullen (2008b). ©Crown Copyright, Met Office.

The results from the UM are shown in Fig. 5.6. They are plotted for different Rossby numbers and the expectation is that they will converge to the semi-geostrophic solution as the Rossby number tends to zero. Convergence is seen from the results, but these contain much sharper discontinuities penetrating into the fluid. It is likely that the semi-geostrophic solution is discontinuous in this case, but a direct computation of the semi-geostrophic solution by finite difference methods is less accurate than a solution of the Euler equations with a small Rossby number. This probably reflects the fact that the advection terms in the Lagrangian time derivatives are absorbed into the **Q** matrix. Regularisation of this matrix may then seriously degrade the accuracy of the advection. When solving the Euler equations, the advection is treated explicitly, and geostrophic and hydrostatic balance are approximately enforced by semi-implicit time integration. While the latter also involves regularisation, the accuracy of the advection is not lost. Fig. 5.19 in section 5.3.4 shows that the Euler solutions are converging to a solution satisfying geostrophic and hydrostatic balance, which may be a more accurate solution of the semi-geostrophic equations than that shown in Fig. 5.5.

5.2 Shallow water solutions

5.2.1 *Nature of solutions*

In section 2.3, the dimensionless shallow water model was used to exemplify the importance of different asympotic limits in determining the nature of 'slow' solutions of the equations. In section 2.4 it was shown how much of this carries over to the three-dimensional case. In section 2.6 the existence of the expected asymptotic behaviour in real atmospheric data was demonstrated. Sections 2.3.4, 2.3.5 and 2.3.6 considered the cases $\varepsilon = Fr, \varepsilon = Ro = Fr$, and $\varepsilon = Ro = Fr^2$ respectively. The different behaviours were explained in terms of the relation of the length scale L of the flow to the radius of deformation L_R noting that $L_R/L = Ro/Fr$.

Take $L = 1$ throughout. Then in the case $\varepsilon = Fr$, $Ro = O(1)$, $L_R \to \infty$ as $\varepsilon \to 0$. Section 2.3.4 shows that the limiting solutions describe two-dimensional incompressible flow. The behaviour of the solutions to these equations have been widely studied and an overview is given by Vallis (2017), p.423 *et seq.*. In general, the energy cascades to the largest available scale, while the vorticity cascades to the smallest scale. However, with a variable Coriolis parameter included in the absolute vorticity, the asymmetry in the absolute vorticity means that the solution converges to a zonal flow, Wirosoetismo (2015). The north-south scale is given by the *Rhines* scale, which is derived from the theory of two-dimensional turbulence, see Vallis (2017), p.446.

In the case $\varepsilon = Ro = Fr$, $L_R = 1$ and so is independent of ε, so there is no change of 'slow' behaviour as $\varepsilon \to 0$. This is described by quasi-geostrophic theory as in section 2.3.5. Inertia-gravity waves are increasingly suppressed in the limit.

In the case $\varepsilon = Ro = Fr^2$, $L_R = \varepsilon^{\frac{1}{2}}$, and so tends to zero as $\varepsilon \to 0$. The limiting behaviour with constant Coriolis parameter is controlled by the geostrophic degeneracy discussed in section 2.3.6 which means that the nonlinear evolution of the coherent vortices will be small. The vortices have to satisfy the inertial stability condition that the potential (3.91) is convex. This is a constraint on the depth field, which means that it cannot oscillate as the evolution proceeds, hence preserving coherence of the vortices. With variable Coriolis parameter, the nonlinear terms in the evolution of h are still small because of the geostrophic degeneracy, and positive definiteness of the matrix $\mathbf{Q}$, (3.74), prevents oscillatory behaviour of h. The vortices will, however, propagate with the 'slow' Rossby wave speed (2.81) derived in section 2.3.5. This gives non-dispersive propagation for $L \gg L_R$, so that the vortices will remain coherent if they are large enough.

5.2.2 *Examples of the asymptotic behaviour*

The shallow water equations (2.33) are written in dimensional form in spherical polar coordinates (λ, ϕ) on a sphere of radius a with rotation rate Ω. The Coriolis parameter $f = 2\Omega \sin\phi$. The velocity $\mathbf{u} = (u, v)$ and the depth is h. The equations are

$$\begin{aligned}
\frac{\mathrm{D}u}{\mathrm{D}t} + \frac{g}{a\cos\phi}\frac{\partial h}{\partial \lambda} - fv &= 0, \\
\frac{\mathrm{D}v}{\mathrm{D}t} + \frac{g}{a}\frac{\partial h}{\partial \phi} + fu &= 0, \\
\frac{\partial h}{\partial t} + \nabla\cdot(h\mathbf{u}) &= 0.
\end{aligned} \tag{5.20}$$

The first example is taken from Hayashi *et al.* (2007). An initial random distribution of vorticity is used as shown in Fig. 5.7(f). This has a much smaller scale than typical atmospheric anomalies, but is closer to that of ocean eddies. The depth and divergence fields are set to zero. The initial Rossby adjustment will thus decrease the amplitude of the relative vorticity and create a depth field in balance with it. On scales larger than L_R the resulting vorticity will be much smaller because, as shown in section 2.3.2, the balanced state is controlled by the depth field. The rotation rate Ω is chosen so that the Rossby number Ro based on the Earth's radius and a wind speed of 1ms^{-1} is 0.00125. These choices give a Rhines scale of about 250km. The radius of deformation at the pole is aRo/Fr where the

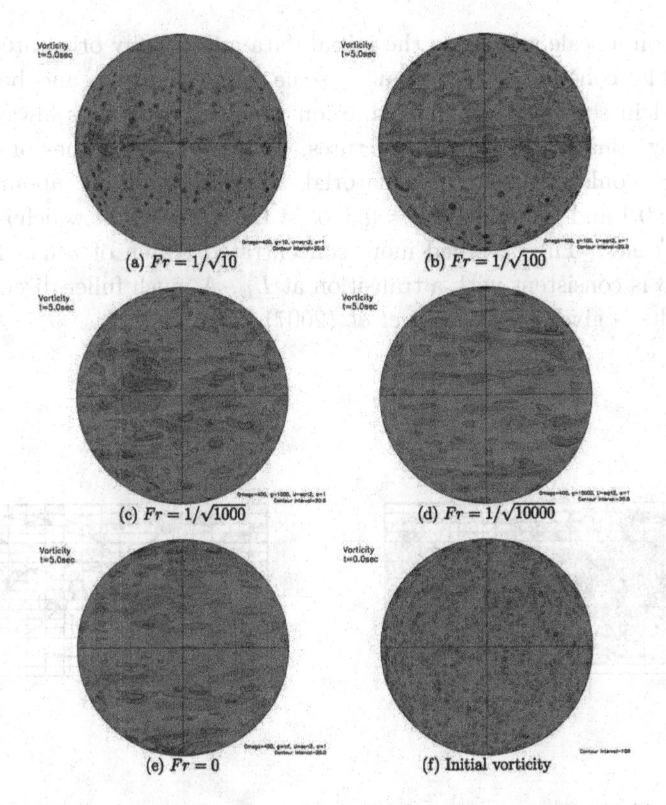

Fig. 5.7 Vorticity at $t = 5$ for various values of Fr and $\Omega = 400$, and in (f) the initial vorticity. Contour intervals are 20 in (a) to (e) and 100 in (f). The zero contour is suppressed. From Hayashi *et al.* (2007). ©American Meteorological Society. Used with permission.

Froude number Fr takes the values shown in Fig. 5.7. These are fixed by different choices of the mean depth h_0. The case $Fr = 0.1$ shown in Fig. 5.7(b) corresponds to L_R =80km at the pole, less than a typical atmospheric value but somewhat larger than a typical ocean value. The equivalent equatorial deformation radius L_E is about 700km.

For the stratification dominated case $Fr = 0$ shown in Fig. 5.1(e), the initial random vortices coalesce into zonally orientated coherent structures. The north-south scale is similar to the Rhines scale, about 250km. The solutions for $Fr < 0.1$ are similar. For instance, if $Fr = 0.03$, L_R is larger than the Rhines scale for latitudes less than 50°. The solution over most of the sphere is then controlled by vortex dynamics.

In the remaining experiments, there is a transition between coherent vortices on a scale similar to the initial data and zonally orientated structures. The coherent vortices have a scale larger than L_R and behave as discussed in section 5.2.1. The solution near the equator is always dominated by zonally orientated structures. For the larger values of Fr, the transition could occur at the equatorial deformation radius, about 700km for $Fr = 0.1$ and 400km for $Fr = 0.3$, or at the Rhines scale which is 250km in both cases. The observed more concentrated region of zonal flows for $Fr = 0.3$ is consistent with a transition at L_E. A much fuller discussion of the results is given by Hayashi *et al.* (2007).

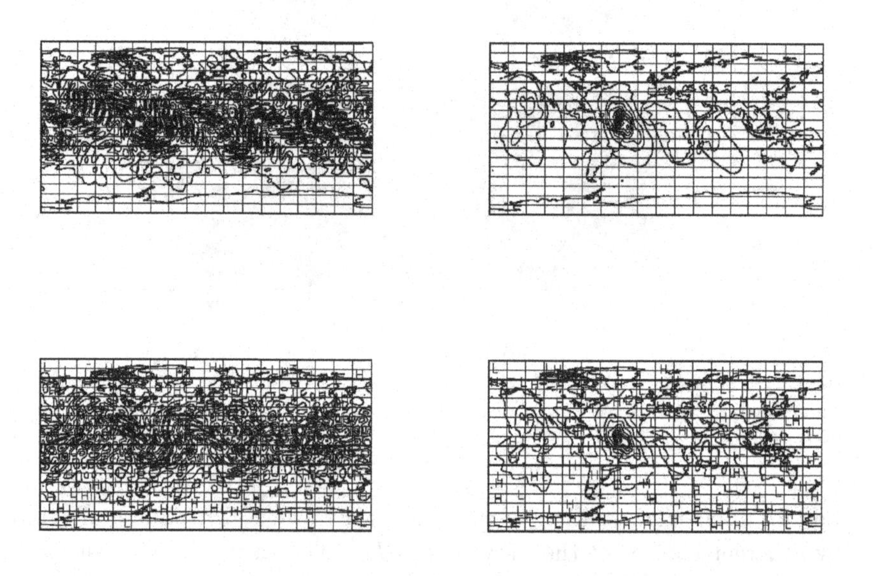

Fig. 5.8 Left: 2 day forecast of potential vorticity, units $(\mathrm{ms})^{-1}$, contour interval 0.3×10^{-9}, from initial data shown in Fig. 2.1 with $gh_0 = 10^5 m^2 s^{-2}$. Top: shallow water solution. Bottom: semi-geostrophic shallow water solution. Right: 2 day forecast of potential vorticity, units $(\mathrm{ms})^{-1}$, contour interval 0.8×10^{-7}, from initial data shown in Fig. 2.2 with $gh_0 = 5000 m^2 s^{-2}$. Top: shallow water solution. Bottom: semi-geostrophic shallow water solution. From Cullen (2002). ©Royal Meteorological Society, Reading, U.K., 2002.

The second example uses (5.20) with the Coriolis parameter f set equal to $1.458 \times 10^{-4}\mathrm{s}^{-1}$ everywhere. This allows the dependence of the solution on ε to be demonstrated uniformly over the sphere. The initial depth field is shown in Fig. 2.1 and the velocities are calculated from it using (5.13), which is the condition that geostrophic balance is maintained in time. The results are taken from Cullen (2002).

Two parameter regimes are demonstrated by making different choices of the mean depth h_0. The first has $gh_0 = 10^5\text{m}^2\text{s}^{-2}$, giving a radius of deformation $L_R = 2148$km. The second has $gh_0 = 5000\text{m}^2\text{s}^{-2}$, giving $L_R = 480$km. The initial potential vorticities (PV), (2.35), implied by the depth field shown in Fig. 2.1 are shown in Figs. 2.1 and 2.2. They are very different in the two regimes. As noted in sections 2.3.4 and 2.3.6 the PV is similar to the relative vorticity in the first regime, and is thus similar to the second derivative of the depth field. The features are much smaller scale than L_R. In the second regime the PV is similar to the depth field itself and is on a larger scale than L_R.

It is not surprising that the subsequent evolution, which is essentially generated by advection of potential vorticity, is completely different in the two cases. This is shown in the top row of Fig. 5.8. In the first case, the dynamics is essentially two-dimensional vortex dynamics, and the potential vorticity features collapse to an even smaller scale than that in the initial data. In the second case, the dynamics is essentially semi-geostrophic. Since the Coriolis parameter is constant in this experiment, the solution hardly varies in time at all, as discussed in section 5.2.1. Results are also shown from a semi-geostrophic shallow water model. These are accurate for the second case, as would be expected. In the first case they are completely unlike the correct solution, though there has been some degree of scale collapse from the initial data.

The third example follows Wheadon (2018), which should be consulted for greater detail. As in the second example, equations (5.20) are integrated on a sphere with a constant Coriolis parameter f so that the dependence of the solution on the chosen ε is uniform in space. The test problem is based on the initial data of Galewsky *et al.* (2004). This has a zonal jet of maximum speed $u_0 = 80\text{ms}^{-1}$, centred at 45°N and extending over a total of 20° latitude, which is in geostrophic balance with the depth field. A localised perturbation extending over 60° longitude is added to the depth field over a period of time to promote instability of the jet without exciting gravity waves.

In order to study the asymptotic behaviour, the Coriolis parameter f and mean fluid depth h_0 are varied. In the initial reference solution, $Ro = Fr = 1, h_0 = u_0^2/g, f = 9u_0/a\pi$. This gives approximately h_0 =640m and $f = 0.365 \times 10^{-4}\text{s}^{-1}$. Then the quasi-geostrophic limit $\varepsilon = Ro = Fr \to 0$

is obtained by increasing f and $\sqrt{h_0}$ in the same proportion. The semi-geostrophic limit $\varepsilon = Ro = Fr^2 \to 0$ is obtained by increasing f and h_0 in the same proportion. In both cases the perturbations to h_0 needed to be in balance with the jet and to provide a consistent local perturbation are increased proportionally to f.

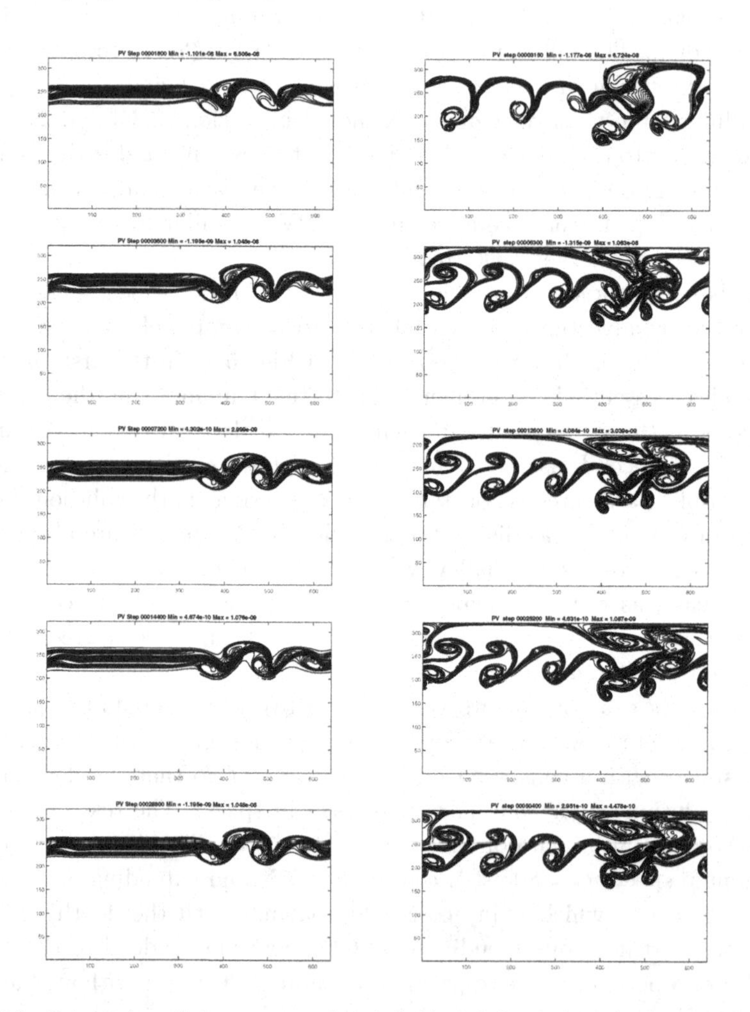

Fig. 5.9 Potential vorticity maps at day 4 (left) and day 7 (right) approaching the quasi-geostrophic limit. The values of ε from top to bottom are 1, 0.5, 0.25, 0.125 and 0.0625. From Wheadon (2018). ©Andrew Wheadon.

Fig. 5.9 shows the evolution of the potential vorticity (2.35) as the quasi-geostrophic limit is approached. Even at the later time (day 7), the solution has essentially converged for $\varepsilon = 0.25$. This shows that with this choice of scaling, the solution is independent of ε for ε sufficiently small. As discussed in secton 5.2.1, this results from the invariance of the deformation radius with ε. The solution demonstrates continued growth of instability in the potential vorticity, which is consistent with the cascade of potential enstrophy to small scales as expected in geostrophic turbulence as discussed in section 2.6.1.

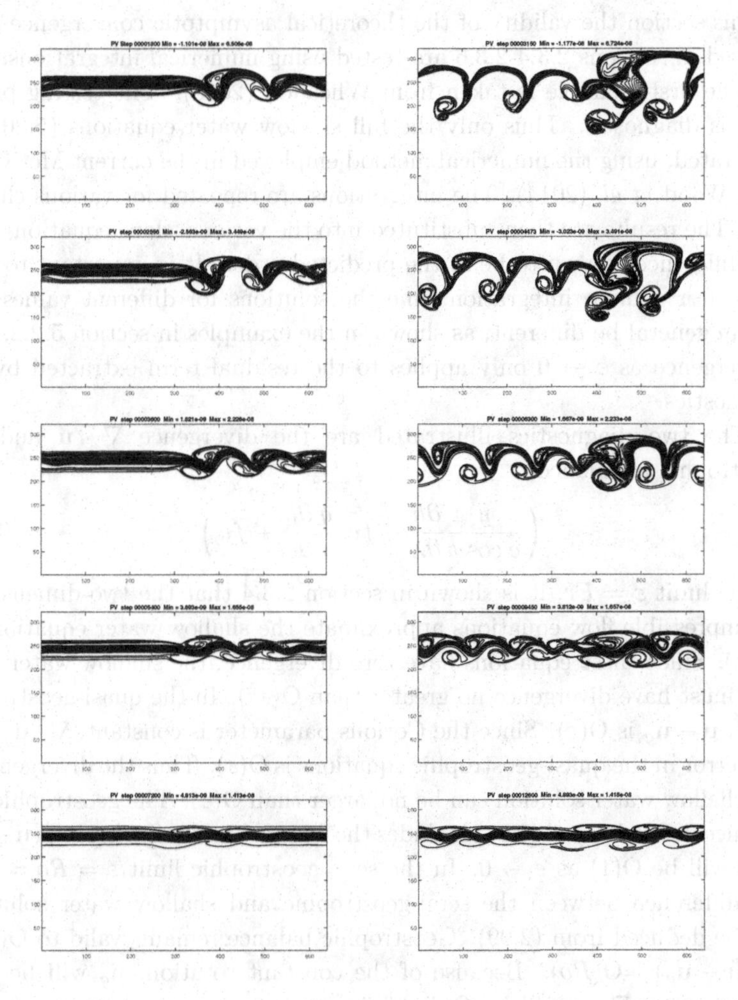

Fig. 5.10 As Fig. 5.9 approaching the semi-geostrophic limit. From Wheadon (2018). ©Andrew Wheadon.

Fig. 5.10 illustrates the solutions in the semi-geostrophic limit. In this case the qualitative nature of the solution changes as ϵ is reduced. This is because, as shown in section 5.2.1, the deformation radius is proportional to $\varepsilon^{\frac{1}{2}}$. Thus for small ε the solution becomes more and more like the initial data, and the instabilities are suppressed.

5.2.3 *Numerical demonstration of the asymptotic convergence*

In this section the validity of the theoretical asymptotic convergence rates derived in sections 2.3.4-2.3.6 are tested using numerical integrations.

The first example is taken from Wheadon (2018). The testing procedure is diagnostic. Thus only the full shallow water equations (5.20) are integrated, using the numerical method employed in the current Met Office UM, Wood *et al.* (2014). The integrations are repeated for various choices of ε. The results are then substituted into the various 'slow' equations, and the differences shown to be of the predicted order. It is important to note that after a finite integration time the solutions for different values of ε will in general be different, as shown in the examples in section 5.2.2. The convergence as $\varepsilon \to 0$ only applies to the residual term extracted by the diagnostics.

The two diagnostics illustrated are the divergence $\nabla \cdot \mathbf{u}$ and the geostrophic imbalance

$$\left(\frac{g}{a\cos\phi}\frac{\partial h}{\partial \lambda} - fv, \frac{g}{a}\frac{\partial h}{\partial \phi} + fu\right). \tag{5.21}$$

In the limit $\varepsilon = Fr$, it is shown in section 2.3.4 that the two-dimensional incompressible flow equations approximate the shallow water equations to $O(\varepsilon^2)$. Since these equations have zero divergence, the shallow water solution must have divergence no greater than $O(\varepsilon^2)$. In the quasi-geostrophic limit, $\mathbf{u} - \mathbf{u}_g$ is $O(\varepsilon)$. Since the Coriolis parameter is constant, $\nabla \cdot \mathbf{u}_g = 0$. The error in the quasi-geostrophic equations is $O(\varepsilon)$. Thus the divergence in the shallow water solution can be no larger than $O(\varepsilon)$. The geostrophic imbalance defined in eq. (5.21) includes the factor f, so is effectively $f(\mathbf{u}-\mathbf{u}_g)$. This will be $O(1)$ as $\varepsilon \to 0$. In the semi-geostrophic limit $\varepsilon = Ro = Fr^2$, the difference between the semi-geostrophic and shallow water solutions can be deduced from (2.99). Geostrophic balance remains valid to $O(Ro)$, so $(\mathbf{u} - \mathbf{u}_g) = O(Ro)$. Because of the constant rotation, $\mathbf{u}_g$ will be non-divergent, so $\nabla \cdot \mathbf{u}$ will be $O(Ro)$. As in the quasi-geostrophic case, the geostrophic imbalance (5.21) will be $O(1)$.

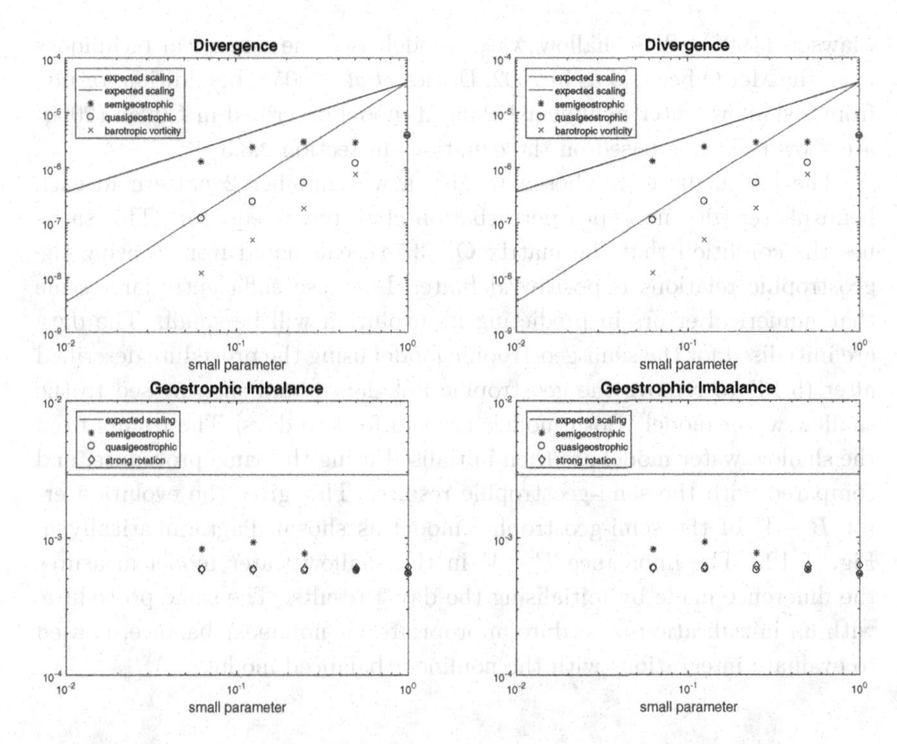

Fig. 5.11 Convergence diagrams for divergence and geostrophic imbalance (5.21) in the jet region. Left: small timestep; right: timestep 10 times larger. From Wheadon (2018). ©Andrew Wheadon.

Fig. 5.11 illustrates the behaviour of the divergence in the three limits. The integrations use two different timesteps. In the left hand column, the timestep is chosen to give accurate solutions of (5.20). In the right hand column a larger value is used which can solve potential vorticity advection accurately, but treats the inertia-gravity waves implicitly and thus eliminates high frequencies. Eq. (2.99) shows that the difference between the semi-geostrophic and shallow water solutions is a second time derivative and may thus be affected by using a long timestep. However, with both choices of timestep, all the differences are as expected for sufficiently small ε. The figure also illustrates the behaviour of the geostrophic imbalance. This is again as expected for the quasi-geostrophic and semi-geostrophic limits. The behaviour for the 'strong rotation' limit $\varepsilon = Ro, Fr = 1$ is also illustrated. This is almost the same as the quasi-geostrophic case.

The second example compares results from a shallow water model and a shallow water semi-geostrophic model. Both models are described in

Mawson (1996). The shallow water model uses the numerical techniques from the Met Office UM as in 2002, Davies *et al.* (2005). In addition, results from a shallow water nonlinear balanced model described in Cullen (2000) are shown. This is based on the equations in section 2.3.3.

The initial data are chosen to give a wavenumber 2 pattern in each hemisphere, with no depth perturbation close to the equator. This satisfies the condition that the matrix $\mathbf{Q}$, (3.74), calculated from h using the geostrophic relations is positive definite. It is also sufficiently large-scale that numerical errors in predicting its evolution will be small. The data are initialised for the semi-geostrophic model using the procedure described after (5.16) to remove the geostrophic imbalance, and then passed to the shallow water model. Both models are run for two days. The results from the shallow water model are then initialised using the same procedure, and compared with the semi-geostrophic results. This gives the evolution error $B - V$ of the semi-geostrophic model as shown diagrammatically in Fig. 5.12. The imbalance $P - V$ in the shallow water model measures the difference made by initialising the day 2 results. The same procedure, with an initialisation procedure appropriate for nonlinear balance, is used to evaluate integrations with the nonlinear balanced model.

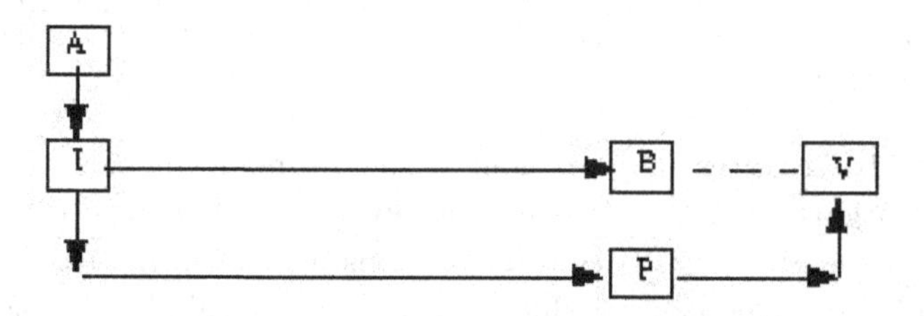

Fig. 5.12 Experimental set-up: A-analytic initial data, I-data initialised for balanced model, B-forecast using balanced model, P-forecast using shallow water model, V-initialised end state from shallow water model. From Cullen (2000). ©Royal Meteorological Society, Reading, U.K., 2000.

The base resolution for the experiments was a latitude-longitude grid with 96 points around latitude circles and 65 points between the poles. The results were also generated using a higher resolution of 192×129 points. Though the general behaviour as the parameters were varied was the same for both resolutions, in the semi-geostrophic tests there were significant differences in individual results. The complete semi-geostrophic experiment was thus also run at the higher resolution, and a further check carried out using a 288×193 grid for one set of parameters. The nonlinear balance model is more compatible with the shallow water model. Sample runs with the 192×129 grid showed that it was unnecessary to repeat the whole experiment at higher resolution.

Table 5.1 Parameters used.

ϕ_0 m	$Ro = U/fL$	$Fr = U/\sqrt(gh_0)$
5760	0.04	0.04
2880	0.04	0.06
1440	0.04	0.08
720	0.04	0.11
360	0.04	0.16
180	0.04	0.22

The experiments were designed to test the effect of varying the Burger number Ro/Fr for fixed Ro. Since the Froude number for shallow water flow is $U/\sqrt{gh_0}$, this was achieved by using the same perturbation depth field for all runs, but varying the mean depth h_0 from 5760m down to 182.5m. The amplitude of the superposed wave was ±170m, so that the lowest mean value used is just sufficient to avoid the depth becoming zero. The horizontal velocity had a maximum value of about 10ms^{-1}. The gravity wave speed implied by the choice of h_0 varied from 240ms^{-1} to 42 ms^{-1}. Table 5.1 lists the values used, together with typical Froude and Rossby numbers.

The results for the depth evolution errors are shown in Fig. 5.7. The imbalance calculated from the depth fields is shown in Fig. 5.8. Similar results for the wind fields are shown in Cullen (2000). The partitioning of the error in the semi-geostrophic model could not be carried out at later times in high resolution runs because the shallow water model developed large depth gradients at the equator which could not be handled by the initialisation procedure. Results for individual cases run at higher resolution were also plotted by Cullen (2000) to validate the results.

The expected proportionality of the evolution error of the semi-geostrophic model to $(Ro/Fr)^2$ is clearly demonstrated. The nonlinear balanced model shows similarly small errors for $Ro \ll Fr$. The evolution errors for the nonlinear balanced model for larger values of mean depth are consistent with the expected accuracy of $O(\varepsilon^2)$, where ε is defined by

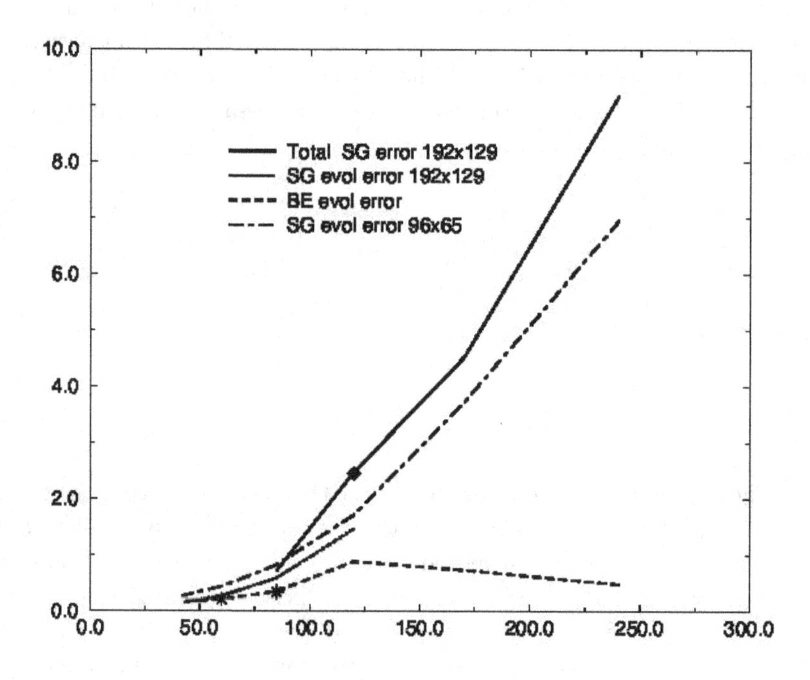

Fig. 5.13 Root-mean-square depth-evolution errors (m) after 48 hours plotted against gravity-wave speed (ms^{-1}). Stars indicate nonlinear balance-equation results on a 192×129 grid, the diamond a semi-geostrophic result on a 288×193 grid. From Cullen (2000). ©Royal Meteorological Society, Reading, U.K., 2000.

(2.55), and the results of Allen *et al.* (1990). Solvability issues did not arise because of the choice of very smooth initial data and the inherent regularisation in the numerical algorithm. The two models have comparable errors once the gravity wave speed is of the order of $50ms^{-1}$ implying a radius of deformation $L_R \simeq$500km in middle latitudes. This is smaller than a typical atmospheric value.

The measures of imbalance also show the expected behaviour. In the semi-geostrophic case, the $O((Ro/Fr)^2)$ dependence is clearly seen at high resolution, though the low resolution results are less reliable in this measure. In the nonlinear balance equation results, the rate of generation of imbalance clearly reduces for small Ro/Fr. The rate of growth of imbalance from balanced initial data, and the errors in the potential vorticity

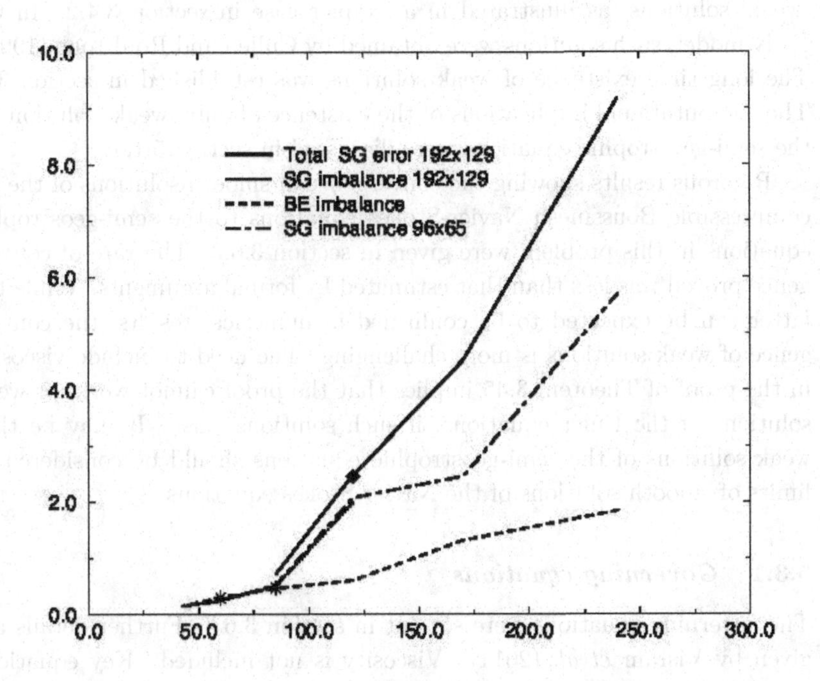

Fig. 5.14 Root-mean-square depth imbalances (m) after 48 hours plotted against gravity-wave speed (ms^{-1}). Notation as Fig. 5.13. From Cullen (2000). ©Royal Meteorological Society, Reading, U.K., 2000.

evolution are of comparable size for all cases tested.

5.3 The Eady wave

In this section, solutions to a vertical slice version of the incompressible Boussinesq Euler equations are studied. This is a classical model for the formation of atmospheric fronts. The background is described by Vallis (2017), p.351 *et seq.* The finite difference results given here are taken from Visram *et al.* (2014). Further work on this problem using different Eulerian numerical methods is described by Yamazaki *et al.* (2016). Results using Lagrangian numerical methods, as discussed in section 5.1.2, are taken from Cullen and Roulstone (1993), Cullen (2007a) and Egan *et al.* (2021).

As expected from a model of frontogenesis, the vertical slice problem described in this section develops discontinuities in finite time. These solutions can be continued under the semi-geostrophic approximation as 'weak' solutions, as illustrated in a simpler case in section 3.4.2. In the Eady model, such solutions were obtained by Cullen and Roulstone (1993). The long time existence of weak solutions was established in section 3.5. The computational implications of the existence of only 'weak' solutions of the semi-geostrophic equations were discussed in section 5.1.4.

Rigorous results showing the convergence of smooth solutions of the incompressible Boussinesq Navier-Stokes equations to the semi-geostrophic equations in this problem were given in section 3.6.3. The rate of convergence proved was less than that estimated by formal arguments. While the latter can be expected to be confirmed in numerical results, the convergence of weak solutions is more challenging. The need to include viscosity in the proof of Theorem 3.45 implies that the proof cannot work for weak solutions of the Euler equations, if such solutions exist. It may be that weak solutions of the semi-geostrophic equations should be considered as limits of smooth solutions of the Navier-Stokes equations.

5.3.1 *Governing equations*

The governing equations were set out in section 3.6.3. Further details are given by Visram *et al.* (2014). Viscosity is not included. Key equations are repeated here for convenience. The potential temperature Θ and zonal wind U are decomposed as

$$\begin{aligned} \Theta(x,y,z,t) &= \theta_0 + \bar{\theta}(y) + \theta(x,z,t), \\ U(x,y,z,t) &= \bar{u}(z) + u(x,z,t), \\ \bar{u}(z) &= -\frac{g}{f\theta_0}\frac{\partial\bar{\theta}}{\partial y}\left(z - \frac{H}{2}\right). \end{aligned} \tag{5.22}$$

The incompressible non-hydrostatic Boussinesq Euler equations are then

$$\begin{aligned} \frac{\mathrm{D}u}{\mathrm{D}t} - fv + \frac{\partial\varphi}{\partial x} &= 0, \\ \frac{\mathrm{D}v}{\mathrm{D}t} + fu + \frac{g}{f\theta_0}\frac{\partial\bar{\theta}}{\partial y}\left(z - \frac{H}{2}\right) &= 0, \\ \frac{\mathrm{D}w}{\mathrm{D}t} - \frac{g\theta}{\theta_0} + \frac{\partial\varphi}{\partial z} &= 0, \\ \frac{\mathrm{D}\theta}{\mathrm{D}t} + v\frac{\partial\bar{\theta}}{\partial y} &= 0, \\ \frac{\partial u}{\partial x} + \frac{\partial w}{\partial z} &= 0. \end{aligned} \tag{5.23}$$

The semi-geostrophic and hydrostatic approximation to (5.23) is given by (3.206). It can be rewritten by using (3.27), thus setting $X = f^{-1}v + x$, $Z = \frac{g\Theta}{f^2\theta_0}$. This gives (3.209), which is repeated below:

$$
\begin{aligned}
-fv + \frac{\partial \varphi}{\partial x} &= 0, \\
\frac{\mathrm{D}X}{\mathrm{D}t} + \frac{g}{f\theta_0}\frac{\partial \bar{\theta}}{\partial y}\left(z - \frac{H}{2}\right) &= 0, \\
\frac{\mathrm{D}Z}{\mathrm{D}t} + v\frac{g}{f^2\theta_0}\frac{\partial \bar{\theta}}{\partial y} &= 0, \\
\frac{\partial u}{\partial x} + \frac{\partial w}{\partial z} &= 0. \\
P &= \frac{1}{2}x^2 + f^{-2}\Phi, \\
\nabla P &= (X, Z).
\end{aligned}
\tag{5.24}
$$

The solutions of eqs. (5.24) are invariant under the rescaling $x \to \beta x_0, u \to \beta u_0, f \to f_0/\beta$. If the Rossby number Ro for eqn. (3.206) is defined as U/fL, where U is a typical velocity in the $x-$direction and L is a length-scale in the $x-$direction, then the rescaling replaces Ro by βRo. Thus taking the limit as $Ro \to 0$ can be replaced by taking the limit as $\beta \to 0$. Applying this rescaling to the terms omitted in the semi-geostrophic approximation, the Lagrangian Rossby number, which is the ratio of $\mathrm{D}u/\mathrm{D}t$ to fv, is decreased by β^2. Thus the semi-geostrophic system (5.24) approximates a hydrostatic version of (5.23) to $\mathrm{O}(Ro^2)$. It is possible to extend this argument to the non-hydrostatic case.

5.3.2 *Numerical experiments*

Solutions of (5.23) for initial data of the form (5.22) are illustrated. The data used are taken from Nakamura (1994), so that L = 1000km, H = 10km, $N_0^2 = 2.5 \times 10^{-5}\mathrm{s}^{-2}$, $f = 10^{-4}\mathrm{s}^{-1}$, $g = 10\mathrm{ms}^{-1}$, $\theta_0 = 300\mathrm{K}$, $\frac{\partial \bar{\theta}}{\partial y} = -3 \times 10^{-6}\mathrm{m}^{-1}\mathrm{K}$. A perturbation corresponding to an unstable mode of the linearised equations (5.23) is added, as described in Visram *et al.* (2014). As discussed in Gill (1982), p. 556, if the isentropes have negative slope $\mathrm{d}x/\mathrm{d}z$, then v_g will increase with z, and the evolution equation for θ will increase the vertical gradient of θ, giving a positive feedback. It represents conversion of potential energy from the infinite reservoir implied by the imposed basic state $\partial\bar{\theta}/\partial y$ into kinetic energy.

The construction of the solutions in the semi-geostrophic limit is discussed in section 5.1.2 and demonstrated by Egan *et al.* (2021). In that limit the solution is described by the Lagrangian evolution equations for (X, Z) in (5.24). The solution can then be constructed in physical space by requiring that $\nabla P = (X, Z)$ with P convex together with mass conservation. It is desirable to reflect this behaviour in the numerical method used to solve (5.23). This is described in Visram *et al.* (2014). It is similar to that used in the current Met Office UM, Wood *et al.* (2014). Thus, as in the compressible version of the problem solved in section 5.1.4, semi-Lagrangian methods which respect the property that advection cannot create new values are used, and a semi-implicit method with slight decentering is used to advance the solution in time. This will ensure that the condition $\nabla P = (X, Z)$, which characterises the 'slow' solution, is adequately represented. No explicit smoothing is applied. However, the semi-Lagrangian advection uses limiters to enforce the Lagrangian advection constraint, and these introduce some smoothing where steep gradients are simulated, such as at the fronts in the Eady model solution.

The standard resolution in the results shown is 121×61 and the high resolution is 481×241. In the Lagrangian integration of Egan *et al.* (2021), the data are represented by 1556 Dirac masses .

5.3.3 *Nature of the solution*

Fig. 5.15 shows the time evolution of the r.m.s. meridional velocity v for various values of the rescaling parameter β. The value $\beta = 1$ corresponds to the data defined in section 5.3.2. The results of Nakamura and Held (1989) were obtained by a different finite-difference model with artificial smoothing included. The results of Egan *et al.* (2021) were obtained with a fully Lagrangian model as described in section 5.1.2. This model does not require any smoothing and obeys the condition that advection cannot create new values.

Initially there is normal mode growth of the initial perturbation. After 7 days the growth saturates in all the integrations with $\beta \leq 1$ and also in the results of Nakamura and Held (1989) and Egan *et al.* (2021). After the growth saturates, the vertical shear in the basic state zonal velocity $\bar{u}$, eq. (5.22) will tilt the isentropes so that they have a positive slope, and the kinetic energy of the perturbation will be converted back into potential energy. This decrease is shown in the integrations between days 7 and 11. Continued advection by the vertically sheared zonal velocity and the periodic boundary conditions then recreates an unstable state and a second growth phase occurs. Repeating this leads to a quasi-periodic solution

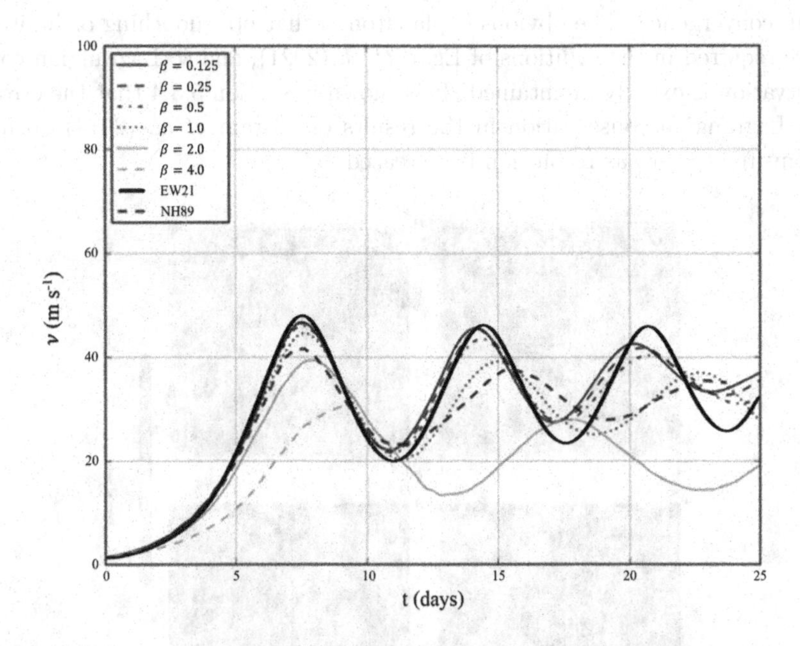

Fig. 5.15 Evolution of the r.m.s. meridional velocity using high resolution with varying β. EW21 results are taken from Egan *et al.* (2021). NH89 are results from Nakamura and Held (1989). From Visram *et al.* (2014), ©Royal Meteorological Society, Reading, U.K., 2013; and from Egan *et al.* (2021), ©Charles Egan.

as shown. Note that purely passive advection of a lower and upper level anomaly by the basic state zonal wind would give a period of 2×10^5s, about 2.3 days. The much longer period observed is because of the phase-locking during the growth phase.

Convergence of the solutions as $\beta \to 0$ is clearly observed. For $\beta \leq 1/2$ the results are close to those for all smaller values of β. The results of Nakamura and Held (1989) are for the standard resolution and $\beta = 1$. After the first lifecycle they tend to have lower amplitude and a longer period. The discussion in Visram *et al.* (2014) shows that this is partly due to resolution and partly to the use of artificial diffusion. The results of Egan *et al.* (2021) have slightly larger amplitude in the first lifecycle, but maintain this amplitide almost unchanged through two further lifecycles. The results of Visram *et al.* (2014) lose amplitude, particularly in the third lifecycle, and the period becomes shorter. While the differences between the results and those of Egan *et al.* (2021) are smallest for small β, they do not achieve full convergence. The obvious explanaton is that no smoothing or limiters are required in the solutions of Egan *et al.* (2021), and so Lagrangian conservation is exactly maintained. It is shown in section 5.3.4 that the errors in Lagrangian conservation in the results of Visram *et al.* (2014) do not converge to zero as resolution is increased.

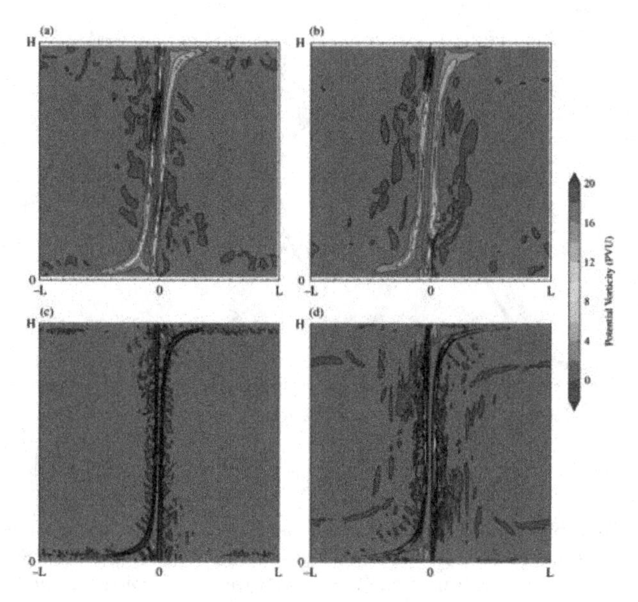

Fig. 5.16 Comparison of the potential vorticity (5.25) at day 8 for standard resolution and (a) $\beta = 1/8$, (b) $\beta = 1$; and for high resolution and (c) $\beta = 1/8$, (d) $\beta = 1$. Negative regions are dashed. From Visram *et al.* (2014). ©Royal Meteorological Society, Reading, U.K., 2013.

This behaviour can be illustrated by plots of the potential vorticity q which will be conserved by smooth solutions of (5.23) where

$$q = \left(\frac{\partial \bar{\theta}}{\partial y} \left(\frac{\partial u}{\partial z} - \frac{\partial w}{\partial x} \right) + f \left(\frac{\partial X}{\partial x} \frac{\partial \Theta}{\partial z} - \frac{\partial X}{\partial z} \frac{\partial \Theta}{\partial x} \right) \right). \tag{5.25}$$

At the initial time, the potential vorticity is almost uniform. However, when the front forms, the potential vorticity increases in the frontal zone because a larger region in (X, Z) space is mapped into the same physical space, as discussed in section 3.2.3. In the Eulerian finite difference calculation, the artificial limiters applied in the advection scheme will be dissipative, and create a potential vorticity source as illustrated. At the time shown in Fig. 5.16, the slope of the isentropes defining the front and the slope of the region of high potential vorticity are both slightly positive. This corresponds to decay of the perturbation which starts at this time as shown in Fig. 5.15.

Comparing the different panels of Fig. 5.16 shows that the region of enhanced potential vorticity is narrower using $\beta = 1/8$, which should be closer to the semi-geostrophic limit. It is also narrower using the high resolution. Both these features are consistent with convergence to a discontinuous semi-geostrophic limit.

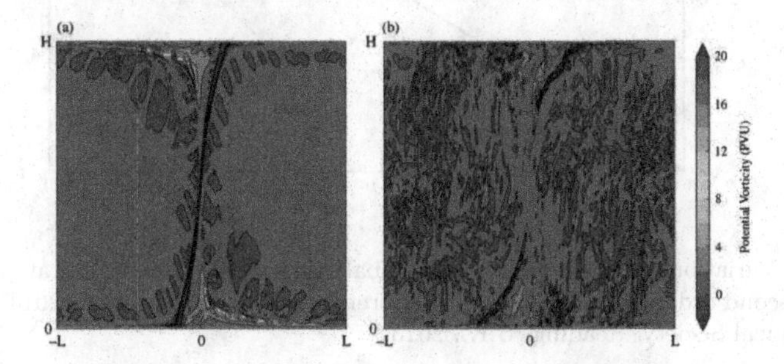

Fig. 5.17 Potential vorticity using high resolution, (a) for $\beta = 1/8$ at day 11 and (b) for $\beta = 1$ at day 12. These correspond to the first minimum in meridional amplitude. From Visram *et al.* (2014). ©Royal Meteorological Society, Reading, U.K., 2013.

Fig. 5.17 shows solutions at the first minimum in meridional amplitude, which occurs at day 11 with $\beta = 1/8$ and day 12 for $\beta = 1$. The region of anomalous potential vorticity with $\beta = 1/8$ is much narrower than at day 8, shown in Fig. 5.16. With $\beta = 1$ the frontal zone has almost lost its identity at day 12. However, anomalies remain at the upper and lower boundaries which will be able to start the next lifecycle.

5.3.4 *Convergence of the Euler solution to the semi-geostrophic solution*

In this subsection two quantitative measures are used to assess the convergence illustrated in the previous subsection. The first measure is the geostrophic imbalance

$$\eta = fv - \frac{\partial \varphi}{\partial x}, \tag{5.26}$$

which will be zero for the semi-geostrophic solution.

Fig. 5.18 shows the geostrophic imbalance at different times for the standard resolution and different values of β. The front forms at about day

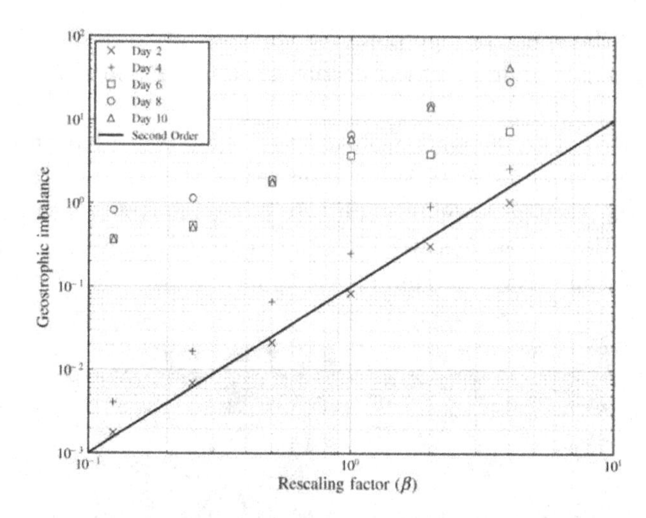

Fig. 5.18 Variations of r.m.s. geostrophic imbalance with rescaling factor at days 2,4,6,8 and 10. Second order slope β^2 shown for reference. From Visram *et al.* (2014). ©Royal Meteorological Society, Reading, U.K., 2013.

6. At days 2 and 4 the solution is smooth, and the geostrophic imbalance reduces at a rate $O(\beta^2)$ as expected. At day 6, the imbalance reduces with β for $\beta > 2$. The front has not yet formed at day 6 for these values of β. It then remains constant till $\beta = 1$ and then reduces more slowly for smaller values of β. This is due to the generation of inertia-gravity waves as the front forms as illustrated in Visram *et al.* (2014) and also in similar integrations by Snyder *et al.* (1991). At day 8, the imbalance reduces as $O(\beta^2)$ for $\beta \geq 0.5$ and at day 10 it reduces as $O(\beta^2)$ for all β plotted. The value of the imbalance is much greater at these times than it is at days 2 and 4. This reflects the continued presence of inertia-gravity waves excited by the frontogenesis.

Similar rates of convergence were obtained with the solutions of the compressible Eady problem by Cullen (2008b) described in section 5.1.4. The results are shown in Fig. 5.19. At the earlier times, the imbalance reduces as Ro^2. At day 7, when the fronts have formed, the imbalance is much larger and only reduces at a rate $O(Ro)$ for large Ro, but $O(Ro^2)$ for $Ro < 0.3$. At day 8, the rate $O(Ro^2)$ is only achieved for $Ro < 0.1$.

However, the accuracy of the solution will depend on the enforcement of the Lagrangian conservation properties that determine the semi-geostrophic solution. The semi-Lagrangian advection scheme determines the value of an advected quantity at an arrival point by interpolating values at the pre-

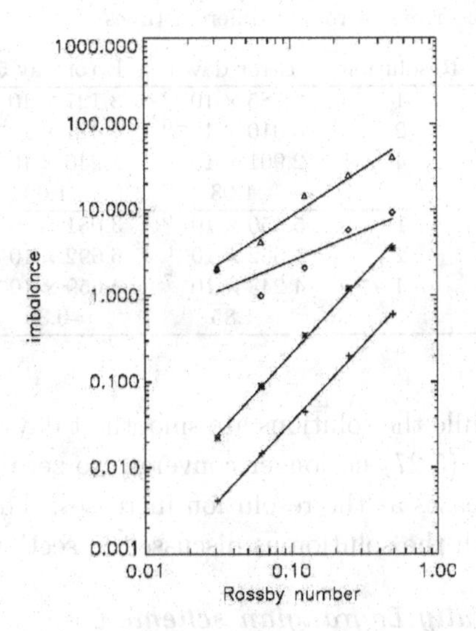

Fig. 5.19 (a) Plot of the r.m.s. imbalance in a vertical slice UM integration as in section 5.1.4 with quasi-monotone interpolation, decentering parameter 0.6, and the specified Rossby number. Both axes are plotted on a logarithmic scale. Plus signs are results after 3 days, asterisks are results after 6 days, diamonds are results after 7 days and triangles are results after 8 days. For each time the best linear fit is shown. From Cullen (2008b). ©Crown Copyright, Met Office.

vious timestep to a departure point, and then transferring the value to the arrival point. Table 5.2 illustrates this by calculating, for the integrations of Visram *et al.* (2014),

$$\left((\chi_a)^2 - (\chi^2)_a\right)/\chi^2, \tag{5.27}$$

where χ is a model variable and $_a$ indicates a value at an arrival point. The value of (5.27) would be zero if there were no interpolation errors. It thus measures the failure of Lagrangian conservation. Table 5.2 shows the r.m.s. value of (5.27) before and after frontal formation for the two variables v and θ whose advection determines the semi-geostrophic solution using various resolutions. Resolutions 1, 2 and 4 are $121 \times 61, 241 \times 121$ and 481×241 respectively. Quintic quasi-monotone advection is used. Visram *et al.* (2014) show more results with different interpolation schemes.

While the solutions are smooth, Lagrangian conservation and the value of (5.27) in particular should be satisfied to the order of accuracy of the finite difference scheme. Table 5.2 shows that the interpolation error is close

Table 5.2 r.m.s. errors at different times

Variable	Resolution	Error day 4	Error day 5.5
v	1	2.785×10^{-5}	3.127×10^{-3}
	2	6.310×10^{-6}	1.421×10^{-2}
	4	2.901×10^{-6}	1.246×10^{-2}
Convergence order		1.63	-1.00
θ	1	5.550×10^{-6}	3.081×10^{-4}
	2	1.562×10^{-6}	6.692×10^{-4}
	4	4.248×10^{-7}	4.659×10^{-4}
Convergence order		1.85	-0.30

to second order accurate while the solutions are smooth at day 4. Once the front is forming at day 5.5, (5.27) no longer converges to zero for either v or θ. In fact the error increases as the resolution increases. This indicates a fundamental problem with the solution as discussed in section 5.1.4.

5.3.5 *Results from a fully Lagrangian scheme*

In this section the results of the fully Lagrangian scheme used by Cullen (2007a) are discussed. The method is outlined in section 5.1.2. The efficient geometric algorithm described there is modified to work with periodic boundary conditions and used to solve (5.24). The solutions update the original solution given in Cullen and Roulstone (1993) which used the version of the geometric algorithm developed by Chynoweth (1987). The initial data are chosen as piecewise constant in X and Z. However the data is different from that described in section 5.3.2 and Nakamura and Held (1989) in that $N_0^2 = 1.3 \times 10^{-4}$. Two resolutions are used, 21×13 elements and 40×16 elements. These are much coarser than the finite difference discretisation used in section 5.3.2 and somewhat coarser than the Lagrangian integrations of Egan *et al.* (2021). This works because eq. (5.24) contains no spatial derivatives of the prognostic variables, as they are all subsumed into the Lagrangian time derivative. The remaining terms can be represented adequately on quite a coarse mesh. The results are illustrated in Fig. 5.20 after 8 days integration.

At this point there is strong frontogenesis, as can be inferred from the element geometry. The slope of the isentropes is negative, indicating growth of the perturbation as discussed in section 5.3.3. A graph of the kinetic energy against time is shown in Fig. 5.21 for the two resolutions used. This shows that the maximum kinetic energy is reached after 8 days, which is the time illustrated in Fig. 5.20. It then illustrates the subsequent

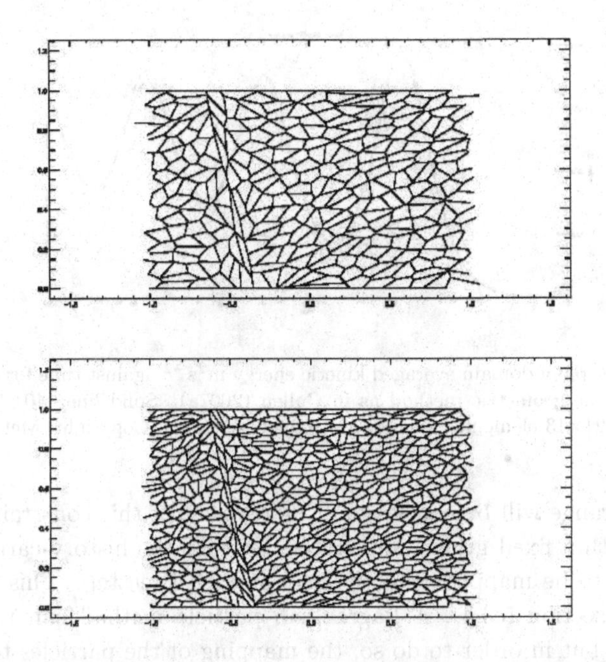

Fig. 5.20 Distribution of elements after 8 days integration of (5.24). Top: 21×13 elements, Bottom: 40×16 elements. Source: Met Office. ©Crown Copyright, Met Office.

cycles, which continue to day 30. The graphs for the two resolutions are almost identical, showing that the solution is highly predictable, despite the formation of fronts. As in section 5.3.2, the period is much longer than the natural period of the system with this data also, so that the prediction of the same period by two different discretisations is a non-trivial achievement.

5.3.6 *Summary of Eady problem*

These results show a fundamental difference between the Eulerian and Lagrangian discretisations. The Eulerian model will conserve potential vorticity while there is no dissipation. When the front forms, the model is unable to maintain Lagrangian conservation while staying close to geostrophic balance. The dissipation used to allow the integration to continue acts as a source of potential vorticity, as shown in section 3.2.3, which will then be conserved by the subsequent evolution.

If c is a conserved scalar, then the mass of fluid with values of c in

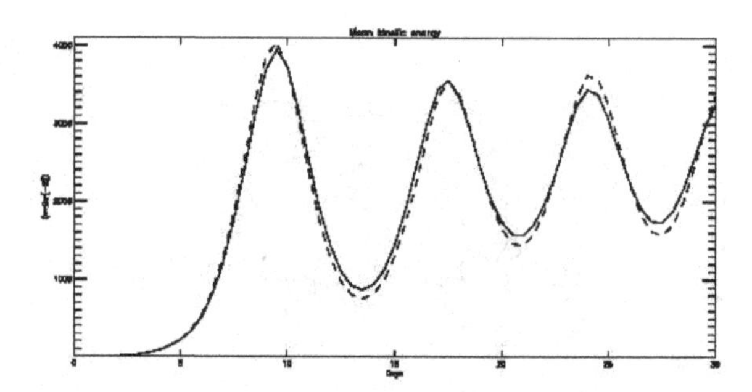

Fig. 5.21 Graph of domain averaged kinetic energy m^2s^{-2} against time for the solution of (5.24) by the geometric method as in Cullen (2007a). Solid line: 40×16 elements, dashed line: 21×13 elements. Source: Met Office. ©Crown Copyright, Met Office.

any given range will be conserved. Enforcement of this constraint using a method with a fixed grid is impossible, because the history-carrying variables have to be mapped onto the grid at each timestep. This is demonstrated in section 5.3.4. A Lagrangian particle method can respect the constraint, but in order to do so, the mapping of the particles to physical space has to conserve a control volume associated with each particle. The control volumes have to exactly cover the physical domain. Constructing such a map is an optimal transport problem, as demonstrated in section 5.1.2 and used by Egan *et al.* (2021) and Cullen (2007a).

The Lagrangian method is built on the knowledge that balance and Lagrangian conservation can be simultaneously maintained if the solution is allowed to be discontinuous. This is proved in Theorem 3.13 using optimal transport theory. The methods illustrated here have been proved to converge to weak solutions of the semi-geostrophic equations by Cullen *et al.* (2007c) in Theorem 3.25. The discretisation handles discontinuities naturally, as it is based on piecewise constant data. While a potential vorticity source will appear when there is frontogenesis, for the reasons discussed in section 3.2.3, this can disappear again as the perturbation amplitude decreases.

The results shown in section 5.3.4 show that a regularised finite difference method for the Euler equations almost achieves convergence to the weak semi-geostrophic solution given by the optimal transport method for the first lifecycle, but increasingly fails to do so in subsequent lifecycles. Since the diagnostics shown in section 5.3.4 show that Lagrangian conservation properties are not maintained in the Euler solution, the most likely deduction is that the discontinuous semi-geostrophic solution is the correct limit of the Euler solutions, but that standard Eulerian numerical methods cannot reach it. That would be a significant issue for numerical modelling of atmosphere and ocean flows.

This example also suggests that using a Lagrangian discretisation based on optimal transport to solve the Euler Boussinesq equations could be advantageous, if practical. Such methods have been illustrated by Gallouët and Mérigot (2018).

5.4 Simulations of baroclinic waves

5.4.1 *Diagnosis of vertical motion*

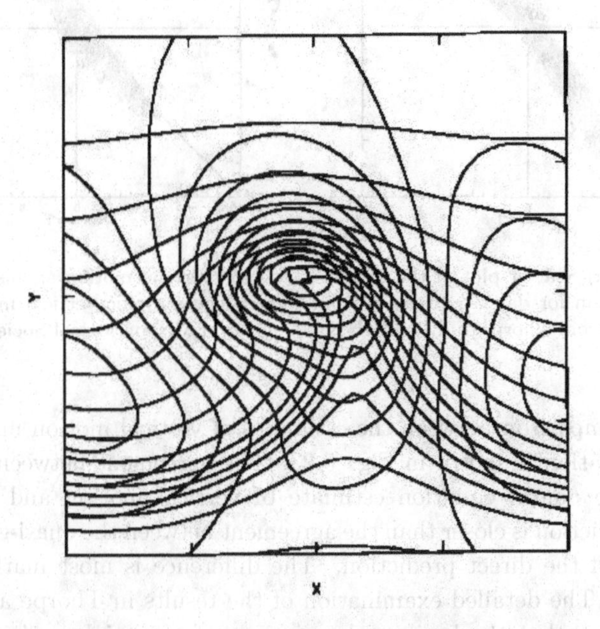

Fig. 5.22 Horizontal section at the surface showing contours of pressure (full lines; contour interval 3hpa) and temperature (dashed lines; contour interval 3°K) at day 7. From Thorpe and Pedder (1999). ©Royal Meteorological Society, Reading, U.K., 1999.

This subsection shows results from three-dimensional simulations by Thorpe and Pedder (1999) using the non-hydrostatic Euler equations (3.20) to provide a simulation of a growing baroclinic wave. They then diagnosed the vertical motion using the quasi-geostrophic omega equation (5.6) and compared the result with the vertical motion predicted directly from (3.20). The procedure was then repeated using the semi-geostrophic omega equation (5.7), which they solved using the methods outlined in section 5.1.1. This forms an idealised test of the diagnostic use of semi-geostrophic theory illustrated in section 2.7.4.

The data used for the comparisons shown here is shown in Fig. 5.15. It is an idealised baroclinic wave at the stage of maximum development. There has been marked frontogenesis at the surface.

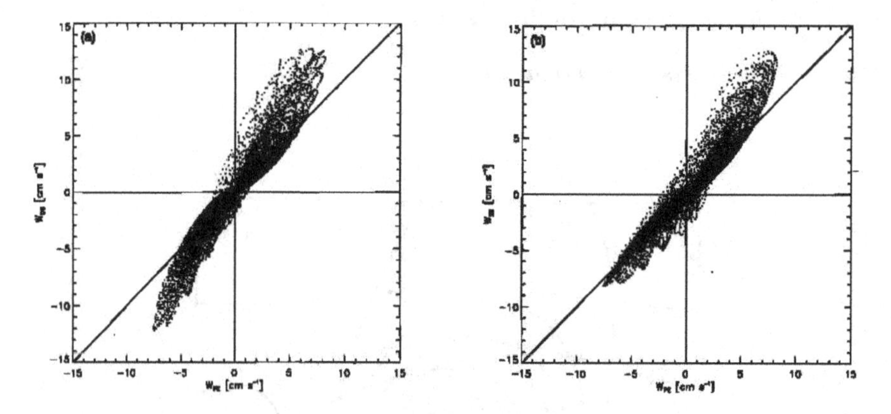

Fig. 5.23 (a) Scatter plot of the quasi-geostrophic vertical motion against the model vertical motion for day 7 ; (b) as (a) but for semi-geostrophic vertical motion versus the model. From Thorpe and Pedder (1999). ©Royal Meteorological Society, Reading, U.K., 1999.

The comparison between the estimates of vertical motion made by the omega equation is shown in Fig. 5.23. The agreement between the semi-geostrophic omega equation estimate of vertical motion and the direct model prediction is closer than the agreement between the quasi-geostrophic version and the direct prediction. The difference is most marked at upper levels. The detailed examination of the results in Thorpe and Pedder (1999) shows that the horizontal variations of potential vorticity and the variations in static stability included in the left hand side of (5.7) are important. There are also important effects from the calculation of the forcing on the right hand side, because the geostrophic coordinate transformation has a significant effect. The main disagreements between the semi-geostrophic diagnosis and the model prediction are in regions of ascent. Further examination in Thorpe and Pedder (1999) shows that these are mostly at low levels, and the semi-geostrophic potential vorticity calculated from the model data is negative in this region. The discussion following equation (2.175) in section 2.7.4 shows that the displacement required to maintain geostrophic balance becomes large if the potential vorticity matrix $\mathbf{Q}$ has a small eigenvalue, implying large ageostrophic velocities. The overestimate by the semi-geostrophic diagnosis is thus not surprising. The model predictions in this region will be strongly influenced by the artificial viscosity needed to maintain stability. The effect of artificial viscosity on frontogenesis was illustrated in sections 5.3.3 and 5.3.4, and the model predictions may not be reliable there.

5.4.2 *Stability of frontal zones*

The next example shows results from Malardel *et al.* (1997) on the stability of frontal zones in geostrophic balance. The fronts are characterised by a potential vorticity anomaly as shown in Figure 5.16. Two cases are studied. Front 1 has a width of 330km. The aspect ratio of the anomaly is less than f/N, so that the effect of the anomaly is therefore mainly seen in the potential temperature. Front 2 has a narrower anomaly, about 150km, with the same vertical extent. The aspect ratio is now greater than f/N and so the effect of the anomaly is mainly seen in the vorticity field.

In the Eady model considered in section 5.3, the initial potential vorticity is independent of x, so the only instability is baroclinic. However, Theorem 4.20 shows that a state with the potential vorticity depending on x, which arises in the Eady model after a discontinuity has formed, is unstable to horizontal perturbations according to semi-geostrophic theory. It is also unstable according to the Charney-Stern theorem for two-dimensional incompressible flow, Vallis (2017), p.351. The detailed analysis in Malardel *et al.* (1997) shows that both types of instability occur, but in different ranges of horizontal wavelength. Baroclinic instability requires a wavelength greater than about 2000km. Barotropic instability predominates on smaller scales. Observations of instabilities on fronts, such as the weather systems illustrated in Fig. 1.2, suggest that instabilities which grow to a significant amplitude are baroclinic. Barotropic instability is likely to be important in preventing the long time existence of fronts with large aspect ratios. Thus fronts with aspect ratios greater than f/N are not often observed.

The error estimate $Ro_L Bu^2$ given by (2.157) for semi-geostrophic solutions uses the Lagrangian Rossby number. This will depend on the horizontal wavelength of the instability. The additional accuracy obtained if $Ro < Fr$ will apply if the cross-frontal structure has an aspect ratio smaller than f/N, so that the errors for Front 1 are expected to be up to 5 times smaller than for Front 2.

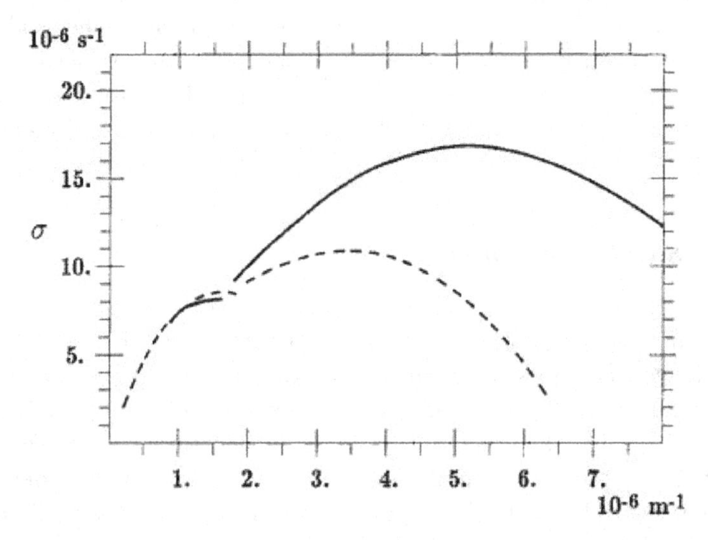

Fig. 5.24 Growth rates ($10^{-6}s^{-1}$) as a function of the along-front wave-number ($10^{-6}m^{-1}$) for Front 1. Dashed curve, semi-geostrophic results; solid curve-results for exact equations. From Malardel *et al.* (1997). ©American Meteorological Society. Used with permission.

The results show the growth rates computed from an 'exact' model similar to (5.1) and a semi-geostrophic model similar to (5.2). Fig. 5.24 shows the results for Front 1. There is fairly good agreement for wavelengths greater than 3000km, but a large under-estimation of the growth of smaller-scale perturbations. It is shown in Malardel *et al.* (1997) that this mainly reflects an inaccurate treatment of the barotropic instability which is governed by vortex dynamics. The result is thus consistent with the behaviour shown in section 5.2.2 and with the error estimate (2.157). Though such a flow is still unstable according to semi-geostrophic theory, the results shown in 5.2.2 demonstrate that the semi-geostrophic model underestimates the evolution when $Fr < Ro$ and vortex dynamics dominates. However, baroclinic instability is more accurately represented because the Lagrangian Rossby number Ro_L computed from the wave-number of the disturbance is smaller. Fig. 5.25 shows the result for Front 2. The differ-

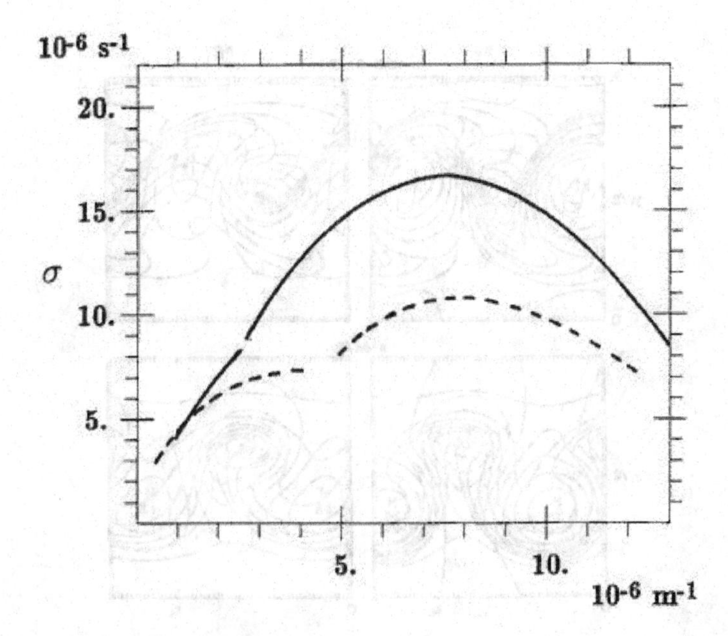

Fig. 5.25 As Fig. 5.24 for Front 2. From Malardel *et al.* (1997). ©American Meteorological Society. Used with permission.

ence in the solutions is now large for wavelengths less than 5000km, and increases more rapidly as a function of wave-number. This is consistent with (2.157).

5.4.3 *Evolution of a baroclinic wave*

Next illustrate the difference between semi-geostrophic simulations of a baroclinic wave and those made by the non-hydrostatic Boussinesq Euler equations (3.20). A comparison by Snyder *et al.* (1991), after 6 days of integration from initial data similar to that used by Thorpe and Pedder (1999), is shown in Fig. 5.26. The integrations were carried out in a domain $\Gamma : (0, x_L) \times (0, y_L) \times (0, H)$ with periodic boundary conditions in (x, y), and rigid boundaries at $z = 0, H$. The numerical method used employs the geostrophic coordinate transformation as described in section 5.1.1. The results shown had $x_L = 4090$km, $y_L = 5623$km and $H = 9$km. As discussed by Gill (1982), p. 556, baroclinic instability has to occur with an aspect ratio comparable to f/N, so the accuracy of semi-geostrophic theory as given by (2.157) will be $\mathrm{O}(Ro_L)$. There are significant differences in the structure of the wave by 6.3 days, the time illustrated.

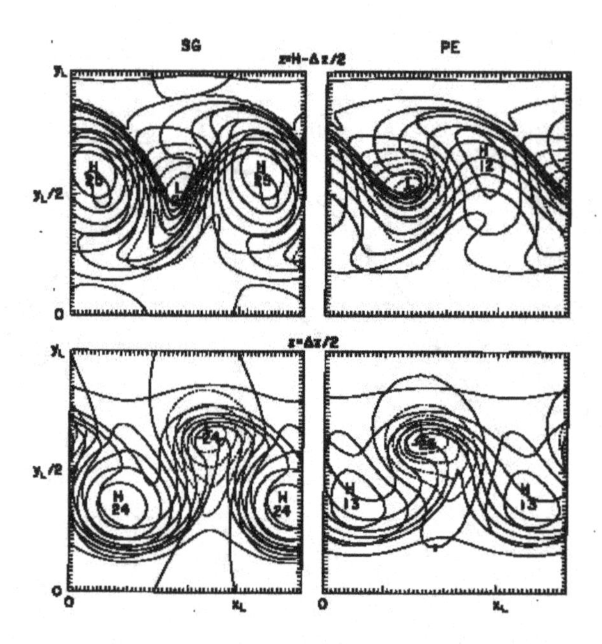

Fig. 5.26 Semi-geostrophic solutions (left) and Euler solutions (right) at day 6.3. Solid lines, potential temperature θ, contour interval 5°K. Dashed lines, geopotential φ, contour interval $500\text{m}^2\text{s}^{-2}$. Solutions are shown as a function of (x, y) for $z = 0.225$ km (bottom) and 8.775 km (top). From Snyder *et al.* (1991). ©American Meteorological Society. Used with permission.

This is consistent with the expected error. There is more distortion of the potential temperature contours in the 'exact' solution. This difference is consistent with the reduced growth of vorticity perturbations illustrated in section 5.2.2. As discussed in section 5.1.1, the results shown neglected the nonlinear term in the geostrophic coordinate transformation. They also ignore the effect of frontogenesis in computing the transformation. Both were estimated as being smaller than the differences shown in Fig. 5.26.

Another study, by Schär and Wernli (1993), showed that the structure of semi-geostrophic baroclinic waves was strongly influenced by the initial data. Their study showed that all the main features of observed waves could be reproduced qualitatively, including simultaneous warm and cold fronts, well marked cold and warm 'conveyor belts' associated with the fronts, and concentrated ascent within the warm front. In Fig. 5.27 the results after 4 days are shown for one of their experiments. The potential temperature is now highly distorted, as in the 'exact' solution of Snyder *et al.* (1991).

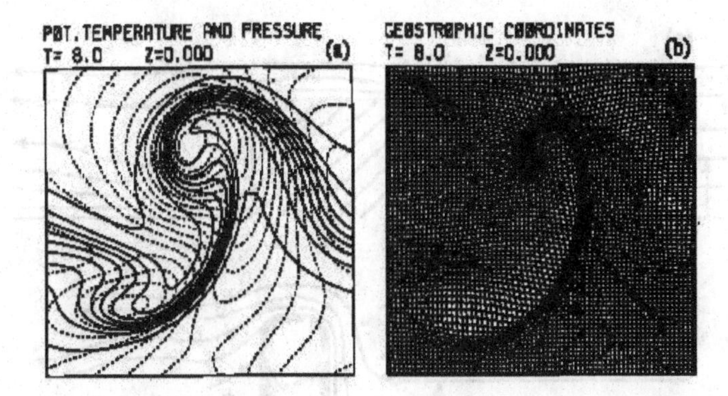

Fig. 5.27 Left: surface potential temperature (solid lines) contour interval 2.1°K, and surface pressure (dashed lines), contour interval 3.2hpa, after 4 days. The domain size is 14640×7000 km. Right: Geostrophic space as viewed from physical space. From Schär and Wernli (1993). ©Royal Meteorological Society, Reading, U.K., 1993.

These results suggest that semi-geostrophic theory is sufficiently accurate to describe the qualitative features of observed extra-tropical weather systems. However it loses accuracy when the aspect ratio is close to f/N so the horizontal scale is comparable to the Rossby radius of deformation. The demonstration illustrated in section 2.7.4 shows that semi-geostrophic dynamics together with realistic representations of physical effects can reproduce the large-scale behaviour of real systems in the extratropics quite well.

5.5 Orographic flows

The first step is to consider a schematic of the flow past a mountain ridge which is longer than the Rossby radius of deformation based on the ridge height, so that the flow is controlled by geostrophic balance. This is illustrated in Fig. 5.28 and discussed more fully in Shutts (1998). The main characteristics of the solution in the central part of the ridge are shown in the top part of Fig. 5.28 which shows the flow in an (x, z) cross-section. A particular feature is the blocking of cold air near the surface on the upstream side of the ridge and the associated barrier jet.

The behaviour in the central part of the ridge can be described by solutions of the semi-geostrophic equations in a vertical slice, as in the Eady model (5.24) of section 5.3. In this case the flow is assumed to be driven by a pressure gradient in the y-direction which is independent of z. Thus

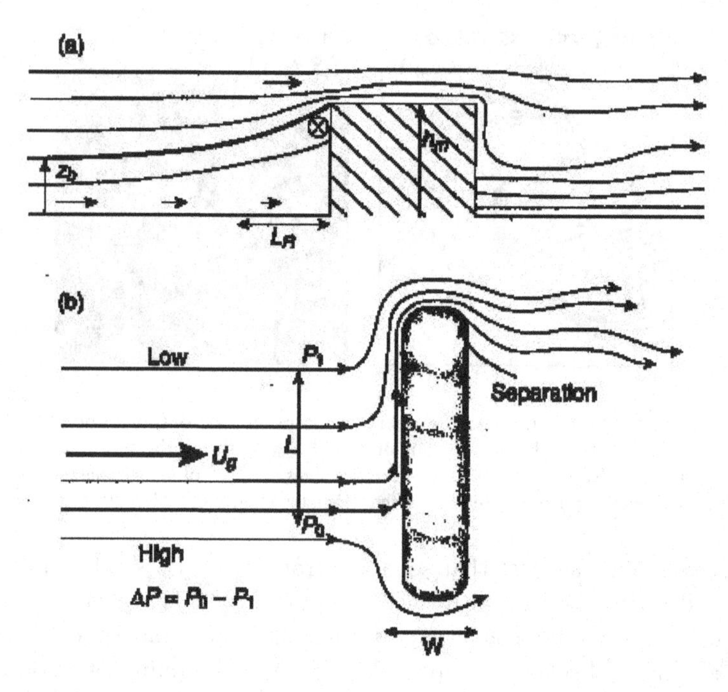

Fig. 5.28 Schematic view of the flow around a mesoscale mountain ridge. (a) Elevation view showing contours of potential temperature in relation to a rectangular block mountain. Arrows indicate the strength of the cross-mountain flow component, and the circle containing a cross marks the location of a barrier jet into the picture. (b) Plan view of the flow streamlines splitting around the mesoscale mountain ridge. The oncoming geostrophic flow is diverted to the left in the Northern hemisphere, forming a barrier jet. This mesoscale jet-stream eventually separates from the northern tip of the ridge. From Shutts (1998). ©Crown Copyright, Met Office.

there is no basic state θ gradient in the y direction. The geostrophically balanced velocity in the x direction is U. The equations are then

$$\begin{aligned} -fv_g + \frac{\partial \varphi}{\partial x} &= 0, \\ \frac{\mathrm{D}v_g}{\mathrm{D}t} + fu - fU &= 0, \\ \frac{\mathrm{D}\theta}{\mathrm{D}t} &= 0, \\ \frac{\partial \varphi}{\partial z} - g\frac{\theta}{\theta_0} &= 0, \\ \nabla \cdot \mathbf{u} &= 0. \end{aligned} \tag{5.28}$$

All variables are assumed to be functions of (x, z) only. The equations are solved on a region $\Gamma = [-L, L] \times [h(x), H]$ with periodic boundary conditions in x and $\mathbf{u} \cdot \mathbf{n} = 0$ on the upper and lower boundaries. The orography is defined by the function $z = h(x)$. This problem was first solved by Cullen *et al.* (1987), and then by Shutts (1987a) and Shutts (1998). The results here are taken from Cullen (2007b).

The blocking of the cold air on the upstream side of the ridge occurs because of the admissibility condition for semi-geostrophic solutions. Definition 3.2 requires in this context that $\varphi + \frac{1}{2}f^2x^2$ is a convex function, so that

$$\left(f^2 + \frac{\partial^2 \varphi}{\partial x^2}\right)\frac{\partial^2 \varphi}{\partial z^2} \geq \left(\frac{\partial^2 \varphi}{\partial x \partial z}\right)^2. \tag{5.29}$$

For a state at rest, with Brunt-Väisälä frequency given by $\frac{g}{\theta_0}\frac{\partial \theta}{\partial z} = N^2$, this means that

$$\left|\frac{\partial \theta}{\partial x}\right| \leq \sqrt{(\theta_0/g)} fN. \tag{5.30}$$

Thus if the ridge $z = h(x)$ is an isentrope, it follows that

$$\left|\frac{\mathrm{d}h}{\mathrm{d}x}\right| \leq f/N. \tag{5.31}$$

If the ridge is steeper than this, it cannot be an isentrope. However, isentropes can extend from the top of the ridge with slope f/N. This can be considered as defining a broader ridge, over which the flow is smooth, with blocking occurring within the region $x = x_0 \pm Nh(x_0)/f$, where x_0 is the position of the ridge top. This states that the influence of the ridge occurs for a horizontal distance equal to the Rossby radius of deformation based on the ridge height. Further discussion is given in Purser and Cullen (1987).

Consider the solution procedure as set out in section 3.1.2. This defines solutions as minimum energy states in the sense of Theorems 3.1 and 3.2. If the rectangular block in Fig. 5.21 is replaced by a vertical barrier with zero width, the solution does not need to change significantly from that shown in Fig. 5.21. However, it is clearly not a global energy minimiser, because if the barrier were removed, the solution would have a discontinuity in φ and would not be admissible in the sense of Definition 3.1. The solution with the barrier is admissible because φ does satisfy the conditions of Definition 3.1 within the domain which excludes the barrier. It is thus an energy minimiser in the sense of Theorem 3.2, but not a global energy minimiser in the sense of the theorems in section 3.5.2 which show that there is an (essentially) unique energy minimiser for this problem. It is therefore necessary to seek local rather than global energy minimisers. The results of section 3.5.2 do not address this issue, and it will be necessary to generalise the theory significantly.

The solution is therefore described in a formal way. Assume that the flow is always in a locally energy minimising state, and can be constructed by a solution procedure similar to that described in section 5.1.3. Given admissible initial data, the solution is stepped forward in time for a time δt to give a first guess to the solution which is not in geostrophic and hydrostatic balance. The final solution is then found by minimising the energy using the iteration (5.15). This requires the energy to decrease through the iteration, and would prevent the fluid temporarily increasing its energy by rising to the top of the ridge before achieving a lower energy on the downstream side.

Apply this procedure to solve (5.28) with initial data $\theta(0,\cdot) = \theta_1(z)$, $v_g(0,\cdot) = 0$. This is clearly admissible. The first guess solution will have $\theta^*(\delta t,\cdot) = \theta_1(z)$, $v_g^*(\delta t,\cdot) = fU\delta t$. In the absence of orography, the energy minimising solution would be given by choosing $u = U$, so that $v_g(\delta t,\cdot) = 0$. However, with orography, this requires transport through the ridge and is not possible. A solution like that shown in the top panel of Figure 5.28 is therefore obtained with $v_g > 0$, the barrier jet, on the upstream side. Observational support for this picture is provided by Schwerdtfeger (1975). Since $\partial v_g/\partial z < 0$ in this region, geostrophic balance requires $\partial\theta/\partial x < 0$, so the air next to the ridge is colder than the upstream air at a given height. On the downstream side, the solution $u = U$ is allowed. However, it would leave a vacuum on the downstream side of the ridge. The actual solution thus has to look like that shown in Fig. 5.28.

The solution has a pressure force acting on the ridge, which is a model of orographic drag. This can easily be estimated for a vertical barrier by integrating the hydrostatic equation down from the top of the barrier and noting the difference in potential temperature at given heights across the barrier. The model as formulated in (5.28) conserves energy. This is because the pressure gradient fU in the y direction is imposed. In a real case, the drag would act to reduce this gradient and the associated geostrophic wind. Air trapped on the upstream side of the ridge will have v_g increasing with time, as u is constrained to be zero. Therefore the slopes of the isentropes, which is related to $\partial v_g/\partial z$, will increase with time and trapped air will reach the top of the ridge. At this point, the minimum energy position of such a parcel will be near $x = x + f^{-1}v_g$, so the parcel will 'jump' downstream. This will correspond to a measure-valued velocity, though it will still fall within the scope of Definition 3.12. In reality, there would be a rapid down-slope wind not described by semi-geostrophic theory. Such winds often occur downstream of mountain ranges. The rate of loss of energy in these jumps has to equal the rate of working by the ridge on the fluid.

In the three-dimensional case, the flow will be like the lower part of Fig. 5.28. Since the trapped air has $v_g > 0$, the air can reach the end of the ridge before it reaches the top. At this point there will again be a jump in the parcel position to near $x = x + f^{-1}v_g$. The preferred deflection of the upstream flow to the left in the Northern hemisphere is regularly seen in observations. Now estimate when this happens. Since the upstream influence of the ridge extends for a distance Nh/f, the maximum displacement due to the ridge of a parcel with given $X = x + f^{-1}v_g$ will be of order Nh/f, and so the maximum barrier jet velocity will be of order Nh. Using a typical tropospheric value 10^{-2}s^{-1} of N gives a jet velocity of 10ms^{-1} for a ridge height of 1km.

Since (5.28) implies that $\mathrm{D}X/\mathrm{D}t = fU$, the y coordinate of a trapped parcel obeys the equation

$$\frac{\mathrm{d}^2 y}{\mathrm{d}t^2} = fU. \tag{5.32}$$

Thus the parcel will reach the end of a ridge of length D in a time $\sqrt{(2D/fU)}$ with a velocity $\sqrt{(2fUD)}$. This is less than the maximum barrier jet velocity if

$$D < N^2h^2/(2fU). \tag{5.33}$$

If this condition is satisfied, the parcel will flow round the end of the ridge. Using the data above and $U = 10\text{ms}^{-1}$ gives $D < 50$km. This problem is discussed more fully in Shutts (1998). D will increase as the square of the ridge height, so a 2km ridge woud give a maximum deflection of 200km. If the ridge is longer than the maximum D given by (5.33), then the parcel will cross the ridge after reaching the top. Thus the maximum north-south distance that a parcel can travel is equal to the right hand side of (5.33). The associated Lagrangian Rossby number Ro_L is $2Fr^2$. Thus the asymptotic behaviour has $Ro \propto Fr^2$ so that semi-geostrophic theory will be accurate to $\mathrm{O}(Ro_L^2)$ as shown in section 2.4.6.

Now consider the solution to the two-dimensional ridge problem. As discussed after (5.31), trapping of the air upstream of the ridge means that the effective aspect ratio of the ridge is f/N, so that the Lagrangian Rossby number Ro_L for the two-dimensional ridge problem will be $U/Nh = Fr$ regardless of the width of the ridge. The error in the semi-geostrophic solution will then be proportional to Ro_L, which will be proportional to U. A test where the cross-ridge velocity U is varied is shown, with f, N and the definition of the ridge $z = h(x)$ fixed.

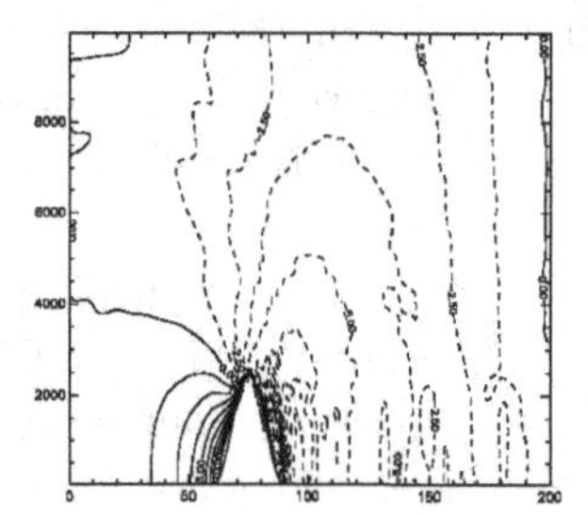

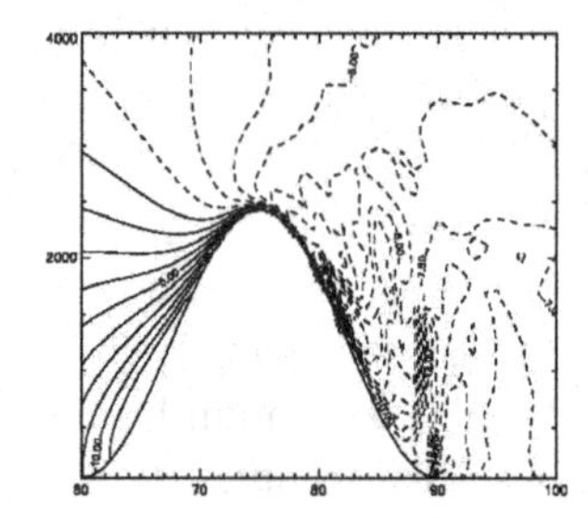

Fig. 5.29 Left: UM solution for the $v-$ component of the wind (ms^{-1} with solid(dashed) contours positive(negative) and contour interval 1.25 ms^{-1}) at 40 days with U =0.083 ms^{-1} and using a 201×121 grid, plotted against height (m) and grid-point number. Right: enlarged extract from this solution with the same plotting conventions. From Cullen (2007b). ©Crown Copyright, Met Office.

The results illustrated are taken from Cullen (2007b). As in section 5.1.4, they were obtained by solving the full compressible Navier-Stokes equations (4.1) from the Met Office UM in a vertical cross-section. The solutions were compared with those from a semi-geostrophic version, (4.4), of the same model. The solutions of the semi-geostrophic model will include measure-valued velocities as noted above. Therefore the success of finite difference methods is not guaranteed, though an iteration equivalent to (5.15) as discussed in section 5.1.3 should give the correct solution. Both models were integrated on a 121×61 grid. The UM integrations were also carried out on a 210×121 grid, since a greater variety of solutions is expected on small scales.

The domain used for the experiments had length 2000km and height 10km. The results shown are for a ridge 150km wide and 2400m high centred at x =750km. The Coriolis parameter is set to $f = 10^{-4}\text{s}^{-1}$ and the Brunt-Väisälä frequency is set to $N = 10^{-2}\text{s}^{-1}$. This would give $Ro/Fr =$ 1.6 if the Rossby number was based on the ridge width, so it is expected that there will be upstream trapping of air and a barrier jet. The equations are integrated for several inertial periods to achieve a steady state. The steady-state solution given by the semi-geostrophic model is independent of the wind speed, since the solution after a time-step is determined by $U\delta t$. If U is increased, then the same solution will be obtained but after a shorter time. The steady-state solution must therefore be the same. The UM solutions depend strongly on U. The values chosen range from 0.625ms^{-1} to 10ms^{-1}. This gives Froude numbers ranging from 0.026 to 0.42 and Rossby numbers ranging from 0.042 to 0.67. Two verification times are illustrated, corresponding to 45 and 60 hours with U =10ms^{-1}. These represent 2-3 inertial periods and are regarded as separate estimates of the steady state solution.

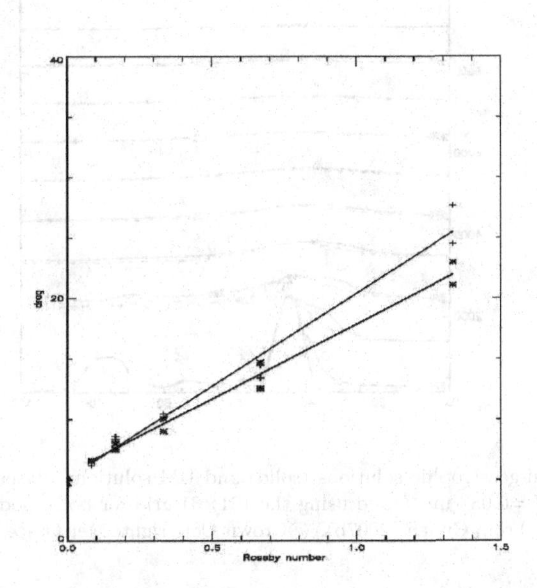

Fig. 5.30 Convergence of the drag on the ridge between UM and semi-geostrophic solutions for fixed L_R plotted against Rossby number (*Ro*). The drag is measured per unit length in the y-direction, and units are 10^5 Pa m^{-1}. Semi-geostrophic values are plotted as triangles at *Ro*=0. UM values using a 121×61 grid are plotted as + symbols and values using a 210×121 grid are plotted as $*$ symbols. The thin (thick) line shows the best linear fit in *Ro* using the 121×61(210×121) grid. Two verification times are given for each *Ro*. From Cullen (2007b). ©Crown Copyright, Met Office.

Fig. 5.29 shows the along-ridge velocity v from the higher resolution UM integration using U = 0.083ms^{-1} and verifying at 40 days. This is expected to match the semi-geostrophic solution most closely among the values tested. Many features shown in the schematic cross-section in Fig. 5.28 can be seen. In particular, there is an upstream barrier jet reaching the top of the ridge with a maximum velocity of about 12ms^{-1}, There is a downstream region of overturning circulations extending to the bottom of the ridge where there is a discontinuity near the point when the slope stops. Inspection of the right panel of Fig. 5.29 shows that this is about 110km downstream of the largest values of v on the upwind side. This is where the parcel 'jump' in the semi-geostrophic solution should terminate because, as explained above, the distance 'jumped' should be $f^{-1}v$. Beyond this, inertia-gravity waves continue further downstream.

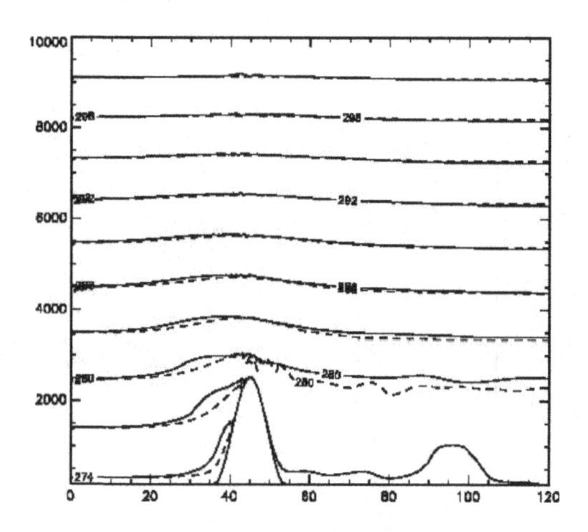

Fig. 5.31 Semi-geostrophic solutions (solid) and UM solutions (dashed) for θ (°K) at 40 days with U =0.083 ms^{-1} and using the 121×61 grid for both models. The contour interval is 3°K. From Cullen (2007b). ©Crown Copyright, Met Office.

Fig. 5.30 shows the convergence of the drag. As discussed above, only linear convergence in Ro is expected. This is shown for both resolutions, with slightly faster convergence for the higher UM resolution.

Fig. 5.31 compares the potential temperature fields from the semi-geostrophic model and the UM for the integrations with the smallest Rossby number. The upstream effect of the ridge is slightly greater in the semi-geostrophic model. The extent is consistent with the expected maximum slope f/N of an isentrope. The potential temperature near the surface is about 3°K higher downstream of the ridge in the semi-geostrophic model, the well-known Föhn effect. In the UM integration, the downstream region

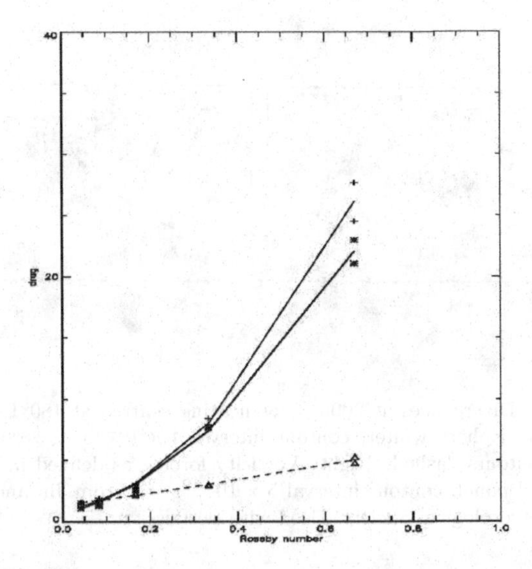

Fig. 5.32 Convergence of the drag on the ridge between UM and semi-geostrophic solutions with $N \propto U^{\frac{1}{2}}$ plotted against Rossby number (Ro). The resolutions and plotting conventions are as in Fig. 5.30 except that the lines represent the best quadratic fit and the triangles represent semi-geostrophic values. From Cullen (2007b). ©Crown Copyright, Met Office.

is a statically unstable mixed layer which is not resolved by the contour interval in the figure. These features are consistent with the schematic in Fig. 5.28.

Fig. 5.32 shows the convergence of the drag where N is reduced in proportion to $U^{\frac{1}{2}}$ so that $Ro \propto Fr^2$. In this case the semi-geostrophic solution is no longer independent of Ro. Second order convergence in Ro is expected according to (2.157). This means that condition (5.31) will be satisfied, and it is possible to obtain smooth semi-geostrophic solutions which agree much more closely with the UM solutions. Fig. 5.32 shows a small discrepancy between the semi-geostrophic solutions and the UM solutions for very small Ro which reflect numerical errors.

5.6 Tropical-extratropical interaction

Another diagnostic application of the semi-geostrophic model is to show the effect of tropical convection on extra-tropical flow. This is very important because, on a long time-scale, the amount of convection is strongly related to the sea-surface temperature, and on shorter time-scales, the amount of

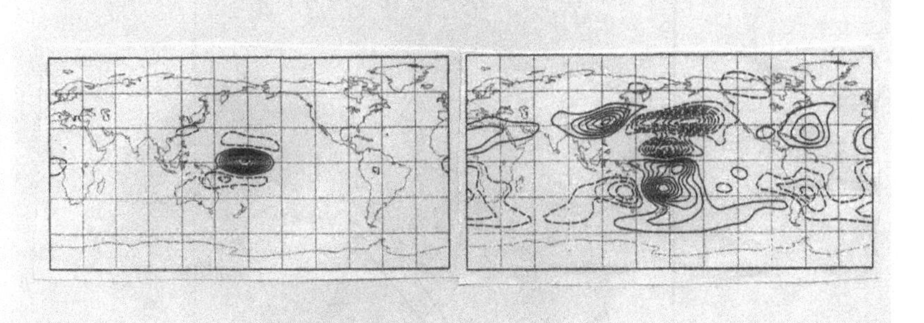

Fig. 5.33 Left: Divergences at 200hpa for heating centred at 180°E using basic state for Northern hemisphere winter, contour interval $4 \times 10^{-7}s^{-1}$, zero contour omitted and negative contours dashed. Right: Vorticity forcing as defined in eq. (5.34). Same case as left hand panel, contour interval $5 \times 10^{-12}s^{-2}$. From Jin and Hoskins (1995). ©American Meteorological Society. Used with permission.

convection shows strong variability which propagates in space, particularly the Madden-Julian oscillation (MJO), Wheeler and Kiladis (1999). In both situations, it is believed that anomalies in the intensity of tropical convection have an effect on extratropical flow. An example is the paper by Matthews *et al.* (2004), where the time dependent heating deduced from reanalyses is used to determine the response to forcing of a simple general circulation model (GCM) using a climatological mean state. This is compared with the observed evolution deduced from satellite data. This is supported by idealised studies, for instance Jin and Hoskins (1995).

As explained in section 2.4.7, the semi-geostrophic model can only describe states satisfying the weak temperature gradient approximation outside the boundary layer. Convection represents local upward motion and thus a divergent circulation which spreads out the associated latent heating to be consistent with the weak temperature gradient, as in Beare and Cullen (2019). This circulation spreads out to an equatorial deformation radius and can thus affect the subtropical jet. Once this has happened, the effect propagates downstream and affects the rest of the circulation. Since the semi-geostrophic model cannot represent tropical circulations directly, the anomaly can be created by imposing a local heat source which generates upward motion and the required divergent circulation. This technique was also used by Jin and Hoskins (1995).

This was studied using the semi-geostrophic diagnostic, Cullen (2018), described in section 2.7.4 and also used in sections 6.1.6 and 6.2.2. The procedure can be applied using artificial physical forcing as well as that derived from the model. Semi-geostrophic dynamics calculates the 'instant' response of the geostrophic flow to forcing at a particular time. It can thus be compared with the study of Jin and Hoskins (1995). That study used an idealised general circulation model solving (2.23) in dimensional form. Either an idealised or an actual atmospheric state could be used. In the former case, an artificial forcing term was included to maintain a steady state. 'Physical' forcing was then added and maintained over a long timescale, so that the divergent circulation in the tropics could spin up. The divergent response is shown in the left panel of Fig. 5.33. This does not propagate much downstream from the forcing. The right panel is a plot of

$$(\zeta + f)\Delta' + \mathbf{u}_\chi \cdot \nabla\zeta', \tag{5.34}$$

where $\mathbf{u}_\chi$ is the horizontal divergent wind and Δ' and ζ' the perturbations to the horizontal divergence and the vertical component of the relative vorticity. In the source region it is dominated by Δ', but changes sign across the equator because of the multiplication by f. Away from the source region, this term is dominated by ζ'. It can be seen that this is significant at 30°N after five days. This is in the location of the subtropical jet.

The semi-geostrophic diagnostic results are shown in Fig. 5.34. The heating region is centred on the dateline as in the study of Jin and Hoskins (1995) shown in Fig. 5.33. This extends through the troposphere. The zonal geostrophic wind in the upper troposphere is shown in the top right panel. This comes from a Met Office UM analysis, filtered to remove non semi-geostrophic motions as discussed in section 2.7.4. The diagnosed geostrophic wind tendency due to the forcing is shown in the bottom left panel. A region of strong response to the forcing of about $0.5\text{ms}^{-1}\text{day}^{-1}$ is seen around 30°N. This is in a position to extend the subtropical jet, as can be seen from the top right panel. This will reinforce the forward propagation of the jet-stream expected under quasi-geostrophic dynamics, and so influence the rest of the circulation. If this response were maintained over 15 days, it would give a similar impact to that shown in Figure 12a of Jin and Hoskins (1995). There are also impacts at lower latitudes, but these are not likely to be physically correct as they are outside the validity of semi-geostrophic theory.

The bottom right panel of Fig. 5.34 shows the second diagonal element of the matrix $\mathbf{BQ}'$ derived from the global semi-geostrophic equations solved by the Met Office UM. It is defined in Cullen (2018), equations (17)

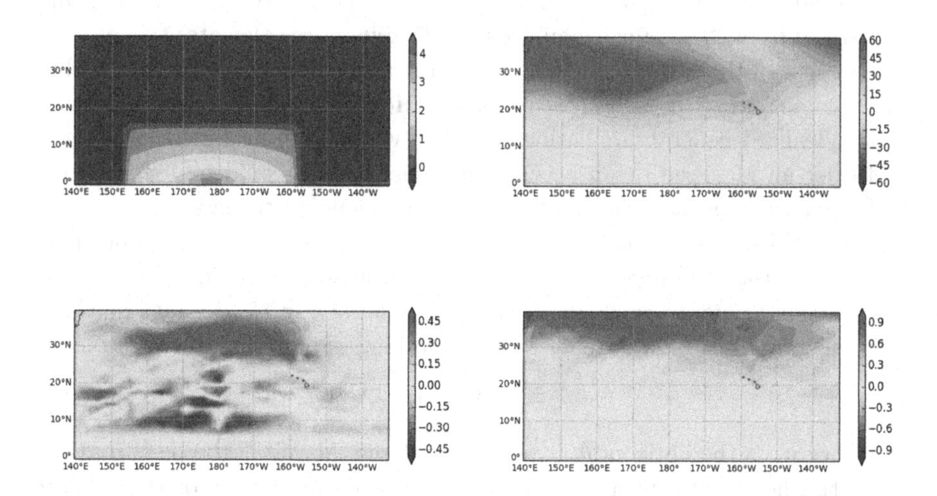

Fig. 5.34 Top: (left) Heating rate at 4000m, units °K day^{-1}; (right) zonal geostrophic wind, units ms^{-1}. Bottom: (left) diagnosed zonal geostrophic wind tendency at 11,500m units ms^{-1}day^{-1}; and (right) (22) coefficient of **Q** matrix, eq. (2.151) at 11,500m, units 10^{-8}s^{-2}. From Cullen (2018). ©Crown Copyright, Met Office.

and (21). This is a generalisation of the matrix **Q** defined in (2.151). It corresponds to the inertial stability of the model state. In the absence of a non-trivial model state, it would take the value f^2, about 0.5×10^{-8} at 30°N. As shown in section 2.7.4, eq. (2.175), small values of this coefficient will be associated with large values of $\frac{\partial}{\partial t}\left(\frac{\partial \phi}{\partial y}\right)$, which is the zonal geostrophic wind tendency shown in the bottom left panel. The plots show that $\mathbf{Q}_{22}$ decreases rapidly on the anticyclonic side of the jet-stream. This allows the effect of the forcing to propagate efficiently northward to the subtropical jet, where further progression is prevented by large values of $\mathbf{Q}_{22}$. These results are consistent with the results of Jin and Hoskins (1995) shown in the right panel of Fig. 5.33, which also shows the downstream propagation over the five day period used.

Chapter 6

Inclusion of Additional Physics

As noted in section 2.1 and section 2.2.3, the governing equations (2.1) have a no-slip boundary condition at the Earth's surface. Even though the Reynolds number is very large, the no-slip condition still exerts control over the solution through a turbulent boundary layer. The turbulence is highly state-dependent, and its bulk effect has to be represented by a vertical diffusivity. This term is therefore included in the large-scale approximation set out in eq. (2.17). Similar terms also have to be included in the thermodynamic and moisture equations once sources and sinks of heat and moisture are included.

Most significant weather phenomena involve moisture. The changes of phase to and from water vapour, liquid water, and ice are associated with large heat sources and sinks which have a big effect on the atmospheric circulation. While these effects are included in production atmospheric models, there has been relatively little work on incorporating them into theories of the large-scale circulation.

There are equivalent additional effects in the ocean as well, but they are quite different from those in the atmosphere and are not discussed in this book.

6.1 Inclusion of friction

6.1.1 *Semi-geotriptic theory*

A major deficiency in representing large-scale atmospheric dynamics by a semi-geostrophic model is the neglect of frictional drag. Though this neglect makes it much easier to study the problem, actual predictions made without including frictional drag give completely unrealistic results. Beare (2007) shows that the depth of an extratropical cyclone could be increased by 25hpa over 60 hours by switching off the frictional drag. In the atmospheric boundary layer, frictional drag is of similar magnitude to the pressure gradient and the Coriolis acceleration. In the tropics, frictional drag can balance the pressure gradient.

In the presentation of the governing equations (2.1), a kinematic viscosity term was used to enforce the no-slip boundary condition. As noted above, kinematic viscosity acts on a scale far too small to relate to large-scale solutions. It is necessary to introduce a turbulent boundary layer, and then represent the bulk effect of this layer by a vertical diffusion term which has to be highly state-dependent to reproduce the observed behaviour. Since frictional drag is small above the boundary layer, the vertical diffusion term should be written in conservation form so that the effect on the momentum budget is well-defined.

It is important to develop a method for incorporating the vertical diffusion in the boundary layer in a way that is consistent with the semi-geostrophic approximation above the boundary layer. This problem has been studied by Cullen (1989a) and Ostdiek and Blumen (1997). More recent work includes Holtslag and Boville (1993) and a series of papers by Beare and Cullen (2010), Beare and Cullen (2012), Beare and Cullen (2013) and Beare and Cullen (2016). The latter papers have further developed and tested a semi-geotriptic model that combines standard boundary layer theory with semi-geostrophic dynamics in a consistent way, following on from Cullen (1989a). This model was introduced in section 2.5.2 and used in the diagnostic studies illustrated in section 2.7.4. These papers form the basis of the rest of this section.

The conceptual structure of the theory is shown in Fig. 6.1. The steady-state balance without friction is geostrophic balance, where the Coriolis acceleration balances the pressure gradient and the wind blows along isobars. In the presence of friction there is a three-way balance between the pressure gradient, Coriolis acceleration and friction. This is called *Ekman balance.* The Ekman balanced wind is called the geotriptic wind as in section 2.5.2. Fig. 6.1 shows that this leads to the wind blowing down the pressure gradient rather than along the isobars. Thus energy is dissipated. Near the equator the Coriolis term is negligible and there is a balance between the pressure gradient and frictional drag. The wind now blows straight down

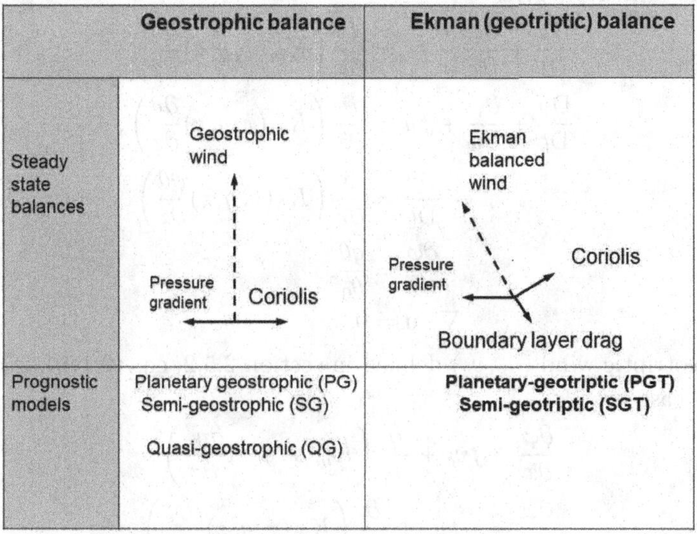

	Geostrophic balance	Ekman (geotriptic) balance
Steady state balances	Geostrophic wind Pressure gradient Coriolis	Ekman balanced wind Pressure gradient Coriolis Boundary layer drag
Prognostic models	Planetary geostrophic (PG) Semi-geostrophic (SG) Quasi-geostrophic (QG)	**Planetary-geotriptic (PGT)** **Semi-geotriptic (SGT)**

Fig. 6.1 The balances (a) above and (b) within the boundary layer. (i) The steady-state balances, and (ii) the associated time-varying balanced models described in chapter 2, sections 2.4.5 and 2.4.6. From Beare and Cullen (2013).

the pressure gradient.

The corresponding evolution equations are also shown in Fig. 6.1. The three types of geostrophic evolution equations are the quasi-geostrophic equations described in section 2.4.5 and the planetary geostrophic and semi-geostrophic equations described in section 2.4.6. The last two become the planetary geotriptic and semi-geotriptic equations. The extra assumptions of quasi-geostrophic theory mean that only a simpler representation of friction can be included in that model, as discussed in Vallis (2017) p. 210.

6.1.2 *The semi-geotriptic equations*

The derivation and demonstration of the semi-geotriptic equations in this section is carried out using Cartesian coordinates (x, y, z) and a constant Coriolis parameter f following Beare and Cullen (2010). Start from the hydrostatic Boussineq Euler equations (3.1). Including a friction term and thermal diffusivity gives

$$
\begin{aligned}
\frac{\mathrm{D}u}{\mathrm{D}t} + \frac{\partial \varphi}{\partial x} - fv &= \frac{\partial}{\partial z}\left(K_m(x,y,z)\frac{\partial u}{\partial z}\right), \\
\frac{\mathrm{D}v}{\mathrm{D}t} + \frac{\partial \varphi}{\partial y} + fu &= \frac{\partial}{\partial z}\left(K_m(x,y,z)\frac{\partial v}{\partial z}\right), \\
\frac{\mathrm{D}\theta}{\mathrm{D}t} &= \frac{\partial}{\partial z}\left(K_h(x,y,z)\frac{\partial \theta}{\partial z}\right), \\
\frac{\partial \varphi}{\partial z} &= \frac{g\theta}{\theta_0}, \\
\nabla \cdot \mathbf{u} &= 0.
\end{aligned}
\tag{6.1}
$$

The geotriptic wind (u_e, v_e) defined in section 2.5.2, eq. (2.170), is defined in this case by

$$
\begin{aligned}
\frac{\partial \varphi}{\partial x} &= fv_e + \frac{\partial}{\partial z}\left(K_m(x,y,z)\frac{\partial u_e}{\partial z}\right), \\
\frac{\partial \varphi}{\partial y} &= -fu_e + \frac{\partial}{\partial z}\left(K_m(x,y,z)\frac{\partial v_e}{\partial z}\right).
\end{aligned}
\tag{6.2}
$$

Friction is added to the three-dimensional semi-geostrophic equations (3.2) in the same way as in (2.171), giving

$$
\begin{aligned}
\frac{\mathrm{D}u_e}{\mathrm{D}t} + \frac{\partial \varphi}{\partial x} - fv &= \frac{\partial}{\partial z}\left(K_m(x,y,z)\frac{\partial}{\partial z}(2u_e - u)\right), \\
\frac{\mathrm{D}v_e}{\mathrm{D}t} + \frac{\partial \varphi}{\partial y} + fu &= \frac{\partial}{\partial z}\left(K_m(x,y,z)\frac{\partial}{\partial z}(2v_e - v)\right), \\
\frac{\mathrm{D}\theta}{\mathrm{D}t} &= \frac{\partial}{\partial z}\left(K_h(x,y,z)\frac{\partial \theta}{\partial z}\right), \\
\frac{\partial \varphi}{\partial z} - g\frac{\theta}{\theta_0} &= 0, \\
\nabla \cdot \mathbf{u} &= 0.
\end{aligned}
\tag{6.3}
$$

Note that (u_e, v_e) are now prognostic variables, and (u, v, w) are diagnostic. The kinetic energy is defined in terms of $\mathbf{u}_e$ and the appearance of $(2u_e - u)$ and $(2v_e - v)$ on the right hand side of the first two equations rather than u and v is necessary to ensure that the energy strictly decreases with time in the absence of heat sources. This can be seen to work by writing the first two equations of (6.3), using (6.2), as

$$
\begin{aligned}
\frac{\mathrm{D}u_e}{\mathrm{D}t} + f(v_e - v) &= \frac{\partial}{\partial z}\left(K_m(x,y,z)\frac{\partial}{\partial z}(u_e - u)\right), \\
\frac{\mathrm{D}v_e}{\mathrm{D}t} + f(u - u_e) &= \frac{\partial}{\partial z}\left(K_m(x,y,z)\frac{\partial}{\partial z}(v_e - v)\right).
\end{aligned}
$$

The detailed argument is given by Beare and Cullen (2010) and a demonstration that this is essential is given by Beare and Cullen (2012).

The equations are solved in a domain $\Gamma = [-L, L] \times [-D, D] \times [0, H]$, with $\mathbf{u} = \mathbf{u}_e = 0$ at $z = 0$ and $w = 0$ at $z = H$. The lateral boundary conditions are that $w = \frac{\partial(u,v)}{\partial x} = 0$ at $x = \pm L$, $w = \frac{\partial(u,v)}{\partial y} = 0$ at $y = \pm D$. This choice avoids difficulties with boundary layers at lateral boundaries, which would be important if these equations were applied to the ocean. It is assumed that $K_m = K_h = 0$ above some value $z = z_{bl} < H$. The energy E_e is defined similarly to (3.7):

$$E_e = \int_\Gamma \left(\frac{1}{2}(u_e^2 + v_e^2) - g\theta z/\theta_0 \right) \mathrm{d}x\mathrm{d}y\mathrm{d}z. \tag{6.4}$$

Using the boundary conditions and the vanishing of K_m for large z, in the case $K_h = 0$ the evolution equation for the energy can be shown to be

$$\frac{\mathrm{d}E_e}{\mathrm{d}t} = -\int_\Gamma K_m(x,y,z) \left(\left(\frac{\partial u_e}{\partial z} \right)^2 + \left(\frac{\partial v_e}{\partial z} \right)^2 \right) \mathrm{d}x\mathrm{d}y\mathrm{d}z, \tag{6.5}$$

so that the energy decreases with time in the absence of thermal diffusion.

6.1.3 *Sea-breeze circulations*

In this section it is shown how the friction acts in this model in a very simple case. Solutions in a two-dimensional cross-section $(x, z) \in [-100, 100] \times [0, 10]$ km^2 are shown, as in sections 3.4.2, 5.3 and 5.5. Numerical solutions are obtained using a finite difference method similar to that described in section 5.1.3. This method requires that the matrix $\mathbf{Q}$ that appears in (5.16) is strictly positive definite, which will not be the case in a well-mixed boundary layer. The equations are obtained by omitting the term $\partial\varphi/\partial y$ from (6.3). In addition, the fourth equation of (6.3) is replaced by

$$\frac{\partial \varphi}{\partial z} - g\frac{\theta}{\theta_0} = d_h \frac{\partial^2 w}{\partial x^2}. \tag{6.6}$$

The extra diffusive term regularises the equations when the boundary layer is well-mixed so that $\frac{\partial \theta}{\partial z}$ is small and $\mathbf{Q}$ would not otherwise be strictly positive definite.

The coefficients K_m and K_h depend on the Froude number defined by

$$Fr^2 = \left| \frac{\partial \mathbf{u}_e}{\partial z} \right|^2 / N^2, \tag{6.7}$$

where N^2 is the Brunt-Väisälä frequency defined in (2.21). Fr^{-2} is called the *Richardson number*. The formulae used for K_m and K_h are taken from the meteorological literature as described in Beare and Cullen (2010).

The solutions are compared with solutions of the Euler model (6.1) with the term $\partial\varphi/\partial y$ omitted and the same form of the frictional drag and thermal conductivity. The problem is forced by a prescribed evolution of the surface temperature given by

$$\theta_{surf} = \theta_{init} + \frac{\theta_{amp}}{4}\left(1 - \cos\left(\frac{2\pi t}{\tau}\right)\right)\left(1 + \tanh\left(-\frac{x}{L_x}\right)\right). \quad (6.8)$$

τ is a time-scale, set to 24 hours, and L_x a space scale which strongly restricts the time variation of θ_{surf} for x >50km. Thus these values of x corresponds to the sea while the remainder correspond to land. The initial θ in the atmosphere is a function of z only with $\partial\theta/\partial z = 2 \times 10^{-3}$ °Km^{-1}. The models are run for 36 hours and the Coriolis parameter f is set to 10^{-4}s^{-1}.

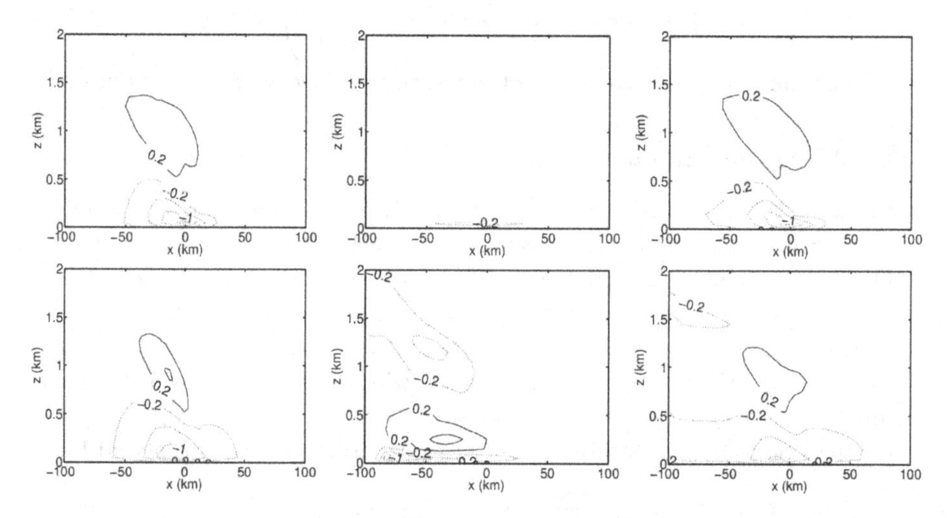

Fig. 6.2 Top: Vertical cross-sections of the u component of the wind, ms^{-1} after 12, 24 and 36 hours from the semi-geotriptic model. Bottom: the same for the Euler model. Contour interval 0.4ms^{-1} with non-zero base value of 0.2 chosen for clarity. Solid contours positive, dotted negative. From Beare and Cullen (2010). ©Royal Meteorological Society, Reading, U.K., 2010.

Figs. 6.2 and 6.3 illustrate the results. Only the bottom 2km of the domain is shown. The thermal diffusivity term in the thermodynamic equation in 6.3 transfers the heat into the atmosphere. If the surface is heated, N is reduced, K_h becomes large and heat is transferred to the atmosphere. The formulation is designed so that N cannot become negative. If the

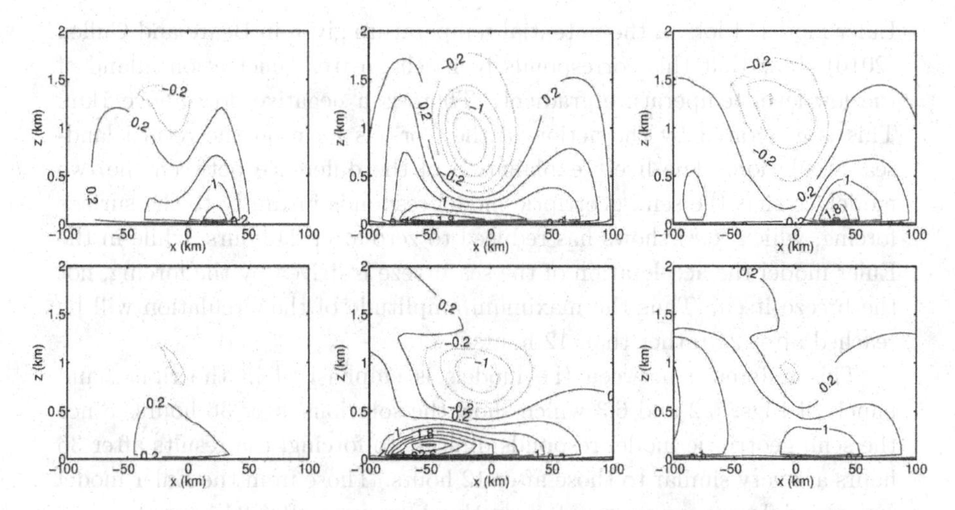

Fig. 6.3 As Fig. 6.2 for the v component of the wind. From Beare and Cullen (2010). ©Royal Meteorological Society, Reading, U.K., 2010.

surface is cooled, N is increased and the heat transfer reduced. With this data, the land x <50km becomes warmer at low levels during the day. The geopotential remains fairly uniform in x at upper levels, so creating lower values at low levels by hydrostatic balance. Over short time-scales, this generates a sea breeze, i.e. a region of negative u at low levels and a return circulation with a larger vertical scale above. This is seen in the left-hand panels of Fig. 6.2. It is similar in both models, showing that this aspect of the semi-geotriptic approximation is valid. In the absence of rotation, the sea breeze would propagate downstream with the friction term balancing the geopotential gradient. In the presence of rotation, a geostrophic wind is generated in the y direction on a time-scale of f^{-1}. The bottom left panel of Fig. 6.3 show that this is happening. In the semi-geotriptic model solution, shown in the top left panel, the effect is much stronger. This is because balance is imposed as a strong constraint.

After 24 hours, the middle panels of Fig. 6.2 show that the sea breeze has almost disappeared in the semi-geotriptic solution, while it has propagated downstream in the Euler solution. In the latter solution, both the sea breeze and the return circulation are confined to low levels. In order to understand this, it is necessary to look at the v component as well, as shown in the middle panels of Fig. 6.3. In the semi-geotriptic solution, there is now a significant positive geostrophic flow at low levels over the land $-75 < x < 50$km. This flow is much stronger and further inland in the

Euler model. Plots of the potential temperature given in Beare and Cullen (2010) show that this corresponds to much greater penetration inland of the low-level temperature gradient. There is a negative flow above 1km. This is generated by the action of the Coriolis term on the return land-sea circulation. The likely explanation of the difference between the two models is that the semi-geotriptic model responds instantly to the surface forcing, which (6.8) shows has reduced to zero after 24 hours, while in the Euler model the acceleration of the sea breeze is driven by the forcing, not the breeze itself. Thus the maximum amplitude of the circulation will be reached after 24 rather than 12 hours.

This difference between the models is emphasised in the right hand panels of Figs. 6.2 and 6.3 which show the solutions after 36 hours. Since the semi-geotriptic model responds directly to forcing, the results after 36 hours are very similar to those after 12 hours. Those from the Euler model are very different, as expected given the differences after 24 hours.

Stull (1988) states that observed sea breezes tend to decay following the surface forcing and not persist into the following day, so the slow decay shown by the Euler model is physically unrealistic. Probably this is because the use of a two-dimensional model eliminates many physical instabilities which would enhance the decay. In addition, accurate treatment of these instabilities is likely to require very much higher resolution, and non-hydrostatic dynamics.

6.1.4 *Modelling low-level jets*

In this section, the semi-geotriptic equations are used as a diagnostic in order to understand the low-level jets which are often observed in the boundary layer. It is based on Beare and Cullen (2013) which should be consulted for further detail. The diagnostic procedure was described and illustrated in section 2.7.4, and a fuller description is given by Cullen (2018). A similar procedure was followed by Thorpe and Pedder (1999), who used the semi-geostrophic omega equation (5.7) to understand the behaviour of an idealised model as described in section 5.4.1.

As in section 6.1.3 a two-dimensional version of the semi-geotriptic model is used. Thus equations (6.4) and the last three equations of (6.3) are solved. Thermal diffusivity is omitted, but a forcing term is included, so that the third equation of (6.3) becomes

$$\frac{\mathrm{D}\theta}{\mathrm{D}t} = S_h. \tag{6.9}$$

$\frac{\partial \varphi}{\partial y}$ is assumed to be zero, and the derivative operators D and ∇ act only in the (x,z) plane. Thus the continuity equation means that $\nabla \cdot (u,w) = 0$, so that (u,w) can be written in terms of a stream-function ψ as $\left(\frac{\partial \psi}{\partial z}, -\frac{\partial \psi}{\partial x}\right)$. Write $\mathcal{D}\xi \equiv \frac{\partial^2}{\partial z^2}(K_m \xi)$ for any function ξ.

The diagnostic equation for Ψ derived from the evolution equations is

$$\left[f\left(f + \frac{\partial v_e}{\partial x}\right) + \mathcal{D}^2 + \frac{\partial u_e}{\partial x}\mathcal{D}\right]\frac{\partial^2 \psi}{\partial z^2} + \frac{g}{\theta_0}\frac{\partial \theta}{\partial z}\frac{\partial^2 \psi}{\partial x^2}$$
$$-\left(\frac{\partial u_e}{\partial z}\mathcal{D} - \mathcal{D}\frac{\partial u_e}{\partial z} + 2\frac{g}{\theta_0}\frac{\partial^2 \theta}{\partial x \partial z}\right)\frac{\partial^2 \psi}{\partial x \partial z} - \frac{\partial \mathcal{D}}{\partial x}\left(\frac{\partial u_e}{\partial z}\right)\frac{\partial \psi}{\partial z} +$$
$$\frac{\partial \mathcal{D}}{\partial z}\left(\frac{\partial u_e}{\partial z}\right)\frac{\partial \psi}{\partial x} = \left(f^2 + \mathcal{D}^2 + \frac{\partial \mathcal{D}}{\partial t}\right)\frac{\partial u_e}{\partial z} - \frac{\partial}{\partial x}\left(\frac{g S_h}{\theta_0}\right). \tag{6.10}$$

As explained in Beare and Cullen (2013), this is solved in a region $[-L, L] \times [0, H]$ with the boundary conditions

$$\frac{\partial \psi}{\partial x} = 0 \quad \text{at } x = \pm L, \tag{6.11}$$
$$\psi = 0 \quad \text{at } z = 0, H.$$

Note that the diffusive term $\mathcal{D}$ appears as $\mathcal{D}^2$ in the coefficient of $\frac{\partial^2}{\partial z^2}$ in (6.10). This improves the ellipticity and hence solvability of the equation. Near the equator, this allows the inertial stability condition $f + \frac{\partial v_e}{\partial x}$ to be relaxed to $f\left(f + \frac{\partial v_e}{\partial x}\right) + \mathcal{D}^2 > 0$. As explained in section 2.4.7, this condition is the main restriction on the applicability of semi-geotriptic theory in the tropics.

The diagnostic equation (6.10) is a generalisation of the classical *Sawyer-Eliassen equation* originally proposed to solve for the vertical circulation at fronts, Eliassen (1962), Sawyer (1956). In this case, the north-south velocity v can be diagnosed separately as

$$(f^2 + d^2)v = \tag{6.12}$$
$$\left(f^2 + d^2 + \frac{\partial d}{\partial t}\right)v_e - d\left(u\frac{\partial v_e}{\partial x} + w\frac{\partial v_e}{\partial z}\right) + f\left(u\frac{\partial u_e}{\partial x} + w\frac{\partial u_e}{\partial z}\right),$$

where $d\xi \equiv \frac{\partial}{\partial z}\left(K_m \frac{\partial \xi}{\partial z}\right)$ for a general function ξ, $v = 0$ at $z = 0$ and $v = v_g$ at $z = H$.

The case used for illustration is taken from Beare and Cullen (2013). It includes a low-level jet parallel to a frontal zone, where there is strong buoyancy and momentum advection in the boundary layer. Such a case is illustrated in Fig. 6.4, taken from an ERA-interim analysis. This shows a

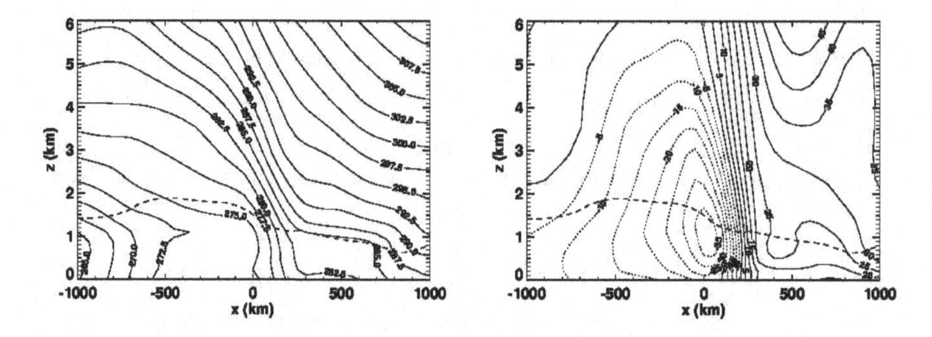

Fig. 6.4 Vertical cross-sections of a mid-latitude cyclone low-level jet from ERA-interim analyses. Cross-sections shown at constant latitude with the origin at latitude 52°N and longitude 41°W for 00UTC on 22 December 2008: (left) potential temperature (contour interval 2.5°K). Dashed line represents the boundary-layer depth; (right) wind in the north-south direction, contour interval 5 ms^{-1}, negative values are represented by dotted lines. From Beare and Cullen (2013).

frontal zone with a 'nose' at the top of the advancing cold air at a height of about 1km. This is associated with with a low level jet exceeding $35\mathrm{ms}^{-1}$ which peaks in the boundary layer. The temperature gradient in the boundary layer is concentrated in a distance of about 300m. The jet extends upwards along the frontal zone and becomes purely geostrophic above the boundary layer.

Since the necessary diagnostics cannot be extracted from an archived analysis, an idealised model was constructed to reproduce the main features of Fig. 6.4. This was achieved by using a geostrophic wind defined by:

$$v_g(\hat{x}, \hat{z}) = V_0 \exp\left[-\left(\frac{c}{c_j}\right)^2\right] \exp\left(-\frac{\hat{z}}{d_j}\right)\left[1 + \tanh\left(-\frac{\hat{x}}{e_j}\right)\right], \tag{6.13}$$

where $\hat{x} = x/L, \hat{z} = z/H$, and $c = \hat{x} + \alpha\hat{z}$. The constants are set as $V_0 = -30\mathrm{ms}^{-1}, c_j = 0.15, d_j = 0.225, e_j = 0.0625$ and $\alpha = 0.3$. The potential temperature field was then created using thermal wind balance. The boundary layer structure and the values of the state-dependent diffusion coefficients K_m and K_h were created by integrating the boundary-layer equations for 6 hours, holding the geostrophic wind and surface potential temperature fixed.

The resulting north-south winds and potential temperatures are shown in Fig. 6.5 for comparison with Fig 6.4. The boundary layer depth is diagnosed as the level where the Froude number decreases to less than 1. The geostrophic wind v_g specified by (6.13) peaks at the surface. The

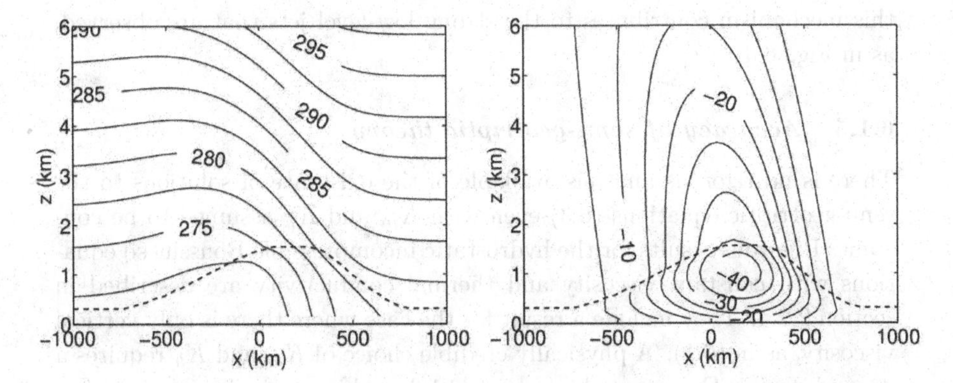

Fig. 6.5 Vertical cross-sections of (left) the potential temperature (contour interval 2.5°K) using the state defined in (6.13). Dashed line represents the boundary-layer depth; (right) the Ekman-balanced wind in the north-south direction (contour interval 5ms^{-1}). From Beare and Cullen (2013).

geotriptic wind v_e found by the initialisation procedure does not, thus there is a low-level jet which peaks below the boundary layer top. The main features of the analysed frontal zone with low-level jet shown in Fig 6.4 have thus been reproduced. It is then possible to solve eq. (6.10) for the total circulation (u, w) and eq. (6.12) for the north-south wind v.

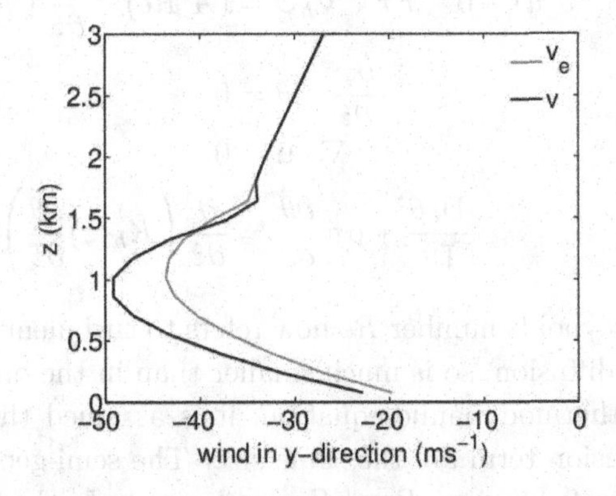

Fig. 6.6 Vertical profiles of v_e (grey) and v (black) at the horizontal location of the jet maximum. From Beare and Cullen (2013).

Fig. 6.6 shows the vertical profiles of v_e and v at the location of the jet maximum. It can be seen that the geotriptic boundary-layer dynamics have increased the jet speed beyond the geotriptic balanced value. This reflects the inclusion of the momentum advection terms in (6.12) and suggests that this mechanism contributes to the strong low-level jets that are observed, as in Fig. 6.4.

6.1.5 *Accuracy of semi-geotriptic theory*

There is no rigorous analysis available of the existence of solutions to the semi-geotriptic equations (6.3) even with K_m and K_h assumed to be constant. Rigorous results for the hydrostatic incompressible Boussinesq equations with constant viscosity and thermal conductivity are described in section 2.2.4. They include a result for the case where there is only vertical viscosity, as in (6.1). A physically credible choice of K_m and K_h requires a dependence on Fr, as noted in section 6.1.3, and in particular requires that K_h tends to zero as $\frac{\partial\theta}{\partial z}$ increases. This almost certainly makes the equations ill-posed, as the term $\frac{\partial}{\partial z}\left(K_h\frac{\partial\theta}{\partial z}\right)$ no longer behaves like a diffusion term. This has always caused practical difficulties.

In order to analyse the convergence of solutions of the Euler equations with friction, (6.1) to those of the semi-geotriptic equations (6.3), it is necessary to write (6.1) in dimensionless form as in (2.116), but assuming constant rotation. Including the scaling of the viscous terms from (2.17) gives

$$\begin{aligned}
\frac{\mathrm{D}(\hat{u},\hat{v})}{\mathrm{D}\hat{t}} + Ro^{-1}(-\hat{v},\hat{u}) + \alpha^{-1}Fr^{-2}\nabla_h\hat{\varphi} &= (A^2Re)^{-1}\frac{\partial}{\partial\hat{z}}\left(\hat{K}_m(\hat{z})\frac{\partial(\hat{u},\hat{v})}{\partial\hat{z}}\right),\\
\frac{\partial\hat{\varphi}}{\partial\hat{z}} - \hat{\theta} &= 0, \\
\nabla\cdot\hat{\mathbf{u}} &= 0,\\
\frac{\mathrm{D}_h\hat{\theta}}{\mathrm{D}\hat{t}} + \alpha\hat{w}\frac{\partial\hat{\theta}}{\partial\hat{z}} &= \frac{\partial}{\partial\hat{z}}\left(\hat{K}_h(\hat{z})\frac{\partial\hat{\theta}}{\partial\hat{z}}\right).
\end{aligned} \tag{6.14}$$

Note that the Reynolds number Re now refers to turbulent diffusion rather than molecular diffusion, so is much smaller than in the original equations (2.17). In the thermodynamic equation, it is assumed that all terms including the diffusion term are the same size. The semi-geostrophic scaling defined in section 2.4.6 gives $Ro = Fr^2$ and $\alpha = 1$. In the boundary layer, a natural scaling is to assume that the rotation, pressure gradient and friction are all the same magnitude. Thus $Re = RoA^{-2}$. The vertical scale which defines the aspect ratio A will be the boundary layer depth H_B. The assumption $Ro = Fr^2$ means that the aspect ratio $A = \frac{f}{N}Ro^{\frac{1}{2}}$. If f, N and the horizontal scale L are held fixed as $Ro \to 0$, that means that $H_B \propto Ro^{\frac{1}{2}}$.

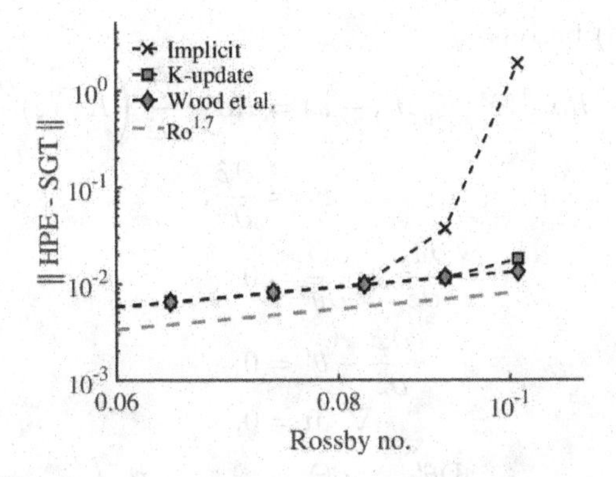

Fig. 6.7 The r.m.s. difference between solutions of (6.14) with boundary-layer timestep 15min and those of the dimensionless form of (6.3) averaged over the domain and plotted against *Ro*. Three methods of boundary-layer time-stepping are used as described in the text and plotted as shown on the figure. The $Ro^{1.7}$ line is shown grey-dashed for reference. From Beare and Cullen (2016). ©Royal Meteorological Society, Reading, U.K., 2016.

Beare and Cullen (2016) tested the convergence of solutions of (6.14) to those of the dimensionless version of (6.3). The semi-geostrophic approximation is accurate to $O(Ro^2)$ with this scaling, as shown in section 2.4.6, but the accuracy of the friction term is only $O(Ro)$ because it uses the geotriptic wind $\mathbf{u}_e$ rather than the total wind $\mathbf{u}$. However, the relation of the boundary layer depth to *Ro* derived above means that the error averaged over a whole domain $[-L, L] \times [0, H]$ will be $O(Ro^{1.5})$. Beare and Cullen (2016) consider a more general power law dependence of H_B on *Ro*.

The test problem used a vertical slice domain $[-1, 1] \times [0, 1]$ in the dimensionless coordinates. The data used was similar to that used for the Eady problem in section 5.3. The geostrophic wind in the $x-$ direction was defined as

$$u_g = \frac{2u_0}{H}\left(z - \frac{H}{2}\right), \tag{6.15}$$

so that the large-scale wave remains stationary. Define $\hat{\theta} = 1 + \Theta(\hat{y}) + \hat{\theta}'$.

Equations (6.14) become

$$\begin{aligned}
\frac{\mathrm{D}(\hat{u},\hat{v})}{\mathrm{D}\hat{t}} - Ro^{-1}(\hat{v}-\hat{v}_g,\hat{u}_g-\hat{u}) &= Ro^{-1}\frac{\partial}{\partial\hat{z}}\left(\hat{K}_m(\hat{z})\frac{\partial(\hat{u},\hat{v})}{\partial\hat{z}}\right),\\
v_g &= \frac{\partial\hat{\varphi}}{\partial\hat{x}},\\
\frac{\partial\hat{u}_g}{\partial\hat{z}} + \frac{\partial\Theta}{\partial\hat{y}} &= 0,\\
\frac{\partial\hat{\varphi}}{\partial\hat{z}} - \hat{\theta}' &= 0,\\
\nabla\cdot\hat{\mathbf{u}} &= 0,\\
\frac{\mathrm{D}\hat{\theta}'}{\mathrm{D}\hat{t}} + \hat{v}\frac{\partial\Theta}{\partial\hat{y}} &= (Ro)^{-1}\frac{\partial}{\partial\hat{z}}\left(\hat{K}_h(\hat{z})\frac{\partial\hat{\theta}'}{\partial\hat{z}}\right).
\end{aligned} \tag{6.16}$$

The boundary conditions were periodic in $\hat{x}$ and

$$\hat{u}-\hat{u}_g = \hat{v} = \frac{\partial\hat{\theta}'}{\partial\hat{z}} = 0$$

at $z=0$, and $\hat{w}=0$ at $z=0,1$. Formulations of $\hat{K}_m(\hat{z})$ and $\hat{K}_h(\hat{z})$ were obtained from the meteorological literature as described in Beare and Cullen (2016).

The initial data used was a uniform $\hat{\theta}'(\hat{z}) = 1+0.09\hat{z}$ with a perturbation added to it with the structure of the fastest growing Eady mode, as in section 5.3.2. This defines $\hat{v}_g$, and $\hat{v}$ was then set equal to $\hat{v}_g$. $(\hat{u},\hat{w})$ were initialised by solving (6.10) in dimensionless form with the friction terms excluded. Solutions were obtained for various values of Ro. Equations (6.16) were then integrated for a time Ro^{-1} so that the solutions remained smooth and the asymptotic convergence could be assessed.

Convergence to the asymptotic limit given by equations (6.3) should be aided by using a numerical method which respects the geotriptic balance (6.2). Thus in a two time level method which advances the solution from time t to $t+\delta t$, all the terms that appear in the first equation of (6.16) should be computed at time $t+\delta t$. This principle is described by Beljaars (1991). The methods of Cullen and Salmond (2003), Diamantakis *et al.* (2006) and Wood *et al.* (2007) all follow this principle, and the tests evaluated the ability of their schemes to achieve the expected convergence rate of $Ro^{1.5}$ to solutions of (6.3).

The convergence was evaluated by performing small timestep simulations for a sequence of values of Ro using (6.16), and using a form of (6.10) to diagnose the semi-geotriptic winds from the predicted pressures. The exact form of (6.10) used is described by Beare and Cullen (2016). The r.m.s.

difference between the winds computed from (6.16) and those diagnosed using (6.10) was calculated and plotted as a function of *Ro*.

The methods used for integrating (6.16) can be described in terms of the discrete equation

$$u_{t+\delta t} - u_t = \delta t F(K(u), u), \tag{6.17}$$

where u represents a model variable and K a diffusion coefficient. The curve labelled 'implicit' in Fig. 6.7 uses an implicit time-stepping scheme with $F(K(u), u)$ calculated as $F(K(u_t), u_{t+\delta t})$, as in Diamantakis *et al.* (2006). Note that $K(u)$ is highly nonlinear, and so cannot be easily calculated at time $t + \delta t$. Also note that because K is state-dependent and can be very large, an implicit method of time-stepping is necessary to ensure computational stability.

The curve labelled $K-$update, uses the iterated scheme described by Cullen and Salmond (2003)

$$\begin{aligned} u_* - u_t &= \delta t F(K(u_t), u_*), \\ u_{t+\delta t} - u_t &= \delta t F(K(u_*), u_{t+\delta t}). \end{aligned} \tag{6.18}$$

This scheme is more accurate because it allows for possible large variations in $K(u)$ during the timestep. However there is significant cost in evaluating $K(u)$ twice per timestep.

The curve labelled 'Wood' uses the method of Wood *et al.* (2007)

$$u_{t+\delta t} - u_t = \delta t\{(F(K(u_t), u_{t+\delta t}) - F(K(u_t), u_t))^{2.5} + F(K(u_t), u_t)\}. \tag{6.19}$$

This works better than the implicit scheme if K increases rapidly during the timestep and avoids the re-evaluation of $K(u)$.

Fig. 6.7 shows that all the schemes show convergence of solutions of the Euler model (6.16) to the semi-geotriptic model (6.3). The convergence rate is closer to $Ro^{1.7}$ than $Ro^{1.5}$, showing that the reduction in boundary layer thickness as Ro decreases is faster than the $Ro^{\frac{1}{2}}$ estimated above. All the schemes give similar results for $Ro \leq 0.08$. At larger Ro, where the wind speeds are greater, the implicit scheme diverges from the others because of the lack of implicit treatment of $K(u)$. The scheme (6.19) performs slightly better than (6.18) for large Ro because it inflates the variation of K during the timestep more than the recalculation in (6.18). However, (6.18) should be more accurate in general situations.

6.1.6 *Tropical performance*

The semi-geostrophic model cannot describe cross-equatorial geostrophic flow, because this is prevented by the inertial stability condition (2.159).

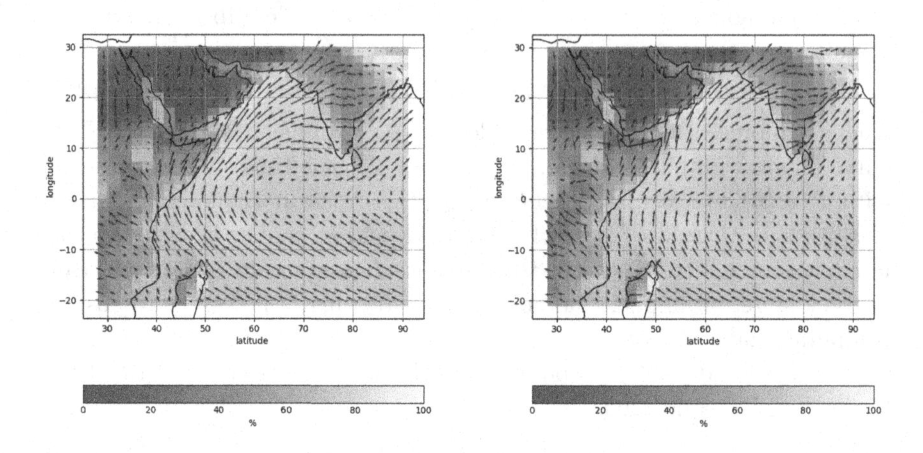

Fig. 6.8 Average winds (arrows) at 277m and relative humidity (%) for June-August of Met Office UM climate simulation. Left: Direct UM prediction. Right: winds calculated from UM pressure using semi-geotriptic diagnostic. Source: Model Evaluation and Diagnostics group, Met Office. ©Crown Copyright, Met Office.

An ageostrophic cross-equatorial flow is possible, and a likely solution if there is more heating in one hemisphere than the other. This will be the case in the Northern hemisphere summer. Such a solution is given in section 4.5.5, fig. 4.5, for a zonal mean model. It was shown in Cullen and Mawson (1992) that the semi-geotriptic model with a representation of the heating over the Tibetan plateau, and a simple representation of the East African orography, could represent the observed cross-equatorial jet.

As discussed in section 2.7.4, a diagnostic procedure consisting of solving the Eulerian form of the semi-geostrophic equations (2.175) will calculate the circulation $\mathbf{u}$ needed to maintain geostrophic and hydrostatic balance against forcing. Thus, in particular, the cross-equatorial flow generated in response to heating over the Tibetan plateau should be represented. The semi-geotriptic model can additionally simulate the Met Office UM boundary layer flow provided that the mixing coefficients are taken from the UM, and the same lower boundary condition is applied, essentially that the surface momentum flux is given by $c_M \mathbf{u}$ where c_M is an exchange coefficient. Thus the low-level jet near East Africa should be represented.

To demonstrate this, the diagnostic procedure described in section 2.7.4 was applied to a climate integration of the UM with a 60km horizontal grid and 70 levels in the vertical. The results presented are averages over June, July and August. The model was run in atmosphere-only mode using sea surface temperatures for 2008. Fig. 6.8 shows the UM output at about 280m, in the middle of the marine boundary layer. The strong cross-equatorial flow is seen in the western Indian Ocean. It extends and strengthens towards the north-western Indian coast, which is consistent with being driven by a heat source over Tibet. Further east there is an area of very light winds along the equator, light westerly winds to the north and stronger south-easterly winds to the south.

The diagnosed winds are also shown in Fig. 6.8. The cross-equatorial flow and the acceleration towards north-west India are well represented. In general the winds are very accurate north of 15°N and south of 15°S. However, the diagnostic does not capture the full magnitude of the winds between 15°N and 15°S. This is probably because the diagnostic cannot capture the rotational tropical circulation above the boundary layer, as shown in section 2.7.5.

6.2 Inclusion of moisture

6.2.1 *Moist semi-geostrophic equations*

The effects of moisture on semi-geostrophic dynamics can be modelled by including additional terms in the thermodynamic equations in equations (6.3). Thus modify (6.3) by setting

$$\begin{aligned}
\frac{\mathrm{D}\theta}{\mathrm{D}t} &= S : q < q_{sat} \text{ or } q = q_{sat} \text{ and } \frac{\mathrm{D}q_{sat}}{\mathrm{D}t} \geq 0,\\
&= S - L\frac{\mathrm{D}q_{sat}}{\mathrm{D}t} : \text{if } q = q_{sat} \text{ and } \frac{\mathrm{D}q_{sat}}{\mathrm{D}t} < 0, \qquad (6.20)\\
\frac{\mathrm{D}q}{\mathrm{D}t} &= 0 : q < q_{sat} \text{ or } q = q_{sat} \text{ and } \frac{\mathrm{D}q_{sat}}{\mathrm{D}t} \geq 0,\\
&= \frac{\mathrm{D}q_{sat}}{\mathrm{D}t} : \text{ if } q = q_{sat} \text{ and } \frac{\mathrm{D}q_{sat}}{\mathrm{D}t} < 0.
\end{aligned}$$

q represents the moisture content, q_{sat} is a prescribed smooth function of θ and z, L is the coefficient of latent heating and S is a source term used to ensure the possibility of non-trivial steady state solutions. In the atmosphere it could represent radiative cooling which compensates the convective heating. The physics behind (6.20), described in Haltiner and Williams (1980), is that q_{sat} is the maximum moisture content which an air parcel can hold. It is a strongly monotonically increasing function of temperature, and is thus a monotonically increasing function of θ and a monotonically decreasing function of z. As an air parcel rises while conserving its potential temperature, its temperature decreases and so q_{sat} decreases while q is conserved. Thus it is common for the condition $q = q_{sat}$ to be met, leading to condensation and release of latent heat. Equations (6.20) represent the simplest model which contains this physics. A rather similar method is used by Majda and Stechmann (2009) to represent moisture in a simple form of the Euler equations with the aim of understanding the Madden-Julian oscillation in the tropics.

It is clear that there will not be a unique global energy minimiser in the sense of Theorem 3.2 when these effects are included. If $q < q_{sat}$ everywhere, then the theorems of section 3.5.1 still hold. Another scenario is that $q = q_{sat}$ everywhere and there is also cloud everywhere so that the dependence of (6.20) on the condition $\mathrm{D}q_{sat}/\mathrm{D}t$ is removed. It is also assumed that $\frac{\partial q_{sat}}{\partial z}$ is a (negative) constant ϱ, so independent of θ. Then (6.20) becomes

$$\begin{aligned} \frac{\mathrm{D}\theta}{\mathrm{D}t} &= S - L\frac{\mathrm{D}q_{sat}}{\mathrm{D}t}, \\ \frac{\mathrm{D}q}{\mathrm{D}t} &= \frac{\mathrm{D}q_{sat}}{\mathrm{D}t}. \end{aligned} \tag{6.21}$$

Then the conserved thermodynamic quantity becomes

$$\theta + L\varrho z. \tag{6.22}$$

Since $L\varrho z$ is independent of x and y, Theorem 3.2 can be extended to this case by using the new thermodynamic variable defined by (6.22). The vertical component of the condition for the matrix $\mathbf{Q}$ of section 3.1 to be positive definite, which was $\partial\theta/\partial z \geq 0$, is replaced by

$$\frac{\partial\theta}{\partial z} \geq -L\varrho. \tag{6.23}$$

Since the right hand side of (6.23) is positive, this is a more severe condition. This is called the *moist stability condition*

Thus, while the global energy minimisation methods will apply to cases where q is either less than q_{sat} everywhere or equal to q_{sat} everywhere, all realistic cases have a mixture of the two. The effect of moisture on the dynamics is then highly non-trivial, giving rise to the complex phenomena observed in reality. This is illustrated in the next subsection. The only rigorous mathematics that has been done in such cases is for the one-dimensional moist rearrangement problem described in section 6.2.4.

6.2.2 *Solutions for stable data*

The effects of latent heating on semi-geostrophic dynamics can be illustrated by using the diagnostic procedure described in section 2.7.4 and used in sections 6.1.4 and 6.1.6. The method is described fully in Cullen (2018). The simplified description here applies to the Euler equations (3.1) with no boundary layer included, and the semi-geostrophic approximation to them, (3.2). Forcing terms are included as in (2.175). Equations (3.2) can be written in Eulerian form as (3.4). Adding the forcing terms in dimensional form gives

$$\nabla \cdot \mathbf{Q}^{-1} \frac{\partial}{\partial t} \nabla \varphi = \nabla \cdot \mathbf{Q}^{-1} \begin{pmatrix} f^2 u_g - S_y \\ f^2 v_g + S_x \\ S_h \end{pmatrix}. \tag{6.24}$$

where

$$\mathbf{Q} = \begin{pmatrix} f^2 + f\frac{\partial v_g}{\partial x} & f\frac{\partial v_g}{\partial y} & f\frac{\partial v_g}{\partial z} \\ -f\frac{\partial u_g}{\partial x} & f^2 - f\frac{\partial u_g}{\partial y} & -f\frac{\partial u_g}{\partial z} \\ \frac{g}{\theta_0}\frac{\partial \theta}{\partial x} & \frac{g}{\theta_0}\frac{\partial \theta}{\partial y} & \frac{g}{\theta_0}\frac{\partial \theta}{\partial z} \end{pmatrix}. \tag{6.25}$$

In the Met Office UM, the effects of moisture on the dynamics are represented by including an appropriate representation of S_h. In the simple formulation of (6.21), which assumes that $q = q_{sat}$ implies the presence of cloud, this would be achieved by defining

$$\begin{aligned} S_h &= 0 : q < q_{sat}(\theta, z), \\ &= -L\frac{\mathrm{D}q_{sat}}{\mathrm{D}t} : q = q_{sat}. \end{aligned} \tag{6.26}$$

Making the same assumption as in deriving (6.22) that $\frac{\partial q_{sat}}{\partial z}$ is a constant ϱ, the same answer could be achieved by replacing the term $Q_{33} = \frac{g}{\theta_0}\frac{\partial \theta}{\partial z}$ in the last row and column of the matrix $\mathbf{Q}$ defined in (6.25) by

$$\begin{aligned} Q_{33} &= \frac{g}{\theta_0}\frac{\partial \theta}{\partial z} : q < q_{sat}(\theta, z), \\ &= \frac{g}{\theta_0}\frac{\partial}{\partial z}(\theta + L\varrho z) : q = q_{sat}, \end{aligned} \tag{6.27}$$

as in (6.23). This is because

$$\begin{aligned} w\frac{g}{\theta_0}\frac{\partial}{\partial z}(\theta + L\varrho z) = w\frac{g}{\theta_0}\frac{\partial \theta}{\partial z} + \frac{g}{\theta_0}L\varrho\frac{\mathrm{D}z}{\mathrm{D}t} = \\ \frac{g}{\theta_0}L\frac{\mathrm{D}q_{sat}}{\mathrm{D}t} = S_h. \end{aligned} \tag{6.28}$$

Thus the dry static stability $\frac{g}{\theta_0}\frac{\partial\theta}{\partial z}$ is replaced by a smaller moist static stability. The effect of latent heating can thus be represented by replacing the dry static stability by the moist static stability. However, if the original equations (6.20) are used, this can only be done where $\frac{\mathrm{D}q_{sat}}{\mathrm{D}t} < 0$, so that the condition $q = q_{sat}$ is maintained. This makes (6.24) fully nonlinear, and it is not at all clear whether it is well-posed. This is why no rigorous results have been obtained for this case.

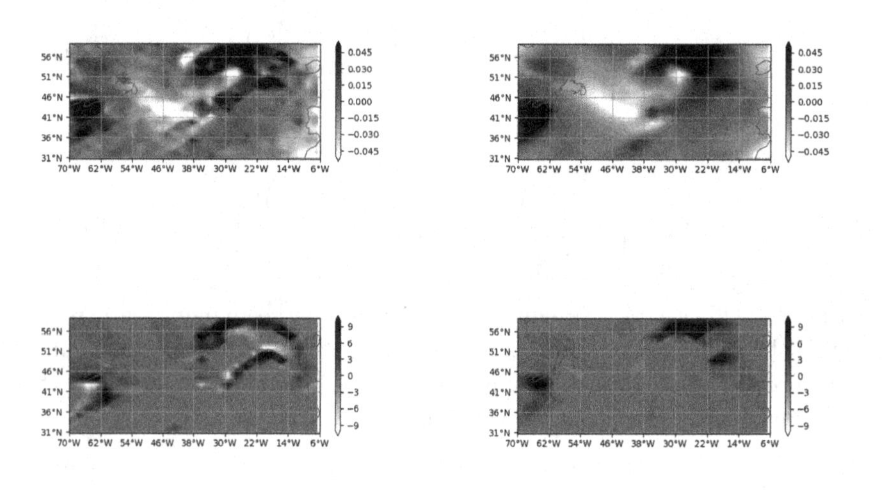

Fig. 6.9 Vertical velocity at 4000m, units ms^{-1}: top left, directly output from UM; top right, calculated from equation (6.24) with heating rates at 4000m; bottom left, heating from large-scale precipitation scheme in UM, units °K day^{-1}; bottom right, latent heating deduced as in (6.28). ©Crown Copyright, Met Office.

Now apply this method to the Met Office UM as in Cullen (2018). The diagnostic procedure described in section 2.7.4 is equivalent to solving (6.24) with forcing terms representing all relevant physical effects. The forcing due to clouds and precipitation can be extracted separately. This is dominated by the latent heating effect, but also includes conversions to and from the ice phase and non-local effects due to clouds and precipitation. The solution is compared with that obtained by omitting this forcing term and replacing the dry static stability by the moist stability as in (6.27). The latter is diagnosed from the UM and includes the effects of cloud water and ice.

Fig. 6.9 shows first the comparison between the vertical velocity w predicted by the UM and that diagnosed by solving the equivalent of (6.24) with the forcing output from the UM. The comparison is over the Atlantic Ocean in the extratropics, so semi-geostrophic theory should be accurate on large scales and is at 4000m, so the representation of the boundary layer does not dominate. This shows very good agreement on large scales, but the direct UM output also includes a lot of small-scale detail which cannot be described by semi-geostrophic theory. The lower left panel shows the heating from clouds and precipitation output from the UM. This shows regions of cooling as well as heating which will usually represent evaporation. This is not allowed for in the very simple model of section 6.2.1. The inferred latent heating is calculated by the equivalent of (6.28) modified to enforce the condition $\frac{\mathrm{D}q_{sat}}{\mathrm{D}t} < 0$. This is shown in the bottom right panel. This captures most of the areas of strong heating, but cannot describe the areas of cooling. Note that this comparison is not exactly the same as that described by Cullen (2018), which did not make this modification and so could describe the cooling.

6.2.3 *Moist instability and moist rearrangements*

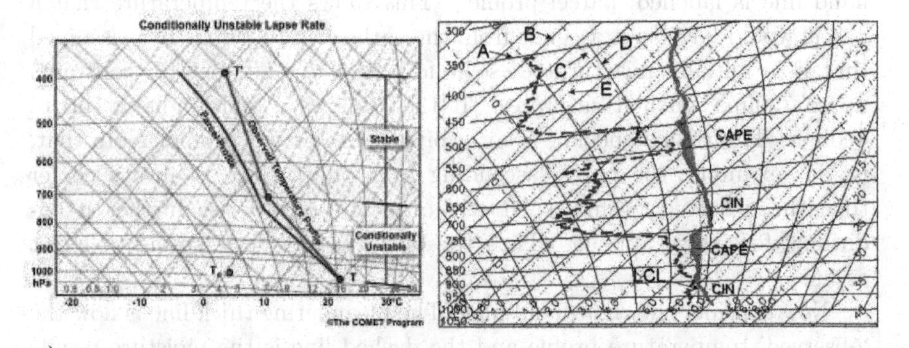

Fig. 6.10 Left: vertical profiles of temperature ($^\circ K$) plotted against pressure (hpa). Fainter curves are lines of constant θ and of θ_E^* as defined below (6.29) (both $^\circ$ K). Thick curves are defined in the text. Right: Same type of plot. A indicates a line of constant p, B a line of constant T, C a line of constant θ, D a line of constant θ_E^*, and E a line of constant q. The thick line is an example temperature profile and the dashed line an example dewpoint profile. Other annotations are described in the text. Left hand plot ©The COMET Program. Right hand plot ©Crown Copyright, Met Office.

The solution of (6.24) and (6.25), with the moist static stability given by (6.27), is only possible if the moist static stability is positive. However, the moist stability condition is only applied if $q = q_{sat}$, which means that some highly non-trivial behaviour becomes possible, as illustrated using Fig. 6.10.

Fig. 6.10 shows example plots of temperature and moisture against pressure on a *tephigram* which is fully explained in Met Office (1975). The reference lines are lines of constant θ and of equivalent potential temperature θ_E^*, which is conserved when moisture is conserved, even when there is evaporation or condensation. The conserved quantity defined in (6.22) achieves this for the simplified equations used in section 6.2.1. A reasonably accurate definition in the real case given in Stull (1988) is

$$\theta_E = \left(T + \frac{L}{C_p}q\right)\left(\frac{p_{ref}}{p}\right)^{(R/C_p)}, \tag{6.29}$$

where C_p, R, p_{ref} are as defined in section 2.1, T is the temperature, q the water vapour mixing ratio, and L is the latent heat defined in section 6.2.1. The equivalent potential temperature at saturation, θ_E^*, is obtained by setting $q = q_{sat}(T,p)$ in (6.29). The lines of constant θ in the tephigram are called *dry adiabats* and the curves of constant θ_E^* are called *moist adiabats*.

Now consider the thick lines in the left-hand plot in Fig. 6.10. The left hand line is labelled 'parcel profile'. This shows the temperature that a parcel would have as it ascends from the surface at 1000hpa. It is assumed that it is initially unsaturated, so it follows a dry adiabat (θ=constant) up to 730hpa when it saturates. It then follows a moist adiabat up to 400hpa. The right hand line is an arbitrary 'observed' profile. Note that, in this example, the parcel ascending from the surface is always cooler then the observed profile, so it will only be able to rise further if additional forcing is applied. Thus the observed profile is stable to moist ascent.

Now consider the right hand plot. The meandering thick line is now the 'observed' temperature profile and the dashed line is the moisture profile as represented by the dewpoint. The dewpoint is the temperature at which the air becomes saturated if it is cooled at constant pressure. Thus if the dashed line coinides with the solid line, the air is saturated. At 1000hpa the observed profile is unsaturated. At 950hpa it become saturated. Thus a parcel ascending from this level, called the LCL (lifting condensation level) will follow a moist adiabat. This moist adiabat is plotted as a solid line.

It can be seen that this moist adiabat is sometimes to the left (colder) and sometimes to the right (warmer) of the observed profile. The difference is shaded. If the moist adiabat is warmer, the ascending parcel will be warmer and so more buoyant than the observed profile, so it will be unstable and 'jump' to the level where the observed profile next crosses the moist adiabat. The tephigram is designed so that areas represent energy, so the size of the shaded region where the parcel jumps indicates the 'CAPE', which means 'convective available potential energy'. The jump releases the CAPE. While in reality this does not happen instantly, usually the time-scale is about 2 hours, this mechanism is responsible for much severe weather. In the layer above, labelled CIN (convective inhibition), the observed profile is warmer than the parcel, so the parcel is less buoyant and cannot rise. At 600hpa, the observed profile again goes to the left of the moist adiabat, so CAPE is indicated. However, the air is very dry at this level so no convection can be initiated.

It can be seen that there are multiple minimum energy states that can be obtained by rearranging a given θ and q profile, even in one dimension. A minimum energy state corresponds to $\frac{\partial \theta}{\partial z} \geq 0$ at points where the air is unsaturated, but $\frac{\partial \theta_E^*}{\partial z} \geq 0$ at points where the air is saturated. However, there is no *a priori* control over when saturation occurs.

6.2.4 *The moist rearrangement problem*

Given the physical background in the previous subsection, the hope is that the time evolution of a moist single column of air acted on by prescribed forcing can be predicted in a rigorous way. The physical analysis above suggests that a Lagrangian parcel approach should be used, and the issue will be to show that this converges to a well-defined solution as the number of parcels is increased. Such a result was proved by Cheng *et al.* (2017a). A description of the result with computational illustrations is given in Cheng *et al.* (2017b). These papers are summarised in this and the following subsections.

The equations solved for the moist single column of air are written in pressure coordinates, since the Boussinesq approximation used in (6.20) gives very unrealistic results when fluid parcels move large distances in the vertical, which is the case here. q_{sat} is therefore now a specified function of θ and p which is monotonically increasing in both variables. The equations are

$$\begin{aligned}
\frac{\mathrm{D}\theta}{\mathrm{D}t} &= 0 : q < q_{sat} \text{ or } q = q_{sat} \text{ and } \frac{\mathrm{D}q_{sat}}{\mathrm{D}t} \geq 0, \\
&= -L\frac{\mathrm{D}q_{sat}}{\mathrm{D}t} : \text{if } q = q_{sat} \text{ and } \frac{\mathrm{D}q_{sat}}{\mathrm{D}t} < 0, \\
\frac{\mathrm{D}q}{\mathrm{D}t} &= 0 : q < q_{sat} \text{ or } q = q_{sat} \text{ and } \frac{\mathrm{D}q_{sat}}{\mathrm{D}t} \geq 0, \\
&= \frac{\mathrm{D}q_{sat}}{\mathrm{D}t} : \text{ if } q = q_{sat} \text{ and } \frac{\mathrm{D}q_{sat}}{\mathrm{D}t} < 0, \\
\frac{\partial \omega}{\partial p} &= 0.
\end{aligned} \tag{6.30}$$

ω is the vertical velocity in pressure coordinates.

The equations are solved in a domain $\Gamma = [p_0, p_\mathcal{T}]$ with $\omega = 0$ at $p = p_0, p_\mathcal{T}$. Some forcing has to be added to create a non-trivial evolution. This is achieved by requiring q_{sat} to be a specified monotonically decreasing function of time $q_{sat}(\theta, p, t)$.

Following (6.23), the condition for a local energy minimiser is now that

$$\begin{aligned}
\frac{\partial \theta}{\partial p} &\leq 0 \text{ if } q < q_{sat} \\
\frac{\partial \theta}{\partial p} &\leq -L\frac{\partial q_{sat}}{\partial p} \text{ if } q = q_{sat}.
\end{aligned} \tag{6.31}$$

Note that the inequalities are reversed because p decreases with physical height. Refering back to Fig. 6.10, this requires the increase in θ with height to be steeper than a dry adiabat if $q < q_{sat}$ and steeper than a moist adiabat if $q = q_{sat}$. It is now possible to define an *admissible* solution of (6.30) by

Definition 6.1. A admissible solution of the moist single column equations (6.30) is a pair $(\theta(p), q(p))$ satisfying (6.31) and with $q(p) \leq q_{sat}(\theta, p, t)$ for all p, t.

Now note that the only classical solution of (6.30) is $\omega = 0$ for all p, so that given admissible initial data and a specified q_{sat} that decreases with time, q will become equal to q_{sat} at some points. At these points (6.30) means that θ will increase and q will decrease to remain equal to q_{sat}. Thus the admissibility condition may be violated. This is unphysical, because such a state would be unstable under the action of the full three-dimensional equations.

This can be resolved by seeking weak solutions which satisfy the admissibility condition. This uses the concepts of a Lagrangian flow and push forward of measures which were introduced in chapter 3. Thus write

$$\begin{aligned}
\frac{\mathrm{D}(\theta + Lq)}{\mathrm{D}t} &= 0, \\
\frac{\mathrm{D}\theta}{\mathrm{D}t} &= 0 : q < q_{sat} \text{ or } q = q_{sat} \text{ and } \frac{\mathrm{D}q_{sat}}{\mathrm{D}t} \geq 0, \\
&= -L\frac{\mathrm{D}q_{sat}}{\mathrm{D}t} : \text{if } q = q_{sat} \text{ and } \frac{\mathrm{D}q_{sat}}{\mathrm{D}t} < 0, \\
F_{t\#}\mathcal{L}^1_{[p_0,p_\tau]} &= \mathcal{L}^1_{[p_0,p_\tau]} \text{ for any } t.
\end{aligned} \tag{6.32}$$

F_t is the Lagrangian flow between particle positions at the initial time and time t and $\mathcal{L}$ is the Lebesgue measure.

The approach taken is to represent the initial data by N parcels of equal mass, thus equal size in p coordinates. The map F_t then describes a rearrangement of the parcels, as defined in chapter 3. In the cases treated in chapter 3, there was a global energy minimiser, and hence a unique rearrangement. In the present case, like the mountain ridge case of section 5.5, it is necessary to find an accessible local energy minimiser. The difficulty in defining the method is to establish the rules for finding a local minimiser. These have to come from the physics of the problem. It is then necessary to prove that the procedure converges to some form of weak solution of (6.32).

6.2.5 *Discrete solutions of the moist rearrangement problem*

Given a set of N parcels with associated potential temperature and moisture content θ_n, q_n, the first step in the rearrangement algorithm is to arrange the parcels so that θ_n is monotonically decreasing in pressure, so increasing in height. An example is shown in Fig. 6.11 of an initially unstable choice of θ_n, with q_n set equal to zero.

The result of the rearrangement is shown in Fig. 6.12. θ now increases monotonically in height. The right hand plot shows that the layer of parcels starting below 2000m has been turned upside down, and the parcels from this layer with higher values of θ have been mixed in with pre-existing parcels with the same θ. A similar effect is seen with the layer starting between 5000m and 8000m. This behaviour reflects the large regions of instability in the initial data shown in Fig. 6.11. In a realistic time-dependent solution, such a state could never be achieved, and the rearrangements to maintain stability would be more like a localised mixing.

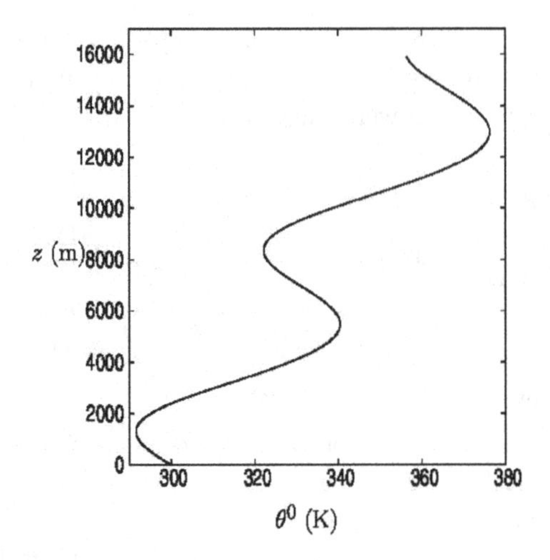

Fig. 6.11 Plot of the initial unstable potential temperature distribution (°K) as a function of height z (m). From Cheng *et al.* (2017b).

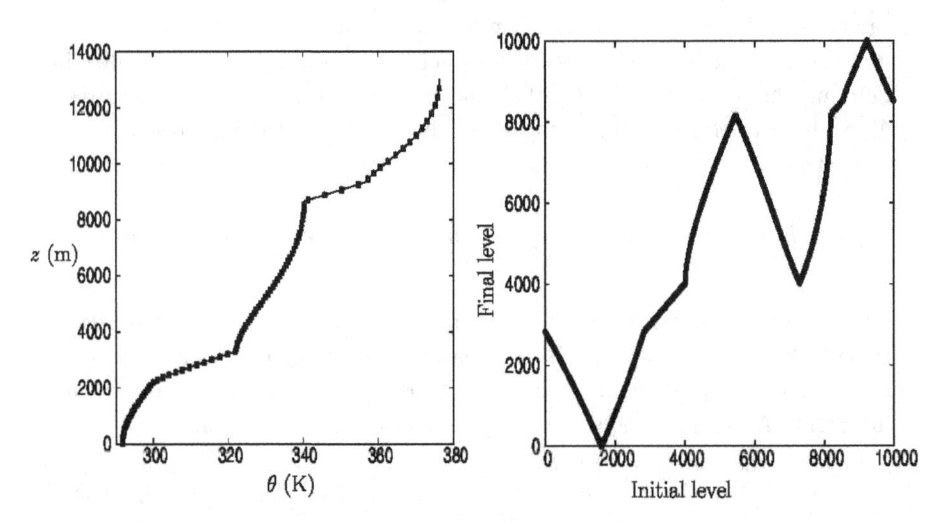

Fig. 6.12 Left: Plot of the final potential temperature distribution as a function of height, using 10000 parcels. Right: Plot of the final Eulerian position of each parcel as a function of its initial position. From Cheng *et al.* (2017b).

Now consider general data with q non-zero. Consider also a simple but realistic time evolution modelled by time-stepping. The admissibility condition, Definition 6.1, has to be satisfied at the end of each timestep. In order to obtain a non-trivial solution, it is assumed that q_{sat} is monotonically decreasing in time, as stated before (6.32). To take the next timestep, q_{sat} is first reduced by the specified amount. Then it is possible that the condition $q_n \leq q_{sat_n}$ will now be violated for some values of n. The right hand panel of Fig. 6.10 shows that there can be more than one layer close to saturation. In the case shown, these are at 900hpa and 550hpa. These may both become supersaturated in the same timestep. The rearrangement algorithm has to handle this situation in a way that gives a converged solution as the timestep tends to zero.

In order to explain the difficulties, consider the example temperature profile in the right hand plot of Fig. 6.10. If a parcel at 740hpa bcame saturated, it would rise, conserving $\theta + Lq$, to the level at which the moist adiabat starting from 740hpa next intersects the profile. This condition is referred to as the *convective inhibition criterion* (CIC) which in this case is at 375hpa. Similarly, if a parcel at 600hpa became saturated, then it would rise to 500hpa. Thus the first parcel would 'overtake' the second. A further complication is that if a parcel ascends from 740hpa to 375hpa, all the other parcels between 740 and 375hpa are pushed down, so that even if a parcel at 600hpa had become saturated, it would no longer be saturated after being pushed down. In order to obtain a well-defined solution as the timestep tends to zero, it is necessary to impose the rule that the unstable parcel with the highest $\theta + Lq$, and thus on the warmest moist adiabat, is moved first. This is addressed in the algorithm below, taken from Cheng *et al.* (2017b).

—START—

In the moist rearrangement algorithm, the superscripts 0 denote the initial pre-adjusted states, the superscripts 1 denote the final adjusted state, the superscripts U denote initially unsaturated parcels, and the superscripts T denote a temporary value while the sort is occurring. A_i is a Lagrangian position label of the ith parcel. The moist rearrangement algorithm proceeds from the top of the column as follows. For the sorting height $p = \tilde{p}_{N_{\text{sort}}}$ (with $N_{\text{sort}} = N$ initially) and $\tilde{\theta}_i^0$, $\tilde{q}_i^0$, A_i^0 given for $i = 1, ..., N$:

(1) Temporarily move all unsorted parcels from their current level to this new sorting level for $j = \{1, ..., N\} \backslash \{\text{sorted parcels}\}$.

(2) (a) **If** $j < N_{\text{sort}}$ **and** $\tilde{q}_j^0 \geq q_{sat}(\tilde{\theta}_j^0, \tilde{p}_{N_{\text{sort}}})$ (i.e. the parcel is saturated or supersaturated) then excess moisture will condense and its potential temperature will increase by the latent heat exchange. Thus its new potential temperature $\widehat{\theta}_j$ is found by solving the implicit

relation

$$\widehat{\theta}_j + Lq_{sat}(\widehat{\theta}_j, \tilde{p}_{N_{\text{sort}}}) - \tilde{\theta}^0_j - L\tilde{q}^0_j = 0.$$

The hat on $\widehat{\theta}_j$ denotes its Eulerian value during the adjustment sorting phase. The corresponding moisture level is then found via

$$\widehat{q}_j = q_{sat}(\widehat{\theta}_j, \tilde{p}_{N_{\text{sort}}}). \tag{6.33}$$

(b) **If** $j < N_{\text{sort}}$ **and** $\tilde{q}^0_j < q_{sat}(\tilde{\theta}^0_j, \tilde{p}_{N_{\text{sort}}})$ (i.e. the parcel is unsaturated) then the parcel should not move. Thus assign $\widehat{\theta}_j = \tilde{\theta}^0_j$.

(c) **If** $j \geq N_{\text{sort}}$ **and** $\tilde{q}^0_j \leq q_{sat}(\tilde{\theta}^0_j, \tilde{p}_j)$ then these parcels move down adiabatically so $\widehat{\theta}_j = \tilde{\theta}^0_j$ and $\widehat{q}_j = \tilde{q}^0_j$.

(d) **If** $j \geq N_{\text{sort}}$ **and** $\tilde{q}^0_j > q_{sat}(\tilde{\theta}^0_j, \tilde{p}_j)$ then the excess moisture is converted into latent heat at level j and then the parcel is moved adiabatically to level N_{sort}. Thus $\widehat{\theta}_j$ is a solution to

$$\widehat{\theta}_j + Lq_{sat}(\widehat{\theta}_j, \tilde{p}_j) - \tilde{\theta}^0_j - L\tilde{q}^0_j = 0,$$

and $\widehat{q}_j$ is given by (6.33).

(3) The process in (2) generates an array of potential temperatures $\Phi(N_{\text{sort}}) = \{\widehat{\theta}_k : k \text{ is an unsorted parcel}\}$ temporarily moved to level N_{sort}. The largest of these potential temperatures, $\widehat{\theta}_J$ from $\tilde{p}_J$ say, is then checked to see whether it satisfies the CIC.

If this parcel is unsaturated that completes the algorithm for this N_{sort}. If it is saturated, check that its potential temperature is greater than that of all the unsaturated parcels it had to pass to get to level N_{sort}. This means checking that

$$\theta^{\text{T}}_J(\tilde{p}^U_k) > \theta^0(\tilde{p}^U_k), \tag{6.34}$$

where θ^{T}_J is a solution of

$$\theta^{\text{T}}_J + Lq_{sat}(\theta^{\text{T}}_J, \tilde{p}^U_k) - \theta^0_J - L\tilde{q}_J = 0,$$

for all $\tilde{p}^U_k$ $k \in (J, N_{\text{sort}})$. If (6.34) holds for all $k \in (J, N_{\text{sort}})$, then the parcel from pressure level $\tilde{p}_J$ is installed at the level $\tilde{p}_{N_{\text{sort}}}$ with

$$\tilde{\theta}^1_{N_{\text{sort}}} = \widehat{\theta}_J, \quad \tilde{q}^1_{N_{\text{sort}}} = \widehat{q}_J, \quad A^1_{N_{\text{sort}}} = A^0_J.$$

The parcel from level $\tilde{p}_J$ is then eliminated from the sort, and the algorithm moves to level $\tilde{p}_{N_{\text{sort}}-1}$.

If (6.34) does not hold for all k in this range then this parcel cannot rise to $\tilde{p}_{N_{\text{sort}}}$. Then check the CIC for the parcel with the second largest potential temperature in $\Phi(N_{\text{sort}})$ at the current sorting level. This process continues until either an unsaturated parcel is found or (6.34) holds for all $k \in (J, N_{\text{sort}})$. Thus either a single parcel is installed at $\tilde{p}_{N_{\text{sort}}}$ or no updating takes place.

In order to ensure convergence of the procedure as $\delta t \to 0$ when used to solve a time-dependent problem, it is essential to show that the rate of mass transfer by the convection is bounded. When a parcel is removed from the column and re-installed at a higher level, the remaining parcels move down and the vertical gradient of θ is increased. Therefore the moist instability condition, which is the reverse of (6.23), will no longer be satisfied and the CIC will no longer be satisfied. This effect is exploited by Cheng *et al.* (2017a) to prove that, if parcels are only moved one at a time, the amount of mass that can be transferred in a given time interval δt is bounded.

Thus the algorithm then takes two forms. Either

(4) As in Cheng *et al.* (2017b). Steps 1-3 are repeated at each decreasing level $N_{\text{sort}} < N$ until $N_{\text{sort}} = 1$ and every parcel is assigned a level. Then the sorted (adjusted) values are $\tilde{\theta}_i^1$, $\tilde{q}_i^1$, A_i^1 for $i = 1, ..., N$. This is much more numerically efficient, and works for data where there are no finite regions of constant θ and q. The results shown below use this algorithm.

(5) As in Cheng *et al.* (2017a). After installing a parcel from level p_J at level $\tilde{p}_{N_{\text{sort}}}$, the values $\tilde{\theta}_i^1$, $\tilde{q}_i^1$, $A_i^1, i \neq J$ are applied to the levels $p_1, ..., N_{\text{sort}} - 1$ and the algorithm is started again from the beginning. This is much more expensive, but ensures that the rate of mass transfer is controlled and the solution will thus converge as $\delta t \to 0$.

—END—

6.2.6 *Measure-valued solutions of the moist rearrangement problem*

This subsection shows that solutions of (6.32) can be obtained by proving convergence of the second version of the discrete algorithm above. These results are due to Cheng *et al.* (2017a), further detail is given in that paper. Assume (6.32) is to be solved with initial data θ_0, q_0 on $p \in [p_0, p_T]$ for a time interval $[0, T]$. F is the Lagrangian map as in (6.32) and $F_t^n(p)$ gives the Eulerian position of the nth particle with initial pressure p at time t.

The first result is for the discrete version of the problem with a fixed N. Write $\theta^M = \theta + Lq$ and $\mathcal{M} = \|\theta_0^n\|_{\mathcal{L}^\infty_{(p_0,p_T)}} + L\|q_0^n\|_{\mathcal{L}^\infty_{(p_0,p_T)}}$. Write

$\hat{\theta}^n(t,p) = \theta^n(t, F^n(t,p))$, $\hat{q}^n(t,p) = q^n(t, F^n(t,p))$. Define $\Theta(\varphi, p, t)$ as the solution of the equation

$$\theta + Lq_{sat}(\theta, p, t) = \varphi,$$

This is equivalent to (6.33). A constant in the theorem that follows is called 'universal' if it depends only on $\mathcal{M}, T$ and q_{sat}.

Theorem 6.1.

(i) $\|\theta^n\|_{\mathcal{L}^\infty_{((p_0,p_\mathcal{T})\times(0,T))}} + L\|q^n\|_{\mathcal{L}^\infty_{((p_0,p_\mathcal{T})\times(0,T))}} \leq C_1$ *for some universal constant* C_1.

(ii) There exists a universal constant $C_2 > 0$, *such that for any* $\epsilon > C_2/n$ *and any* $t \in [0,T)$, *if* $\hat{\theta}^n(t,z) > \Theta(\theta^{M,n}(t,p), F_t^n(p), t) + \epsilon$, *then* $\hat{\theta}^n(t',p) = \hat{\theta}^n(t,p)$ *and* $F_{t'}^n(p) \leq F_t^n(p)$ *for any* $t' - t < \epsilon/C_2$.

(iii) For any $p \in [p_0, p_\mathcal{T}], TV_{t\in[0,T)}(F_t^n(p)) \leq C_3$, *where TV is the total variation.*

(iv) Given any $\epsilon > 0$ *and any* $[t-\epsilon, t+\epsilon] \subset [0,T)$, *then* $|\hat{\theta}^n(t+\epsilon,p) - \hat{\theta}^n(t-\epsilon,p) - (\Theta(\theta^{M,n}(p), F_{t+\epsilon}^n(p), t+\epsilon) - \Theta(\theta^{M,n}(p), F_{t-\epsilon}^n(p), t-\epsilon))^+| \leq 2C_4(\epsilon + \delta t)$.

(v) Given any $s < t \in [0,T)$, *then* $\|\theta^n(t,\cdot) - \theta^n(s,\cdot)\|_{\mathcal{L}^1([p_0,p_\mathcal{T}])} = \|\hat{\theta}^n(t,\cdot) - \hat{\theta}^n(s,\cdot)\|_{\mathcal{L}^1([p_0,p_\mathcal{T}])} \leq C_5\sqrt{t-s+\delta t}$.

(vi) Given any $s < t \in [0,T)$, *then* $\|F^n(t,\cdot) - F^n(s,\cdot)\|_{\mathcal{L}^1(p_0,p_\mathcal{T})} \leq C_6\sqrt{t-s+\delta t}$ *for some universal constant* C_6.

Most of these properties concern regularity of the solution, which is required to prove convergence. Item (iii) states that the length of the discrete trajectory is bounded independently of the number of parcels. This prevents oscillatory behaviour. These properties allow a proof that the mass flux is bounded. This is a key result as stated below:

Theorem 6.2. *Let* $p_{i_0} \in [p_0, p_\mathcal{T}]$. *Chose* $n, \delta t$ *such that* $n\delta t = \frac{1}{2C_1}$, *where* $C_1 = \frac{\sup|\partial_t\Theta|}{\inf(\partial_z\Theta)}$. *Let* $t_0 \in [0,T)$, $\epsilon > 0$ *and define the set*

$$J = \{p \in [p_0, p_\mathcal{T}) : F_{t_0}(p) < F_{t_0}(p_{t_0}), F_{t_0+\epsilon}(p) > F_{t_0+\epsilon}(p_{t_0})\}$$

Then $\mathcal{L}_1(J) \leq C(\epsilon + \delta t)$ *for some universal constant* C.

Next define the function spaces which will contain the converged solution. The first contains all the possible trajectories of an arbitrary parcel. According to item (iii) of Theorem 6.1, these have finite length. So take $B_1 > 0$ and define X_{B_1} as the set of functions $f : (0,T) \to [p_0, p_\mathcal{T}]$ which is left continuous and has total variation no greater than B_1. Next define a space which contains all possible profiles of θ. So take $B_2 > 0$ and define Y_{B_2} as the space of monotone decreasing functions on $[p_0, p_\tau]$ which are right continuous on $[p_0, p_\tau]$ and with absolute bound $\leq B_2$. The admissibility constraint from Definition 6.1 that $\partial_p \theta \leq 0$ is incorporated. Then take $B_3 > 0$ and set $Y = C([0,T); Y_{B_3})$, so Y is the space of continuous maps from $[0,T)$ to Y_{B_3}. This the space of possible evolutions of the potential temperature. Denote a generic trajectory from the space X as γ, a generic profile from the space Y_B as $\theta(p)$, and a generic evolution from the space Y as $\tilde{\theta}(t,p)$.

The non-smooth nature of the problem means that the solution is expected to be probabilistic, as in the measure-valued solutions of the semi-geostrophic equations discussed in sections 3.5.3 and 3.5.4. Thus define a space of probability measures $\mathcal{P}(Y_{B_2} \times \mathbb{R} \times [p_0, p_\tau])$, as introduced in section 3.2.2, whose members define the probability distribution of $\{\theta(p), \theta(p) + Lq(p)\}$ for $p \in [p_0, p_\tau]$. For $\zeta \in \mathcal{P}$, write $\pi_{ij}\zeta$ for the projection to the ith and jth argument.

Then suitable initial data will satisfy the following definition, where $B_2 > 0$ is an arbitrary constant.

Definition 6.2. Let $\zeta_0 \in \mathcal{P}(Y_{B_2} \times \mathbb{R} \times [p_0, p_\tau])$.Then ζ_0 is admissible initial data for (6.32) if the following hold:

(i) $\pi_{13\#}\zeta_0 = \mu_0 \times \mathcal{L}^1_{p_0,p_\tau]}$ for some $\mu_0 \in \mathcal{P}(Y_{B_2})$, and $\pi_{2\#}\zeta_0$ has compact support.

(ii) $Lq_{sat}(\theta(p), p, p_0) \geq s - \theta(p)$ for $\zeta_0 - a.e.(\theta, s, p)$.

The second point says that $(\theta + Lq) - \theta \leq Lq_{sat}$ for all p as required by Definition 6.1.

Now define evaluation maps for the spaces X and Y. For any $t \in [0,T)$, define $e_t : X \to [p_0, p_\tau]$ by $\gamma \mapsto \gamma(t)$ and $e'(\tilde{\theta}) : Y \to Y_{B_3}$ by $\tilde{\theta} \mapsto \tilde{\theta}(t)$. The solution which is a limit of the discrete solutions for large N is now characterised as follows:

Definition 6.3. Let $\lambda \in \mathcal{P}(Y_B \times \mathbb{R} \times [p_0, p_\tau] \times X_{B_1})$, and write the Lagrangian flow (F_t in (6.32)) as $\eta_t = (e'_t \times \mathrm{id} \times \mathrm{id} \times e_t)_{\#}\lambda$, $\zeta_t = \pi_{124\#}\eta_t \in \mathcal{P}(Y_{B_3} \times \mathbb{R} \times [p_0, p_\tau])$. Then λ is a measure-valued solution of (6.32) for admissible initial data ζ_0 if the following hold:

(i) $\zeta_t \to \zeta_0, \pi_{34\#}\eta_t \in \Gamma(\mathcal{L}^1_{[p_0,p_T]}, \mathcal{L}^1_{[p_0,p_T]}) \to (\mathrm{id} \times \mathrm{id}_{\#}\mathcal{L}^1_{[p_0,p_T]})$ narrowly as $t \to 0$, and $t \mapsto \eta_t$ is narrowly continuous.

(ii) For any $t \in [0,T)$, $\pi_{13\#}\zeta_t = \mu_t \times \mathcal{L}^1_{[p_0,p_T]} \in (Y_{B_3} \times [p_0,p_T])$, $\pi_{2\#}\zeta_t$ has compact support. In addition, $Lq_{sat}(\theta(p),p,t) \geq s - \theta(p)$ for $\zeta_t -$ a.e.(θ,s,p).

(iii) For $\lambda -$ a.e.$(\tilde{\theta},s,p,\gamma)$, $\tilde{\theta}_t(\gamma_t) \leq \tilde{\theta}_{t'}(\gamma_{t'})$ for $\mathcal{L}^2 -$ a.e.$(t,t') \in (0,T)^2$ and $t < t'$.

(iv) For $\lambda -$ a.e.$(\tilde{\theta},s,p,\gamma)$, there is equality of the measures

$$\partial_t(\tilde{\theta}_t(\gamma_t)) = L[\partial_t(q_{sat}(\tilde{\theta}_t(\gamma_t),\gamma_t,t))]^E,$$

where E is the 'wet set' defined by

$$\{t \in (0,T) : (\tilde{\theta}(\gamma))^*(t) = s - q_{sat}((\tilde{\theta}(\gamma)^*(t),\gamma_t,t)\}$$

and $\tilde{\theta}(\gamma)^*$ is the monotone decreasing, left continuous version of $\tilde{\theta}_t(\gamma_t)$.

In (i), $\Gamma(\mathcal{L}^1_{[p_0,p_T]},\mathcal{L}^1_{[p_0,p_T]})$ is the set of probability measures on $[p_0,p_T]^2$ whose projections on both components are equal to $\mathcal{L}^1_{[p_0,p_T]}$.

Point (i) specifies in what sense the initial data is satisfied. Thus the probability distribution ζ_t narrowly converges to ζ_0. The second convergence statement means that the Lagrangian flow converges to the identity as $t \to 0$. The statement $\pi_{34\#}\eta_t \in \Gamma(\mathcal{L}^1_{[p_0,p_T]},\mathcal{L}^1_{[p_0,p_T]})$ is a statement that the Lagrangian flow is measure-preserving. Point (ii) shows that ζ_t satisfies the same conditions as for admissibility of the initial data. Therefore any ζ_t can be taken as initial data for subsequent integration. Point (iii) shows that $t \mapsto \tilde{\theta}_t(\gamma_t)$ is almost always monotonically increasing in t. Point (iv) shows that for all choices of evolution of $\tilde{\theta}$ and the trajectory γ, the correct equation is satisfied.

Then the existence theorem states that

Theorem 6.3. *Let ζ_0 be admissible initial data. Then there exists a measure-valued solution of (6.32) with initial data ζ_0.*

6.2.7 *Examples of solutions of the moist rearrangement problem*

Because all saturated parcels are temporarily ascended to every height until they are sorted (with the algorithm choosing the largest θ value satisfying the CIC for the post-rearrangement θ at that level), the algorithm described in section 6.2.5 produces a rearrangement which generates as much latent heating as the initial conditions will allow, thus condensing as much moisture as possible. Solutions of the first version of the algorithm are shown, as much finer discretisations are affordable. The data is chosen to avoid

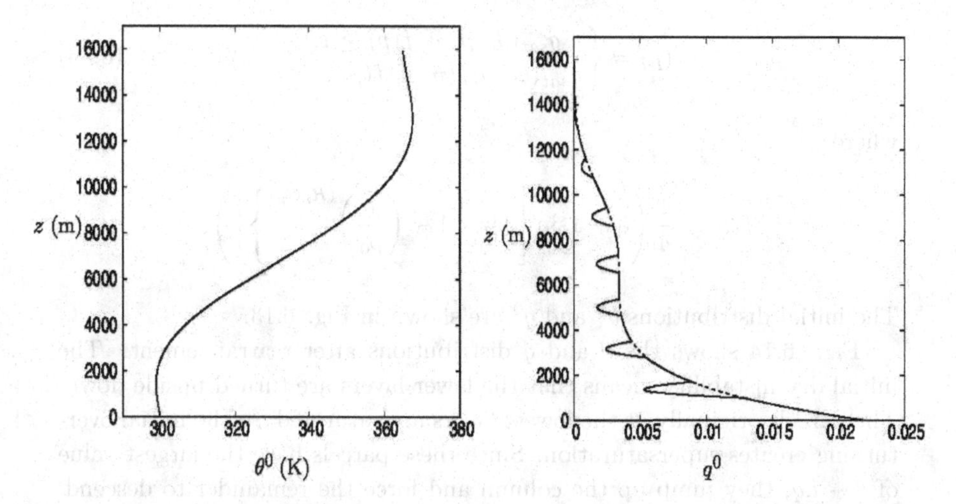

Fig. 6.13 (left) The initial potential temperature distribution $\theta^0(z)$ and (right) the initial moisture distribution $q^0(z)$ given by (6.35) and (6.36). The right panel also shows (dashed) $q_{sat}(\theta, z)$. From Cheng *et al.* (2017b).

the situations of well-mixed $\theta + Lq$ where the algorithm may not give the correct solution. The results demonstrate convergence to either smooth solutions with jumps (this means smooth solutions in adjoining regions), or non-smooth solutions where saturated and unsaturated parcels are interleaved. The latter situation is an example of a measure-valued solution as defined in section 6.2.6.

In section 6.2.6, (6.32) is integrated in time, and a rearrangement is applied at each timestep to ensure that the data remains admissible. In the example shown in this subsection, there is no time integration but an intentionally non-admissible initial state is chosen to force the rearrangement algorithm to make non-trivial changes. Thus the initial potential temperature distribution was given by

$$\theta(p) = 300 \exp\left[\frac{7}{15}\left\{1 - \left(\frac{p}{p_0}\right)^{(R/C_p)}\right\}\right] \times \tag{6.35}$$
$$\left(1 - \frac{1}{25}\sin\left[\frac{14\pi}{3}\left\{1 - \left(\frac{p}{p_0}\right)^{(R/C_p)}\right\}\right]\right).$$

This increases with p, so decreases with z, for p slightly greater than p_0 as shown in Fig. 6.13. The initial moisture distribution was chosen to give regions of saturated and unsturated air. This was

$$q^0(p) = \begin{cases} q_{sat}(\theta_0, p) \text{ if } f(p) \geq 1, \\ q_{sat}(\theta_0, p) f(p) \text{ if } f(p) < 1 \end{cases}, \tag{6.36}$$

where

$$f(p) = \frac{1}{4}\left(5 + 3\sin\left[34\pi\left\{1 - \left(\frac{p}{p_0}\right)^{(R/C_p)}\right\}\right]\right).$$

The initial distributions θ^0 and q^0 are shown in Fig. 6.13.

Fig. 6.14 shows the θ and q distributions after rearrangement. The initial dry instability means that the lower layers are turned upside down. The parcels originally at the lowest layers are saturated, so the initial overturning creates supersaturation. Since these parcels have the largest value of $\theta + Lq$, they jump up the column and force the remainder to descend. They blend with pre-existing parcels with similar θ. The discrete representation creates a rapidly oscillating solution as shown in the right hand plots of Fig. 6.14. Between 4000 and 7000m, the alternation of saturated and unsaturated parcels with the same θ is clearly seen. This is why only measure-valued solutions can be obtained in the limit as the number of parcels increases. Such a solution has also been observed in an oceanographic context by Hieronymus and Nycander (2015). The initial overturning also creates an unsaturated layer below 1000m at the bottom of the column.

6.2.8 *Examples of large-scale flows including moist instability*

In this subsection, the semi-geostrophic solutions of the moist problem solved in section 6.2.1 for stable data are extended to unstable data. The basic numerical time-stepping approach for equations (6.24) is described in section 5.1.3. Given admissible data at each timestep, a predictor step adds increments to θ and $\mathbf{u}_g$, giving fields which are not gradients of φ. These can include increments S_h to θ from latent heat release. A corrector step is then carried out to restore admissibility of the data, given by (5.15) in the dry case. This depends on the matrix $\mathbf{Q}$ defined in (6.25) being strictly positive definite. Alternatively, the geometric method described in section 5.1.2 could be used for this purpose. In either case, the corrector step may result in further vertical motion which will create extra supersaturation. In section 6.2.2 this was addressed by replacing the explicit forcing term by a

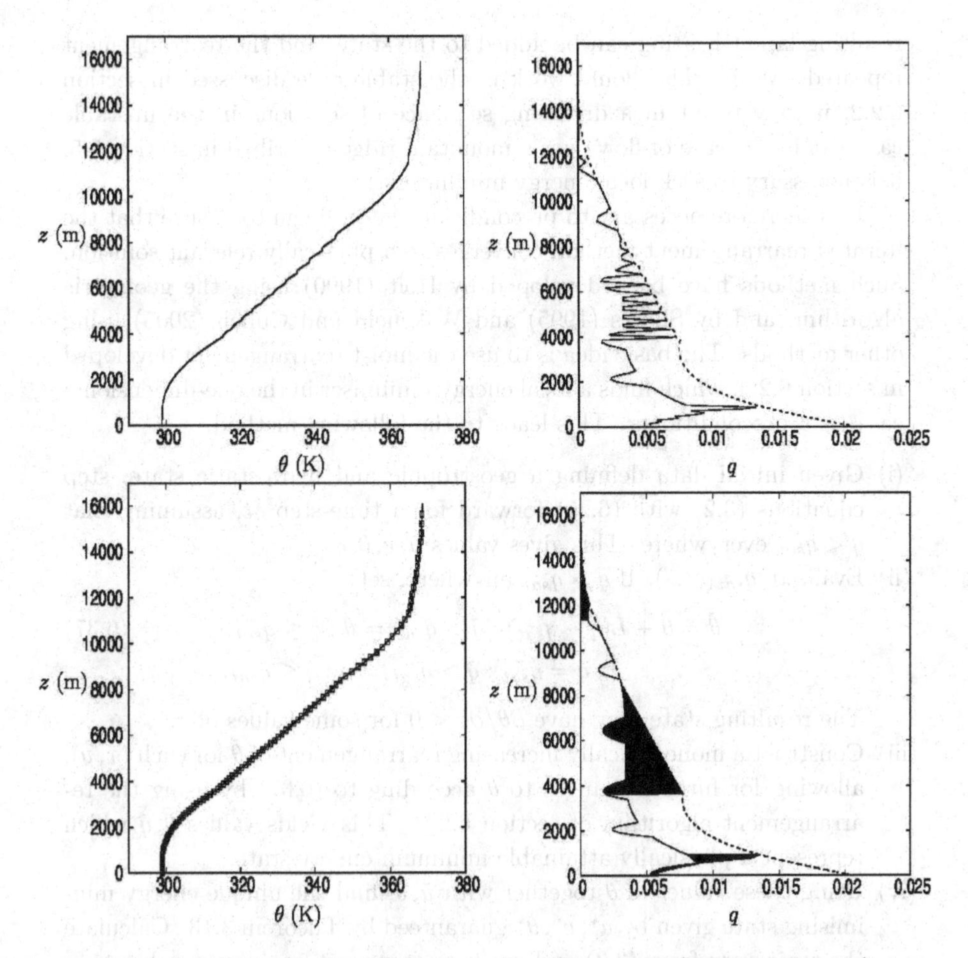

Fig. 6.14 (left) Final potential temperature profiles using (top) 100 parcels and (bottom) 1000 parcels. (right) Final moisture distribution using (top) 100 parcels and (bottom) 1000 parcels. In the right hand plots, the dashed line gives the value of $q_{sat}(\theta, z)$. From Cheng *et al.* (2017b).

modified **Q** given in (6.27). However, this will be less positive definite, and if the moist stability condition (6.23) fails, the method will not work.

The geometric method described in section 5.1.2 and demonstrated in section 5.3.5 represents the state at the end of each timestep as a global energy minimiser. After the prognostic variables are updated in the predictor step, the results will no longer be an energy minimiser and the geometric algorithm will find one by a global rearrangement. However, this cannot take account of extra supersaturation generated by the rearrangement. The resulting latent heating can be added to the state, and the rearrangement repeated. While this should work in the stable case discussed in section 6.2.2, it may result in a diverging sequence of solutions in the unstable case. As in the case of flow over a mountain ridge described in section 5.5, it is necessary to seek local energy minimisers.

It is therefore necessary to precondition the problem to ensure that the iterated rearrangement method converges to a physically relevant solution. Such methods have been developed by Holt (1990) using the geometric algorithm, and by Shutts (1995) and Wakefield and Cullen (2005) using other methods. The basic idea is to use the moist rearrangement developed in section 6.2.4, which finds a local energy minimiser in the one-dimensional case, as a preconditioner. This leads to the following method.

(i) Given initial data defining a geostrophic and hydrostatic state, step equations (3.2) with (6.20) forward for a time-step δt, assuming that $q < q_{sat}$ everywhere. This gives values $\tilde{u}, \tilde{v}, \tilde{\theta}, \tilde{q}$.

(ii) Evaluate $q_{sat}(\delta t, \cdot)$. If $q > q_{sat}$ anywhere, set

$$\begin{aligned} \breve{\theta} = \tilde{\theta} + L(q - q_{sat}) : \tilde{q} > q_{sat}, = \tilde{\theta} : \tilde{q} \leq q_{sat}, \\ \breve{q} = q_{sat} : \tilde{q} > q_{sat}, = \tilde{q} : \tilde{q} \leq q_{sat}. \end{aligned} \tag{6.37}$$

The resulting state may have $\partial\breve{\theta}/\partial z < 0$ for some values of z.

(iii) Construct a monotonically increasing rearrangement of $\breve{\theta}$ for each (x, y), allowing for further changes to $\breve{\theta}$ according to (6.37) by using the rearrangement algorithm of section 6.2.4. This yields values $\hat{\theta}, \hat{q}$ which represent a physically attainable minimum energy state.

(iv) Using these values of $\hat{\theta}$ together with $\tilde{u}, \tilde{v}$, find the unique energy minimising state given by u^*, v^*, θ^* guaranteed by Theorem 3.13. Calculate the trajectory from (3.3) and use it to transport $\breve{q}$, giving a value q^*.

(v) If this state has $q^* \geq q_{sat}$, the procedure has to be iterated starting from step (ii). It is found that step (iii) is a very effective preconditioner for this iteration.

(vi) The result is then $(u_g(\delta t, \cdot), v_g(\delta t, \cdot), \theta(\delta t, \cdot), q(\delta t, \cdot))$, which is an admissible state in the sense of Definition 3.1, satisfies $q \leq q_{sat}$ everywhere and so can be used for the next timestep.

While it is proved that the rigorous moist rearrangement algorithm described in section 6.2.5 gives a physically relevant solution, there is no guarantee that the overall iteration above will converge. The convergence proof in section 6.2.5 relies on all the rearrangement being confined to a single column, so that the convective inhibition generated by the parcels displaced downwards controls the rate of upward mass transfer. In a three-dimensional case, or a two-dimensional case as used for illustration here, it is likely that the descent of parcels to balance the moist ascending parcels will be over a larger area, This motivated the scheme of Kuell and Bott (2008) which represents the upward 'jumps' by mass sources and sinks. However, it makes a proof of convergence of the algorithm above very unlikely as there is no control over where the stable layers are, nor any control over the horizontal moisture distribution. Thus it is unlikely that a large-scale solution can be computed independently of knowledge of the small-scale behaviour, so there is no longer a 'slow manifold' once moisture is taken into account.

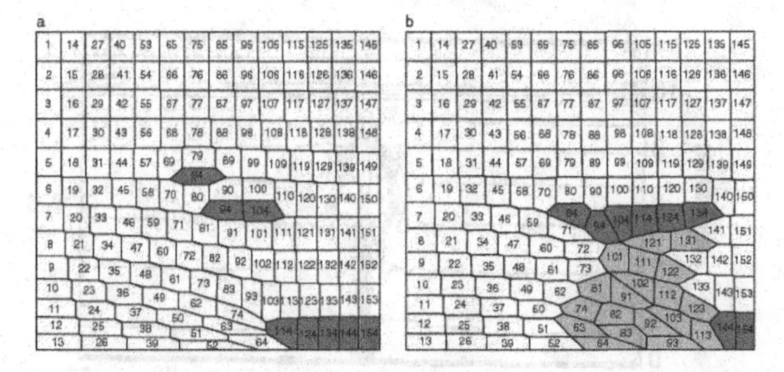

Fig. 6.15 Fluid element pictures showing a vertical cross-section of a frontal zone with moisture. Left: the hatched elements are moist. Right: the striped elements have been cooled by precipitation falling from the convecting hatched elements. From Holt (1990). ©Crown Copyright, Met Office.

Now illustrate the results obtained using the algorithm above. Fig. 6.15 shows a solution of the frontogenesis problem (3.134), with moisture included according to (6.20), obtained by Holt (1990) using the geometric algorithm. Only the hatched elements contain moisture. The effect of the frontogenesis is that the air is forced up at the front, as shown in section 3.4.2. Some of the air becomes saturated, and therefore θ is increased, leading to further upward motion. This gives a positive feedback which strengthens the frontogenesis. This is the reason why weather systems are intensified by latent heat release. Fig. 6.15 shows a further effect, not included in (6.20). If the excess moisture falls out as rain, some of it will re-evaporate into the air below, thus cooling it. This effect occurs in the striped elements in the right-hand panel of Fig. 6.15, and gives a further positive feedback as the convergence into the front is increased. The position of the rainfall relative to the front is critical for whether the feedback is positive or negative.

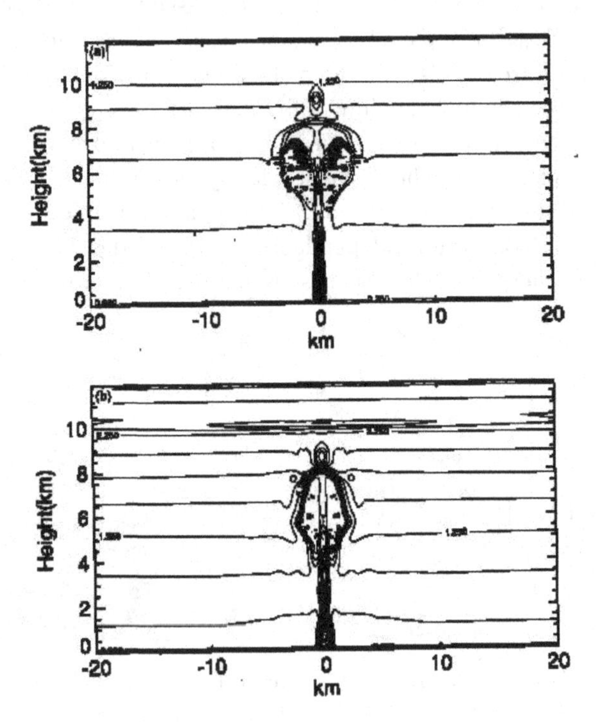

Fig. 6.16 The potential vorticity field for experiments with (a) $f = 1 \times 10^{-3}\mathrm{s}^{-1}$ and (b) $f = 2 \times 10^{-3}\mathrm{s}^{-1}$. Contour interval $0.25 \times 10^{-5}\mathrm{m}^2\mathrm{s}^{-1}{}^\circ\mathrm{K}\mathrm{kg}^{-1}$. Regions of negative potential vorticity are shaded. From Shutts and Gray (1994). ©Crown Copyright, Met Office.

The effect of solving the moist semi-geostrophic equations (3.2) with (6.20) can also be studied using the dual formulation (3.45). Assuming a thermal forcing S from other effects, this becomes

$$\begin{aligned} &\frac{\partial \sigma}{\partial t} + \mathbf{U} \cdot \nabla \sigma = 0, \\ &\mathbf{U} = (U, V, W), \\ &\quad = \left(f(y - Y), f(X - x), S : q < q_{sat}; S - L\frac{\mathrm{D}q_{sat}}{\mathrm{D}t} : q = q_{sat} \right). \end{aligned} \tag{6.38}$$

Thus the latent heat release term creates a positive W if there is upward vertical motion. If the moist stability condition (6.23) is violated, parcels will jump in the vertical to new stable positions as shown in Figs. 6.15. As in the mountain ridge case, section 5.5, the jumps will result in energy loss. In that case W becomes measure-valued as does the physical space velocity w. The potential density σ represents a mass density in (X, Y, Z) coordinates. The effect of the jump is to create a mass source at large Z and a corresponding sink at small Z, as in Kuell and Bott (2008). Since the potential vorticity is the inverse of the potential density, the effect can be seen in the potential vorticity calculated from the solutions in physical space as a source at low levels and a sink at high levels.

An example is shown in Fig. 6.16, from Shutts and Gray (1994). The solutions were obtained using a non-hydrostatic Euler model similar to (3.20). Moist effects were included but with a more accurate representation than the simple one used in (6.20). The initial potential vorticity was independent of x and y, and increasing with z, as shown away from the convecting region in Fig. 6.16. The domain size was 20km square in (x, y) and 12km deep. The Brunt-Väisälä frequency was $1.2{\times}10^{-2}\text{s}^{-1}$. Two values of rotation rate were used, $1 \times 10^{-3}\text{s}^{-1}$ giving $f/N = 0.08$ and $2 \times 10^{-5}\text{s}^{-1}$ giving $f/N = 0.16$. Convection was initiated by creating a warm bubble at the surface, which convects to about 7km. For the parameters chosen, the horizontal spreading of the effect of this convection on the semi-geostrophic solution would be about 80km and 40km respectively. The actual spreading is much less, as explained using an analytic solution in Shutts and Gray (1994). The potential vorticity source at low levels and sink at high levels, leading to weakly negative values, are easily seen. Other diagnostics calculated by Shutts and Gray (1994), their Fig. 15, show that about 35% of the total production of energy by latent heat release is retained in the balanced flow, the rest being dissipated or dispersed as gravity waves. In a semi-geostrophic solution, this dissipation is represented by the energy lost in the convective jumps.

Semi-geostrophic theory can be expected to give useful predictions of convection if the instability is created by large-scale effects which lead to violations of (6.23). Fig. 6.17 shows a case where a storm developed over eastern England on the north side of a vortex in the upper troposphere. In Fig. 6.18 the vertical motion diagnosed by the omega equation (5.6) is shown by the + and − signs. It indicates upward motion associated with the vortex. Fig. 6.18 also shows the warming and moistening of the low level troposphere by the large-scale flow. The effect is to create instability according to (6.23), because the upward motion creates saturation and brings the more stringent condition into effect

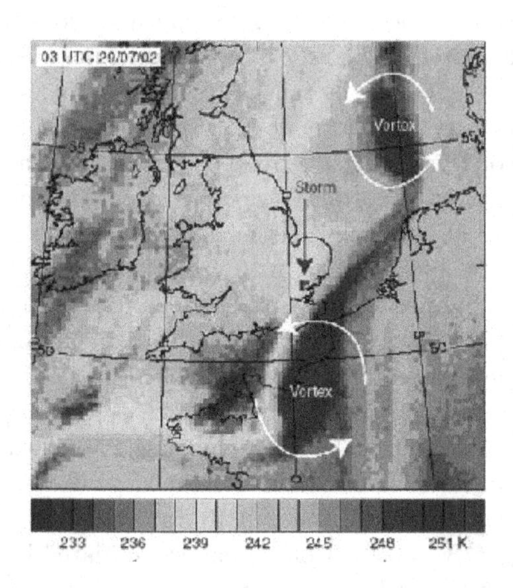

Fig. 6.17 Case study for 29 July 2002 over the U.K. The shading indicates cloud-top heights, the darker shading is high cloud. The position of an upper level vortex is marked, together with the position of a severe convective storm. Source: N. Roberts, JCMM, Met Office. ©Crown Copyright, Met Office.

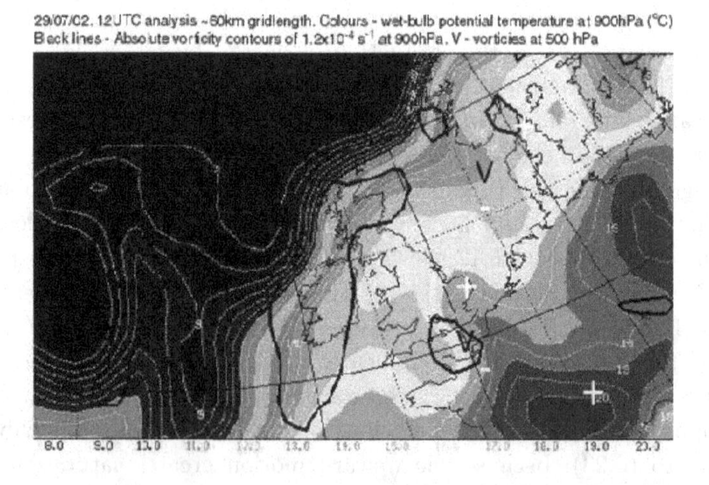

Fig. 6.18 The same case as in Fig. 6.17. Shading indicates wet-bulb potential temperature at 900hpa. High values indicate almost saturated air. The plus and minus signs indicate upward and downward motion diagnosed using (5.6). Source: N. Roberts, JCMM, Met Office. ©Crown Copyright, Met Office.

Chapter 7

Summary

In this volume it has been demonstrated that the semi-geostrophic model is an appropriate model for the large-scale evolution of the atmosphere and ocean, where 'large-scale' means scales dominated by the Earth's rotation. The solutions are typically long-lived stable structures which explain the often observed long-lived periods of anomalous weather. These structures may be anisotropic, thus including fronts and jet-streams which are characteristic of atmospheric behaviour in the extratropics. In the ocean, these structures correspond to long-lived eddies. It has also been demonstrated that most smaller scale regimes are dominated by instabilities which inhibit the existence of long-lived structures. An exception is the axisymmetric vortex, which describes tropical cyclones. In the deep tropics, only the zonal mean state can be described by the semi-geostrophic model.

The solution can be regarded as an evolution through a sequence of minimum energy states, where the maintenance of this state against dynamical and physical forcing is achieved by the ageostrophic circulaton. In the tropics, maintenance of the zonal mean state is achieved by the Hadley and Walker circulations. Mathematically, the existence of minimum energy states can be established by solving an 'optimal transport' problem. The solutions of such a problem are characterised by a convexity condition. In the atmosphere, applying the convexity condition to the pressure ensures that the solutions consist of large-scale structures. Small-scale oscillations are prevented. This corresponds to the observed large-scale behaviour.

The application of optimal transport methods to the semi-geostrophic equations yields a time-dependent pressure field, with associated geostrophic wind, which satisfies a convexity condition. The ageostrophic trajectory is also predicted. In some cases, however, this trajectory can only be described as a probability measure. This is because the Lagrangian formulation, which underpins semi-geostrophic theory, attempts to track the history of individual fluid particles. However, this is not possible in a well-mixed fluid, which can easily be generated as a result of external forcing. Thus non-classical mathematical techniques are required.

Additional physics can be built into the model, which is essential if it is to describe the observed large-scale behavior. Boundary layer friction can be included, giving the semi-geotriptic model. This is shown to give an accurate representation of the boundary layer behaviour in an operationl weather forecasting model. Moist effects can also be included by making scaling assumptions consistent with the semi-geostrophic assumptions. These have been demonstrated to give useful information, as shown in this volume. However, neither the boundary layer behaviour or the moist evolution can yet be analysed rigorously.

On present evidence, the semi-geostrophic model is the most general model which can be solved for large times and is asymptotically valid on large scales. Idealised tests described in this volume demonstrate that the expected asymptotic behaviour is observed in practice. The diagnostics illustrated in this volume show that the accuracy of the semi-geostrophic model is sufficient for giving a useful description of the large-scale behaviour of real data. It is thus the most general 'slow manifold', which characterises atmospheric states that can be predicted without knowledge of the smaller-scale behaviour. The existence of such manifolds underpins the practical success of weather forecasting.

Bibliography

Adams, R. A. (1975) *Sobolev spaces*, (Academic Press, New York).

Allen, J. S., Barth, J. A. and Newberger, P. (1990) On intermediate models for barotropic continental shelf and slope flow fields. Part III: Comparison of numerical model solutions in periodic channels. *J. Phys. Oceanog.* **20**, pp. 1949–1973.

Ambrosio, L. (2004) Transport equation and Cauchy problem for BV vector fields. *Invent. Math.* **158**, pp. 227–260.

Ambrosio, L. and Gangbo, W. (2008) Hamiltonian ODEs in the Wasserstein space of probability measures. *Comm. Pure. Appl. Math.* **61**, pp. 18–53.

Ambrosio, L., Gigli, N. and Savaré, G. (2008) *Gradient Flows: In Metric Spaces and in the Space of Probability Measures.* Lectures in Mathematics. (ETH Zrich, Birkhäuser Basel).

Ambrosio, L., Colombo, M., de Philippis, G. and Figalli, A. (2012) Existence of Eulerian solutions to the semigeostrophic equations in physical space: the 2-dimensional periodic case. *Comm. PDE.* **37**, pp. 2209–2227.

Ambrosio, L., Colombo, M., de Philippis, G. and Figalli, A. (2014) A global existence result for the semigeostrophic equations in three-dimensional convex domains. *Disc. Cont. Dyn. Syst.* **A**.

Angenent, S., Haker, S. and Tannenbaum, A. (2003) Minimising flows for the Monge-Kantorovich problem. *SIAM. J. Math. Anal.* **359**, pp. 61–97.

Babin, A., Mahalov, A. and Nicolaenko, B. (1999) Global regularity of 3D rotating Navier-Stokes equations for resonant domains. *Indiana University Mathematics Journal.* **48**, pp. 1133–1176.

Babin, A., Mahalov, A. and Nicolaenko, B. (2002) Fast singular oscillating limits of stably stratified 3-D Euler and Navier-Stokes equations and ageostrophic wave fronts. In *Large-Scale Atmosphere-Ocean Dynamics*, vol. I, J. Norbury and I. Roulstone eds., (Cambridge University Press), pp. 126–201.

Bannon, P. R. (1996) On the anelastic approximation for a compressible atmosphere. *J. Atmos. Sci.* **53**, pp. 3618–3628.

Beare, R. J. (2007) Boundary layer mechanisms in extratropical cyclones. *Quart. J. Roy. Meteorol. Soc.* **133**, pp. 503–515.

Beare, R. J. and Cullen, M. J. P. (2010) A semi-geostrophic model incorporating well-mixed boundary layers. *Quart. J. Roy. Meteorol. Soc.* **136**, pp. 906–917.

Beare, R. J. and Cullen, M. J. P. (2012) Balanced models of boundary-layer convergence. *Quart. J. Roy. Meteorol. Soc.* **138**, pp. 1452–1464.

Beare, R. J. and Cullen, M. J. P. (2013) Diagnosis of boundary-layer circulations. *Phil. Trans. Roy. Soc.* **A** (371), 20110474.

Beare, R. J. and Cullen, M. J. P. (2016) Validating weather and climate models at small Rossby numbers: Including a boundary layer. *Quart. J. Roy. Meteorol. Soc.* **142**, pp. 2636–2645.

Beare, R. J. and Cullen, M. J. P. (2019) A simple model of a balanced boundary layer coupled to a large-scale convective circulation. *J. Atmos. Sci.* **76**, pp. 837–849.

Beljaars, A. C. M. (1991) Numerical schemes for parametrizations. In *Proc. Seminar on Numerical Methods in Atmospheric Models*, (ECMWF, Reading, U.K.), pp. 308–334.

Benamou, J.-D. and Brenier, Y. (1997) A numerical method for the optimal time-continuous mass transport problem and related problems. In *Monge-Ampère equations: applications to geometry and optimisation. (Deerfield Beach, FL. 1997).* Contemporary Math., Amer. Math. Soc. **226**, pp. 1–11.

Benamou, J.-D. and Brenier, Y. (1998) Weak existence for the semi-geostrophic equations formulated as a coupled Monge-Ampère/transport problem. *SIAM J. Appl. Math.* **58**, pp. 1450–1461.

Benamou, J.-D. and Brenier, Y. (2000) A computational fluid mechanics solution to the Monge-Kantorovich mass transfer problem. *Numer. Math.* **84**, pp. 375–393.

Bennetts, D. A. and Hoskins, B. J. (1979) Conditional symmetric instability-a possible explanation for frontal rainbands. *Quart. J. Roy. Meteorol. Soc.* **105**, pp. 945–962.

Bogue, N. M., Huang, R. X. and Bryan, K. (1986) Verification experiments with an isopycnal coordinate ocean model. *J. Phys. Oceanog.* **16**, pp. 985–990.

Bourgeois, A. J. and Beale, J. T. (1994) Validity of the quasigeostrophic model for large-scale flow in the atmosphere and ocean. *SIAM J. Math. Anal.* **25**, pp. 1023–1068.

Brenier, Y. (1991) Polar factorisation and monotone rearrangement of vector-valued functions. *Commun. Pure Appl. Math.* **44**, pp. 375–417.

Brenier, Y. and Cullen, M. J. P. (2009) Rigorous derivation of the $x-z$ semi-geostrophic equations. *Commun. Math. Sci.* **7**, pp. 779–784.

Brenier, Y. (2013) Rearrangement, convection, convexity and entropy. *Phil. Trans. Roy. Soc.* **A** (371), 20120343.

Burton, G. R. and Douglas, R. J. (1998) Rearrangements and polar factorisation of countably degenerate functions. *Proc. Roy. Soc. Edin.* **128A**, pp. 671–681.

Burton, G. R. and McLeod, J. B. (1991) Maximisation and minimisation on classes of rearrangements. *Proc. Roy. Soc. Edin.* **119A**, pp. 287–300.

Burton, G. R. and Nycander, J. (1999) Stationary vortices in three-dimensional quasi-geostrophic flow. *J. Fluid Mech.* **389**, pp. 255–274.

Caffarelli, L. A. (1992a) The regularity of mappings with a convex potential. *J. Amer. Math. Soc.* **5**, pp. 99–104.

Caffarelli, L. A. (1992b) Boundary regularity of maps with convex potentials. *Comm. Pure Appl. Math.* **45**, pp. 1141–1151.

Caffarelli, L. A. (1996) Boundary regularity of maps with convex potentials-II. *Ann. Math.* **144**, pp. 453–496.

Callies, J. and Ferrari, R. (2013) Interpreting energy and tracer spectra of upper-ocean turbulence in the submesoscale range (1-200km). *J. Phys. Oceanog.* **43**, pp. 2456–2474.

Cao, C. and Titi, E. S. (2007) Global well-posedness of the three-dimensional viscous primitive equations of large scale ocean and atmospheric dynamics. *Ann. Math.* **166**, pp. 245–267.

Cao, C., Ibrahim, S., Nakanishi, K. and Titi, E. S. (2015) Finite-time blowup for the inviscid primitive equations of oceanic and atmospheric dynamics. *Comm. Math. Phys.* **337**, pp. 473–482.

Chan, I. H. and Shepherd, T. G. (2014) Diabatic balance model for the equatorial atmosphere. *J. Atmos. Sci.* **71**, pp. 985–1001.

Charney, J. G. (1948) On the scale of atmospheric motions. *Geofys. Publ.* **17** (2), pp. 4–17.

Charney, J. G. (1971) Geostrophic turbulence. *J. Atmos. Sci.* **28**, pp. 1087–1095.

Charney, J. G. and Devore, J. (1979) Multiple flow equilibria in the atmosphere and blocking. *J. Atmos. Sci.* **36**, pp. 1205–1216.

Charney, J. G., Fjortoft, R. and von Neumann, J. (1950) Numerical integration of the barotropic vorticity equations. *Tellus.* **2**, pp. 237–254.

Chemin, J-Y. (2000) *Perfect incompressible fluids.* (Oxford University Press), 185pp.

Cheng, J. (2017) Semigeostrophic equations with free upper boundary. *Calc. Var. Partial Differential Equations.* **55**,(6) DOI 10.1007/s00526-016-1072-x.

Cheng, J., Cullen, M. J. P. and Feldman, M. (2018) Semi-geostrophic system with variable Coriolis parameter. *Arch. Rat. Mech. Anal.* **227**, pp. 215–272.

Cheng, B., Cheng, J., Cullen, M. J. P., Norbury, J. and Turner, M. (2017a) A rigorous treatment of moist convection in a single column. *SIAM J. Math. Anal.* **49**, pp. 3854–3892.

Cheng, B., Cullen, M. J. P., Esler, G., Norbury, J., Turner, M., Vanneste, J. and Cheng, J. (2017b) A Model for Moist Convection in an Ascending Atmospheric Column. *Quart. J. Roy. Meteorol. Soc.* **143**, pp. 2925–2939.

Chynoweth, S. (1987) *The semi-geostrophic equations and the Legendre transform.* Ph. D thesis. (University of Reading, U.K.), 248pp.

Chynoweth, S. and Sewell, M. J. (1989) Dual variables in semi-geostrophic theory. *Proc. Roy. Soc. Lond.* **A** (424), pp. 155–186.

Cloke, P. and Cullen, M. J. P. (1994) A semi-geostrophic ocean model with outcropping. *Dyn. Atmos. Ocean.* **21**, pp. 23–48.

Cordero-Erausquin, D. (1999) Sur le transport de mesures périodiques. *C. R. Acad. Sci. Paris Ser. I Math.* **329**, pp. 199–202.

Craig, G. C. (1991) A three-dimensional generalisation of Eliassen's balanced vortex equations derived from Hamilton's principle. *Quart. J. Roy. Meteorol.*

Soc. **117**, pp. 435–448.

Cullen, M. J. P. and Purser, R. J. (1984) An extended Lagrangian theory of semi-geostrophic frontogenesis. *J. Atmos. Sci.* **41**, pp. 1477–1497.

Cullen, M. J. P., Chynoweth, S. and Purser, R. J. (1987) On semi-geostrophic flow over synoptic-scale topography. *Quart. J. Roy. Meteorol. Soc.* **113**, pp. 163–180.

Cullen, M. J. P. (1989a) On the incorporation of atmospheric boundary layer effects into a balanced model. *Quart. J. Roy. Meteorol. Soc.* **115**, pp. 1109–1131.

Cullen, M. J. P. (1989b) Implicit finite difference methods for modelling discontinuous atmospheric flows. *J. Comp. Phys.* **81**, pp. 319–348.

Cullen, M. J. P. and Purser, R. J. (1989) Properties of the Lagrangian semi-geostrophic equations. *J. Atmos. Sci.* **46**, pp. 2684–2697.

Cullen, M. J. P. and Mawson, M. H. (1992) An idealised simulation of the Indian monsoon using primitive-equation and quasi-equilibrium models. *Quart. J. Roy. Meteorol. Soc.* **118**, pp. 153–164.

Cullen, M. J. P. and Roulstone, I. (1993) A geometric model of the nonlinear equilibration of two-dimensional Eady waves. *J. Atmos. Sci.* **50**, pp. 328–332.

Cullen, M. J. P. (2000) On the accuracy of the semi-geostrophic approximation. *Quart. J. Roy. Meteorol. Soc.* **126**, pp. 1099–1115.

Cullen, M. J. P. and Gangbo, W. (2001) A variational approach for the 2-D semi-geostrophic shallow water equations. *Arch. Rat. Mech. Anal.* **156**, pp. 241–273.

Cullen, M. J. P. and Maroofi, H. (2003) The fully compressible semi-geostrophic system from meteorology. *Arch. Rat. Mech. Anal.* **167**, pp. 309–336.

Cullen, M. J. P. (2002) Large scale non-turbulent dynamics in the atmosphere. *Quart. J. Roy. Meteorol. Soc.* **128**, pp. 2623–2640.

Cullen, M. J. P. and Douglas, R. J. (2003) Large amplitude nonlinear stability results for atmospheric circulations. *Quart. J. Roy. Meteorol. Soc.* **129**, pp. 1969–1988.

Cullen, M. J. P. and Salmond, D. J. (2003) On the use of a predictor-corrector scheme to couple the dynamics with the physical parametrizations in the ECMWF model. *Quart. J. Roy. Meteorol. Soc.* **129**, pp. 1217–1236.

Cullen, M. J. P., Douglas, R. J., Roulstone, I. and Sewell, M. J. (2005) Generalised semi-geostrophic theory on a sphere. *J. Fluid Mech.* **531**, pp. 123–157.

Cullen, M. J. P. and Feldman, M. (2006) Lagrangian solutions of semi-geostrophic equations in physical space. *SIAM J. Math. Anal.* **37**, pp. 1371–1395.

Cullen, M. J. P. (2007a) Modelling atmospheric flows. *Acta Numer.* **16**, pp. 67–154.

Cullen, M. J. P. (2007b) Semigeostrophic solutions for flow over a ridge. *Quart. J. Roy. Meteorol. Soc.* **133**, pp. 491–501.

Cullen, M. J. P., Gangbo, W. and Pisante, G. (2007c) The semigeostrophic equations discretized in reference and dual variables. *Arch. Rat. Mech. Anal.* **185**, pp. 341–363.

Cullen, M. J. P. (2008a) Analysis of the semi-geostrophic shallow water equations.

Physica **D. 237**, pp. 1461–1465.

Cullen, M. J. P. (2008b) A comparison of numerical solutions to the Eady frontogenesis problem. *Quart. J. Roy. Meteorol. Soc.* **134**, pp. 2143–2155.

Cullen, M. J. P., Gilbert, D. K. and Pelloni, B. (2014) Solution of the fully compressible semi-geostrophic system. *Comm. PDE.* **39**, pp. 591–625.

Cullen, M. J. P. and Sedjro, M. (2014) On a model of forced axisymmetric flows. *SIAM J. Math. Anal.* **46**, pp. 3983–4013.

Cullen, M. J. P., Gangbo, W. and Sedjro, M. (2015a) A mathematically rigorous analysis of forced axisymmetric flows in the atmosphere. *Quart. J. Roy. Meteorol. Soc.* **141**, pp. 1836–1844.

Cullen, M. J. P., Gilbert, D. K., Kuna, T. and Pelloni, B. (2015b) Free upper boundary value problems for the semi-geostrophic equations. *arXiv* preprint arXiv:1409.8560.

Cullen, M. J. P. (2017) The impact of high vertical resolution in the Met. Office Unified Model. *Quart. J. Roy. Meteorol. Soc.* **143**, pp. 278–287.

Cullen, M. J. P. (2018) The use of semi-geostrophic theory to diagnose the behaviour of an atmospheric GCM. *MDPI Fluids.* www.mdpi.com/2311-5521/3/4/72/pdf.

Cullen, M. J. P., Kuna, T., Pelloni, B. and Wilkinson, M. (2019) The stability principle and global weak solutions of the free surface semi-geostrophic equations in geostrophic coordinates. *Proc. Roy. Soc.* **A**, 47520180787.

Cullen, M. J. P., Feldman, M. and Tudorascu, A. (2020) From Euler to the semi-geostrophic system: convergence under uniform convexity. *Methods and Applications of Analysis*, to appear.

Davies, T., Staniforth, A., Wood, N. and Thuburn, J. (2003) Validity of anelastic and other equation sets as inferred from normal mode analysis. *Quart. J. Roy. Meteorol. Soc.* **129**, pp. 2761–2775.

Davies, T., Cullen, M. J. P., Malcolm, A. J., Mawson, M. H., Staniforth, A., White, A. A. and Wood, N. (2005) A new dynamical core for the Met Office's global and regional modelling of the atmosphere. *Quart. J. Roy. Meteorol. Soc.* **131**, pp. 1759–1782.

De Philippis, G. and Figalli, A. (2012) $W^{2,1}$ regularity for solutions of the Monge-Ampère equation. *Invent. Math.*, DOI 10.1007/s0222-012-0405-4.

Diamantakis, M., Wood, N. and Davies, T. (2006) An improved implicit predictor-corrector scheme for boundary-layer vertical diffusion. *Quart. J. Roy. Meteorol. Soc.* **132**, pp. 959–978.

DiPerna, R. J. and Lions, P. L. (1989) Ordinary differential equations, transport theory, and Sobolev spaces. *Invent. Math.* **98**, pp. 511–547.

Douglas, R. J. (1994) Rearrangements of functions on unbounded domains. *Proc. Roy. Soc. Edin.* **124A**, pp. 621–644.

Douglas, R. J. (2002) Rearrangements of functions with application to meteorology and ideal fluid flow. In *Large-Scale Atmosphere-Ocean Dynamics*, vol. I, J. Norbury and I. Roulstone eds., (Cambridge University Press), pp. 288–341.

Durran, D. R. (1989) Improving the anelastic approximation. *J. Atmos. Sci.* **46**, pp. 1453–1461.

Egan, C., Bourne, D., Cotter, C., Cullen, M. J. P., Pelloni, B., Roper, S. and Wilkinson, M. (2021) A new implementation of the geometric method for solving the Eady slice equations. *Quart. J. Roy. Meteorol. Soc.*, submitted.

Eliassen, A. (1949) The quasi-static equations of motion with pressure as an independent variable. *Geofys. Publ.* **17**, pp. 3–43.

Eliassen, A. and Kleinschmidt, E. (1957) Dynamic meteorology. *Handbuch der Physik, Geophysik II.* (Springer Verlag), p. 154.

Eliassen, A. (1962) On the vertical circulation in frontal zones. *Geofys. Publ.* **24**, pp. 147–160.

Eliassen, A. (1984) Geostrophy. *Quart. J. Roy. Meteorol. Soc.* **110**, pp. 1–12.

Emanuel, K. A. (1983) The Lagrangian parcel dynamics of moist symmetric instability. *J. Atmos. Sci.* **40**, pp. 2368–2376.

Embid, P. F. and Majda, A. J. (1998) Low Froude number limiting dynamics for stably stratified flow with small or finite Rossby numbers. *Geophys. Astrophys. Fluid Dynamics.* **87**, pp. 1–50.

Faria, J. C. O., Lopes Filho, M. C. and Nussenzveig Lopes, H. J. (2009) Weak stability of Lagrangian solutions to the semi-geostrophic equations. *Nonlinearity.* **22**, pp. 2521–2539.

Feldman, M. and Tudorascu, A. (2013) On Lagrangian solutions for the semi-geostrophic equations with singular data. *SIAM J. Math. Anal.* **38**, pp. 1616–1640.

Feldman, M. and Tudorascu, A. (2015) On the semi-geostrophic system in physical space with general initial data. *Arch. Rat. Mech. Anal.* **218**, pp. 527–551.

Feldman, M. and Tudorascu, A. (2015a) Lagrangian solutions for the semi-geostrophic shallow water system in physical space with general initial data. *Algebra i Analiz 27* **3**, pp. 272–300; reprinted in *St. Petersburg Math. J.* **27**, pp. 547–568, (2016).

Feldman, M. and Tudorascu, A. (2016) The semi-geostrophic system: weak-strong uniqueness under uniform convexity. *Calc. Var. PDE's.* **56**, (6), paper no. 158, 22pp.

Fincham, A. M., Maxworthy, T. and Spedding, G. R. (1996) The horizontal and vertical structure of the vorticity field in freely-decaying stratified grid turbulence. *Dyn. Atmos. Ocean.* **23**, pp. 153–160.

Ford, R., McIntyre, M. E. and Norton, W. A. (2000) Balance and the slow quasi-manifold: some explicit results. *J. Atmos. Sci.* **57**, pp. 1236–1254.

Fjortoft, R. (1946) On the frontogenesis and cyclogenesis in the atmosphere. Part I. On the stability of the stationary circular vortex. *Geofys. Publik.* **16**, pp. 1–28.

Fjortoft, R. (1950) Applications of integral theorems in deriving criteria of stability for laminar flows and for the baroclinic circular vortex. *Geofys. Publ.* **17**, (6).

Galewsky, J., Scott, R. K., and Polvani, L. M. (2004) An initial-value problem for testing numerical models of the global shallow-water equations. *Tellus.* **56**, pp. 429–440.

Gallouët, T. and Mérigot, Q. (2018) A Lagrangian scheme à la Brenier for the incompressible Euler equations. *Foundations of Comp. Math.* **18**, (Springer-

Verlag), pp. 835–865.

Gangbo, W. and McCann, R. J. (1996) The geometry of optimal transportation. *Acta Math.* **177**, pp. 113–161.

Gill, A. E. (1982) *Atmosphere-Ocean Dynamics.* (Academic Press). 662pp.

Godson, W. L. (1950) Generalised criteria for dynamic instability. *J. Meteor.* **7**, pp. 268–278.

Gordon, A., Grace, W., Schwertfeger, P. and Byron-Scott, R. (1998) *Dynamic Meteorology: a basic course.* (Routledge/Taylor and Francis, London). 322pp.

Halmos, P. R. (1950) *Measure Theory.* (van Nostrand).

Haltiner, G. J. and Williams, R. T. (1980) *Numerical Prediction and Dynamic Meteorology.* (John Wiley). 477pp.

Hayashi, Y.-Y., Nishizawa, S., Takehiro, S.-I., Yamada, M., Ishioka, K., and Yoden, S. (2007) Rossby waves and jets in two-dimensional decaying turbulence on a rotating sphere. *J. Atmos. Sci.* **65**, pp. 4246–4269.

Haynes, P. H. and McIntyre, M. E. (1990) On the conservation and impermeability theorems for potential vorticity. *J. Atmos. Sci.* **47**, pp. 2021–2031.

Heckley, W. A. and Gill, A. E. (1984) Some simple analytical solutions to the problem of forced equatorial long waves. *Quart. J. Roy. Meteorol. Soc.* **110**, pp. 203–217.

Held, I. M., Pierrehumbert, R. T., Garner, S. T. and Swanson, K. L. (1995) Surface quasi-geostrophic dynamics. *J. Fluid Mech.* **282**, pp. 1–20.

Hieronymus, M. and Nycander, J. (2015) Finding the minimum potential energy state by adiabatic parcel rearrangements with a nonlinear equation of state: an exact solution in polynomial time. *J. Phys. Oceanog.* **45**, pp. 1843–1857.

Holm, D. D. (1996) Hamiltonian balance equations. *Physica* **D**. **98**, pp. 379–414.

Holt, M. W. (1990) Semi-geostrophic moist frontogenesis in a Lagrangian model. *Dyn. Atmos. Ocean.* **14**, pp. 463–481.

Holtslag, A. A. M. and Boville, B. A. (1993). Local versus nonlocal boundary-layer diffusion in a global climate model. *J. Climate.* **6**, pp. 1825–1842.

Hoskins, B. J. and Bretherton, F. P. (1972) Atmospheric frontogenesis models: formulation and solutions. *J. Atmos. Sci.* **29**, pp. 11–37.

Hoskins, B. J. (1975) The geostrophic momentum approximation and the semi-geostrophic equations. *J. Atmos. Sci.* **32**, pp. 233–242.

Hoskins, B. J. and Draghici, I. (1977) The forcing of ageostrophic motion according to the semi-geostrophic equations and in an isentropic coordinate model. *J. Atmos. Sci.* **34**, pp. 1859–1867.

Hoskins, B. J. and West, N. V. (1979) Baroclinic waves and frontogenesis. Part II: uniform potential vorticity jet flows- cold and warm fronts. *J. Atmos. Sci.* **36**, pp. 1663–1680.

Hoskins, B. J. (1982) The mathematical theory of frontogenesis. *Ann. Rev. Fluid Mech.* **14**, pp. 131–151.

Hoskins, B. J., McIntyre, M. E. and Robertson, A. W. (1985) On the use and significance of isentropic potential vorticity maps. *Quart. J. Roy. Meteorol. Soc.* **111**, pp. 877–946. Also **113**, pp. 402–404.

Jian, H.-Y., and X.-J. Wang. (2007) Continuity estimates for the Monge-Ampère equation. *SIAM J. Math. Anal.* **39**, pp. 608–626.

Jin, F. and Hoskins, B. J. (1995) The direct response to tropical heating in a baroclinic atmosphere. *J. Atmos. Sci.* **52**, pp. 307–319.

Kantorovich, L. (1942) On the translocation of masses. *C.R. (Doklady) Acad. Sci. URSS (N.S.)* **37**, pp. 199–201.

Klein, R. (2009) Scale-dependent models for atmospheric flows. *Ann. Rev. Fluid Mech.* **42**, pp. 249–274.

Kuell, V. and Bott, A. (2008) A hybrid convection scheme for use in nonhydrostatic numerical weather prediction models. *Meteorol. Z.* **17**, pp. 775–783.

Kushner, P. J. and Shepherd, T. G. (1995a) Wave-activity conservation laws and stability theorems for semi-geostrophic dynamics: Part 1. Pseudomomentum based theory. *J. Fluid Mech.* **290**, pp. 67–104.

Kushner, P. J. and Shepherd, T. G. (1995b) Wave-activity conservation laws and stability theorems for semi-geostrophic dynamics: Part 2. Pseudo-energy based theory. *J. Fluid Mech.* **290**, pp. 105–129.

Larichev, V. D. and McWilliams, J. C. (1991) Weakly decaying turbulence in an equivalent-barotropic fluid. *Phys. Fluids* **A, 3(5)**, pp. 938–950.

Leith, C. E. and Kraichnan, R. E., (1972) Predictablity of turbulent flows. *J. Atmos. Sci.* **29**, pp. 1041–1052.

Lions, J.-L., Temam, R. and Wang, S. (1992) On the equations of the large-scale ocean. *Nonlinearity.* **5**, pp. 237–288.

Liu, X. and Titi, E. S. (2019) Global existence of weak solutions to the compressible primitive equations of atmospheric dynamics with degenerate viscosities. *SIAM J. Math. Anal.* **51**, pp. 1913–1964.

Loeper, G. (2005) On the regularity of the polar factorisation for time-dependent maps. *Calc. of Variations and PDEs.* **22**, pp. 343–374.

Loeper, G. (2006) A fully nonlinear version of the incompressible Euler equations: the semi-geostrophic system. *SIAM J. Math. Anal.* **38**, pp. 795–823.

Lopes Filho, M. C. and Nussenzveig Lopes, H. J. (2002) Existence of a weak solution for the semi-geostrophic equation with integrable initial data. *Proc. Roy. Soc. Edin.* **132A**, pp. 329–339.

Lorenz, E. N. (1963) Deterministic non-periodic flow. *J. Atmos. Sci.* **20**, pp. 131–141.

Lorenz, E. N. (1992) The slow manifold–What is it? *J. Atmos. Sci.* **49**, pp. 2449–2451.

Lynch, P. (1989) The slow equations. *Quart. J. Roy. Meteorol. Soc.* **115**, pp. 201–219.

McCann, R. J. (1997) A convexity principle for interacting gases. *Adv. Math.* **I**, pp. 153–179.

McCann, R. J. (2001) Polar factorisation of maps on Riemannian manifolds. *Geom. Funct. Anal.* **11**, pp. 589–608.

McIntyre, M. E. and Roulstone, I. (2002) Are there higher-accuracy analogues of semi-geostrophic theory. In *Large-Scale Atmosphere-Ocean Dynamics*, vol. II, J. Norbury and I. Roulstone eds., (Cambridge University Press), pp. 301–364.

McWilliams, J. C. and Gent, P. R. (1980) Intermediate models of planetary circulations in the atmosphere and ocean. *J. Atmos. Sci.* **37**, pp. 1657–1678.

McWilliams, J. C. and Yavneh, I. (1998) Fluctuation growth and instability associated with a singularity of the balance equations. *Phys. Fluids.* **10**, pp. 2587–2596.

McWilliams, J. C., Yavneh, I., Cullen, M. J. P. and Gent, P. R. (1999) The breakdown of large-scale flows in rotating, stratified fluids. *Phys. Fluids.* **10**, pp. 3178–3184.

Magnusdottir, G. and Schubert, W. H. (1991) Semi-geostrophic theory on the hemisphere. *J. Atmos. Sci.* **48**, pp. 1449–1456.

Majda, A. J. (1984) *Compressible fluid flow, and systems of conservation laws in several space variables.* Applied Mathematical Sciences, **53**, (Springer, New York).

Majda, A. J. (2003) *Introduction to PDEs and waves for the atmosphere and ocean.* Courant Lecture Notes, **9**, (American Math. Society).

Majda, A. J. and Klein, R. (2003) Systematic multiscale models for the tropics. *J. Atmos. Sci.*, **60**, pp. 357–372.

Majda, A. J. and Stechmann, S. N. (2009) The skeleton of tropical intraseasonal oscillations. *PNAS.* **106**, pp. 8417–8422.

Malardel, S., Thorpe, A. J. and Joly, A. (1997) Consequences of the geostrophic momentum approximation on barotropic instability. *J. Atmos. Sci.* **54**, pp. 103–112.

Mansfield, L. (2017) *An optimal transport approach to find the background state of the atmosphere.* MSc. thesis, (University of Reading, U.K.).

Matthews, A. J., Hoskins, B. J. and Masutani, M. (2004) The global response to tropical heating in the Madden-Julian oscillation during the northern winter. *Quart. J. Roy. Meteorol. Soc.* **130**, pp. 1991–2011.

Mawson, M. H. (1996) A shallow water semi-geostrophic model on a sphere. *Quart. J. Roy. Meteorol. Soc.* **122**, pp. 267–290.

Met. Office (1975) *Handbook of weather forecasting*, Met.O, No. 875, (HMSO, London).

Methven, J. and Berrisford, P. (2015) The slowly evolving background state of the atmosphere. *Quart. J. Roy. Meteorol. Soc.* **141**, pp. 2237–2258.

Mohebalhojeh, A. R. and Dritschel, D. G. (2001) Hierarchies of balance conditions for the $f-$plane shallow water equations. *J. Atmos. Sci.* **58**, pp. 2411–2426.

Monge, G. (1781) *Mémoire sur la théorie des déblais et de remblais*, Memoires de l'Academie des Sciences, pp. 686–704.

Montgomery, M. T. and Shapiro, L. J. (1995) Generalized Charney-Stern and Fjortoft theorems for rapidly rotating vortices. *J. Atmos. Sci.* **52**, pp. 1829–1833.

Muraki, D. J., Snyder C., and Rotunno R. (1999) The next-order corrections to quasigeostrophic theory. *J. Atmos. Sci.* **56**, pp. 1547–1560.

Nakamura, N. and Held, I. M. (1989) Nonlinear equilibriation of two-dimensional Eady waves. *J. Atmos. Sci.* **46**, pp. 3055–3064.

Nakamura, N. (1994) Nonlinear equilibriation of two-dimensional Eady waves: simulations with viscous geostrophic momentum equations. *J. Atmos. Sci.* **51**, pp. 1023–1035.

Nastrom, G. D. and Gage, K. S. (1985) A climatology of atmospheric wavenumber spectra of wind and temperature observed by commercial aircraft. *J. Atmos. Sci.* **42**, pp. 950–960.

Neumann, C. J. (1993) 1993 Global overview. Chapter 1 in Holland, G. J. (ed.) *Global guide to tropical cyclone forecasting.* WMO/TD 560, TCP Report 31.

Norbury, J. and Roulstone, I. (2002) *Large-scale atmosphere-ocean dynamics*, Vols. I and II, (Cambridge University Press).

O'Neill, T. (2020) *A rigorous analysis, via the Monge-Ampère equation, of the existence of classical solutions of the semi-geostrophic equations with explicit Rossby number scaling.*, Ph D. Thesis, (University of Surrey, U.K.).

Ostdiek, V. and Blumen, W. (1997) A dynamic trio: inertial oscillation, deformation frontogenesis, and the Ekman-Taylor boundary layer. *J. Atmos. Sci.* **54**, pp. 1490–1502.

Parsons, A. T. (1969) A two-layer model of Gulf Stream separation. *J. Fluid Mech.* **39**, pp. 511–528.

Pedder, M. A. and Thorpe, A. J. (1999) The semi-geostrophic diagnosis of vertical motion. I: Formulation and coordinate transformations. *Quart. J. Roy. Meteorol. Soc.* **125**, pp. 1231–1256.

Pedlosky, J. (1987) *Geophysical fluid dynamics.* (Springer-Verlag).

Phillips, N. A. (1963) Geostrophic motion. *Rev. Geophysics*, **1**, pp. 123–176.

Phillips, N. A. (2000) The start of numerical weather prediction in the United States. *Proc. 50th Anniversary of Numerical Weather Prediction.* Deutsche Meteorologische Gesellschaft, pp. 13–28.

Pogorelov, A. V. (1964) *Monge-Ampère equations of elliptic type.* Noordhoff, Groningen, 114pp.

Preparata, F. P., and Hong, S. J. (1977) Convex hulls of finite sets of points in two and three dimensions. *Comm. A. C. M.* **20**, pp. 87–93.

Purser, R. J. and Cullen, M. J. P. (1987) A duality principle in semi-geostrophic theory. *J. Atmos. Sci.* **44**, pp. 3449–3468.

Purser, R. J. (2002) Legendre-transformable semi-geostrophic theories. In *Large-Scale Atmosphere-Ocean Dynamics*, vol. II, J. Norbury and I. Roulstone eds., (Cambridge University Press), pp. 224–250.

Ragone, F. and Badin, G. (2016) A study of surface semi-geostrophic turbulence: Freely decaying dynamics. *J. Fluid Mech.* **792**, pp. 740–774.

Rasmussen, E. (1985) A case study of a polar low development over the Barents Sea. *Tellus.* **37A**, pp. 407–418.

Reed, R. J. and Albright, M. D. (1986) A case study of explosive cyclogenesis in the Eastern Pacific. *Mon. Weather Rev.* **114**, pp. 2297-2319.

Ren, S. (2000) Finite-amplitude wave-activity invariants and nonlinear stability theorems for shallow water semi-geostrophic dynamics. *J. Atmos. Sci.* **57**, pp. 3388–3397.

Reiser, H. (2000) The development of numerical weather prediction in Deutsche Wetterdienst. *Proc. 50th Anniversary of Numerical Weather Prediction.* Deutsche Meteorologische Gesellschaft, pp. 51–80.

Richardson, L. F. (1922) *Weather prediction by numerical process*, (Cambridge University Press), 236pp. (also from Dover Publications Inc., New York,

1965).
Robert, R. and Sommeria, J. (1991) Statistical equilibrium states for two-dimensional flows. *J. Fluid Mech.* **229**, pp. 291–310.
Rockafellar, R. T. (1970) *Convex analysis*, (Princeton University Press), 451pp.
Roulstone, I. and Norbury, J. (1994) A Hamiltonian structure with contact geometry for the semi-geostrophic equations. *J. Fluid Mech.* **272**, pp. 211–233.
Roulstone, I. and Sewell, M. J. (1996) Potential vorticities in semi-geostrophic theory. *Quart. J. Roy. Meteorol. Soc.* **122**, pp. 983–992.
Roulstone, I. and Sewell, M. J. (1997) The mathematical structure of theories of semi-geostrophic type. *Phil. Trans. Roy. Soc. London* **A**, pp. 2489–2517.
Ryff, J. V. (1970) Measure preserving transformations and rearrangements. *J. Math. Anal. and Applics.* **30**, pp. 431–437.
Salmon, R. (1985) New equations for nearly geostrophic flow. *J. Fluid Mech.* **153**, pp. 461–477.
Salmon, R. (1998) *Lectures on Geophyscal Fluid Dynamics*. (Oxford University Press), 378pp.
Sawyer, J. S. (1956) The vertical circulation at meteorological fronts, and its relation to frontogenesis. *Proc. Roy. Soc. London.* **A 234**, pp. 346–362.
Schär, C. and Wernli, H. (1993) Structure and evolution of an isolated semi-geostrophic cyclone. *Quart. J. Roy. Meteorol. Soc.* **119**, pp. 57–90.
Schecter, D. A., and Montgomery, M. T. (2006) Conditions that inhibit the spontaneous radiation of spiral inertia-gravity waves from an intense mesoscale cyclone. *J. Atmos. Sci.* **63**, pp. 435–456.
Schubert, W. H. and Hack, J. J. (1983) Transformed Eliassen balanced vortex model. *J. Atmos. Sci.* **40**, pp. 1571–1583.
Schubert, W. H. (1985) Semi-geostrophic theory. *J. Atmos. Sci.* **42**, pp. 1770–1772.
Schubert, W. H., Ciesieleski, P. E., Stevens, D. E. and Kuo, H-C. (1991) Potential vorticity modelling of the ITCZ and the Hadley circulation. *J. Atmos. Sci.* **48**, pp. 1493–1509.
Schubert, W. H. and Magnusdottir, G. (1994) Vorticity coordinates, transformed primitive equations, and a canonical form for balanced models. *J. Atmos. Sci.* **51**, pp. 3309–3319.
Schubert, W. H., Montgomery, M. T., Taft, R. K., Guinn, T. A., Fulton, S. R., Kossin, J. P. and Edwards, J. P. (1999). Polygonal eyewalls, asymmetric eye contraction, and potential vorticity mixing in hurricanes. *J. Atmos. Sci.* **56**, pp. 1197–1223.
Schwerdtfeger, W. (1975) The effects of the Antarctic peninsula on the temperature regime of the Weddell Sea. *Mon. Weath. Rev.* **103**, pp. 45–51.
Sewell, M. J. (2002) Some applications of transformation theory in mechanics. In *Large-scale Atmosphere-Ocean Dynamics*, vol. II, J. Norbury and I. Roulstone eds., (Cambridge University Press), pp. 143–223.
Shapiro, L. J. and Montgomery, M. T. (1993) A three-dimensional balance theory for rapidly rotating vortices. *J. Atmos. Sci.* **50**, pp. 3322–3335.
Shutts, G. J. (1987a) The semi-geostrophic weir: a simple model of flow over mountain barriers. *J. Atmos. Sci.* **44**, pp. 2018–2030.

Shutts, G. J. (1987b) Balanced flow states resulting from penetrative, slant-wise convection. *J. Atmos. Sci.* **44**, pp. 3363–3376.

Shutts, G. J., Booth, M. W. and Norbury, J. (1988) A geometric model of balanced, axisymmetric flow with embedded penetrative convection. *J. Atmos. Sci.* **45**, pp. 2609–2621.

Shutts, G. J. (1989) Planetary semi-geostrophic theory. *J. Fluid Mech.* **208**, pp. 545–573.

Shutts, G. J. and Cullen, M. J. P. (1987) Parcel stability and its relation to semi-geostrophic theory. *J. Atmos. Sci.* **44**, pp. 1318–1330.

Shutts, G. J. and Gray, M. E. B. (1994) A numerical modelling study of the geostrophic adjustment process following deep convection. *Quart. J. Roy. Meteorol. Soc.* **120**, pp. 1145–1178.

Shutts, G. J. (1995) An analytical model of the balanced flow created by localised convective mass transfer in a rotating fluid. *Dyn. Atmos. Oceans.* **22**, pp. 1–17.

Shutts, G. J. (1998) Idealised models of the pressure drag force on mesoscale mountain ridges. *Contr. Atmos. Phys.* **71**, pp. 303–313.

Skamarock, W. C., Park, S-H., Klemp, J. B. and Snyder, C. (2014) Atmospheric kinetic energy spectra from global high-resolution non-hydrostatic simulations. *J. Atmos. Sci.* **71**, pp. 4369–4381.

Smith, G. B. II, Montgomery, M. T. (1995) Vortex axisymmetrization: Dependence on azimuthal wave-number or asymmetric radial structure changes. *Quart. J. Roy. Meteorol. Soc.* **121**, pp. 1615–1650.

Smith, R. K. and Montgomery, M. T. (2016) Understanding hurricanes. *Weather* **71**, pp. 219–223.

Smith, R. K., Montgomery, M. T. and Rui, H. (2018) Axisymmetric balance dynamics of tropical cyclone intensification and its breakdown revisited. *J. Atmos. Sci.* **75**, pp. 3169–3189.

Smyth, W. D. and McWilliams, J. C. (1998) Instability of an axisymmetric vortex in a stably stratified, rotating environment. *Theoret. Comput. Fluid Dynamics.* **11**, pp. 305–322.

Snyder, C., Skamarock, W. C. and Rotunno, R. (1991) A comparison of primitive-equation and semi-geostrophic simulations of baroclinic waves. *J. Atmos. Sci.* **48**, pp. 2179–2194.

Sobel, A. H., Nilsson, J. and Polvani, L. M. (2001) The weak temperature gradient approximation and balanced tropical moisture waves. *J. Atmos. Sci.* **58**, pp. 3650–3665.

Soldatenko, S. and Tingwell, C. (2013) The sensitivity of characteristics of large scale baroclinic unstable waves in the Southern hemisphere to the underlying climate. *Adv. Meteor.* **2013**, ID 981271, 11pp.

Stechmann, S. N., Majda, A. J. and Khouider, B. (2008) Dynamics of hydrostatic internal gravity waves. *Theoretical and Computational Fluid Dynamics.* **22**, pp. 407–432.

Stull, R. B. (1988). *An Introduction to Boundary Layer Meteorology*, (Kluwer Academic, Dordrecht, The Netherlands), 666 pp.

Thomson, W. (Lord Kelvin) (1910) Maximum and minimum energy in vortex motion. *Mathematical and Physical Papers*, **4**, (Cambridge University Press), pp. 172–183.

Thorpe, A. J. and Pedder, M. A. (1999) The semi-geostrophic diagnosis of vertical motion. II: Results for an idealized baroclinic wave life cycle. *Quart. J. Roy. Meteorol. Soc.* **125**, pp. 1257–1276.

Vallis, G. K. (2017) *Atmospheric and Oceanic Fluid Dynamics*, (Cambridge University Press), 944 pp.

Van Meighem, J. M. (1952) Hydrodynamic stability. *Compendium of meteorology*, (Amer. Met. Soc.), pp. 434–453.

Vanneste, J. (2013) Balance and spontaneous wave generation in geophysical flows. *Ann. Rev. Fluid Mech.* **45**, pp. 147–172.

Veitch, G. and Mawson, M. H. (1993) A comparison of inertial stability conditions in the planetary semi-geostrophic and quasi-equilibrium models. Short Range Forecasting Tech. Report no. 60, (Met. Office, U.K.).

Villani, C. (2003) *Topics in optimal transportation.*, Vol. 58 of *Graduate Studies in Mathematics*, (Amer. Math. Soc., Providence, RI).

Villani, C. (2008) *Optimal transport, old and new*, (Springer Verlag).

Visram, A. R., Cotter, C. J. and Cullen, M. J. P. (2014) A framework for evaluating model error using asymptotic convergence in the Eady model. *Quart. J. Roy. Meteorol. Soc.* **140**, pp. 1629–1639.

Waite, M. L., Bartello, P. 2006. The transition from geostrophic to stratified turbulence. *J. Fluid Mech.* **568**, pp. 89–108.

Wakefield, M. A. and Cullen, M. J. P. (2005) Atmospheric response to equatorial forcing. *Int. J. Numer. Meth. Fluids*, **47**, pp. 1345–1351.

Wang, X.-J. (1995) Some counter-examples to the regularity of Monge-Ampère equations. *Proc. Amer. Math. Soc.* **123**, pp. 841–845.

Warn, T., Bokhove, O., Shepherd, T. G. and Vallis, G. K. (1995) Rossby-number expansions, slaving principles, and balance dynamics. *Quart. J. Roy. Meteorol. Soc.* **121**, pp. 723–739.

Wheadon, A. (2018) *Wave-turbulence interaction in shallow water numerical models: asymptotic limits, and subgrid interactions.* Ph.D thesis, (Dept. Maths. University of Exeter, U.K.).

Wheeler, M., and Kiladis, G. N. (1999) Convectively-coupled equatorial waves: Analysis of clouds and temperature in the wavenumber-frequency domain. *J. Atmos. Sci.* **56**, pp. 374–399.

White, A. A. (2002) The equations of meteorological dynamics and various approximations. In *Large-scale Atmosphere-Ocean Dynamics*, vol. I, J. Norbury and I. Roulstone eds., (Cambridge University Press), pp. 1–100.

White, A. A., Hoskins, B. J., Roulstone, I. and Staniforth, A. (2005) Consistent approximate models of the global atmosphere: shallow, deep, hydrostatic, quasi-hydrostatic and non-hydrostatic. *Quart. J. Roy. Meteorol. Soc.* **131**, pp. 2081–2108.

Wirosoetismo, D. (2015) Navier-Stokes equations on a rapidly rotating sphere *DCDS-B* **20**, pp. 1263–1271.

Wood, N. and Staniforth, A. (2003) The deep atmosphere Euler equations with a mass-based vertical coordinate. *Quart. J. Roy. Meteorol. Soc.* **129**, pp. 1289–1300.

Wood, N., Diamantakis, M. and Staniforth, A. (2007) A monotonically damping second-order-accurate unconditionally stable numerical scheme for diffusion. *Quart. J. Roy. Meteorol. Soc.* **133**, pp. 1559–1573.

Wood, N., Staniforth, A., White, A., Allen, T., Diamantakis, M., Gross, M., Melvin, T., Smith, C., Vosper, S., Zerroukat, M., Thuburn, J. (2014) An inherently mass-conserving semi-implicit semi-Lagrangian discretization of the deep atmosphere global non-hydrostatic equations. *Q. J. R. Meteorol. Soc.* **140**, pp. 1505–1520.

Yamazak, H., Shipton, J., Cullen, M. J. P., Mitchell, L., and Cotter, C. J. (2016) Vertical slice modelling of nonlinear Eady waves using a compatible finite element method. *J. Comp. Phys.* **343**, pp. 130–149.

Index